# The Potential of Food By-Products: Bioprocessing, Bioactive Compounds Extraction and Functional Ingredients Utilization

# The Potential of Food By-Products: Bioprocessing, Bioactive Compounds Extraction and Functional Ingredients Utilization

Editors

**Michela Verni**
**Federico Casanova**

MDPI • Basel • Beijing • Wuhan • Barcelona • Belgrade • Manchester • Tokyo • Cluj • Tianjin

*Editors*
Michela Verni
University of Bari,
Bari, Italy

Federico Casanova
Technical University of Denmark,
Lyngby, Denmark

*Editorial Office*
MDPI
St. Alban-Anlage 66
4052 Basel, Switzerland

This is a reprint of articles from the Special Issue published online in the open access journal *Foods* (ISSN 2304-8158) (available at: https://www.mdpi.com/journal/foods/special_issues/The_Potential_Food_By_Products).

For citation purposes, cite each article independently as indicated on the article page online and as indicated below:

LastName, A.A.; LastName, B.B.; LastName, C.C. Article Title. *Journal Name* **Year**, *Volume Number*, Page Range.

**ISBN 978-3-0365-6876-8 (Hbk)**
**ISBN 978-3-0365-6877-5 (PDF)**

# Contents

*foods* MDPI

*Editorial*

# The Potential of Food By-Products: Bioprocessing, Bioactive Compounds Extraction and Functional Ingredients Utilization

**Michela Verni [1,*] and Federico Casanova [2]**

1 Department of Soil, Plant, and Food Sciences, University of Bari "Aldo Moro", Via Amendola 165/A, 70126 Bari, Italy
2 Research Group for Food Production Engineering, National Food Institute, Technical University of Denmark, Søltofts Plads, 2800 Kongens Lyngby, Denmark
* Correspondence: michela.verni@uniba.it; Tel.: +39-0805442950

**Citation:** Verni, M.; Casanova, F. The Potential of Food By-Products: Bioprocessing, Bioactive Compounds Extraction and Functional Ingredients Utilization. *Foods* **2022**, *11*, 4092. https://doi.org/10.3390/foods11244092

Received: 12 December 2022
Accepted: 16 December 2022
Published: 17 December 2022



Achieving sustainability in the agro-food sector can only be possible with the valorization of food industry waste and side streams, products with an extremely high intrinsic value but often discarded because they are unfit for further processing that meets consumer expectations. Apart from their use as feed, a more practical solution responding to the modern vision of a circular economy must be sought. In this framework, this Special Issue aimed at covering the most recent advances in the valorization of food by-products (of both animal and plant origin), from extraction to bioprocessing, including their application in a variety of food-related industries throughout the entire food supply chain.

Several processes can guarantee that the potential of food industry by-products is fully unleashed, generating high-value ingredients while ensuring nutritional quality. For example, ultrasound technology was successfully used to extract phenolic acids, flavonoids, anthocyanins, and carotenoids from *Morus alba* leaves [1], pomegranate peels and seeds [2], guaraná by-products [3], and horchata, a Spanish beverage obtained from pressing tiger nut by-products [4]. Such side streams, rich in phenolic compounds, are often studied for their beneficial properties on human health, particularly their antioxidant properties determined in vitro [1,2,4–6]. The extraction of olive pomace phenolic compounds, followed by their purification using macroporous absorbing resins was also explored by Zhao et al. [5], confirming the great potential as natural antioxidants, preservatives, and antimicrobials in clean-label foods.

Enzymatic treatments, alone or ultrasound-assisted, are among the processes employed to valorize food industry waste and by-products [7–9]. Ultrasound-assisted enzymatic treatment was often used to extract soluble and insoluble dietary fibers from orange peel [7] and olive pomace [8]. The renowned physiological effects of dietary fibers are varied and often inversely correlated to obesity, type two diabetes, cancer, and cardiovascular disease. Lignocellulosic-based adsorbents from sugarcane bagasse, cornstalk piths, and corn cob were also found to improve the physicochemical properties and quality of fried oils [10]; nevertheless, beyond the technological properties of dietary fibers, their cholesterol adsorption capacity [7] and antioxidant and prebiotic activity [8] suggest that the application of such by-products in functional foods should be recommended. Indeed, L-arabinose, a bio-active compound derived from the processing of many fibrous materials, was studied by Pol et al. [11], who concluded that its addition to sugary drinks used for a clinical trial, despite the large quantities of starch and fat, significantly lowered postprandial glycemic and insulinemic responses in healthy subjects.

Although the sole extraction and characterization of the compounds of interest is the most preferred option, as in the above-reported cases, as well as for *Aloe vera* anthraquinones [12] or sesame oil asarin [13], the use of the whole discarded biomass is the valorization alternative that can prevent the further generation of by-products and should be preferred in terms of sustainability of the overall upcycling process. This option was chosen by Najjar et al. [14] for the valorization of date seed flour, incorporating it in composite cookies and thus leading to a high amount of total polyphenolic content, with flavonoids showing high in vitro antioxidant activity.

Two review articles were also collected in this Special Issue. One of them overviewed the development of starch-based films as environmentally friendly packaging alternatives which could also be prepared by adding antioxidant or anti-bacterial substances, thus extending food shelf-life, and reducing food waste [15]. The other review focused on the interaction of dairy and plant-based proteins, more specifically, the application of pulsed electric field technology to enhance the functional properties of food proteins [16]. Indeed, the colloidal and acid gelling properties of mixed suspensions of pea and milk proteins [17], and the foaming properties of pea protein isolates and fish skin gelatin [18] were the subject of other studies. In addition, Guan et al. [19] exploited the use of aquatic collagen, obtained from by-products of fish processing, concluding that its higher thermal stability compared to that of terrestrial sources of collagen provides promising applications in food, cosmetics, and biomedical fields.

The research papers published in this Special Issue represent some of the novel strategies at our disposal to valorize waste and by-products generated by the food industry. However, much more support is needed in the future, from academia, industry, government, and consumers, to make sure more sustainable approaches can be put in place to minimize or counteract this issue.

**Author Contributions:** M.V. and F.C. contributed equally to the proposal, editorial work of this Special Issue, and to the writing of the editorial. All authors have read and agreed to the published version of the manuscript.

**Funding:** This research received no external funding.

**Institutional Review Board Statement:** Not applicable.

**Informed Consent Statement:** Not applicable.

**Conflicts of Interest:** The authors declare no conflict of interest.

## References

1. Martín-García, B.; Aznar-Ramos, M.; Verardo, V.; Gómez-Caravaca, A. The Establishment of Ultrasonic-Assisted Extraction for the Recovery of Phenolic Compounds and Evaluation of Their Antioxidant Activity from Morus alba Leaves. *Foods* **2022**, *11*, 314. [CrossRef] [PubMed]
2. Campos, L.; Seixas, L.; Dias, S.; Peres, A.; Veloso, A.; Henriques, M. Effect of Extraction Method on the Bioactive Composition, Antimicrobial Activity and Phytotoxicity of Pomegranate By-Products. *Foods* **2022**, *11*, 992. [CrossRef] [PubMed]
3. Pinho, L.; de Lima, P.; de Sá, S.; Chen, D.; Campanella, O.; da Costa Rodrigues, C.; Favaro-Trindade, C.; Pinho, L.S.; de Lima, P.M.; de Sá, S.H.G.; et al. Encapsulation of Rich-Carotenoids Extract from Guaraná (*Paullinia cupana*) Byproduct by a Combination of Spray Drying and Spray Chilling. *Foods* **2022**, *11*, 2557. [CrossRef] [PubMed]
4. Razola-Díaz, M.; Gómez-Caravaca, A.; Guerra-Hernández, E.; Garcia-Villanova, B.; Verardo, V. New Advances in the Phenolic Composition of Tiger Nut (*Cyperus esculentus* L.) by-Products. *Foods* **2022**, *11*, 343. [CrossRef] [PubMed]
5. Zhao, H.; Avena-Bustillos, R.; Wang, S. Extraction, Purification and In Vitro Antioxidant Activity Evaluation of Phenolic Compounds in California Olive Pomace. *Foods* **2022**, *11*, 174. [CrossRef] [PubMed]
6. Jiang, S.; Yu, M.; Wang, Y.; Yin, W.; Jiang, P.; Qiu, B.; Qi, H. Traditional Cooking Methods Affect Color, Texture and Bioactive Nutrients of Undaria pinnatifida. *Foods* **2022**, *11*, 1078. [CrossRef] [PubMed]
7. Sang, J.; Li, L.; Wen, J.; Gu, Q.; Wu, J.; Yu, Y.; Xu, Y.; Fu, M.; Lin, X. Evaluation of the Structural, Physicochemical and Functional Properties of Dietary Fiber Extracted from Newhall Navel Orange By-Products. *Foods* **2021**, *10*, 2772. [CrossRef] [PubMed]
8. Li, Y.; Yu, Y.; Wu, J.; Xu, Y.; Xiao, G.; Li, L.; Liu, H. Comparison the Structural, Physicochemical, and Prebiotic Properties of Litchi Pomace Dietary Fibers before and after Modification. *Foods* **2022**, *11*, 248. [CrossRef] [PubMed]
9. Gruppi, A.; Dermiki, M.; Spigno, G.; FitzGerald, R. Impact of Enzymatic Hydrolysis and Heat Inactivation on the Physicochemical Properties of Milk Protein Hydrolysates. *Foods* **2022**, *11*, 516. [CrossRef] [PubMed]
10. Ahmed, E.; Zeitoun, A.; Hamad, G.; Zeitoun, M.; Taha, A.; Korma, S.; Esatbeyoglu, T. Lignocellulosic Biomasses from Agricultural Wastes Improved the Quality and Physicochemical Properties of Frying Oils. *Foods* **2022**, *11*, 3149. [CrossRef] [PubMed]
11. Pol, K.; Puhlmann, M.; Mars, M. Efficacy of L-Arabinose in Lowering Glycemic and Insulinemic Responses: The Modifying Effect of Starch and Fat. *Foods* **2022**, *11*, 157. [CrossRef] [PubMed]

12. Sadiq, U.; Gill, H.; Chandrapala, J. Temperature and pH Stability of Anthraquinones from Native Aloe vera Gel, Spray-Dried and Freeze-Dried Aloe vera Powders during Storage. *Foods* **2022**, *11*, 1613. [CrossRef] [PubMed]
13. Yu, Q.; Wang, X.; Liu, H.; Ma, Y. Preparation and Characterization of Solid Acid Catalysts for the Conversion of Sesamin into Asarinin in Sesame Oil. *Foods* **2022**, *11*, 1225. [CrossRef] [PubMed]
14. Najjar, Z.; Kizhakkayil, J.; Shakoor, H.; Platat, C.; Stathopoulos, C.; Ranasinghe, M. Antioxidant Potential of Cookies Formulated with Date Seed Powder. *Foods* **2022**, *11*, 448. [CrossRef] [PubMed]
15. Liu, D.; Zhao, P.; Chen, J.; Yan, Y.; Wu, Z. Recent Advances and Applications in Starch for Intelligent Active Food Packaging: A Review. *Foods* **2022**, *11*, 2879. [CrossRef] [PubMed]
16. Taha, A.; Casanova, F.; Šimonis, P.; Stankevič, V.; Gomaa, M.; Stirkė, A. Pulsed Electric Field: Fundamentals and Effects on the Structural and Techno-Functional Properties of Dairy and Plant Proteins. *Foods* **2022**, *11*, 1556. [CrossRef] [PubMed]
17. Oliveira, I.; de Paula Ferreira, I.; Casanova, F.; Cavallieri, A.; Lima Nascimento, L.; de Carvalho, A.; Nogueira Silva, N. Colloidal and Acid Gelling Properties of Mixed Milk and Pea Protein Suspensions. *Foods* **2022**, *11*, 1383. [CrossRef] [PubMed]
18. Odelli, D.; Sarigiannidou, K.; Soliani, A.; Marie, R.; Mohammadifar, M.; Jessen, F.; Spigno, G.; Vall-llosera, M.; de Carvalho, A.; Verni, M.; et al. Interaction between Fish Skin Gelatin and Pea Protein at Air-Water Interface after Ultrasound Treatment. *Foods* **2022**, *11*, 659. [CrossRef] [PubMed]
19. Guan, Y.; He, J.; Chen, J.; Li, Y.; Zhang, X.; Zheng, Y.; Jia, L. Valorization of Fish Processing By-Products: Microstructural, Rheological, Functional, and Properties of Silver Carp Skin Type I Collagen. *Foods* **2022**, *11*, 2985. [CrossRef] [PubMed]

*foods*

MDPI

*Article*

# Lignocellulosic Biomasses from Agricultural Wastes Improved the Quality and Physicochemical Properties of Frying Oils

**Eman Ahmed [1,*], Ashraf Zeitoun [1], Gamal Hamad [2], Mohamed A. M. Zeitoun [1], Ahmed Taha [1,3], Sameh A. Korma [4,5] and Tuba Esatbeyoglu [6,*]**

1 Department of Food Science, Faculty of Agriculture (Saba Basha), Alexandria University, Alexandria 21531, Egypt
2 Department of Food Technology, Arid Lands Cultivation Research Institute (ALCRI), City of Scientific Research and Technological Applications (SRTA City), New Borg El-Arab 21934, Egypt
3 Department of Functional Materials and Electronics, Center for Physical Sciences and Technology, Saulėtekio al. 3, 10257 Vilnius, Lithuania
4 Department of Food Science, Faculty of Agriculture, Zagazig University, Zagazig 44519, Egypt
5 School of Food Science and Engineering, South China University of Technology, Guangzhou 510641, China
6 Department of Food Development and Food Quality, Institute of Food Science and Human Nutrition, Gottfried Wilhelm Leibniz University Hannover, Am Kleinen Felde 30, 30167 Hannover, Germany
* Correspondence: eman-abdelaziz@alexu.edu.eg (E.A.); esatbeyoglu@lw.uni-hannover.de (T.E.); Tel.: +20-1212484089 (E.A.); +49-5117625589 (T.E.)

**Abstract:** In this work, the effects of using natural lignocellulosic-based adsorbents from sugarcane bagasse (SC), cornstalk piths (CP), and corn cob (CC) on the physicochemical properties and quality of fried oils were studied. The properties of lignocellulosic biomasses were examined using X-ray diffraction (XRD), scanning electron microscope (SEM), and Fourier transform infrared spectroscopy (FTIR). Moreover, the changes in the physicochemical properties of fresh, fried oils (for 4, 8, 12, 16 and 20 h) and adsorbents-treated oils were examined. The XRD results revealed that SC and CP biomasses have more amorphous regions than CC biomass, which had the highest crystallinity percentage. The results also showed that lignocellulosic biomasses enhanced the quality of the used oils. SC was the most effective biomass to enhance the properties of the used sunflower oil. For instance, the acid value of oil samples fried for 20 h reduced from $0.63 \pm 0.02$ to $0.51 \pm 0.02$ mg KOH/g oil after SC biomass treatment. For the peroxide value, the SC biomass treatment reduced it from $9.45 \pm 0.56$ (fried oil for 20 h) to $6.91 \pm 0.12$ meq $O_2$/kg. Similarly, SC biomass adsorbent reduced the *p*-Anisidine Value (*p*-AV) of the used oil (20 h) from $98.45 \pm 6.31$ to $77.92 \pm 3.65$. Moreover, SC adsorbents slightly improved the lightness of the used oils (20 h). In conclusion, natural lignocellulosic biomasses, particularly SC, could be utilized as natural adsorbents to improve the oil quality. The results obtained from this study could help in developing sustainable methods to regenerate used oils using natural and cheap adsorbents.

**Keywords:** fried oil; adsorbent; lignocellulosic biomass; fatty acid; food waste; oil regeneration; oil quality; sustainability

**Citation:** Ahmed, E.; Zeitoun, A.; Hamad, G.; Zeitoun, M.A.M.; Taha, A.; Korma, S.A.; Esatbeyoglu, T. Lignocellulosic Biomasses from Agricultural Wastes Improved the Quality and Physicochemical Properties of Frying Oils. *Foods* **2022**, *11*, 3149. https://doi.org/10.3390/foods11193149

Academic Editors: Michela Verni and Federico Casanova

Received: 3 September 2022
Accepted: 8 October 2022
Published: 10 October 2022



## 1. Introduction

Deep-fat frying is an extensively used cooking approach worldwide as fried products have a unique texture, color and flavor [1]. In this cooking approach, the food product is immersed in edible oil at temperatures ranging from 120–190 °C [2]. Frying also involves simultaneous mass and heat transfer in both directions, from the frying medium "edible oil" to the food product and vice versa. For example, in the case of French fries, the oil penetrates potato strips while a few soluble materials, water and starch, escape from the strips [3,4]. The physicochemical changes of the frying medium and the fried product are influenced by several process factors, including frying time, frying oil temperature and nature of the fried products (i.e., thermal characteristics and moisture content) [4]. The presence of oxygen with high temperatures and moisture can cause oil degradation and decrease the quality of the fried products. Oil degradation due to oxidation [5], hydrolysis [6], and polymerization [7] can form undesirable components such as free fatty acids, dimers, cyclic compounds, alcohols, polymers and acrylamides (a chemical carcinogen formed via the Maillard reaction between reducing sugars and asparagine amino acid) [8,9]. The consumption of a diet produced by repeatedly heated oil could increase the risk of severe health issues, including cardiovascular diseases [10], cancers [11], impaired renal function [12], hepatorenal toxicity [13], hypertension and atherosclerosis [14].

Recent studies estimated that the generation of the used frying oils is about 20–32% of the total vegetable oils consumption. Thus, the global cooking oil production is around 42 million tons per year [15]. Discarding the used oils within solid residues or through siphons and sinks causes many environmental and economic issues. This could increase the maintenance costs of waste treatment plants and sewage, encourage the proliferation of pests and vectors, trigger pollution and public health issues and damage the infrastructure. Moreover, discarding the used oils to soil resources can damage the ecosystem as oil isolates soil from water and air, killing useful natural organisms such as bacteria and worms. In the water resources, waste oil pollution could block the sunlight and reduce the oxygen content, damaging aquatic fauna and flora [15,16]. Thus, the reutilization of waste frying oils is highly recommended to mitigate the mentioned negative effects.

Passive and active filtrations have been used to regenerate used oil. Passive filtration via filter paper and cloth only removed coarse particles; however, fine particles and water remained in the oil. In comparison, active filtration via adsorptive properties of natural and synthetic adsorbents removes chemical substances, including free fatty acids, peroxides, dimers, dark color matter and aldehydes [9,17]. There are many advantages of adsorption-based filtration of the used oil, such as low generation of wastewater, low energy consumption and suitability for any quality of oil [15]. On the other hand, the main drawbacks of active filtration are the high consumption of adsorbents, the unavailability of effective filtration equipment and the leaching of some metals into oils [18]. Several types of natural (i.e., clays, ceramics, etc.) and chemical or synthetic materials (bentonite, calcium silicate, magnesium oxide and aluminum hydroxides) can be used for active filtration of the used cooking oils [19]. There is a rising interest in natural and sustainable materials due to the higher costs and toxicity concerns of synthetic adsorbents. Thus, raw materials from agricultural wastes could be a suitable alternative to conventional chemical adsorbents thanks to their low cost, availability and biodegradability [9,20]. The lignocellulosic biomass from agricultural wastes, including sugarcane bagasse, corn cobs, rice straw, corn piths, orange peels and wheat bran, can be utilized as natural adsorbents because of their extensive production and availability [21–24].

Corn cobs (CC) are solid residues from corn production; they contain lignin (20.3%), hemicellulose (48.7%), cellulose (~32%) and acetyl groups (3.40%) [25,26]. Several studies used CC as an effective adsorbent in different applications. For example, Alves et al. [27] used modified CC to adsorb phenylalanine and tyrosine from an aqueous solution. Bavaresco et al. [24] evaluated the potential of using CC to reduce the acidity of the used oils to produce biodiesel. Cornstalk piths (CP) are found in the center of cornstalk, forming approximately 50% of cornstalk by weight. CP comprises around 37% cellulose, 13% lignin and 24% pentosane and contains ash and other traces [28]. CP was applied as an affordable adsorbent during wastewater treatment to remove dyes, spilled oil and toxic metal [29–31].

Moreover, malic acid-modified CP was successfully used to adsorb methylene blue and crystal violet dyes [31]. Sugarcane bagasse piths (SC) is the byproduct of biomass residue after extracting juice from sugarcane, and it is mainly used in sugar industry factories to run boilers [32]. SC consists of around 40% cellulose and almost 32% of hemicellulose, and 5% lignin and other traces [33]. Wannahari et al. [34] utilized SC to recover the used palm oil; 5 $w/v$ of bagasse caused a reduction in the free fatty acids to 82.14% and reduced the color density to 75.67%.

Therefore, we hypothesized that lignocellulosic materials could restore some original attributes of the used frying oils because of their morphological properties, such as the spores at the surface of their particles. Moreover, these materials could have functional groups that act as active sites to adsorb some chemical residues resulting from oil deterioration. To date, few studies investigated the potential of using natural lignocellulosic biomass to enhance the quality of the used frying oils [9,35,36].

The effects of using CC, CP and SC as natural lignocellulosic-based adsorbents on the quality of the used sunflower oils are yet to be studied. Thus, this work aims to evaluate the effects of adding using some lignocellulosic biomass materials (CC, CP and SC) on the quality of sunflower used oil. Standard sunflower oil mainly consisted of linoleic acid (C18:2, 48.3–74.0%), oleic acid (C18:1, 14.0–39.4), palmitic acid (C16:0, 5–7.6%) and stearic acid (C:18:0, 2.7–6.5%) [37].

In this work, the physicochemical and morphological properties of the lignocellulosic materials were investigated using XRD, FTIR and SEM. Lignocellulosic biomass mainly comprised lignin (10–25%), hemicellulose (20–35%), and cellulose (35–50%) with traces of ash and extractives [38]. The FTIR, SEM and XRD techniques are usually used to investigate the physicochemical and microstructural properties of lignocellulosic biomasses [39]. The results obtained from these techniques could help to interpret the physicochemical properties of the used frying oils before and after treatment with adsorbents. Furthermore, the physicochemical properties of the fried sunflower oil were examined after using the lignocellulosic materials as natural adsorbents.

## 2. Materials and Methods

### 2.1. Materials

The corn cobs (CC) and cornstalk piths (CP) from corn stover (*Zea maize*, cultivar 30 K 8, produced by Corteva Agriscience LLC., New Cairo, Egypt) were collected from a local corn farm in Beheira governorate, Egypt. Sugarcane bagasse (SC) wastes were collected from a local sugarcane juice producer immediately after squeezing for juice production. Sunflower oil (Crystal brand from Arma for the food industry, Al-Sharqia Governorate, Egypt) was purchased from a local market in Alexandria, Egypt. Frozen potato strips (10 × 10 × 50 mm) were bought from Farm Frites (Tenth of Ramadan City, Egypt). Ethanol, NaOH, KOH, phenolphthalein, potassium iodide, glacial acetic acid, chloroform, sodium thiosulfate and starch were purchased from Elnasr chemical company and Elgomhouria pharmaceutical company (Cairo, Egypt). Isooctane was purchased from the Alfa chemical group (Cairo, Egypt). *n*-Hexane, sodium hydrogen sulfate, *p*-AV and methyl esters fatty acid standards were obtained from Sigma Chemical Co., Ltd. (St. Louis, MO, USA). All reagents used in this investigation were of analytical grade.

### 2.2. Methods

#### 2.2.1. Adsorbents Preparation

The preparation of lignocellulosic adsorbents (SC, CP and CC) was performed once, according to Schneider et al. [35], with minor modifications. The corn wastes (CC and CP) were collected a month after cultivation directly from the corn farm. After collection, SC wastes were dried in a sunny area for 24 h (daytime only). Then, all biomasses were kept in a closed package in the refrigerator until further experiments. After that, 500 g of each adsorbent material was washed using distilled water to remove the dust particles and left in a sunny and dry area for two days, then dried in the oven for 24 h at 90 °C. Then, the dried products were ground in a grinder and sieved using sieves of 50 mesh (~0.3 mm). These three types of adsorbents were selected based on preliminary results, which showed that these adsorbents had better performance than other tested absorbents (including orange and banana peel biomasses). The released pigments from orange and banana peel biomasses changed the color of oil samples to an unacceptable color for consumers.

#### 2.2.2. Scanning Electron Microscope (SEM)

An SEM (Tescan Vega 3, Tescan Company, Brno, Czech Republic) at 20 kV was applied to examine the morphological properties. The SEM was combined with Energy Dispersive X-ray spectroscopy (EDX) to investigate the element contents of the adsorbents [40]. A high vacuum sputtering instrument (Quoroum150T, UK) was used to sputter samples (at 40 mA sputter current for 120 s) with gold.

#### 2.2.3. Fourier Transform Infrared Spectroscopy (FTIR)

The Fourier transform infrared-attenuated total reflectance (FTIR-ATR) spectroscopy was performed to identify the main functional groups of wastes and to get an overview of their chemical structure. The FTIR transmittance spectra were recorded in the frequency range 4000–400 $cm^{-1}$ with 64 scans and 4 $cm^{-1}$ resolution using a VERTEX 70v FT-IR Spectrometer (Bruker Optics, Ettlingen, Germany) [41]. The FTIR spectra were further analyzed and fitted using OriginPro 2021 software (OriginLab Corporation, Northampton, MA, USA) and PeakFit 4.12 software (SeaSolve Software Inc., San Jose, CA, USA). The Levenberg-Marquardt algorithm was used during the peak analysis to perform non-linear fitting of the peaks. Baseline correction was made using a second derivative method for finding anchor points and detecting the baseline.

#### 2.2.4. X-ray Diffraction (XRD)

The crystallinity behavior of the adsorbents was identified using a PW 1390 Philips X-ray diffractometer with Ni-filtered Cu-K$\alpha$ radiation ($\lambda$ = 0.15418 nm) energized at 45 kV. The XRD patterns were recorded in reflection mode from 0° to 80° in a diffraction angle range of 2θ. The crystallinity index (*CrI%*) was calculated based on the method of Karthika et al. [42] and Equation (1). Where (*I am*) is the intensity at 2θ = 18.4° and (*I cr*) is the intensity of the main peak around 2θ = 21.6°.

$$Crystallinity\ index\ (CrI\%) = \frac{(I\ cr - I\ am)}{I\ cr} \times 100 \quad (1)$$

#### 2.2.5. Frying Process

An electric deep fryer (Moulinex, Alençon, France) was used for deep frying. Sunflower oil (2.5 L) was poured into the fryer and heated at 180 °C. Fifteen minutes were required to heat the oil to the frying temperature; one batch (150 g potato strips) was fried for 5 min every hour. The frying process lasted for 5 days (4 h/day). About 600 mL of fried oil was withdrawn (each day). The fried oils were allowed to cool to room temperature [9] and mixed with different adsorbents (150 mL for each adsorbent). About 200 mL of fresh oil was added to the frying oil each day (after 4 h of frying) to replace part of the withdrawn fried oil.

#### 2.2.6. Filtration

An amount of 150 mL of the used oil was mixed with 5% (*w*/*w*) adsorbent + 3% (*w*/*w*, citric acid/adsorbent), then mechanically stirred for 30 min and filtered with cloths then with filter papers (Whatman number 41). After filtration, the used (control) and treated oil samples were kept at −18 °C for the physiochemical analysis. Citric acid, as a metal chelator, was added to prevent the oxidation of oil samples during the filtration and adsorption processes, as lignocellulosic biomasses could release some minerals during the adsorption process [43]. Moreover, previous studies showed that citric acid did not reduce the acid value of fried oils [17,44].

#### 2.2.7. Acid Value (AV)

The AV was calculated following the method of the American Oil Chemists' Society (AOCS Method Cc 5a-40) [45]. Briefly, 5 g of oil was added to hot, neutral ethyl alcohol in a

25 mL elementary. Then, 0.1 N NaOH was used to titrate the mixture and phenolphthalein was used as an indicator. Equation (2) was used to calculate the AV.

$$\text{Acid value (AV)} = \frac{\text{mL} \times \text{N} \times 56.1}{\text{sample weight}} \quad (2)$$

where 56.1 = molecular weight of KOH, N = normality of NaOH and mL = milliliter of required NaOH.

### 2.2.8. Peroxide Value (PV)

The PV was measured based on the method of Nuru and Getachew [46]. Five gram of each sample was poured into a 250 mL Erlenmeyer flask, 0.5 mL of saturated potassium iodide solution, and 30 mL of chloroform-glacial acetic acid (2:3, $v/v$) were mixed and kept in a dark place for 1 min at room temperature (~25 °C). Then, distilled water (30 mL) was added to terminate the reaction then saturated starch (2 mL) was added as an indicator. Subsequently, the mixture (with dark brown to purple color) was titrated with standardized 0.12 M sodium thiosulfate solution until the disappearance of the color. The PV was expressed as milliequivalent of oxygen/kilogram of oil (meq $O_2$/kg) following Equation (3).

$$\text{Peroxide value (PV)} = \frac{(\text{S} - \text{B}) \times \text{M} \times 1000}{\text{m}} \quad (3)$$

where m = sample weight, M = Molarity of $Na_2S_2O_3$, S = titration of sample and B = titration of the blank.

### 2.2.9. *p*-AV and the Total Oxidation Value (TOTOX)

The *p*-AV was calculated based on the IUPAC method [47] with a minor modification. Ten milliliters of isooctane were used to dilute a 0.1–0.3 g sample. The adsorption at 350 nm ($A_b$) was measured against isooctane as a blank using a UV-VIS spectrophotometer (T80, PG Instruments, Wibtoft, UK). Then, 1 mL of *p*-AV solution was added to 5 mL of fat solution in a tube. The *p*-AV solution (1 mL) was mixed with 5 mL of the solvent in a second tube and shaken. The adsorption was then measured after 10 min ($A_s$), and the second tube was used as a blank. The *p*-AV was calculated according to Equation (4). Where m is the mass of the oil sample.

$$(p - \text{AV}) = \frac{10 \times (1.2 \times A_S - A_b)}{\text{m}} \quad (4)$$

The TOTOX value was estimated based on the values of *p*-AV and PV following Equation (5) [48].

$$\text{TOTOX value} = 2 \times \text{PV} + p - \text{AV} \quad (5)$$

### 2.2.10. Fatty Acids Analysis

Fatty acid compositions of fresh, treated and untreated oil samples were determined according to Yang et al. [49] with minor modifications. An amount of 5 mL of *n*-hexane was used to dissolve 100 mg of an oil sample. 11.2 g of KOH was added to 100 mL ethanol (as a transesterification agent) and poured into the mixture. The tube was vortexed for 1 min, and solid sodium hydrogen sulfate (0.5 g) was added to the mixture. The mixture was then centrifuged at 4000× *g* at 25 °C for 3 min then the supernatant was injected for analysis. The obtained methyl esters were examined using an Agilent Technologies 7890A GC (LTD Corporation, Mundelein, IL, USA) combined with a flame ionization detector and equipped with a fused silica capillary column (0.4 mm diameter; 0.25 µm film thickness; 30 m length). The analysis was operated at 0.5 bar pressure, a nitrogen flow rate of 3 mL $min^{-1}$ and an oven temperature of 195 °C. The column and detector temperatures were set at 230 and 250 °C, respectively. The fatty acid methyl esters were identified based on retention time using commercial methyl esters fatty acid standards (Sigma Chemical Co., St. Louis, MO, USA). The peaks were identified using Agilent Technologies ChemStation software.

2.2.11. Color

The color was measured according to the method of Caiping Su and Pamela White [50] using the Hunter Lab instrument (Easy Match QC, Hunter Associates Laboratory, Inc., Reston, VA, USA). For the $L$ value, the value closer to 0 reflect black (darkness), and the value closer to 100 reflects whiter (lighter). While $a^*$ values indicate green ($-a$) to red ($+a$) colors, and $b^*$ represent blue ($-b$) to yellow ($+b$) colors.

2.2.12. Statistical Analysis

Most experiments were performed in triplicates, while the color and fatty acids analysis were performed in duplicates. The data are displayed as means and standard deviations. Data analysis was completed using SPSS version 25 (IBM SPSS Statistics®, Chicago, IL, USA). One-way ANOVA and Duncan's tests (at $p < 0.05$) were employed to test statistical differences among means.

## 3. Results and Discussion

### 3.1. Characterizations of Adsorbents

3.1.1. FTIR Analysis

FTIR analysis was also employed to confirm the presence of CC, CP and SC biomass components such as lignin, cellulose and hemicellulose. The FTIR spectra of CC, CP and SC biomasses are shown in Figure 1. The presence of cellulose, hemicelluloses and lignin (phenolic+$CH_2OH$) was confirmed by the bands 3309–3324 $cm^{-1}$, corresponding to -OH axial deformation and intermolecular hydrogen-bonded H–O–H stretching [51]. Moreover, the deconvolution of broad bands in the 3500–3300 $cm^{-1}$ range was performed to show the differences in hydrogen interactions (Figure S1 and Table S1). The analysis showed that SC biomass had a strong hydrogen bonding interaction (higher intensity), followed by CP biomass, while CC biomass had the lowest density of hydrogen bonding. The fitted peaks centred at 3460–3480 $cm^{-1}$ are assigned to H-bonded -OH group (intermolecular H-bonds) [52]. The characteristic spectral band of lignin assigned to the skeleton vibrations in the phenolic ring (vC=C) at 1517 $cm^{-1}$ was detected [53,54]. The hemicellulose bands at 1173 $cm^{-1}$ referred to acetate groups.

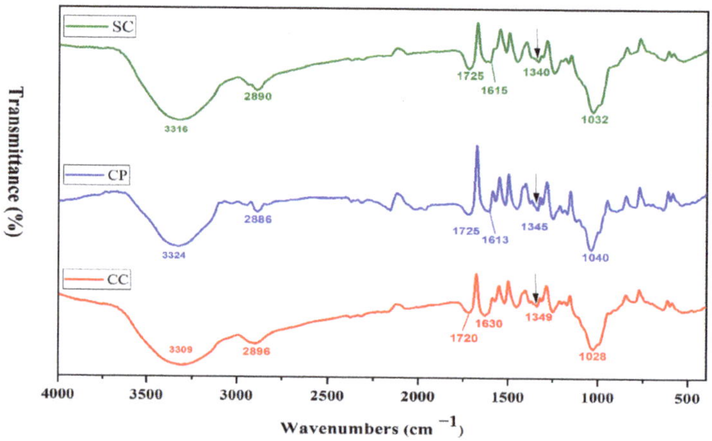


**Figure 1.** The FTIR spectra of lignocellulosic biomasses, including corn cob (CC), cornstalk piths (CP) and sugarcane bagasse (SC).

The IR bands at 1609–1630 $cm^{-1}$ related to C=O axial deformation. While the IR band in the region between 1180–950 $cm^{-1}$ presented the fingerprint region of hemicelluloses and cellulose; the bands positioned at 1028–1040 $cm^{-1}$ were assigned to C-O-C skeletal vibration of the polysaccharides ring [51,55]. The total crystalline index (TCI) was calculated (Table 1) from the ratio of peak areas at 1372 $cm^{-1}$ and 2900 $cm^{-1}$ from the FTIR spectrum [56,57]. TCI ($A_{1372}/A_{2900}$ $cm^{-1}$) indicates the crystallinity degree of cellulose. The TCI value of CP samples was the lowest among all adsorbents, indicating a higher percentage of amorphous structures in CP. CC samples had a higher TCI ratio than CP and SC, meaning that CC had a more ordered crystalline structure. These findings are matched with the findings of crystallinity behavior collected from XRD results.

**Table 1.** The total crystalline index (TCI) from FTIR parameters for crystallinity assessment and the crystallinity index (CrI%) from XRD data of adsorbents; including corn cob (CC), cornstalk piths (CP) and sugarcane bagasse (SC). Means with different letters ([a], [b], [c]) within each column indicate statistically significant differences between the samples after Duncan's analysis ($p < 0.05$).

| Samples | TCI ($A_{1372}/A_{2900}$) | CrI (%) |
|---|---|---|
| CC | 0.53 ± 0.08 [a] | 53.32 ± 1.13 [a] |
| CP | 0.40 ± 0.05 [c] | 48.53 ± 0.92 [b] |
| SC | 0.45 ± 0.06 [b] | 49.24 ± 0.24 [a,b] |

3.1.2. Crystallinity Analysis

Generally, cellulose molecules have both amorphous and crystalline regions, while lignin and hemicellulose only contain amorphous areas. XRD can examine the crystal structure of cellulose and conclude the CrI (%) [54]. Crystallinity is considered the percentage of an ordered crystalline structure in a material. The CrI (%) is related to cellulose, as lignin and hemicellulose only have amorphous structures [58]. XRD analysis of lignocellulosic biomass is challenging because this method measures the crystalline materials' crystal structure. Notwithstanding, the lignocellulosic material has a complex structure with a high percentage of amorphous structures. Materials with high amorphous contributions exhibit a huge background signal, superimposing the crystalline regions' clear signals; this phenomenon was observed in our XRD results (Figure 2). As shown in Figure 2, SC and CC biomasses had similar crystalline peaks at 21.5° 2θ, while CP had a wide, non-sharp peak at 21.5°. It was also found that CC had significantly higher CrI (%) compared to SC and CP biomasses (Table 1). These findings were matched with crystallinity properties confirmed by TCI results from FTIR analysis (Table 1). Thus, it could be concluded that the cellulose of CC biomass has more crystalline regions in its structure compared to the cellulose structure of CP and SC.

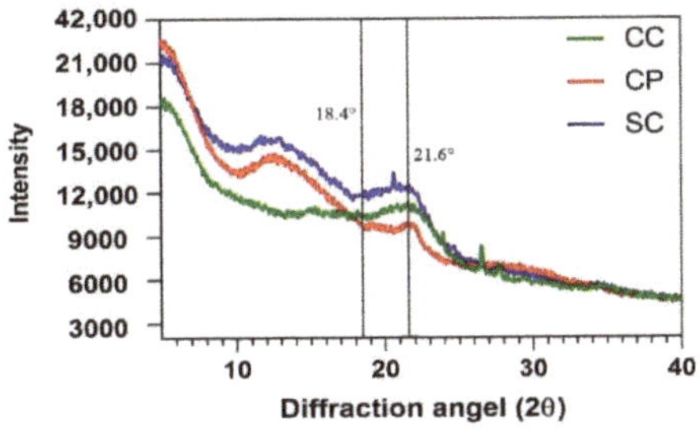


**Figure 2.** The X-ray diffraction (XRD) spectra of lignocellulosic biomasses, including corn cob (CC), cornstalk piths (CP) and sugarcane bagasse (SC).

### 3.1.3. Microstructure of Adsorbents

The differences in microstructure and element content of the adsorbents were observed using SEM and EDX analysis. The surface morphology and elements content could help in understanding the effects of the lignocellulosic adsorbents on the physicochemical properties of the used edible oils. As shown in Figure 3, CC biomass had an irregular surface and hollow tubular structure with different sizes and shapes of the particles. It was noticed that CC biomass is like wood in terms of mechanical and microstructure properties [59]. Moreover, the microstructure of CP and SC had a more defined structure, and homogenous sponge-like tissues with visible pores were observed in the SEM images of SC (Figure 3). The EDX analysis (Table 2) revealed that CC had higher carbon levels than other adsorbents, while SC and CP biomasses had significantly higher oxygen values compared to CC. Higher oxygen levels could indicate the presence of oxygenated functional groups (e.g., carboxyl, carbonyl, hydroxyl, etc.) that might interact with adsorbates [24].

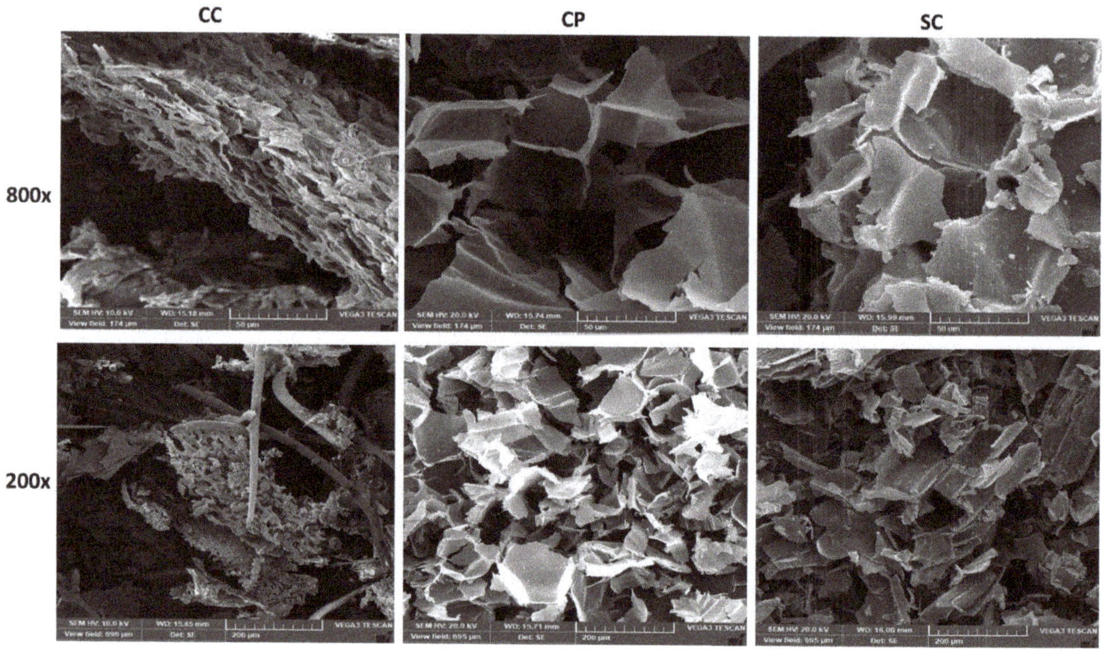


**Figure 3.** The SEM micrographs of lignocellulosic biomasses; including corn cob (CC), cornstalk piths (CP) and sugarcane bagasse (SC).

**Table 2.** Element analysis (mass %) of lignocellulosic biomasses, including corn cob (CC), cornstalk piths (CP) and sugarcane bagasse (SC) using Energy Dispersive X-ray spectroscopy (EDX). Means with different letters ([a], [b]) within each column indicate statistically significant differences between the samples after Duncan's analysis ($p < 0.05$). Post-hoc tests were not performed if there were fewer than three groups.

| Samples | Carbon | Oxygen | Calcium | Potassium | Cobalt |
|---|---|---|---|---|---|
| CC | $69.52 \pm 3.35$ [a] | $27.87 \pm 0.75$ [b] | - | $2.23 \pm 0.95$ [a] | $0.59 \pm 0.02$ |
| CP | $57.53 \pm 2.63$ [a,b] | $35.25 \pm 0.92$ [a,b] | $4.14 \pm 0.13$ | $3.26 \pm 0.23$ [a] | - |
| SC | $53.30 \pm 1.84$ [b] | $39.34 \pm 1.22$ [a] | $5.67 \pm 0.17$ | $1.88 \pm 0.27$ [a] | - |

### 3.2. Acid and Peroxide Values

Acid value (AV) is a widely used parameter to evaluate oil quality; it is a reliable and fast method to check oil acidity during frying. Generally, the AV must be less than 1.0 mg KOH/g oil [36,60]. Table 3 presents the changes in the acidity of the used sunflower oil before and after treatment with lignocellulosic biomasses. The AV of fresh oil was 0.11 mg KOH/g oil and increased after frying for 20 h to 0.63 mg KOH/g oil. The AV increased with the increase of frying hours due to the hydrolysis reaction. The water is released from the potato during frying in the form of vapor, which then interacts with triglycerides, producing di-mono glyceride and free fatty acids [61]. Another reason for the increase in the AV could be the accumulation of secondary oxidation substances [62]. The treatment of fried oils with CC, CP and SC biomasses decreased the AV by 8%, 14% and 19%, respectively, compared to the oil samples fried for 20 h. This could be due to the hydrophilicity nature of lignocellulosic materials, adsorbing vapor from used oil samples and thus slowing the chemical reactions [29,63].

**Table 3.** Acid values (mg KOH/g oil) of the used and treated oils with different lignocellulosic biomasses; corn cob (CC), cornstalk piths (CP) and sugarcane bagasse (SC). Means with different letters ($^{a}$, $^{b}$, $^{c}$) within each column indicate statistically significant differences between the samples after Duncan's analysis ($p < 0.05$). The numbers 4, 8, 12, 16 and 20 indicate the frying period by (h, hours).

| | 0 h | 4 h | 8 h | 12 h | 16 h | 20 h |
|---|---|---|---|---|---|---|
| Fresh oil | 0.11 ± 0.00 | - | - | - | - | - |
| Used oil | - | 0.32 ± 0.01 $^{a}$ | 0.39 ± 0.05 $^{a}$ | 0.44 ± 0.02 $^{a}$ | 0.59 ± 0.03 $^{a}$ | 0.63 ± 0.02 $^{a}$ |
| CC-treated oil | - | 0.22 ± 0.01 $^{b}$ | 0.27 ± 0.00 $^{b}$ | 0.37 ± 0.05 $^{b}$ | 0.50 ± 0.08 $^{a,b}$ | 0.58 ± 0.03 $^{a,b}$ |
| CP-treated oil | - | 0.17 ± 0.01 $^{c}$ | 0.28 ± 0.02 $^{b}$ | 0.35 ± 0.01 $^{b}$ | 0.49 ± 0.04 $^{a,b}$ | 0.54 ± 0.00 $^{b,c}$ |
| SC-treated oil | - | 0.14 ± 0.03 $^{c}$ | 0.23 ± 0.03 $^{b}$ | 0.34 ± 0.02 $^{b}$ | 0.41 ± 0.09 $^{b}$ | 0.51 ± 0.02 $^{c}$ |

Moreover, the SC biomass showed a more porous structure, as shown in SEM images (Figure 3). This could result in more adsorption positions, a larger specific surface area and a higher binding possibility with free fatty acids [31]. These features could improve the adsorption capacity of these adsorbents and reduce acid values

The peroxide value (PV) is a widely used analytical method to measure hydroperoxides formed at the primary oxidation stage; it was determined by iodometric titration assay [64]. As shown in Table 4, the initial PV of fresh oil was 4.63, then increased to 11.10 after 12 frying hours and thereafter declined to 9.45 meq $O_2$/kg after frying for 20 h. The decline in the PV after 12 h of frying could be due to the decomposition of hydroperoxides under the high-temperature condition, forming carbonyl and aldehydic components [65]. The results exhibited the ability of lignocellulosic biomasses to reduce the PV of different oil samples. After applying CC, CP and SC to used oils (20 h), the PVs were reduced to 8.23, 8.96 and 6.91 meq $O_2$/kg, respectively. SC biomass adsorbents induced a significant ($p < 0.05$) reduction in PV with lower PVs than CC- and CP-treated oils. This agreed with the results of Ali and El-Anany [36], who found that using 3% sugarcane bagasse ash adsorbent reduced the sunflower oil's PV.

**Table 4.** Peroxide values (meq $O_2$/kg) of the used and treated oils with different lignocellulosic biomasses; corn cob (CC), cornstalk piths (CP) and sugarcane bagasse (SC). Means with different letters ([a], [b], [c]) within each column indicate statistically significant differences between the samples after Duncan's analysis ($p < 0.05$). The numbers 4, 8, 12, 16 and 20 indicate the frying period by (h, hours).

| | 0 h | 4 h | 8 h | 12 h | 16 h | 20 h |
|---|---|---|---|---|---|---|
| Fresh oil | 4.63 ± 1.03 | - | - | - | - | - |
| Used oil | - | 9.11 ± 0.13 [a] | 8.72 ± 2.84 [a] | 11.10 ± 0.89 [a] | 10.96 ± 1.39 [a] | 9.45 ± 0.56 [a] |
| CC-treated oil | - | 3.43 ± 0.29 [c] | 5.22 ± 0.77 [b] | 9.37 ± 1.45 [a] | 6.21 ± 0.96 [b] | 8.23 ± 1.76 [a] |
| CP-treated oil | - | 5.26 ± 1.30 [b] | 6.40 ± 0.76 [a,b] | 9.02 ± 0.80 [a] | 4.99 ± 0.93 [b] | 8.96 ± 0.61 [a] |
| SC-treated oil | - | 2.55 ± 0.41 [c] | 5.19 ± 0.44 [b] | 5.97 ± 1.03 [b] | 5.01 ± 0.92 [b] | 6.91 ± 0.12 [b] |

### 3.3. *p*-AV and TOTOX

PV measures hydroperoxides (initial oxidation products), which decompose to carbonyl components and aldehyde at a longer frying or higher temperature. Thus, PV does not provide a complete evaluation of oil oxidation. Unsaturated aldehydes induce the oils' off-flavors. *p*-AV determine the secondary oxidation products, especially aldehydes, by color reaction [66,67]. Table 5 displays the *p*-AV of the used oil samples before and after adding SC, CP and CC as natural adsorbents. *p*-AV gradually increased by increasing the frying hours due to the accumulation of secondary oxidation products. *p*-AV of fresh oil was 16.40 and reached 98.45 after 20 h of frying. Using CC, CP and SC adsorbents reduced the *p*-AV of the used oils (20 h) by 9.7, 15.2 and 20.8%, respectively. These findings indicate that lignocellulosic materials have a high affinity with aldehydes and ketones, probably due to the presence of the OH group, which binds with aldehydes and ketones via carbonyl groups [64]. The EDX analysis showed that SC had more oxygen levels than CC and CP, indicating the existence of oxygenated functional groups that could interact with aldehydes, reducing the *p*-AV. Moreover, the FTIR and XRD results showed that the cellulose structure of CC biomass has more crystalline regions than SC and CP. This indicates that more amorphous regions in lignocellulosic biomasses (SC and CP) could improve the absorbance of oxidation residues from fried oils. However, more studies are needed to investigate the relationship between the crystallinity of lignocellulosic materials and the absorbance capacity and oil quality.

**Table 5.** *p*-AV of the used and treated oils with different lignocellulosic biomasses; corn cob (CC), cornstalk piths (CP) and sugarcane bagasse (SC). Means with different letters ([a], [b], [c], [d]) within each column indicate statistically significant differences between the samples after Duncan's analysis ($p < 0.05$). The numbers 4, 8, 12, 16 and 20 indicate the frying period by (h, hours).

| | 0 h | 4 h | 8 h | 12 h | 16 h | 20 h |
|---|---|---|---|---|---|---|
| Fresh oil | 16.40 ± 1.32 | - | - | - | - | - |
| Used oil | - | 43.23 ± 2.30 [a] | 53.81 ± 5.43 [a] | 69.32 ± 4.75 [a] | 73.89 ± 3.68 [a] | 98.45 ± 6.31 [a] |
| CC-treated oil | - | 36.91 ± 1.42 [b] | 49.71 ± 3.65 [b] | 61.71 ± 6.15 [c] | 71.35 ± 2.43 [a,b] | 88.84 ± 5.74 [b] |
| CP-treated oil | - | 31.86 ± 1.97 [c] | 51.79 ± 4.59 [a,b] | 66.39 ± 3.35 [b] | 70.59 ± 2.67 [a,b] | 83.56 ± 4.53 [c] |
| SC-treated oil | - | 32.17 ± 2.23 [c] | 48.99 ± 2.36 [b] | 60.10 ± 2.84 [c] | 69.63 ± 4.21 [b] | 77.92 ± 3.65 [d] |

The TOTEX value is an indicator of the total oxidation state in oils; it is calculated from PV and *p*-AV [67]. The changes in the TOTOX values are displayed in Table 6. Like *p*-AV, the TOTOX value steadily increased with frying and reached 117.35 after 20 h of frying. Furthermore, the results indicated that the TOTOX value of the used oils (20 h) treated with CC, CP and SC decreased to 105.3, 101.5 and 91.74, respectively. CC biomass was shown the highest TOTOX value, followed by CP, while SC treatment had the lowest TOTOX value. These findings are matched with the PV and *p*-AV results.

**Table 6.** Total oxidation value (TOTOX) of the used and treated oils with different lignocellulosic biomasses; corn cob (CC), cornstalk piths (CP) and sugarcane bagasse (SC). Means with different letters ([a], [b], [c], [d]) within each column indicate statistically significant differences between the samples after Duncan's analysis ($p < 0.05$). The numbers 4, 8, 12, 16 and 20 indicate the frying period by (h, hours).

| | 0 h | 4 h | 8 h | 12 h | 16 h | 20 h |
|---|---|---|---|---|---|---|
| Fresh oil | 25.66 ± 2.32 | - | - | - | - | - |
| Used oil | - | 61.45 ± 5.31 [a] | 71.25 ± 6.25 [a] | 91.52 ± 4.32 [a] | 95.81 ± 3.81 [a] | 117.35 ± 9.15 [a] |
| CC-treated oil | - | 43.77 ± 3.84 [b] | 60.15 ± 3.34 [d] | 80.45 ± 5.26 [b] | 83.77 ± 4.43 [b] | 105.3 ± 5.32 [b] |
| CP-treated oil | - | 42.38 ± 4.19 [b] | 64.59 ± 4.65 [b] | 84.43 ± 3.79 [b] | 80.57 ± 3.32 [b] | 101.5 ± 6.62 [b] |
| SC-treated oil | - | 38.27 ± 2.27 [c] | 59.37 ± 3.43 [c] | 72.04 ± 4.54 [c] | 79.65 ± 2.73 [b] | 91.74 ± 4.76 [c] |

### 3.4. Fatty Acid Composition

The fatty acid composition of oil samples is listed in Table 7. The fatty acid profile of oil changed after 20 h of frying, the proportion of unsaturated fatty acids (USFA) decreased to 84.02%, and the proportion of saturated fatty acids (SFA) increased to 15.98%. These changes may be due to the destruction of USFA by polymerization and oxidation of oil at high temperatures [68,69]. The ratio of linoleic acid to palmitic acid indicates the degree of deterioration of frying oils [60,70]. In this study, the ratio of linoleic acid to palmitic acid decreased with increased frying time. Similar findings were reported by Sharoba and Ramadan [70], who found that the ratio of linoleic acid to palmitic acid of palm olein, cottonseed and sunflower oils decreased after increasing the frying time up to 16 h. All adsorbents induced changes in the fatty acid profile of the used oil. In the case of the used oils (20 h), the linoleic acid (C18:2) increased from 58.84% (before treatment) to 61.39%, 61.99% and 61.05% for CC, CP and SC treatments, respectively. Oleic acid decreased in oil treated with CC, CP and SC by 6.99%, 7.11% and 5.21%, respectively. Lignocellulosic adsorbents (CC, CP and SC) reduced the content of USFA in the case of the fried oils for 12 h compared to used oils (12 h). These findings were consistent with the findings of Udomkun et al. [9], who found that cellulose acetate combined with commercial adsorbents reduced the content of USFA.

### 3.5. Color Analysis

Color is another important indicator of frying oil quality. Table 8 shows the effect of adsorption treatments using lignocellulosic biomasses on the oil color. The $L^*$ value refers to the brightness from zero value (full dark) to 100 value (full brightness), $a^*$ value indicates the level of greenness with a negative and the level of redness with a positive value, while the $b^*$ value ranges from blueness as negative and yellowness as a positive value [19]. Compared to fresh oils, after 20 h of potato frying, the color of the used oils tends to be dark ($L^* = 35.96$, $a^* = 1.78$ and $b^* = 18.60$). These changes in the oil color can occur because of the Maillard reaction that occurred between the potato carbohydrates and their amino acids. In addition, the increase in the release of some pigments from frying products. Another reason could be the accumulation of conjugated oxidation products induced by polymerization, oxidation and pyrolysis [19,71]. There was a slight, but not significant, improvement in $L^*$ value in SC-treated oil, while other adsorbents were ineffective in recovering the color of the used oils.

**Table 7.** Fatty acids profile of fresh, fried (for 12 and 20 h), and adsorbents-treated oils with different lignocellulosic biomasses adsorbents, including corn cob (CC), cornstalk piths (CP), and sugarcane bagasse (SC). Means with different letters ($^{a}$, $^{b}$) within each row indicate statistically significant differences between the samples after Duncan's analysis ($p < 0.05$).

| Fatty Acids (%) | 0 h | 12 h | | | | 20 h | | | |
|---|---|---|---|---|---|---|---|---|---|
| | Fresh Oil | Used | CC-Treated Oil | CP-Treated Oil | SC-Treated Oil | Used | CC-Treated Oil | CP-Treated Oil | SC-Treated Oil |
| C16:0 | 6.84 ± 0.14 $^{b}$ | 7.92 ± 0.22 $^{b}$ | 13.92 ± 0.36 $^{a}$ | 9.83 ± 0.69 $^{b}$ | 9.45 ± 0.84 $^{b}$ | 8.78 ± 0.54 $^{b}$ | 7.98 ± 1.59 $^{b}$ | 8.04 ± 0.54 $^{b}$ | 8.32 ± 0.73 $^{b}$ |
| C18:0 | 3.41 ± 0.11 $^{a}$ | 3.75 ± 0.19 $^{a}$ | 4.30 ± 0.66 $^{a}$ | 4.28 ± 0.43 $^{a}$ | 4.49 ± 0.18 $^{a}$ | 4.27 ± 0.62 $^{a}$ | 3.52 ± 0.97 $^{a}$ | 3.58 ± 0.61 $^{a}$ | 3.45 ± 0.84 $^{a}$ |
| C18:1 | 21.64 ± 0.90 $^{a}$ | 23.63 ± 0.37 $^{a}$ | 22.93 ± 0.72 $^{a}$ | 25.81 ± 1.21 $^{a}$ | 24.44 ± 0.95 $^{a}$ | 24.72 ± 0.85 $^{a}$ | 22.99 ± 1.10 $^{a}$ | 22.7 ± 0.92 $^{a}$ | 23.43 ± 1.18 $^{a}$ |
| C18:2 | 65.30 ± 1.12 $^{a}$ | 62.39 ± 1.13 $^{a,b}$ | 52.89 ± 2.31 $^{b}$ | 56.68 ± 1.23 $^{a,b}$ | 57.27 ± 1.52 $^{a,b}$ | 58.84 ± 1.18 $^{a,b}$ | 61.39 ± 1.14 $^{a,b}$ | 61.99 ± 2.25 $^{a,b}$ | 61.05 ± 1.93 $^{a,b}$ |
| SFA (%) ** | 12.31 ± 0.63 $^{b}$ | 13.78 ± 0.82 $^{b}$ | 23.07 ± 2.17 $^{a}$ | 17.51 ± 0.65 $^{a,b}$ | 17.46 ± 0.74 $^{a,b}$ | 15.98 ± 0.59 $^{b}$ | 15.39 ± 0.89 $^{b}$ | 15.31 ± 0.64 $^{b}$ | 15.14 ± 0.54 $^{b}$ |
| USFA (%) *** | 87.69 ± 1.89 $^{a}$ | 86.22 ± 1.21 $^{a}$ | 76.93 ± 2.46 $^{a}$ | 82.49 ± 1.13 $^{a}$ | 82.53 ± 1.78 $^{a}$ | 84.02 ± 2.15 $^{a}$ | 84.61 ± 1.76 $^{a}$ | 84.69 ± 2.32 $^{a}$ | 84.86 ± 1.86 $^{a}$ |

** SFA = total saturated fatty acids. *** USFA = total unsaturated fatty acids.

**Table 8.** Color analysis ($L^*$, $a^*$, and $b^*$ values) of fresh, used and treated oils with different lignocellulosic biomasses; including corn cob (CC), cornstalk piths (CP), and sugarcane bagasse (SC); 4, 12, and 20 means the frying period by (h, hours). Means with different letters ([a], [b], [c], [d]) within each column indicate statistically significant differences between the samples after Duncan's analysis ($p < 0.05$).

| Frying Time (h) | Samples | $L^*$ | $a^*$ | $b^*$ |
|---|---|---|---|---|
| 0 | Fresh oil | 40.91 ± 0.32 [a] | −1.69 ± 0.03 [c] | 7.38 ± 1.21 [d] |
| 4 | Used oil | 38.79 ± 0.41 [a] | −1.74 ± 0.04 [c] | 13.27 ± 2.34 [c] |
| | CC-treated oil | 39.47 ± 0.65 [a] | −1.70 ± 0.02 [c] | 13.32 ± 2.12 [c] |
| | CP-treated oil | 39.29 ± 0.43 [a] | −1.98 ± 0.02 [c] | 14.57 ± 1.84 [b,c] |
| | SC-treated oil | 40.30 ± 1.23 [a] | −1.96 ± 0.03 [c] | 14.37 ± 2.47 [c] |
| 12 | Used oil | 36.82 ± 0.95 [a] | 0.49 ± 0.06 [a,b,c] | 17.65 ± 1.59 [a,b] |
| | CC-treated oil | 36.94 ± 0.56 [a] | 0.78 ± 0.05 [a,b,c] | 19.23 ± 0.12 [a] |
| | CP-treated oil | 36.68 ± 0.77 [a] | 0.56 ± 0.06 [a,b,c] | 17.68 ± 0.36 [a,b] |
| | SC-treated oil | 36.62 ± 0.84 [a] | 0.25 ± 0.02 [b,c] | 17.90 ± 0.83 [a,b] |
| 20 | Used oil | 35.96 ± 0.92 [a] | 1.78 ± 0.08 [a,b] | 18.60 ± 0.32 [a] |
| | CC-treated oil | 35.29 ± 0.75 [a] | 3.31 ± 0.21 [a,b] | 17.76 ± 0.89 [a,b] |
| | CP-treated oil | 35.55 ± 0.54 [a] | 3.71 ± 0.33 [a] | 18.52 ± 1.10 [a] |
| | SC-treated oil | 38.51 ± 0.42 [a] | 2.38 ± 0.16 [a,b] | 20.84 ± 0.97 [a] |

## 4. Conclusions

Lignocellulosic biomasses can be considered effective natural adsorbents to increase the quality of edible oils. However, the effects of these adsorbents on the physicochemical properties and quality of fried sunflower oils depended on their type and structure. The XRD analysis of adsorbents showed that CC biomass had the highest crystallinity index (CI, %), followed by SC and CP. In most cases, SC biomass was the most effective adsorbent to improve the used oils' quality, followed by CP, then CC. SC biomass reduced the acid, peroxide and *p*-AV of fried oils.

Moreover, the lightness of the used oils slightly improved after using SC biomass. The fatty acid analysis showed that using SC and CP biomasses did not significantly alter the fatty acid composition of the used oils. Based on our results, it is recommended to extract cellulose from SC biomass and examine its effect on the physicochemical properties of the used oils. It could also be suggested to study the correlation between the crystallinity degree of lignocellulosic materials and the absorbance capacity and oil quality. Moreover, the infrared spectroscopy technique could be used in future studies to investigate the structural changes of oils.

**Supplementary Materials:** The following supporting information can be downloaded at: https://www.mdpi.com/article/10.3390/foods11193149/s1, Figure S1: Fitted peaks of FTIR spectra at 3300–3500 $cm^{-1}$ of lignocellulosic biomasses, including corn cob (CC), cornstalk piths (CP) and sugarcane bagasse (SC); Table S1: Lorentz components of fitted peaks of lignocellulosic biomasses, including corn cob (CC), cornstalk piths (CP) and sugarcane bagasse (SC).

**Author Contributions:** Conceptualization, M.A.M.Z. and E.A.; methodology, E.A. and M.A.M.Z.; software, A.T.; validation, E.A., M.A.M.Z. and A.Z.; formal analysis, E.A.; investigation, E.A.; resources, G.H.; data curation, S.A.K. and T.E.; writing—original draft preparation, E.A.; writing—review and editing, A.T., M.A.M.Z., G.H., A.Z., S.A.K. and T.E.; visualization, E.A.; supervision, M.A.M.Z., A.Z. and G.H.; project administration, E.A., A.T., S.A.K. and T.E.; funding acquisition, A.T., S.A.K., T.E. and E.A. All authors have read and agreed to the published version of the manuscript.

**Funding:** The publication of this article was funded by the Open Access Fund of Leibniz Universität Hannover.

**Institutional Review Board Statement:** Not applicable.

**Informed Consent Statement:** Not applicable.

**Data Availability Statement:** The data presented in this study are available on request from the corresponding author.

**Acknowledgments:** The graphical abstract in the online version was created with Biorender.com with a publication license. The authors thank the editors and reviewers for their critical reading of the manuscript.

**Conflicts of Interest:** The authors declare no conflict of interest.

## References

1. Ogata, F.; Kawasaki, N. Regeneration of Waste Edible Oil by the Use of Virgin and Calcined Magnesium Hydroxide as Adsorbents. *J. Oleo Sci.* **2016**, *65*, 941–948. [CrossRef] [PubMed]
2. Aşkın, B.; Kaya, Y. Effect of Deep Frying Process on the Quality of the Refined Oleic/Linoleic Sunflower Seed Oil and Olive Oil. *J. Food Sci. Technol.* **2020**, *57*, 4716–4725. [CrossRef]
3. Mohammadalinejhad, S.; Dehghannya, J. Effects of Ultrasound Frequency and Application Time Prior to Deep-Fat Frying on Quality Aspects of Fried Potato Strips. *Innov. Food Sci. Emerg. Technol.* **2018**, *47*, 493–503. [CrossRef]
4. Asokapandian, S.; Swamy, G.J.; Hajjul, H. Deep Fat Frying of Foods: A Critical Review on Process and Product Parameters. *Crit. Rev. Food Sci. Nutr.* **2020**, *60*, 3400–3413. [CrossRef] [PubMed]
5. Zhao, L.; Zhang, M.; Wang, H.; Devahastin, S. Effects of Carbon Dots in Combination with Rosemary-Inspired Carnosic Acid on Oxidative Stability of Deep Frying Oils. *Food Control* **2021**, *125*, 107968. [CrossRef]
6. Yuan, Y.; Cui, C.; Liu, H.; Li, X.; Cao, Y.; Zhang, Y.; Yan, H. Effects of Oxidation and Hydrolysis of Frying Oil on MCPD Esters Formation in Chinese Fried Dough Sticks. *LWT* **2022**, *154*, 112576. [CrossRef]
7. Hwang, H.S.; Ball, J.C.; Doll, K.M.; Anderson, J.E.; Vermillion, K. Investigation of Polymers and Alcohols Produced in Oxidized Soybean Oil at Frying Temperatures. *Food Chem.* **2020**, *317*, 126379. [CrossRef] [PubMed]
8. Li, P.; Yang, X.; Lee, W.J.; Huang, F.; Wang, Y.; Li, Y. Comparison between Synthetic and Rosemary-Based Antioxidants for the Deep Frying of French Fries in Refined Soybean Oils Evaluated by Chemical and Non-Destructive Rapid Methods. *Food Chem.* **2021**, *335*, 127638. [CrossRef]
9. Udomkun, P.; Innawong, B.; Jumrusjumroendee, N. Cellulose Acetate and Adsorbents Supported on Cellulose Fiber Extracted from Waxy Corn Husks for Improving Shelf Life of Frying Oil. *LWT* **2018**, *97*, 45–52. [CrossRef]
10. Ganesan, K.; Sukalingam, K.; Xu, B. Impact of Consumption and Cooking Manners of Vegetable Oils on Cardiovascular Diseases- A Critical Review. *Trends Food Sci. Technol.* **2018**, *71*, 132–154. [CrossRef]
11. Ganesan, K.; Sukalingam, K.; Xu, B. Impact of Consumption of Repeatedly Heated Cooking Oils on the Incidence of Various Cancers- A Critical Review. *Crit. Rev. Food Sci. Nutr.* **2019**, *59*, 488–505. [CrossRef]
12. Perumalla Venkata, R.; Subramanyam, R. Evaluation of the Deleterious Health Effects of Consumption of Repeatedly Heated Vegetable Oil. *Toxicol. Rep.* **2016**, *3*, 636–643. [CrossRef]
13. Ahmad Nazri, K.A.; Fauzi, N.M.; Buang, F.; Mohd Saad, Q.H.; Husain, K.; Jantan, I.; Jubri, Z. Gynura Procumbens Standardised Extract Reduces Cholesterol Levels and Modulates Oxidative Status in Postmenopausal Rats Fed with Cholesterol Diet Enriched with Repeatedly Heated Palm Oil. *Evid. Based Complement. Altern. Med.* **2019**, *2019*, 7246756. [CrossRef]
14. Kummerow, F.A. Interaction between Sphingomyelin and Oxysterols Contributes to Atherosclerosis and Sudden Death. *Am. J. Cardiovasc. Dis.* **2013**, *3*, 17. [PubMed]
15. Cárdenas, J.; Orjuela, A.; Sánchez, D.L.; Narváez, P.C.; Katryniok, B.; Clark, J. Pre-Treatment of Used Cooking Oils for the Production of Green Chemicals: A Review. *J. Clean. Prod.* **2021**, *289*, 125129. [CrossRef]
16. Hosseinzadeh-Bandbafha, H.; Nizami, A.S.; Kalogirou, S.A.; Gupta, V.K.; Park, Y.K.; Fallahi, A.; Sulaiman, A.; Ranjbari, M.; Rahnama, H.; Aghbashlo, M.; et al. Environmental Life Cycle Assessment of Biodiesel Production from Waste Cooking Oil: A Systematic Review. *Renew. Sustain. Energy Rev.* **2022**, *161*, 112411. [CrossRef]
17. Bhattacharya, A.B.; Sajilata, M.G.; Tiwari, S.R.; Singhal, R.S. Regeneration of Thermally Polymerized Frying Oils with Adsorbents. *Food Chem.* **2008**, *110*, 562–570. [CrossRef]
18. Subramanian, R.; Nandini, K.E.; Sheila, P.M.; Gopalakrishna, A.G.; Raghavarao, K.S.M.S.; Nakajima, M.; Kimura, T.; Maekawa, T. Membrane Processing of Used Frying Oils. *J. Am. Oil Chem. Soc.* **2000**, *77*, 323. [CrossRef]
19. Yılmaz, E.; Güner, M. Regeneration of Used Frying Oils by Selected Metal–Organic Frameworks as Adsorbents. *J. Am. Oil Chem. Soc.* **2018**, *95*, 1497–1508. [CrossRef]
20. Somnuk, C.; Innawong, B.; Tirawattanawanich, C. Cytotoxicity of Used Frying Oil Recovered by Different Adsorbents. *Kasetsart J. Natural Sci.* **2013**, *47*, 874–884.
21. Adejumo, A.L.; Azeez, L.; Oyedeji, A.O.; Adetoro, R.O.; Aderibigbe, F.A. Nanostructured and Surface Functionalized Corncob as Unique Adsorbents for Anionic Dye Remediation. *SN Appl. Sci.* **2020**, *2*, 301. [CrossRef]
22. Herrera-Barros, A.; Tejada-Tovar, C.; Villabona-Ortiz, A.D.; Gonzalez-Delgado, A.D.; Alvarez-Calderon, J. Adsorption of Nickel and Cadmium by Corn Cob Biomass Chemically Modified with Alumina Nanoparticles. *Indian J. Sci. Technol.* **2018**, *11*, 1–11. [CrossRef]
23. Patel, S. Potential of Fruit and Vegetable Wastes as Novel Biosorbents: Summarizing the Recent Studies. *Rev. Environ. Sci. Biotechnol.* **2012**, *11*, 365–380. [CrossRef]

24. Bavaresco, A.; Fonseca, J.M.; Scheufele, F.B.; da Silva, C.; Teleken, J.G. Use of Carbonized Corn Cob Biomass to Reduce Acidity of Residual Frying Oil. *Acta Sci. Technol.* **2021**, *43*, e51303. [CrossRef]
25. Torre, P.; Aliakbarian, B.; Rivas, B.; Domínguez, J.M.; Converti, A. Release of Ferulic Acid from Corn Cobs by Alkaline Hydrolysis. *Biochem. Eng. J.* **2008**, *40*, 500–506. [CrossRef]
26. Singh, H.K.; Patil, T.; Vineeth, S.K.; Das, S.; Pramanik, A.; Mhaske, S.T. Isolation of Microcrystalline Cellulose from Corn Stover with Emphasis on Its Constituents: Corn Cover and Corn Cob. *Mater. Today Proc.* **2020**, *27*, 589–594. [CrossRef]
27. Alves, C.C.O.; Franca, A.S.; Oliveira, L.S. Evaluation of an Adsorbent Based on Agricultural Waste (Corn Cobs) for Removal of Tyrosine and Phenylalanine from Aqueous Solutions. *BioMed Res. Int.* **2013**, *2013*, 978256. [CrossRef]
28. Han, Q.; Gao, X.; Zhang, H.; Chen, K.; Peng, L.; Jia, Q. Preparation and Comparative Assessment of Regenerated Cellulose Films from Corn (Zea Mays) Stalk Pulp Fines in DMAc/LiCl Solution. *Carbohydr. Polym.* **2019**, *218*, 315–323. [CrossRef]
29. Peng, D.; Li, W.; Zheng, L. Laccase-Modified Cornstalk Pith for Cleanup of Spilled Diesel Oil. *Cellulose* **2021**, *28*, 7123–7142. [CrossRef]
30. Fan, T.; Cheng, S.; Peng, D.; Kong, S. A Comparative Study on Sorption Characteristics by Corn Stalk Pith Using Green Modification with Different Polycarboxylic Acids. *Fibers Polym.* **2021**, *22*, 862–876. [CrossRef]
31. Peng, D.; Cheng, S.; Li, H.; Guo, X. Effective Multi-Functional Biosorbent Derived from Corn Stalk Pith for Dyes and Oils Removal. *Chemosphere* **2021**, *272*, 129963. [CrossRef] [PubMed]
32. Karnitz, O.; Gurgel, L.V.A.; de Melo, J.C.P.; Botaro, V.R.; Melo, T.M.S.; de Freitas Gil, R.P.; Gil, L.F. Adsorption of Heavy Metal Ion from Aqueous Single Metal Solution by Chemically Modified Sugarcane Bagasse. *Bioresour. Technol.* **2007**, *98*, 1291–1297. [CrossRef] [PubMed]
33. Loh, Y.R.; Sujan, D.; Rahman, M.E.; Das, C.A. Sugarcane Bagasse—The Future Composite Material: A Literature Review. *Resour. Conserv. Recycl.* **2013**, *75*, 14–22. [CrossRef]
34. Wannahari, R.; Firdhaus, M.; Nordin, M. Under a Creative Commons Attribution (CC-BY) 3.0 License The Recovery of Used Palm Cooking Oil Using Bagasse as Adsorbent. *Am. J. Eng. Appl. Sci.* **2012**, *5*, 59–62.
35. Schneider, L.T.; Bonassa, G.; Alves, H.J.; Meier, T.R.W.; Frigo, E.P.; Teleken, J.G. Use of Rice Husk in Waste Cooking Oil Pretreatment. *Environ. Technol.* **2019**, *40*, 594–604. [CrossRef]
36. Ali, R.F.M.; el Anany, A.M. Recovery of Used Frying Sunflower Oil with Sugar Cane Industry Waste and Hot Water. *J. Food Sci. Technol.* **2014**, *51*, 3002–3013. [CrossRef]
37. Gunstone, F.D. *Vegetable Oils in Food Technology: Composition, Properties and Uses*, 2nd ed.; John Wiley and Sons: Hoboken, NJ, USA, 2011; ISBN 9781444332681.
38. Ponnusamy, V.K.; Nguyen, D.D.; Dharmaraja, J.; Shobana, S.; Banu, J.R.; Saratale, R.G.; Chang, S.W.; Kumar, G. A Review on Lignin Structure, Pretreatments, Fermentation Reactions and Biorefinery Potential. *Bioresour. Technol.* **2019**, *271*, 462–472. [CrossRef]
39. Rodrigues, R.C.L.B.; Green Rodrigues, B.; Vieira Canettieri, E.; Acosta Martinez, E.; Palladino, F.; Wisniewski, A.; Rodrigues, D. Comprehensive Approach of Methods for Microstructural Analysis and Analytical Tools in Lignocellulosic Biomass Assessment—A Review. *Bioresour. Technol.* **2022**, *348*, 126627. [CrossRef]
40. Xu, Q.; Chen, Z.; Wu, Z.; Xu, F.; Yang, D.; He, Q.; Li, G.; Chen, Y. Novel Lanthanum Doped Biochars Derived from Lignocellulosic Wastes for Efficient Phosphate Removal and Regeneration. *Bioresour. Technol.* **2019**, *289*, 121600. [CrossRef]
41. Galal, A.M.F.; DiaaAtta; Abouelsayed, A.; Ibrahim, M.A.; Hanna, A.G. Configuration and Molecular Structure of 5-Chloro-N-(4-Sulfamoylbenzyl) Salicylamide Derivatives. *Spectrochim. Acta Part A Mol. Biomol. Spectrosc.* **2019**, *214*, 476–486. [CrossRef]
42. Karthika, K.; Arun, A.B.; Rekha, P.D. Enzymatic Hydrolysis and Characterization of Lignocellulosic Biomass Exposed to Electron Beam Irradiation. *Carbohydr. Polym.* **2012**, *90*, 1038–1045. [CrossRef]
43. Hrebień-Filisińska, A. Application of Natural Antioxidants in the Oxidative Stabilization of Fish Oils: A Mini-Review. *J. Food Process. Preserv.* **2021**, *45*, e15342. [CrossRef]
44. Udomkun, P.; Innawong, B.; Siasakul, C.; Okafor, C. Utilization of Mixed Adsorbents to Extend Frying Oil Life Cycle in Poultry Processing. *Food Chem.* **2018**, *248*, 225–229. [CrossRef]
45. AOCS. *Official Methods and Recommended Practices of the American Oil Chemists' Society*, 4th ed.; AOCS Press: Champaign, IL, USA, 2004.
46. Nuru, Z.; Getachew, P. Improving the Quality of Used Frying Niger Seed Oil with Adsorbent Treatment. *Heliyon* **2021**, *7*, e06748. [CrossRef] [PubMed]
47. IUPAC, C. Method Number 2.504. Determination of the P-Anisidine Value (P-AV). *Stand. Methods Anal. Oils Fats Deriv.* **1987**, *51*, 2–207.
48. Sun-Waterhouse, D.; Zhou, J.; Miskelly, G.M.; Wibisono, R.; Wadhwa, S.S. Stability of Encapsulated Olive Oil in the Presence of Caffeic Acid. *Food Chem.* **2011**, *126*, 1049–1056. [CrossRef]
49. Yang, Y.; Ferro, M.D.; Cavaco, I.; Liang, Y. Detection and Identification of Extra Virgin Olive Oil Adulteration by GC-MS Combined with Chemometrics. *J. Agric. Food Chem.* **2013**, *61*, 3693–3702. [CrossRef]
50. Su, C. Pamela White Frying Stability of High-Oleate and Regular Soybean Oil Blends. *AOCS* **2004**, *81*, 783–788.
51. Boukir, A.; Fellak, S.; Doumenq, P. Structural Characterization of Argania Spinosa Moroccan Wooden Artifacts during Natural Degradation Progress Using Infrared Spectroscopy (ATR-FTIR) and X-ray Diffraction (XRD). *Heliyon* **2019**, *5*, e02477. [CrossRef]

52. Asemani, M.; Rabbani, A.R. Detailed FTIR Spectroscopy Characterization of Crude Oil Extracted Asphaltenes: Curve Resolve of Overlapping Bands. *J. Pet. Sci. Eng.* **2020**, *185*, 106618. [CrossRef]
53. Moura, H.O.M.A.; Campos, L.M.A.; da Silva, V.L.; de Andrade, J.C.F.; de Assumpção, S.M.N.; Pontes, L.A.M.; de Carvalho, L.S. Investigating Acid/Peroxide-Alkali Pretreatment of Sugarcane Bagasse to Isolate High Accessibility Cellulose Applied in Acetylation Reactions. *Cellulose* **2018**, *25*, 5669–5685. [CrossRef]
54. Shao, X.; Wang, J.; Liu, Z.; Hu, N.; Liu, M.; Xu, Y. Preparation and Characterization of Porous Microcrystalline Cellulose from Corncob. *Ind. Crops Prod.* **2020**, *151*, 112457. [CrossRef]
55. Pereira, S.C.; Maehara, L.; Machado, C.M.M.; Farinas, C.S. Physical–Chemical–Morphological Characterization of the Whole Sugarcane Lignocellulosic Biomass Used for 2G Ethanol Production by Spectroscopy and Microscopy Techniques. *Renew. Energy* **2016**, *87*, 607–617. [CrossRef]
56. Ninomiya, K.; Soda, H.; Ogino, C.; Takahashi, K.; Shimizu, N. Effect of Ionic Liquid Weight Ratio on Pretreatment of Bamboo Powder Prior to Enzymatic Saccharification. *Bioresour. Technol.* **2013**, *128*, 188–192. [CrossRef]
57. Mou, H.; Wu, S. Comparison of Hydrothermal, Hydrotropic and Organosolv Pretreatment for Improving the Enzymatic Digestibility of Bamboo. *Cellulose* **2017**, *24*, 85–94. [CrossRef]
58. Yao, W.; Weng, Y.; Catchmark, J.M. Improved Cellulose X-Ray Diffraction Analysis Using Fourier Series Modeling. *Cellulose* **2020**, *27*, 5563–5579. [CrossRef]
59. Zou, Y.; Fu, J.; Chen, Z.; Ren, L. The Effect of Microstructure on Mechanical Properties of Corn Cob. *Micron* **2021**, *146*, 103070. [CrossRef]
60. Nayak, P.K.; Dash, U.; Rayaguru, K.; Krishnan, K.R. Physio-Chemical Changes during Repeated Frying of Cooked Oil: A Review. *J. Food Biochem.* **2016**, *40*, 371–390. [CrossRef]
61. Sun, Y.; Zhang, M.; Fan, D. Effect of Ultrasonic on Deterioration of Oil in Microwave Vacuum Frying and Prediction of Frying Oil Quality Based on Low Field Nuclear Magnetic Resonance (LF-NMR). *Ultrason. Sonochem.* **2019**, *51*, 77–89. [CrossRef]
62. Farag, R.S.; El-Anany, A.M. Improving the Quality of Fried Oils by Using Different Filter Aids. *J. Sci. Food Agric.* **2006**, *86*, 2228–2240. [CrossRef]
63. Nwadiogbu, J.O.; Ajiwe, V.I.E.; Okoye, P.A.C. Removal of Crude Oil from Aqueous Medium by Sorption on Hydrophobic Corncobs: Equilibrium and Kinetic Studies. *J. Taibah Univ. Sci.* **2016**, *10*, 56–63. [CrossRef]
64. Shahidi, F. *Bailey's Industrial Oil and Fat Products, Industrial and Nonedible Products from Oils and Fats*, 6th ed.; John Wiley & Sons, Inc.: Hoboken, NJ, USA, 2005.
65. Choe, E.; Min, D.B. Chemistry of Deep-Fat Frying Oils. *J. Food Sci.* **2007**, *72*, R77–R86. [CrossRef]
66. Ismail, S.A.E.A.; El-Anany, A.M.; Ali, R.F.M. Regeneration of Used Frying Palm Oil with Coffee Silverskin (CS), CS Ash (CSA) and Nanoparticles of CS (NCS). *J. Oleo Sci.* **2017**, *66*, 897–905. [CrossRef] [PubMed]
67. Koohikamali, S.; Alam, M.S. Improvement in Nutritional Quality and Thermal Stability of Palm Olein Blended with Macadamia Oil for Deep-Fat Frying Application. *J. Food Sci. Technol.* **2019**, *56*, 5063–5073. [CrossRef]
68. Solati, Z.; Baharin, B.S. Antioxidant Effect of Supercritical $CO_2$ Extracted *Nigella sativa* L. Seed Extract on Deep Fried Oil Quality Parameters. *J. Food Sci. Technol.* **2015**, *52*, 3475–3484. [CrossRef] [PubMed]
69. Aydinkaptan, E.; Mazi, B.G.; Barutçu Mazi, I. Microwave Heating of Sunflower Oil at Frying Temperatures: Effect of Power Levels on Physicochemical Properties. *J. Food Process. Eng.* **2017**, *40*, e12402. [CrossRef]
70. Sharoba, A.M.; Ramadan, M.F. Impact of Frying on Fatty Acid Profile and Rheological Behaviour of Some Vegetable Oils. *J. Food Process. Technol.* **2017**, *3*, 7. [CrossRef]
71. Udomkun, P.; Niruntasuk, P.; Innawong, B. Impact of Novel Far-Infrared Frying Technique on Quality Aspects of Chicken Nuggets and Frying Medium. *J. Food Process. Preserv.* **2019**, *43*, e13931. [CrossRef]

foods

MDPI

Article

# Valorization of Fish Processing By-Products: Microstructural, Rheological, Functional, and Properties of Silver Carp Skin Type I Collagen

Yongxin Guan [1,2], Jianlin He [2], Junde Chen [2,*], Yushuang Li [2], Xingkun Zhang [2], Yan Zheng [2] and Linyan Jia [1,*]

1 College of Chemistry and Chemical Engineering, Mudanjiang Normal University, Mudanjiang 157011, China
2 Technical Innovation Center for Utilization of Marine Biological Resources, Third Institute of Oceanography, Ministry of Natural Resources, Xiamen 361005, China
* Correspondence: jdchen@tio.org.cn (J.C.); 1002002@mdjnu.edu.cn (L.J.); Tel./Fax: +86-592-215527 (J.C.); +86-453-6426868 (L.J.)

**Abstract:** The objective of this study was to develop aquatic collagen production from fish processing by-product skin as a possible alternative to terrestrial sources. Silver carp skin collagen (SCSC) was isolated and identified as type I collagen, and LC-MS/MS analysis confirmed the SCSC as *Hypophthalmichthys molitrix* type I collagen, where the yield of SCSC was 40.35 ± 0.63% (dry basis weight). The thermal denaturation temperature ($T_d$) value of SCSC was 30.37 °C, which was superior to the collagen of deep-sea fish and freshwater fish. Notably, SCSC had higher thermal stability than human placental collagen, and the rheological experiments showed that the SCSC was a shear-thinning pseudoplastic fluid. Moreover, SCSC was functionally superior to some other collagens from terrestrial sources, such as sheep, chicken cartilage, and pig skin collagen. Additionally, SCSC could provide a suitable environment for MC3T3-E1 cell growth and maintain normal cellular morphology. These results indicated that SCSC could be used for further applications in food, cosmetics, and biomedical fields.

**Keywords:** silver carp; collagen; by-products; functional properties; rheological properties

**Citation:** Guan, Y.; He, J.; Chen, J.; Li, Y.; Zhang, X.; Zheng, Y.; Jia, L. Valorization of Fish Processing By-Products: Microstructural, Rheological, Functional, and Properties of Silver Carp Skin Type I Collagen. *Foods* **2022**, *11*, 2985. https://doi.org/10.3390/foods11192985

Academic Editors: Michela Verni and Federico Casanova

Received: 6 September 2022
Accepted: 21 September 2022
Published: 24 September 2022



## 1. Introduction

Collagen is the main component of the extracellular matrix of animals and is found in bones, skin, and tendons, accounting for about 30% of the total protein content. Collagen consists of three polypeptide chains twisted to create a triple helix structure, providing mechanical stability, flexibility, and strength for the organism [1]. Scholars have identified 29 different types of collagens with different distinct amino acid sequences and molecular structures [2]. Among the discovered collagens, type I collagen has a wide range of food, cosmetic, and biomedical applications due to its excellent biocompatibility, low antigenicity, and high biodegradability [3].

Traditional type I collagen found on the market is primarily derived from the skin and bones of mammals such as cattle and pigs. In recent years, outbreaks of terrestrial infectious diseases such as foot and mouth disease, avian influenza, and Creutzfeldt–Jakob disease have caused restrictions on collagen applications. Religious factors also limit the use of collagen of terrestrial origin [4]. The currently available sources of collagen cannot satisfy the increasing consumer demand, and new sources of collagen are urgently required. Marine collagen has attracted the attention of researchers. Nalinanon et al. (2010) [5] extracted collagen from the Arabesque greenling skin and examined its properties. Sun et al. (2017) [3] extracted Pacific cod skin collagen and evaluated its biocompatibility. Li et al. (2013) [6] extracted collagen using two different methods to extract collagen from the skin and bones of Spanish mackerel. Although fish skin collagen research has focused on marine fish, the poor thermal stability of deep-sea fish collagen has limited its application. Freshwater fish live in a temperature environment very different from deep-sea fish, which may result in the better thermal stability of freshwater fish collagen compared to deep-sea fish collagen.

Silver carp (*Hypophthalmichthys molitrix*) is a filter-feeding freshwater fish of the carp genus. It is widely distributed in the principal waters of China and other Asian countries, and is the primary economic fish species in China, with an annual production value of more than 3.8 million tons. About 80% of the by-products from silver carp are not rationally utilized during processing [7]. Some of these by-products are processed into low-value products such as fertilizers, with most discarded in water bodies. The presence of $NO_3^-$ and $PO4^{3-}$ in by-product skins will alter the dissolved oxygen capacity of water, causing eutrophication of the water bodies and disrupting the ecological balance of aquatic organisms [8]. These by-product fish skins contain 50–70% collagen. Thus, the use of silver carp skin may be an effective way to obtain high-value-added products as an alternative source of traditional collagen. In recent years, researchers have obtained collagen from the skin and scales of silver carp processed by-products and investigated its structural properties. Zhang et al. (2010) identified the potential of collagen from silver carp skin and scales in food processing [9]. Faralizadeh et al. (2021) determined the feasibility of collagen-based biomaterials from the perspective of cytocompatibility [10]. Sionkowska et al. (2020) analyzed the influence of UV light on rheological properties of collagen extracted from silver carp skin [11]. However, there are few reports that systematically study the rheological and functional properties of collagen from silver carp skin. Rheological properties are crucial parameters required for protein product production. In the processing of collagen products, rheological properties will affect the material properties, and the functional properties are also critical indicators for determining collagen applications.

Therefore, the purpose of this investigation was to address the extraction of silver carp skin collagen (SCSC), and characterize its microstructural, rheological, functional, and biological properties to determine its potential as an alternative source of terrestrial collagen.

## 2. Materials and Methods

### 2.1. Materials

Rat tail type I collagen and protein marker (26634) were purchased from the Sigma Chemical Company (St. Louis, MO, USA), while sodium dodecyl sulfate (SDS), *N, N, N, N*-tetramethylethylenediamime (TEMED), and Coomassie Brilliant Blue R-250 were purchased from Bio-Rad Laboratories (Harkles, CA, USA). MC3T3-E1 cell lines (Cat No. CBP60946) were purchased from Cobioer (Nanjing, China). Other reagents used in the experimental process were of analytical-grade purity.

### 2.2. Preparation of SCSC

SCSC was isolated according to a previously described method with slight modifications [12]. Silver carp (2 years of age, 25–50 cm, 1500–2000 g) was purchased from Fujian Fengsheng Food Co., Ltd. (Zhangzhou, China). Silver carp body skins were manually removed using a scalpel and washed with cold water. These skins were incubated in 150 L of 0.1 mol/L sodium bicarbonate for 3 h with continuous stirring, and then rinsed with cold water until a neutral pH was reached. The resulting material was soaked in 0.5 M acetic acid at 1:40 ($w/v$) and stirred for 10 h at 4 °C in a top-mounted electronic stirrer (EUROSTAR 20 digital, IKA, Wilmington Germany). Then, these skins were centrifuged at 9000 rpm at 4 °C for 30 min (Avanti J-26 XP, Beckman, Brea, CA, USA). NaCl was added to the supernatant after centrifugation to achieve a final concentration of 3% ($w/v$) and then salted for 30 min. The sediment was collected after centrifugation at 9000 rpm for 30 min at 4 °C, and the precipitate was redissolved in 0.5 M acetic acid at 1:10 ($w/v$). After allowing for sufficient dissolution, the solution was placed in dialysis bags (MD 77 MM, Viskase, Lombard, IL, USA) in 0.1 M acetic acid for 24 h, followed by distilled water for 48 h. Then, the dialyzed samples were freeze-dried (Telstar, lyoobeta-25, Barcelona, Spain), and the obtained samples were stored at −20 °C. The yield can be calculated as follows:

$$\text{Yield }(\%) = \frac{M}{M_0} \times 100\%, \tag{1}$$

where $M_0$ is the dry weight of the fish skin (g) and $M$ is the weight of the freeze-dried collagen (g).

### 2.3. Sodium Dodecyl Sulfate–Polyacrylamide Gel Electrophoresis (SDS-PAGE)

Collagen samples were mixed with the sample buffer at a rate of 4:1 (*v*/*v*), and the acquired mixture was heated at 100 °C for 3 min. Subsequently, the samples were loaded on 8% separating gel and 3% stacking gel. Electrophoresis gel plates were prepared and electrophoresed at 110 V and 50 mA for 80 min using a mini protein vertical plate electrophoresis system (Bio-Rad Laboratories, Hercules, CA, USA). Electrophoresis was performed after staining the gels using 0.1% (*w/v*) Coomassie Brilliant Blue R-250, 30% (*v*/*v*) $CH_3OH$, and 10% (*v*/*v*) acetic acid, and finally the gels were decolorized with 50% (*v*/*v*) $C_2H_5OH$ and 10% (*v*/*v*) acetic acid. Rat tail type I collagen was utilized as the standard. The protein molecular weights were estimated using protein markers (26634) and Quantity One 4.6.0 (Bio-Rad Laboratories, Hercules, CA, USA). The band intensities were analyzed using ImageJ software (VERSION 1.8.0, National Institute of Mental Health, Bethesda, MD, USA).

### 2.4. Amino Acid Sequence Analysis

The amino acid sequence determination of SCSC was performed according to the method previously reported [13]. The samples were first subjected to electrophoresis experiments. The SDS-PAGE strips were cut, rinsed with double distilled water, and then decolorized. Subsequently, 0.01 μg/μL of proteomics grade trypsin to obtain the appropriate amount, and then 25 mmol/L of $NH_4HCO_3$ solution containing 10% ACN overlay, in a 37 °C water bath. After digestion, the supernatant was transferred to a new centrifuge tube, and 50 μL of extract was added to the remaining gel, which was centrifuged for 15 min after the water bath. Then, the supernatant was combined, concentrated, and centrifuged, and then dried for mass spectrometry analysis.

The lyophilized peptide samples were redissolved in mobile phase A (0.1% formic acid, 5% acetonitrile/water) with a 2 μL/min C18 pre-column (100 μm × 3 cm, C18, 3 μm, 150 Å). Then they were held on an Easy-nLC1200 liquid phase system (HPLC MS/MS, Ultimate 3000-API 4000 Q TRAP, Thermo Fisher Scientific, Dreieich, Germany) for 8 min to rinse and desalt the samples. The column size was a C18 reversed-phase column (75 μm × 15 cm C18, 3 μm, 120 Å, ChromXP Eksigent, San Francisco, CA, USA) and the gradient used for experimentation consisted of 10 min LCMS with mobile phase B (0.1% formic acid, 95% acetonitrile/water), increasing from 5% to 40%. Mass spectrometry was performed using a QEactive system, combined with a nanoliter spray III ion source with a spray voltage of 2.3 kV, an air curtain pressure of 30 PSI, a nebulization air pressure of 14 PSI, and a heating temperature of 150 °C. The mass spectrometry scan mode consisted of information-dependent acquisition in IDA, information-dependent analysis mode, with a scan time of 250 ms for a single spectrum of TOF-MS. Up to 26 secondary spectra with $2^+$ to $5^+$ charges and single counts greater than 200 cps were acquired in each IDA cycle, with a cumulative time of 80 ms for each secondary spectrum. The database was NCBI Mascot, the enzyme was trypsin, the maximum allowed missed cut site was 1, the fixed modification was carbamidomethyl (C), the variable modifications were acetylation (protein N-terminal), deamidation (NQ), and Gln->pyro-Glu (N-terminal Q), Glu->pyro-Glu Glu->pyro-Glu (N-terminal E) and oxidation (M), the MS tolerance was 20 ppm, and the MS/MS tolerance was 0.05 Da. Protein confidence above 95% was considered successful for identification.

### 2.5. Spectral Analysis

#### 2.5.1. Ultraviolet Absorption (UV)

Collagen samples were dissolved in 0.5 M acetic acid at a concentration of 1 mg/mL, and the UV absorption value of the sample solution was calculated using a UV-Vis absorption spectrometer (UV-2550 SHIMADZU, Kyoto, Japan) in the wavelength range of 190–600 nm, with 0.5 M HAC as the blank control.

#### 2.5.2. Fourier Transform Infrared (FTIR)

The collagen and KBr mixtures at 1:100 (*w/w*) were pressed on a translucent sample sheet and analyzed using an FTIR spectrometer (VERTEX 70, Bruker, Ettlingen, Germany). The scan range was from 400 to 4000 $cm^{-1}$ with a resolution of 4 $cm^{-1}$, using KBr as the blank control.

#### 2.5.3. Circular Dichroism (CD)

Collagen samples were dissolved separately in 0.5 M acetic acid to produce 0.1 mg/mL solutions of collagen. Then, the solutions were centrifuged at 18,000 rpm for 10 min at 4 °C. The supernatants were collected and placed in a quartz cell and the CD spectra were measured on a Chirascon CD spectrometer (Applied Photophysics Ltd., London, UK). The spectra were recorded at 260–190 nm wavelengths at 15 °C in 0.1 nm steps with a response time of 1 s.

#### 2.5.4. X-ray Diffraction (XRD)

Collagen samples were measured using an X-ray diffractometer (X'Pert Pro XRD, PANalytical, Almelo, The Netherlands) with CuKa rays, $\lambda$ = 0.154 nm, a scanning range 0–80°, and an angular velocity 0.06 °/s. The d value was calculated from the Bragg equation, Equation (2):

$$\mathrm{d}() = \frac{\lambda}{2sin\theta}, \tag{2}$$

where $\lambda$ is the X-ray wavelength (1.54°) and $\theta$ is the Bragg diffraction angle.

### 2.6. Determination of Denaturation Temperature ($T_d$)

The $T_d$ was according to the method published with slight modifications [12]. Collagen was dissolved in 0.5 M acetic acid to 20 mg/mL, and the Td was measured using a rheometer (MCR 302, Anton Paar, Graz, Austria) with a stainless-steel cone/plate geometry (CP 25-2) (2° cone angle, 25 mm cone diameter). The samples were placed in the gap between the top and bottom plates, which were set at 103 µm. Dimethylsilicone oil was applied around the top and bottom plates to avoid evaporation of moisture from the sample. The samples were measured at temperatures from 10 to 50 °C, where the shear rate was constant at 1 $s^{-1}$ and the temperature was elevated at a rate of 3 °C/min. The temperature at which the relative viscosity was 50% was recorded as the $T_d$ of the samples, which was calculated according to Equation (3):

$$T_d = \frac{A}{B} \times 100\%, \tag{3}$$

where A is the sample viscosity (Pa.s) and B is the sample initial viscosity.

### 2.7. Rheological Properties

The rheological properties of the collagen were examined by dissolving collagen in acetic acid to 5, 10, 15, 20, and 25 mg/mL at 20 °C. The dynamic frequency sweeps were performed for the 15 mg/mL collagen at different temperatures of 20, 25, 30, 35, and 40 °C, while the G′, G″, η*, and tangent angle were determined with a rheometer (MCR 302, Anton Paar, Graz, Austria) using a stainless-steel cone/plate geometry (CP 60-0.5) (0.5° cone angle, 60 mm cone diameter) with a dynamic frequency sweep from 0.01 to 10 Hz in the oscillatory mode. The sample was placed in the gap between the top and bottom plates, which was set at 57 μm. Dimethylsilicone oil was applied around the top and bottom plates to avoid evaporation of moisture from the sample.

### 2.8. Zeta Potential

The Zeta potential was determined according to the method published with slight modifications [14]. The collagen was prepared with 0.1 M acetic acid to 0.2 mg/mL for each 20 mL sample solution. The pH was adjusted to a range of 2–10, with 1.0 M $HNO_3$ and 4.0 M NaOH. These solutions were evaluated with a Zeta potential analyzer (Zetasizer Nano ZS90, Malvern Instr., Melvin, UK).

### 2.9. Functional Properties

#### 2.9.1. Foaming Properties

The foaming properties were determined according to the method published with slight modifications [15]. The collagen was formulated with 0.5 M acetic acid to 0.5 (*w/v*). The pH values were adjusted to 2, 4, 6, 6.98, 8, and 10 with 1 M HCl and 4 M NaOH, respectively. Then, the samples were homogenized at 22,000 rpm (JS25, JUNRUI, Yangzhou, China) for 2 min at room temperature, and the foam volumes at 0 min and 1 h were recorded. The foaming capacity (FC) and foam stability (FS) were calculated according to Equations (4) and (5):

$$\mathrm{FC}\% = \frac{V_2}{V_1} \times 100\%, \tag{4}$$

$$\mathrm{FS}(\%) = \frac{V_3}{V_2} \times 100\%, \tag{5}$$

where $V_1$ is the initial volume (mL), $V_2$ is the foam volume (mL), and $V_3$ is the foam volume after 1 h of rest (mL).

#### 2.9.2. Emulsifying Properties

The emulsifying properties were determined according to the method published with slight modifications [15]. The collagen was formulated with 0.5 M acetic acid to 0.5% (*w/v*), and the pH was adjusted to 2, 4, 6, 6.98, 8, and 10 with 1M HCl and 4 M NaOH. The samples were incorporated with 5 mL of corn oil and homogenized at 22,000 rpm (JS25, JUNRUI, Yangzhou, China) for 1 min. The emulsions were diluted with 0.1% SDS for 0 and 10 min. Then, the oscillator was shaken for 30 s and 0.1% SDS was used as the blank control to evaluate the absorbance at 500 nm. (UV-2550 SHIMADZU, Kyoto, Japan). The emulsification activity index (ESI) and the emulsion stability index (EAI) were calculated according to Equations (6) and (7):

$$\mathrm{EAI}\left(\frac{\mathrm{m}^2}{\mathrm{g}}\right) = \frac{2 \times 2.303 \times A_0}{\phi \times weight\ of\ protein}, \tag{6}$$

$$\mathrm{ESI}(\mathrm{min}) = \frac{A_{10} \times \Delta t}{\Delta A}, \tag{7}$$

where $A_0$ is the absorbance at 0 min, $A_{10}$ is the absorbance at 10 min, $\Delta A$ is $A_{10}$–$A_0$, $\Delta t$ is 10 min, and $\Phi$ is the oil volume fraction.

#### 2.9.3. Water Absorption Capacity

The water absorption capacity (WAC) was determined according to the method published with slight modifications [16]. Collagen (0.2 g) was combined with 20 mL of distilled water and then vortexed for 30 s using a vortex mixer (MX-E, SCILOGEX, Beijing China). Then, the samples were allowed to settle at room temperature for 60 min and were centrifuged at 6000 rpm/min for 10 min at 4 °C. The supernatant was removed and the aspirated collagen was weighed. The water absorption was calculated according to Equation (8):

$$\text{WAC}\left(\frac{\text{g}}{\text{g}}\right) = \frac{W_2 - W_1}{W_0}, \tag{8}$$

where $W_0$ is the collagen weight, $W_1$ is the total mass of the sample and the tube, and $W_2$ is the total mass of the sample and tube after centrifugation.

#### 2.9.4. Oil Absorption Capacity

The oil absorption capacity (OAC) was calculated according to the published method with slight modifications [16]. Collagen (0.2 g) was combined with 5 mL of corn oil and then vortexed for 30 s using a vortex mixer (MX-E, SCILOGEX, Beijing, China). The samples were allowed to settle at room temperature for 60 min and then were centrifuged at 6000 rpm/min for 10 min at 4 °C. The supernatant was removed, and the aspirated collagen was weighed. The oil absorption was calculated according to Equation (9):

$$\text{OAC}\left(\frac{\text{g}}{\text{g}}\right) = \frac{F_2 - F_1}{F_0}, \tag{9}$$

where $F_0$ is the collagen weight, $F_1$ is the total mass of the sample and tube, and $F_2$ is the total mass of the sample and tube after centrifugation.

### 2.10. Biological Properties

The biological properties of collagen on the MC3T3-E1 cells were estimated using a CCK-8 assay with some modifications, according to the method published [10]. The collagen was dissolved in distilled water to 0.5 mg/mL. The bottoms of 48-well plates were coated with the collagen solution and dried under a laminar flow hood before they were sterilized with UV light. The cells were inoculated at a density of $1 \times 10^4$ cells per well and then incubated at 37 °C in a humidified atmosphere containing 5% $CO_2$ for 1, 3, and 5 days. Then, CCK-8 solution was separately added to each sample and incubated for 1.5 h. The absorbance of the samples was measured using a multimode reader (Mithras$^2$ LB 943, Berthold, Germany) at a wavelength of 450 nm. The cell viability was calculated according to Equation (10):

$$\text{Cell Viability\%} = \frac{A_2}{A_1} \times 100\%, \tag{10}$$

where $A_1$ is the absorbance of control; $A_2$ is the absorbance of treatment.

### 2.11. Statistical Analysis

All experiments were performed in triplicate and the results were expressed as the mean ± standard deviation (SD). All data were analyzed by ANOVA using SPSS version 17.0 software (IBM SPSS Statistics, Ehningen, Germany), with values of $p < 0.05$ indicating significant deviation.

## 3. Results and Discussion

### 3.1. Yield

The yield of SCSC was 40.35 ± 0.63% (dry basis weight), while the SCSC yields were higher than the bigeye tuna 16.7 ± 0.7%, (dry basis weight) [2], large fin long barbel catfish 16.8% (dry basis weight) [17], and loach 22.42% (dry basis weight) [18]. The difference in yield was possibly due to the different degrees of cross-linking of the collagenous protofibrils in the different fish raw materials [12]. The degree of cross-linking in the terminal peptide region of the collagen molecule is an essential factor in its yield. The low degree of cross-linking of collagen leads to decreased intermolecular forces of collagen molecules. The decreased intermolecular forces of collagen molecules lead to the higher solubility of collagen, thus, providing a higher collagen yield [2]. Additionally, as silver carp aquaculture is widespread in China, with an annual production value of more than

3.8 million tons, these values indicated that silver carp may provide sufficient skin raw materials for collagen production, and SCSC may have the potential to replace collagen in the market.

### 3.2. SDS-PAGE

The protein patterns from the SCSC and rat tail type I collagen are shown in Figure 1, which all consisted of two distinct $\alpha$ chains ($\alpha_1$ and $\alpha_2$ chains) and their intramolecular cross-linked dimers ($\beta$-chains). The molecular weights of SCSC ($\alpha_1$ 129 kDa, $\alpha_2$ 121 kDa, and $\beta$ 224 kDa) were slightly lower than rat tail collagen ($\alpha_1$ 132 kDa, $\alpha_2$ 121 kDa, and $\beta$ 230 kDa). These differences were possibly caused by the different collagen sources of mammals and aquatic animals [19]. The clear electrophoretic bands of SCSC and the absence of spurious bands also suggested a higher purity of the extracted collagen. In addition, the gray ratio between the $\alpha_1$ and $\alpha_2$ chains of SCSC was 2.07:1, which was close to 2:1. This was consistent with the composition of $[(\alpha_1)_2\alpha_2]$, which was similar to previously published collagen derived from skipjack tuna (2:1) and pufferfish (approx. 2:1), indicating that the prepared collagen was type I collagen. [20,21].

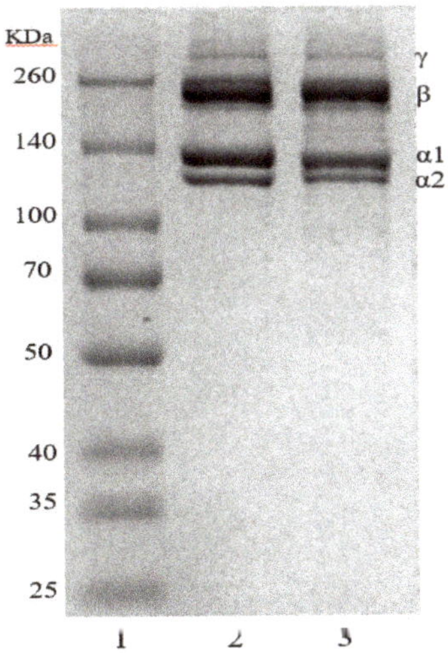


**Figure 1.** SDS-PAGE pattern of SCSC. (1) Marker; (2) rat tail collagen; (3) SCSC.

### 3.3. Amino Acid Sequence Analysis

The amino acid sequences determined using the LC-MS/MS analysis results are presented in the Supplementary Materials. The type of fish skin collagen was determined by ion mass matching of the collagen subunit MS/MS fragments, amino acid sequence coverage, and the fractions calculated from the number of matching peptides of the collagen subunit peptides in the NCBI database.

Among the data detected for $\alpha_1$, five proteins were observed with relatively high scores, with a high number of significant matches. First, the GI was AIL02135.1, the score was 847, the mass was 138,365, the sequence was 19, the coverage was 20%, and the protein species was *Hypophthalmichthys molitrix* with collagen type I alpha 1. Second, GI was AGH32451.1, the fraction was 164, the mass was 42,295, the sequence was 4, and the protein species was *Hypophthalmichthys molitrix* actin alpha 1b. Third, the GI was AUF74474.1, the fraction was 158, the mass was 127,755, the sequence was 5, and the protein species *Hypophthalmichthys molitrix* collagen type I alpha 2. Fourth, the GI was AIE40058.1, the fraction was 50, the mass was 13,041, the sequence was 1, and the protein species was *Hypophthalmichthys molitrix* histone H2A, partial. Fifth, the GI was ACO51127.1, the fraction was 33, the mass was 7835, the sequence was 1, and the protein species was *Hypophthalmichthys nobilis* bactin1 protein, partial.

The identical $\alpha_2$ also obtained five sets of data with high confidence, as shown in Table 1. The first GI was AUF74474.1, with a score of 2009, the mass was 127,755, the sequence was 31, the coverage was 26%, and the protein species was *Hypophthalmichthys molitrix* collagen type I alpha 2. Second, the GI was AGH32451.1, the score was 121, the mass was 42,295, the sequence was 4, and the protein species was *Hypophthalmichthys molitrix* actin alpha 1b. Third, the GI was AHY86406.1, the score was 76, the mass was 55,200, the sequence was 1, and the protein species was *Hypophthalmichthys nobilis* ATP synthase, H+ transport, mitochondrial F1 complex, beta subunit. Fourth, the GI was AIL02135.1, the score was 39, the mass was 138,365, the sequence was 2, and the protein species was *Hypophthalmichthys molitrix* collagen type I alpha 1. Fifth, the GI was ACO51127.1, the score was 33, the mass was 7835, the sequence was 1, and the protein species was *Hypophthalmichthys nobilis* bactin1 protein, partial.

**Table 1.** Matching peptide coverage of SCSC amino acid sequences.

| GI | Score | Mass/pi | Sequences | Coverage [1] | Protein Description |
|---|---|---|---|---|---|
| AIL02135.1 | 847 | 138,365/5.44 | 19 | 20% | *Hypophthalmichthys molitrix* $\alpha_1$ (I) |
| AUF74474.1 | 2009 | 127,755/9.36 | 31 | 26% | *Hypophthalmichthys molitrix* $\alpha_2$ (I) |

[1] Matched peptide coverage of the protein sequence.

The results presented above showed that the extracted collagen had the highest coverage of silver carp type I $\alpha_1$ and the highest coverage of silver carp type I $\alpha_2$ ($\alpha_1$: 20% $\alpha_2$: 26%), where the relative molecular masses ($\alpha_1$: 138,365 Da, $\alpha_2$: 127,755 Da) were similar to the SDS-PAGE results ($\alpha_1$: 129,000 Da, $\alpha_2$: 121,000 Da). This indicated that the experimentally extracted collagen was a typical type I *Hypophthalmichthys molitrix* collagen.

### 3.4. Spectral Analysis

The structure of SCSC was identified using UV absorption spectra, FTIR, CD, and XRD spectroscopy.

As shown in Figure 2A, the UV spectrum of SCSC produced two characteristic absorption peaks, where the maximum appeared at 230 nm. The maximum absorption peak was due to the fact that collagen contains C=O, COOH which was a chromophore and the unsaturated group leaps after the absorption of light radiation [22]. In addition, SCSC contained small amounts of tyrosine, phenylalanine, and tryptophan, which consisted of aromatic amino acids with conjugated double bonds in the phenyl-producing peaks with low absorption at 280 nm. The results of this assay were comparable to hammerhead shark type I collagen and puffer fish type I collagen, showing that the extracted collagen was classic type I collagen [23,24].

As shown in Figure 2B, the FTIR spectrum of SCSC was similar to that of type I collagen with five absorption peaks including Amide A, Amide B, Amide I, Amide II, and Amide III. The Amide A band of SCSC was observed at 3419 $cm^{-1}$, which was related to the N-H stretching vibrations. According to the study, the N-H stretching vibrations ranged from 3400 to 3440 $cm^{-1}$, when the N-H group was bonded to hydrogen the position of Amide A shifted to a much lower frequency [25]. The Amide B band of SCSC was observed at 2931 $cm^{-1}$, which was related to the asymmetric stretching of $CH_2$, which usually occurred at 2900–2950 $cm^{-1}$ [26]. The absorption peaks of Amide I, Amide II, and Amide III bands appeared in the ranges of 1600–1700, 1550–1600, and 1235–1240 $cm^{-1}$, respectively, and their production was related to the C=O, N-H, and C-N stretching vibrations, respectively [6]. The absorption peaks of the Amide I, Amide II, and Amide III bands of SCSC were produced at 1652, 1546, and 1245 $cm^{-1}$. In addition, the absorbance ratio of 1.1 between Amide III and 1400–1454 $cm^{-1}$ was approximated to be

1.0, which indicated that the triple helix structure of collagen was preserved intact [27]. The results of this experiment were similar to the results previously reported for Spanish mackerel and sea cucumber [6,28].

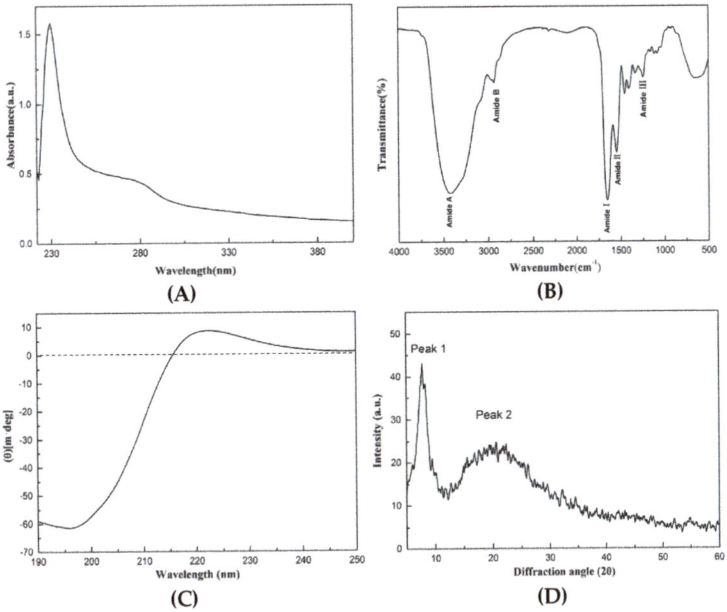


**Figure 2.** Spectral analysis of SCSC. (**A**) UV; (**B**) FTIR; (**C**) CD; (**D**) XRD.

As shown in Figure 2C, the CD of SCSC produced two characteristic absorption peaks that appeared at 196 and 222 nm, respectively. This produced a negative peak at 196 nm and a positive peak at 222 nm with an ellipticity of 0 near 215 nm, which was similar to sliver scale lizardfish scale and pufferfish skin. In addition, the Rpn value of the sample was 0.14, and the study showed that the Rpn value was between 0.12 and 0.15, proving that the triple helix structure of the collagen sample remained intact [18]. The results indicated that SCSC had an intact triple helix structure.

As shown in Figure 2D, the XRD spectra of SCSC produced two characteristic absorption peaks. The first sharp peak appeared near 7.46°. This diffraction peak was produced by the distance between the molecular chains of the collagen fibers [29]. The second broad peak appeared near 20.7° and this peak arose from diffuse scattering caused by the many structural layers of the collagen fibers [30]. The XRD results were evaluated according to the Bragg equation. The distances between the molecular chains of the collagen fibrils $d_1$ = 1.18 nm and collagen skeleton $d_2$ = 0.43 nm were calculated, indicating that the d values were similar to type I collagen diameter, proving the presence of an intact triple helix structure in the samples [12].

### *3.5. Determination of Denaturation Temperature ($T_d$)*

Thermal stability is considered one of the most critical properties of collagen. As the temperature increased, the hydrogen bonds in the collagen molecules were destroyed and the triple helix structure dissociated. This modification of the structure weakened the molecular forces between collagen, resulting in a decrease in viscosity. The temperature corresponding to the viscosity shift to half of the original was known as $T_d$. As shown in Figure 3, the viscosity of the sample gradually decreased with increasing temperature, becoming half of the initial viscosity at 30.37 °C. After continued temperature increase, the viscosity dropped to a minimum, indicating that the $T_d$ of SCSC was 30.37 °C. At this moment, SCSC absorbed enough energy from the outside world and the non-covalent bonds that held the triple helix structure inside rupture. The triple-stranded helical conformation of collagen subsequently converted to an irregular convoluted conformation, and this change caused the physical properties of collagen to change with an increase in temperature [31].

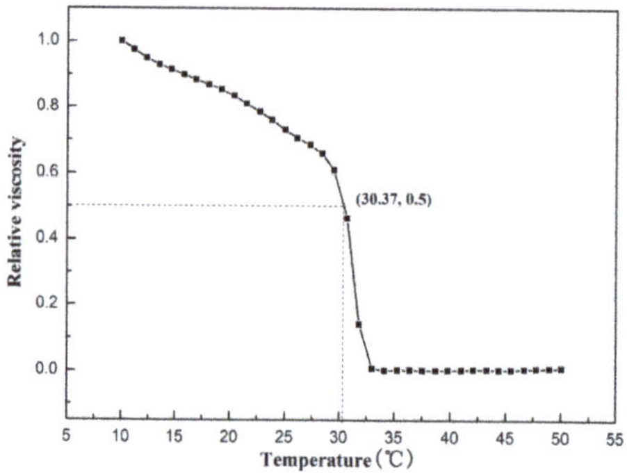


**Figure 3.** Thermal denaturation curve of SCSC.

In addition, the living environment can lead to differences in $T_d$. SCSC showed higher thermal stability than the collagen from deep-sea fish such as Sharpnose skin (28 °C), Pacific cod skin (14.5 °C), and Arabian mackerel (15.5 °C) [1,3,5]. This was possibly due to the fact that deep-sea fish are typically found in environments of high salinity, high pressure, and low temperature. In addition, $T_d$ of SCSC was better than some freshwater fish with similar survival environments, such as sturgeon at 26.86 °C and Mozambique tilapia at 28 °C [32]. In general, the $T_d$ value of fish collagen was lower than that of mammals, surprisingly, and the $T_d$ of SCSC was higher than that of human placental collagen (28.5 °C) [33]. The higher Td made SCSC less susceptible to denaturation during processing at room temperature and it could maintain a stable triple helix structure. This meant that its biological properties (low antigenicity, biocompatibility) remain unchanged. These factors indicated that SCSC might have the potential to be used in food and biological functional material.

### 3.6. Rheological Properties

As shown in Figure 4A,B, the G′ and G″ of SCSC were proportional to the concentration at 0.01 Hz, with the concentration increasing from 5 to 25 mg. G′ increased from 0.0248 to 18.9 Pa, increasing by three magnitudes, and G″ increased from 0.0937 to 15.9 Pa, increasing by three magnitudes. At 10 Hz, G′ and G″ increased from 6.88 and 8.52 Pa to 177 and 96.3 Pa, respectively, for an increase in three magnitudes and two magnitudes, respectively. These results indicated that the effect of concentration on the solution viscoelasticity was more significant at low frequencies. As shown in Figure 4C $\eta^*$ increased linearly with concentration, when the viscosity of 5 mg SCSC was 1.54 Pa.s. The viscosity of 25 mg SCSC reached 393 Pa.s, an increase of two orders of magnitude, and $\eta^*$ decreased with increasing frequency. The viscosity was 1.54–393 Pa.s at 0.1 Hz and decreased to 0.174–3.21 Pa.s at 10 Hz. This decrease in $\eta^*$ with increasing frequency was referred to as shear thinning [34]. As the frequency increased, the physical entanglement points destroyed by shear were too weak to be reconstructed, which was referred to as pseudoplastic fluid. Tanδ was the ratio of G′, G″, which reflected the state of the sample. As shown in Figure 4D, Tanδ decreased with increasing frequency, gradually changing from greater than 1 to less than 1, indicating that the rheological behavior of collagen changed from viscous to elastic properties [35]. The experimental results were similar to the experimental results of collagen from sharp nose stingray skin collagen [1], red stingray skin collagen [36], and grass carp swim bladder collagen [12]. These rheological data can provide useful information for the production of high-quality and stable collagen products.

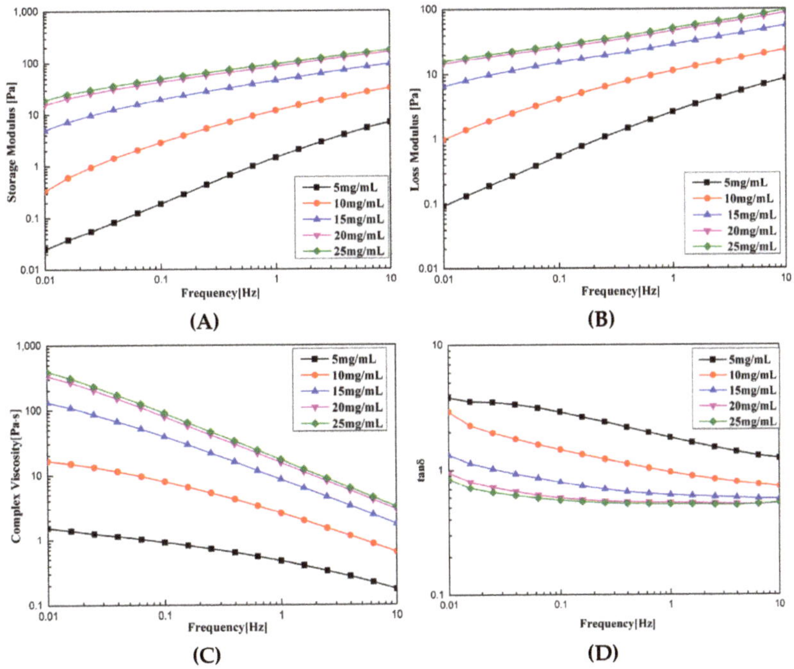


**Figure 4.** Dynamic rheological curve of SCSC at different concentrations. (**A**) Storage modulus (G′); (**B**) loss modulus (G″); (**C**) complex viscosity (η*); (**D**) tanδ.

As shown in Figure 5A,B, under the dynamic rheological curves of SCSC at different temperatures, G′ and G″ were mainly related to the degree of cross-linking and the $T_d$ of SCSC. When the temperature was lower than $T_d$ (30.37 °C), there was very slight variation in G′ and G″ with frequency. However, the temperature was higher than the $T_d$ of SCSC, and the G′ value of collagen decreased rapidly when the shear frequency was higher than 1 Hz. This was caused by the fact that as the temperature increased beyond $T_d$, the thermal motion energy of collagen molecules increased, the leap resistance of collagen molecular chain segments became weak, and collagen gelation occurred. As shown in Figure 5C, η* showed a decreasing trend with increasing temperature, and decreased with increasing frequency, which was related to the $T_d$ and shear thinning of the sample. As the temperature exceeded the $T_d$, the hydrogen bonds were broken and the triple helix structure was deconvoluted into single chains or dimers, which led to denaturing behavior of the collagen. As shown in Figure 5D, collagen exhibited elastic behavior at low frequencies and viscous behavior at high frequencies at temperatures above $T_d$. These results were consistent with the properties of pseudoplastic fluids [35]. In production and processing, different production processes will cause variations in the rheological properties of SCSC during processing, with excessive viscosity adding to the difficulty of transport and processing, and too low viscosity making it difficult to set. Thus, the rheological properties of SCSC could be used to guide the production of collagen products.

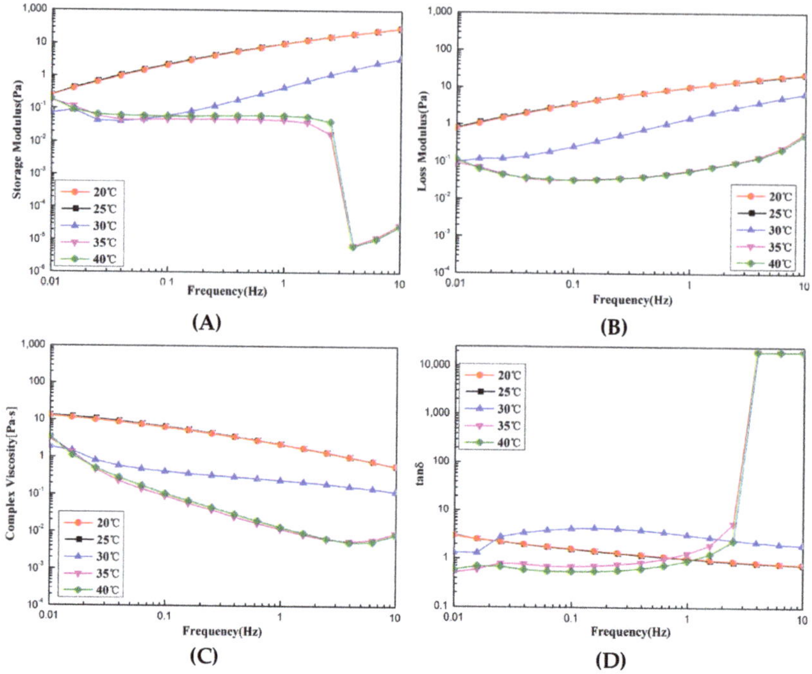


**Figure 5.** Dynamic rheological curve of SCSC at different temperatures. (**A**) Storage modulus (G′); (**B**) loss modulus (G″); (**C**) complex viscosity (η*); (**D**) tanδ.

### *3.7. Zeta Potential*

The polar groups of the proteins tended to dissociate when the pH of the solution reached a specific value, the positive and negative ions of collagen dissociation became equal, the electrostatic charge reached zero, and the electrostatic repulsion between the molecules reached a low value and the stability was poor, where the pH value at this moment was referred to as the isoelectric point (pI) [37]. SCSC was positively charged at pH 2–6, negatively charged at pH 7–10, and the potential value was equal to 0 at a pH of 6.98 (Supplementary Materials Figure S3). The isoelectric point varied slightly among the fish species, with higher values for SCSC (6.98) compared to sea bass (6.46) [38] and yellowfin tuna (6.05) [39]. The phenomenon of this difference was likely due to the distribution of amino acid sequences and the differences among the amino acid residues [40].

### *3.8. Functional Properties*

#### 3.8.1. Foaming Properties

Figure 6A,B shows the FC and FS of SCSC at different pH values, where FC varied from 7.46 ± 0.06% to 23.34 ± 1.22% and FS varied from 57.23 ± 1.99% to 95.39 ± 2.84% with increasing pH. FS and FC both reached their lowest points near an isoelectric point pH 6.98 (FC: 7.46 ± 0.61%, FS: 57.23 ± 1.99%). This phenomenon occurred due to the loss of electrostatic repulsion, which reduced the solubility of the collagen solution at the isoelectric point. The aggregation of undissolved proteins reduced the forces between protein and water, leading to a decrease in foaming capacity. Casein and soy protein are food foaming agents available in the market [41,42]. The FC of SCSC was superior to the FC of casein (3.95 ± 0.07%–14.25 ± 0.35%, pH = 5–9) [43], HBC 19 rice bran protein concentrate (5.2 ± 0.28%–10.03 ± 0.39%, pH = 5–9) [43] in the same pH ranges. While the FS of SCSC was higher than casein (0.17 ± 0.002–0.54 ± 0.61, pH = 5–9) [43], HBC 19 rice bran protein concentrate (3.67 ± 0,09–4.30 ± 0.16% pH = 5–9) [43], and grass carp swim bladder collagen 52.8 ± 2.45%–74.72 ± 0.71%, pH = 2–10) [12] in the same pH ranges. SCSC showed excellent foaming properties and could possibly be applied as a foaming agent in the processing of alcohol, dairy products, and other food products.

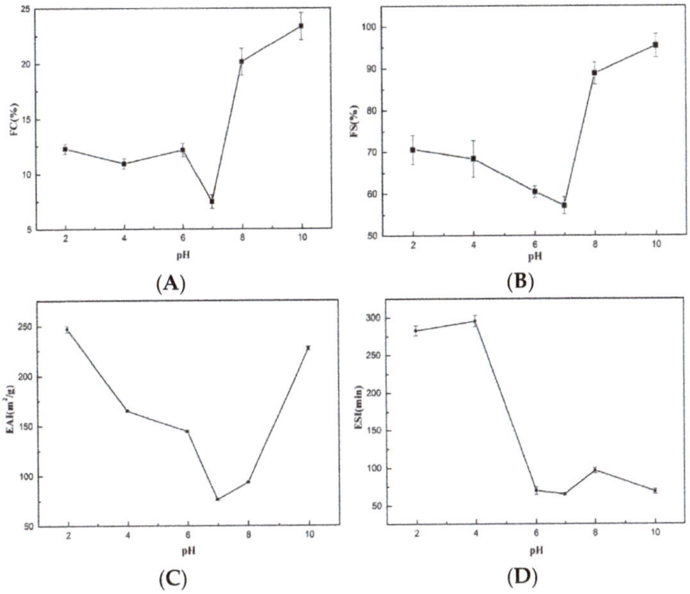


**Figure 6.** Functional properties of SCSC. (**A**) Foam capacity (FC); (**B**) foam stability (FS); (**C**) emulsifying activity index (EAI); (**D**) emulsion stability index (ESI).

3.8.2. Emulsifying Properties

Figure 6C,D shows the EAI and ESI of SCSC at different pH values, where EAI varied between 76.48 ± 0.59 $m^2g$ and 246.73 ± 3.37 $m^2g$ with increasing pH and ESI varied between 64.97 ± 1.23 min and 295.31 ± 7.46 min with increasing pH. Both ESI and EAI reached their minimum values close to the isoelectric point pH 6.98 (ESI: 64.97 ± 1.23 min, EAI: 76.48 ± 0.59 $m^2g$). This phenomenon was consistent with the foaming properties, where the free radical activity in the protein diminished near the isoelectric point, which was not conducive to the emulsification properties [44,45]. Soy protein and casein are food emulsifiers available in the market [41,42]. The ESI of SCSC was better than that of chicken cartilage collagen (25.62–43.3 $m^2g$, pH = 4–10) [46], peanut protein (approx. 52 $m^2g$–175 $m^2g$, pH = 3–7) [47], and soy protein isolate Bosa (25.94 ± 0.3–33.75 ± 1.25 $m^2g$, pH = 3–8) [48] in the same pH ranges. The EAI of SCSC was superior to that of soft-shelled turtle calipash collagen (EAI < 25 min, pH = 4–10) [44], red stingray collagen (1.96 ± 0.05–80.36 ± 0.27 min, pH = 2–10), [36], peanut protein (approx. 5–100 min, pH = 3–9) [47], and soy protein isolate Bosa (16.16 ± 0.12–27.43 ± 0.86 min, pH = 3–8) [48] in the same pH ranges. The experimental results showed that SCSC might be used as supplemental collagen in food and beverages as an emulsion stabilizer [49].

3.8.3. WAC and OAC

WAC refers to the amount of water absorbed and the ability of collagen to retain water, reflecting the capacity of the protein to absorb water and influence the taste and flavor of food during actual production and processing. The process of water absorption by collagen might accumulate around the collagen fibers and diffuse within the triple helix, promoting the formation of hydrogen bonds; thus, enhancing the WAC. The WAC of SCSC was 31.13 ± 0.98 g/g, which was significantly better than walnut protein isolate (3.11 ± 0.28 g/g), sunflower seed isolate (0.985 ± 0.01 g/g), and pig skin collagen

(0.21 ± 0.03 g/g) [50–52]. The excellent WAC allowed SCSC to be applied in food and cosmetic applications. The OAC of proteins has shown to be crucial not only for the taste of the product but also for the functional properties required in the processing of meat and confectionery products. The OAC of SCSC was 26.48 ± 0.51 g/g, which is slightly lower compared to the WAC, indicating that SCSC contained more hydrophilic groups rather than hydrophobic groups. The OAC of SCSC was higher than soybean isolate (8.31 ± 0.36 g/g), green tea water soluble protein (4.5 g/g), and sorghum protein (3.39 g/g) [53–55]. A higher WAC with OAC could be utilized as a food stabilizer to prevent water loss in frozen products during heat treatment or thawing, as well as prevent the separation of fat fraction [56]. This showed that SCSC might have a high development value for various applications in food as well as in cosmetics.

### 3.9. Biocompatibility

The biocompatibility of the MC3T3-E1 fibroblasts was determined using the CCK-8 cell proliferation assay. The optical density (OD) value is a criterion used to determine the cell proliferation rate, where higher OD values are more favorable for cell proliferation. As shown in Table 2, the OD values on days 1, 3, and 5 of culturing were 0.521, 3.179, and 1.281, respectively. The cell proliferation rate reached the highest level on day 3 and decreased on day 5. This was possibly due to apoptosis caused by the decrease in the survival environment of the cells after high density multiplication. In general, the OD values of the cells tended to increase, where the higher cell viability values relative to the blank control indicated that SCSC facilitated cell proliferation.

**Table 2.** Classification of OD relative cell viability and toxicity of SCSC at different days of culture.

| Time (D) | OD Value | Relative Cell Viability (%) | Classification of Toxicity |
|---|---|---|---|
| 1 | 0.521 | 99.30 ± 9.3 | 0 |
| 3 | 3.179 | 95.54 ± 4.2 | 0 |
| 5 | 1.281 | 102.91 ± 8.33 | 0 |

As shown in Figure 7, a CCK-8 assay was performed to assess the effect of different days of culturing of SCSC on the relative cell viability graph of MC3T3-E1. The relative cell viability values of the samples were greater than 95%. All the relative cell viabilities were above 70%, and according to the toxicity grade assessment of IOS, the toxicity grade of all samples with a cell viability greater than 70% was 0 [57]. These results were similar to those of grass carp swim bladder collagen [12] and red stingray skin collagen [36], indicating that SCSC is non-cytotoxic and biocompatible. Thus, SCSC has great potential for application in biomedical materials.

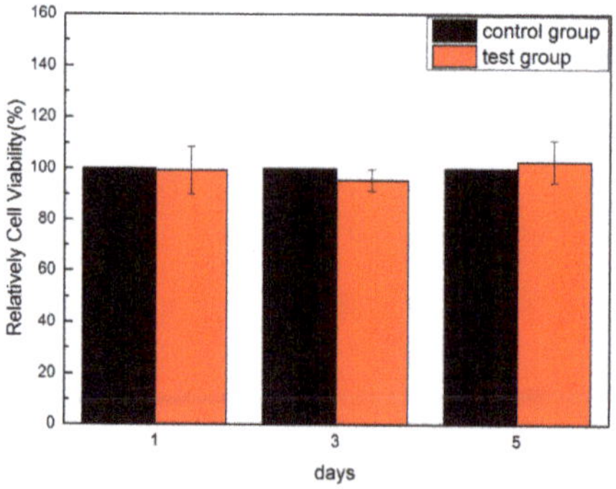


**Figure 7.** Relative cell viability of MC3T3-E1 culture on days 1, 3, and 5 in SCSC.

As shown in Figure 8, the morphology of the cells cultured with SCSC was observed using an inverted microscope (ECLIPSE Ti, Nikon, Japan). On the first day, the MC3T3-E1 cells were sparse, and a shuttle-shaped cell morphology was observed. On the third day, the number of shuttle-shaped cells increased significantly. On the fifth day, the cells were stacked together in a squamous pattern, with no cell lysis or morphological abnormalities observed. These results were consistent with the conclusion of cell proliferation, where SCSC could promote the growth of MC3T3-E1 cells, indicating that SCSC was biocompatible and was a potential biomaterial with great development value.

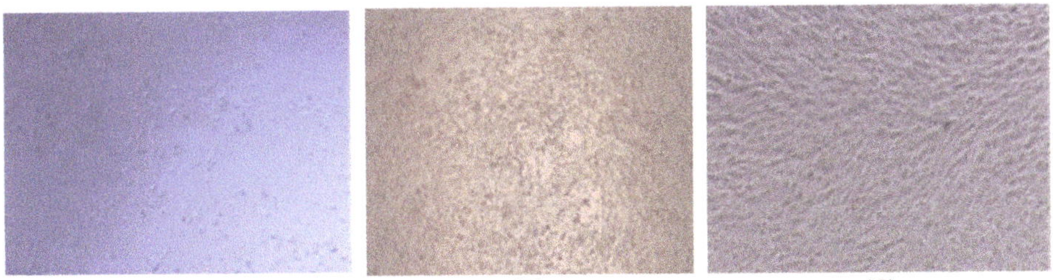

(A) (B) (C)

**Figure 8.** Cell morphology of MC3T3-E1 in SCSC at different days of culture. (**A**) Culture for 1 day; (**B**) culture for 3 days; (**C**) culture for 5 days.

## 4. Conclusions

In this work, SCSC was extracted from the skin of silver carp and identified as type I collagen using electrophoretic experiments and spectroscopic analysis, with a triple helix structure that was intact. The amino acid sequence analysis identified SCSC as *Hypophthalmichthys molitrix*. The $T_d$ tests showed that the $T_d$ value of SCSC was higher than deep-sea fish sources such as Pacific cod, Sharpnose, and Arabian mackerel. The Td value of SCSC was also better than that of some freshwater fish with similar survival environments, such as sturgeon and Mozambique tilapia. Interestingly, the $T_d$ of SCSC was higher than that of human placental collagen. Functional property analyses indicated better functional properties of SCSC compared to other proteins from terrestrial sources in terms of WAC, OAC, foaming properties, and emulsifying properties. Rheological property analyses indicated that SCSC had shear thinning properties and was a pseudoplastic fluid, while the biocompatibility results indicated that the MC3T3-E1 cells could proliferate and maintain normal morphology in the SCSC solutions. Thus, SCSC could be applied as an alternative collagen resource on the market. The collagen extracted from the by-products of silver carp could greatly enhance the economic value of silver carp and realize the high-value utilization of low-value products.

**Supplementary Materials:** The following supporting information can be downloaded at: https://www.mdpi.com/article/10.3390/foods11192985/s1, Figure S1: Amino acid sequence of *Hypophthalmichthys molitrix* type I collagen subunit α1. Bold red peptides: Matched peptide in the NCBI database; Figure S2: Amino acid sequence of *Hypophthalmichthys molitrix* type I collagen subunit α2; Bold red peptides: Matched peptide in the NCBI database. Figure S3: Zeta potential of SCSC.

**Author Contributions:** J.C. conceptualization, validation, resources, writing—review and editing, supervision, project administration, funding acquisition; Y.G. conceptualization, methodology, formal analysis, data curation, writing—original draft preparation; J.H. methodology, software, formal analysis, data curation; Y.L. conceptualization, methodology, software; X.Z. formal analysis, data curation; Y.Z. conceptualization, methodology, software, formal analysis, data curation; L.J. conceptualization, validation, supervision. All authors have read and agreed to the published version of the manuscript.

**Funding:** This work was supported by grants from the National Natural Science Foundation of China (42076120, 41676129, 41106149), Scientific Research Foundation of the Third Institute of Oceanography, SOA (2019010).

**Institutional Review Board Statement:** Not applicable.

**Informed Consent Statement:** Not applicable.

**Data Availability Statement:** Data is contained within the article and Supplementary Materials.

**Conflicts of Interest:** The authors declare no conflict of interest.

## References


1. Shaik, M.I.; Effendi, N.F.A.; Sarbon, N.M. Functional properties of sharpnose stingray (*Dasyatis zugei*) skin collagen by ultrasonication extraction as influenced by organic and inorganic acids. *Biocatal. Agric. Biotechnol.* **2021**, *63*, 102103. [CrossRef]
2. Ahmed, R.; Haq, M.; Chun, B.S. Characterization of marine derived collagen extracted from the by-products of bigeye tuna (*Thunnus obesus*). *Int. J. Biol. Macromol.* **2019**, *135*, 668–676. [CrossRef] [PubMed]
3. Sun, L.L.; Li, B.F.; Song, W.K.; Si, L.L.; Hou, H. Characterization of Pacific cod (*Gadus macrocephalus*) skin collagen and fabrication of collagen sponge as a good biocompatible biomedical material. *Process. Biochem.* **2017**, *35*, 229–235. [CrossRef]
4. Bhagwat, P.K.; Dandge, P.B. Isolation, characterization and valorizable applications of fish scale collagen in food and agriculture industries. *Biocatal. Agric. Biotechnol.* **2016**, *7*, 234–240. [CrossRef]
5. Nalinanon, S.; Benjakul, S.; Kishimura, H. Collagens from the skin of arabesque greenling (*Pleurogrammus azonus*) solubilized with the aid of acetic acid and pepsin from albacore tuna (*Thunnus alalunga*) stomach. *J. Sci. food Agric.* **2010**, *90*, 1492–1500. [CrossRef] [PubMed]
6. Li, Z.R.; Wang, B.; Chi, C.F.; Zhang, Q.H.; Gong, Y.D.; Tang, J.J.; Luo, H.Y.; Ding, G.F. Isolation and characterization of acid soluble collagens and pepsin soluble collagens from the skin and bone of Spanish mackerel (*Scomberomorous niphonius*). *Food Hydrocollold.* **2013**, *31*, 103–113. [CrossRef]
7. Abdollahi, M.; Rezaei, M.; Jafarpour, A.; Undeland, I. Sequential extraction of gel-forming proteins, collagen and collagen hydrolysate from gutted silver carp (*Hypophthalmichthys molitrix*), a biorefinery approach. *Food Chem.* **2018**, *242*, 568–578. [CrossRef] [PubMed]
8. Bhuimbar, M.V.; Bhagwat, P.K.; Dandge, P.B. Extraction and characterization of acid soluble collagen from fish waste: Development of collagen-chitosan blend as food packaging film. *J. Environ. Chem. Eng.* **2019**, *7*, 102983. [CrossRef]
9. Zhang, J.J.; Duan, R.; Ye, C.; Konno, K. Isolation and characterization of collagens from scale of silver carp (*Hypophthalmichthys molitrix*). *J. Food Biochem.* **2010**, *34*, 1343–1354. [CrossRef]
10. Faralizadeh, S.; Rahimabadi, E.Z.; Bahrami, S.H.; Hasannia, S. Extraction, characterization and biocompatibility evaluation of silver carp (*Hypophthalmichthys molitrix*) skin collagen. *Sustain. Chem. Pharm.* **2021**, *22*, 100454. [CrossRef]
11. Sionkowska, A.; Lewandowska, K.; Adamiak, K. The Influence of UV light on rheological properties of collagen extracted from silver carp skin. *Materials* **2020**, *13*, 4453. [CrossRef] [PubMed]
12. Li, Y.S.; Yang, L.H.; Wu, S.J.; Chen, J.D. Structural, functional, rheological, and biological properties of the swim bladder collagen extracted from grass carp (*Ctenopharyngodon idella*). *LWT-Food Sci. Technol.* **2022**, *153*, 112518. [CrossRef]
13. Katayama, H.; Nagasu, T.; Oda, Y. Improvement of in-gel digestion protocol for peptide mass fingerprinting by matrix-assisted laser desorption/ionization time-of-flight mass spectrometry. *Rapid Commun. Mass. Spectrom.* **2001**, *15*, 1416–1421. [CrossRef] [PubMed]
14. Benjakul, S.; Thiansilakul, Y.; Visessanguan, W.; Roytrakul, S.; Kishimura, H.; Prodpran, T.; Meesane, J. Extraction and characterisation of pepsin-solubilised collagens from the skin of bigeye snapper (*Priacanthus tayenus* and *Priacanthus macracanthus*). *J. Sci. food Agric.* **2010**, *90*, 132–138. [CrossRef] [PubMed]
15. Zamorano-Apodaca, J.C.; García-Sifuentes, C.O.; Carvajal-Millán, E.; Vallejo-Galland, B.; Scheuren-Acevedo, S.M.; Lugo-Sánchez, M.E. Biological and functional properties of peptide fractions obtained from collagen hydrolysate derived from mixed by-products of different fish species. *Food Chem.* **2020**, *331*, 127350. [CrossRef] [PubMed]
16. Dhakal, D.; Koomsap, P.; Lamichhane, A.; Sadiq, M.B.; Anal, A.K. Optimization of collagen extraction from chicken feet by papain hydrolysis and synthesis of chicken feet collagen based biopolymeric fibres. *Food Biosci.* **2018**, *23*, 23–30. [CrossRef]
17. Zhang, M.; Liu, W.T.; Li, G.Y. Isolation and characterisation of collagens from the skin of largefin longbarbel catfish (*Mystus macropterus*). *Food Chem.* **2009**, *115*, 826–831. [CrossRef]
18. Wang, J.; Pei, X.L.; Liu, H.Y.; Zhou, D. Extraction and characterization of acid-soluble and pepsin-soluble collagen from skin of loach (*Misgurnus anguillicaudatus*). *Int. J. Biol. Macromol.* **2018**, *106*, 544–550. [CrossRef] [PubMed]
19. Veeruraj, A.; Arumugam, M.; Ajithkumar, T.; Balasubramanian, T. Isolation and characterization of collagen from the outer skin of squid (*Doryteuthis singhalensis*). *Food Hydrocolloid.* **2015**, *43*, 708–716. [CrossRef]
20. Chen, J.D.; Li, M.; Yi, R.Z.; Bai, K.K.; Wang, G.Y.; Tan, R.; Sun, S.S.; Xu, N.H. Electrodialysis extraction of pufferfish skin (*Takifugu flavidus*): A promising source of collagen. *Mar. Drugs* **2019**, *17*, 25. [CrossRef]
21. Yu, D.; Chi, C.F.; Wang, B.; Ding, G.F.; Li, Z.R. The ESI of SCSC was better than that of chicken cartilage collagen (25.62–43.3 $m^2g$, pH = 4–10). *Chin. J. Nat. Med.* **2014**, *35*, 712–720. [CrossRef]
22. Edwards, H.G.M.; Farwell, D.W.; Holder, J.M.; Lawson, E.E. Fourier-transform Raman spectroscopy of ivory: II. Spectroscopic analysis and assignments. *J. Mol. Struct.* **1997**, *435*, 49–58. [CrossRef]

23. Chi, C.F.; Wang, B.; Li, Z.R.; Luo, H.Y.; Ding, G.F. Characterization of acid-soluble collagens from the cartilages of scalloped hammerhead (*Sphyrna lewini*), red stingray (*Dasyatis akajei*), and skate (*Raja porosa*). *Food Sci. Biotechnol.* **2014**, *38*, 236–247. [CrossRef]
24. Iswariya, S.; Velswamy, P.; Uma, T.S. Isolation and Characterization of Biocompatible Collagen from the Skin of Puffer fish (*Lagocephalus inermis*). *J. Polym. Environ.* **2018**, *26*, 2086–2095. [CrossRef]
25. Doyle, B.B.; Bendit, E.G.; Blout, E.R. Infrared spectroscopy of collagen and collagen-like polypeptides. *Biopolymers* **1975**, *14*, 937–957. [CrossRef] [PubMed]
26. Abe, Y.; Krimm, S. Normal vibrations of crystalline polyglycine I. *Biopolymers* **1972**, *11*, 1817–1839. [CrossRef]
27. Liu, H.Y.; Han, J.; Guo, S.D. Characteristics of the gelatin extracted from Channel Catfish (*Ictalurus Punctatus*) head bones. *LWT-Food Sci. Technol.* **2009**, *42*, 540–544. [CrossRef]
28. Li, P.H.; Lu, W.C.; Chan, Y.J.; Ko, W.C.; Jung, C.C.; Huynh, D.T.L.; Ji, Y.X. Extraction and characterization of collagen from sea cucumber (*Holothuria cinerascens*) and its potential application in moisturizing cosmetics. *Aquaculture* **2020**, *515*, 734590. [CrossRef]
29. Giraud-Guille, M.M.; Besseau, L.; Chopin, C.; Durand, P.; Herbage, D. Structural aspects of fish skin collagen which forms ordered arrays via liquid crystalline states. *Biomaterials* **2000**, *21*, 899–906. [CrossRef]
30. Fauzi, M.B.; Lokanathan, Y.; Aminuddin, B.S.; Ruszymah, B.H.I.; Chowdhury, S.R. Ovine tendon collagen: Extraction, characterisation and fabrication of thin films for tissue engineering applications. *Mater. Sci. Eng. C* **2016**, *68*, 163–171. [CrossRef]
31. Usha, R.; Ramasami, T. The effects of urea and n-propanol on collagen denaturation: Using DSC, circular dicroism and viscosity. *Thermochim. Acta* **2004**, *409*, 201–206. [CrossRef]
32. Li, J.; Wang, M.C.; Qiao, Y.Y.; Tian, Y.Y.; Liu, J.H.; Qin, S.; Wu, H. Extraction and characterization of type I collagen from skin of tilapia (*Oreochromis niloticus*) and its potential application in biomedical scaffold material for tissue engineering. *Process Biochem.* **2018**, *74*, 156–163. [CrossRef]
33. Wei, P.; Zheng, H.; Shi, Z.Y.; Li, D.; Xiang, Y.L. Isolation and characterization of Acid-soluble Collagen and Pepsin-soluble Collagen from the Skin of Hybrid Sturgeon. *J. Wuhan Univ. Technol.* **2019**, *34*, 950–959. [CrossRef]
34. Zheng, T.T.; Tang, P.P.; Shen, L.R.; Bu, H.H.; Li, G.Y. Rheological behavior of collagen/chitosan blended solutions. *J. Appl. Polym. Sci.* **2021**, *138*, 50840. [CrossRef]
35. Sharma, M.; Kristo, E.; Corredig, M.; Duizer, L. Effect of hydrocolloid type on texture of pureed carrots: Rheological and sensory measures. *Food Hydrocolloid.* **2017**, *63*, 478–487. [CrossRef]
36. Chen, J.D.; Li, J.Y.; Li, Z.B.; Yi, R.Z.; Shi, S.J.; Wu, K.Y.; Li, Y.S.; Wu, S.J. Physicochemical and functional properties of type I collagens in red stingray (*Dasyatis akajei*) skin. *Mar. Drugs* **2019**, *17*, 558. [CrossRef] [PubMed]
37. Hu, K.; McClements, D.J. Fabrication of surfactant-stabilized zein nanoparticles: A pH modulated antisolvent precipitation method. *Food Res. Int.* **2014**, *64*, 329–335. [CrossRef]
38. Sinthusamran, S.; Benjakul, S.; Kishimura, H. Comparative study on molecular characteristics of acid soluble collagens from skin and swim bladder of seabass (*Lates calcarifer*). *Food Chem.* **2013**, *138*, 2435–2441. [CrossRef]
39. Kaewdang, O.; Benjakul, S.; Kaewmanee, T.; Kishimura, H. Characteristics of collagens from the swim bladders of yellowfin tuna (*Thunnus albacares*). *Food Chem.* **2014**, *155*, 264–270. [CrossRef]
40. Chen, J.D.; Li, L.; Yi, R.Z.; Xu, N.H.; Gao, R.; Hong, B.H. Extraction and characterization of acid-soluble collagen from scales and skin of tilapia (*Oreochromis niloticus*). *LWT-Food Sci. Technol.* **2016**, *66*, 453–459. [CrossRef]
41. Hammam, A.R.A.; Martinez-Monteagudo, S.I.; Metzger, L.E. Progress in micellar casein concentrate: Production and applications. *Compr. Rev. Food Sci. Food Saf.* **2021**, *20*, 4426–4449. [CrossRef] [PubMed]
42. Mune, M.M.; Bouba, A.A.; Minka, S.R. Effects of extraction conditions on the functional properties of bambara bean protein concentrates. *Int. J. Food Eng.* **2016**, *12*, 195–201. [CrossRef]
43. Chandi, G.K.; Sogi, D.S. Functional properties of rice bran protein concentrates. *J. Food Eng.* **2007**, *79*, 592–597. [CrossRef]
44. Zou, Y.; Wang, L.; Cai, P.P.; Li, P.P.; Zhang, M.H.; Sun, Z.L.; Sun, C.; Xu, W.M.; Wang, D.Y. Effect of ultrasound assisted extraction on the physicochemical and functional properties of collagen from soft-shelled turtle calipash. *Int. J. Biol. Macromol.* **2017**, *105*, 1602–1610. [CrossRef] [PubMed]
45. Tang, L.; Sun, J.; Zhang, H.C.; Zhang, C.S.; Yu, L.N.; Bi, J.; Zhu, F.; Liu, S.F.; Yang, Q.L. Evaluation of physicochemical and antioxidant properties of peanut protein hydrolysate. *PLoS ONE* **2012**, *7*, e37863. [CrossRef]
46. Akram, A.N.; Zhang, C.H. Effect of ultrasonication on the yield, functional and physicochemical characteristics of collagen-II from chicken sternal cartilage. *Food Chem.* **2019**, *307*, 125544. [CrossRef]
47. Jamdar, S.N.; Rajalakshmi, V.; Pednekar, M.D.; Juan, F.; Yardi, V.; Sharma, A. Influence of degree of hydrolysis on functional properties, antioxidant activity and ACE inhibitory activity of peanut protein hydrolysate. *Food Chem.* **2010**, *121*, 178–184. [CrossRef]
48. Barac, M.B.; Pesic, M.B.; Stanojevic, S.P.; Kostic, A.Z.; Bivolarevic, V. Comparative study of the functional properties of three legume seed isolates: Adzuki, pea and soy bean. *J. Food Sci. Technol.* **2015**, *52*, 2779–2787. [CrossRef]
49. Xiong, X.Y.; Ho, M.T.; Bhandari, B.; Bansal, N. Foaming properties of milk protein dispersions at different protein content and casein to whey protein ratios. *Int. Dairy J.* **2020**, *109*, 104758. [CrossRef]
50. Mao, X.Y.; Hua, Y.F. Composition, structure and functional properties of protein concentrates and isolates produced from walnut (*Juglans regia* L.). *Int. J. Mol. Sci.* **2012**, *13*, 1561–1581. [CrossRef]

51. Dabbour, M.; He, R.H.; Ma, H.L.; Musa, A. Optimization of ultrasound assisted extraction of protein from sunflower meal and its physicochemical and functional properties. *J. Food Process Eng.* **2018**, *41*, e12799. [CrossRef]
52. Li, F.; Jia, D.; Yao, K. Amino acid composition and functional properties of collagen polypeptide from Yak (*Bos grunniens*) bone. *LWT-Food Sci. Technol.* **2009**, *42*, 945–949. [CrossRef]
53. Tao, X.; Cai, Y.J.; Liu, T.X.; Long, Z.; Huang, L.H.; Deng, X.L.; Zhao, Q.Z.; Zhao, M.M. Effects of pretreatments on the structure and functional properties of okara protein. *Food Hydrocolloid.* **2019**, *90*, 394–402. [CrossRef]
54. Ren, Z.Y.; Chen, Z.Z.; Zhang, Y.Y.; Zhao, T.; Ye, X.G.; Gao, X.; Lin, X.R.; Li, B. Functional properties and structural profiles of water-insoluble proteins from three types of tea residues. *LWT-Food Sci. Technol.* **2019**, *222*, 559–564. [CrossRef]
55. Elkhalifa, A.O.; Bernhardt, R.; Bonomi, F.; Lametti, S.; Pagani, M.A.; Zardi, M. Fermentation modifies protein/protein and protein/starch interactions in sorghum dough. *Eur. Food Res. Technol.* **2006**, *35*, 102103. [CrossRef]
56. Tkaczewska, J.; Wielgosz, M.; Kulawik, P.; Zajac, M. The effect of drying temperature on the properties of gelatin from carps (*Cyprinus carpio*) skin. *Czech J. Food Sci.* **2019**, *37*, 246–251. [CrossRef]
57. Shan, J.; Tang, B.; Liu, L.; Sun, X.B.; Shi, W.; Yuan, T.; Liang, J.; Fan, Y.J.; Zhang, X.D. Development of chitosan/glycerophosphate/collagen thermo-sensitive hydrogel for endoscopic treatment of mucosectomy-induced ulcer. *Mater. Sci. Eng. C* **2019**, *103*, 109870. [CrossRef]

*foods*

MDPI

*Article*

# Encapsulation of Rich-Carotenoids Extract from Guaraná (*Paullinia cupana*) Byproduct by a Combination of Spray Drying and Spray Chilling

**Lorena Silva Pinho [1,2], Priscilla Magalhães de Lima [1], Samuel Henrique Gomes de Sá [1], Da Chen [3], Osvaldo H. Campanella [2], Christianne Elisabete da Costa Rodrigues [1] and Carmen Sílvia Favaro-Trindade [1,*]**

1 Departamento de Engenharia de Alimentos, Faculdade de Zootecnia e Engenharia de Alimentos, Universidade de São Paulo, Pirassununga 13635-900, São Paulo, Brazil

2 Department of Food Science and Technology, Ohio State University, Columbus, OH 43210, USA

3 Department of Animals, Veterinary and Food Sciences, University of Idaho, 875 Perimeter Drive, Moscow, ID 83844, USA

* Correspondence: carmenft@usp.br

**Abstract:** Guaraná byproducts are rich in carotenoids, featuring strong antioxidant capacity and health-promoting benefits. However, these compounds are highly susceptible to oxidation and isomerization, which limits their applications in foods. This research aimed to encapsulate the carotenoid-rich extract from reddish guaraná peels by spray drying (SD), chilling (SC), and their combination (SDC) using gum arabic and vegetable fat as carriers. The carotenoid-rich extract was analyzed as a control, and the formulations were prepared with the following core–carrier ratios: SD20 (20:80), SD25 (25:75), SD33 (33:67), SC20 (20:80), SC30 (30:70), SC40 (40:60), SDC10 (10:90), and SDC20 (20:80). The physicochemical properties of the formed microparticles were characterized, and their storage stability was evaluated over 90 days. Water activity of microparticles formed during the SD process increased during storage, whereas those formed by SC and SDC processes showed no changes in water activity. The formed microparticles exhibited color variation and size increase over time. Carotenoid degradation of the microparticles was described by zero-order kinetics for most treatments. Considering the higher carotenoid content and its stability, the optimum formulation for each process was selected to further analysis. Scanning electron micrographs revealed the spherical shape and absence of cracks on the microparticle surface, as well as size heterogeneity. SD increased the stability to oxidation of the carotenoid-rich extract by at least 52-fold, SC by threefold, and SDC by 545-fold. Analysis of the thermophysical properties suggested that the carrier and the process of encapsulation influence the powder's thermal resistance. Water sorption data of the SDC microparticles depended on the blend of the carrier agents used in the process. Carotenoid encapsulation via an innovative combination of spray drying and spray chilling processes offers technological benefits, which could be applied as a promising alternative to protect valuable bioactive compounds.

**Keywords:** microencapsulation; storage stability; carotenoid degradation; gum arabic; vegetable fat

**Citation:** Pinho, L.S.; de Lima, P.M.; de Sá, S.H.G.; Chen, D.; Campanella, O.H.; da Costa Rodrigues, C.E.; Favaro-Trindade, C.S. Encapsulation of Rich-Carotenoids Extract from Guaraná (*Paullinia cupana*) Byproduct by a Combination of Spray Drying and Spray Chilling. *Foods* **2022**, *11*, 2557. https://doi.org/10.3390/foods11172557

Academic Editors: Michela Verni and Federico Casanova

Received: 20 July 2022
Accepted: 10 August 2022
Published: 24 August 2022



## 1. Introduction

The use of byproducts from agro-industrial processes has been growing over recent years as a strategy to valorize them and reduce environmental impact. Fruit and vegetable peels, seeds, and pomace are byproducts of their processing and show great potential as starting materials for the extraction of bioactive compounds [1]. One fairly representative fruit is guaraná (*Paullinia cupana*), a Brazilian plant native to the Amazon basin. In this region, the flowering of this plant occurs during the dry season, induced by water deficiency. The guaraná fruits become mature 2–3 months after flowering with a peel color of yellow to red [2].

Previous studies have found that guaraná seeds and peels are rich in alkaloids (caffeine, theobromine, and theophylline), polyphenols (catechin, epicatechin, and epicatechin gallate), and carotenoids (β-carotene and lutein) [3,4]. Carotenoids consist of natural pigments that are present in plants and other photosynthetic systems, and they are often used as food colorants. They also possess antioxidant capacity and have been claimed to act as anticancer agents, immune response stimulants, and pro-vitamin A activity promoters [5–10]. However, carotenoids are easily oxidized or isomerized in the presence of oxygen, light, and metals, as well as when exposed to heat during processing and storage, leading to products with reduced color and bioactivity [11]. Considering the potential applications of carotenoid-rich extract in foods, increased stabilization against processing conditions should be guaranteed, which can be fulfilled through microencapsulation.

Microencapsulation is a widespread approach to preserve the bioactivity of oxygen- and light-sensitive compounds, mask unpleasant tastes, and control release of valuable compounds [12]. A process commonly used for microencapsulation of carotenoids is spray drying [13–15]. This process turns a liquid feed material, which is composed of an ingredient (core) and a carrier agent, into a powder. As a consequence of the continuous operation of liquid atomization at high temperature and fast drying, the microparticles are formed. This increases the convenience of further processing and transportation in addition to preservation of active compounds. The efficiency of encapsulation in spray drying depends mainly on the types of carrier and core materials used in addition to the drying operating conditions [16]. Water-soluble polymers such as modified starches, whey protein, maltodextrin, and gum arabic are well-known and widely used carrier materials [17–19]. Gum arabic is a hydrocolloid extensively employed for the spray drying of materials containing bioactive compounds, due to its emulsification capacity, high solubility, low viscosity, nontoxicity, and ideal retention properties.

Spray chilling is an alternative process for stabilizing sensitive compounds such as carotenoids. Unlike spray drying that operates at high temperatures, the formation of lipid microparticles trapping bioactive compounds by spray chilling occurs at low temperatures. The lower processing temperature results in smaller degradation of these compounds [20–22]. However, the technique has relatively low encapsulation efficiency, and degradation of the lipid carrier may occur. Different types of materials, such as vegetable fat, fatty acids, and/or waxes are used as encapsulating agents to produce solid lipid microparticles by spray chilling. Due to its lower price and high availability, the use of vegetable fat is more appropriate than other lipids for this application [23,24].

The combination of spray drying and chilling offers the possibility of using the advantage of the single techniques and has been shown to better preserve bioactive components considering the presence of a double-wall coverage. For example, Arslan-Tontul and Erbas [25] observed higher gastric and thermal resistance of particles containing prebiotics produced by the combination of spray drying and chilling than those observed by processing individually. Fadini et al. [26] reported similar findings on the protection of functional oils when using combined technologies. However, whether this applies to carotenoids remains unknown.

The current study aimed to produce and compare microparticles containing a carotenoid-rich extract from guaraná peels produced by spray drying, chilling, and their combination. The microparticles were monitored during storage in terms of water activity, size, color parameters, and carotenoid stability. The physicochemical properties of microparticles, including morphology, water sorption, and thermal and oxidative stability of carotenoids, were measured. The influence of different carrier materials on the properties of the microparticles was also investigated.

## 2. Material and Methods

### 2.1. Materials

Guaraná fruits were provided by the Executive Commission of the Rural Economic Recuperation Plan in Cacao (Taperoá, Bahia, Brazil). Gum arabic (Acacia gum—Nexira,

Brazil) was used as the carrier in the spray drying process. Microparticles obtained by spray chilling were produced using Al Home P54 vegetable fat with a melting point of 54 °C (Cargill, Itumbiara, Goiás, Brazil) as the carrier material. Microparticles obtained by the combination of spray drying and chilling processes were prepared using gum arabic as the first wall material and vegetable fat as the second coating. Ethanol (purity $\geq$ 99%) was purchased from Êxodo Científica (Sumaré, Brazil). Petroleum ether (30–70 fraction) and magnesium chloride ($MgCl_2 \cdot 6H_2O$) were acquired from Synth (Diadema, Brazil).

### 2.2. Production of Carotenoid-Rich Extract from Guaraná Peels

Guaraná peels were dried in a convective oven (Marconi, MA035/1152) at 50 °C for 18 h and stored at −20 °C until further analysis. For carotenoid extraction, the peels were mixed with ethanol in a ratio of 1:10 (peel/solvent, *w*/*v*). According to Pinho et al. [4] ethanol presented ideal performance when extracting carotenoids from guaraná peels compared to other solvent systems tested (such as hexane and ethyl acetate). Furthermore, the use of ethanol as solvent was based on the fact that it is recognized as safe and is obtained from renewable sources.

The mixture was shaken using an orbital shaker (Orbital Shaker Marconi, MA420, Piracicaba, SP) for 4 h at 50 °C and centrifuged at 7168× *g* for 10 min [4]. Sunflower oil was added to the supernatant with a final concentration of 3% and thoroughly mixed to minimize carotenoid degradation (observed during preliminary experiments). A final concentration of 3% of the oil was selected after considering the liquid–liquid equilibrium for the system composed of sunflower oil and ethanol [27]. Afterward, the material was concentrated using a rotary evaporator (TE-211 Tecnal, Piracicaba, Brazil) at 48 ± 2 °C to 20% of the initial volume. The concentrate was named "carotenoid-rich extract".

### 2.3. Production of Microparticles from the Carotenoid-Rich Extract

Carotenoid-rich extracts were mixed with the carrier materials at different proportions for the production of carotenoid-enriched microparticles by spray drying and chilling. Furthermore, microparticles obtained by the combined spray drying and chilling processes were prepared using spray-dried microparticles as the core with a layer of vegetable fat as a carrier material. Formulations in the processes are shown in Table 1.

**Table 1.** Formulation and composition of microparticles produced by spray drying (SD), chilling (SC), and the combination of spray drying and chilling (SDC).

| Formulation | Core (%) | Carrier Material (%) |
|---|---|---|
| | Extract | Gum Arabic solution (20%, *w*/*v*) |
| SD20 | 20 | 80 |
| SD25 | 25 | 75 |
| SD33 | 33 | 67 |
| | Extract | Vegetable fat |
| SC20 | 20 | 80 |
| SC30 | 30 | 70 |
| SC40 | 40 | 60 |
| | SD33 microparticles | Vegetable fat |
| SDC10 | 10 | 90 |
| SDC20 | 20 | 80 |

#### 2.3.1. Microparticles Obtained by Spray Drying (SD)

An emulsion was prepared by mixing the carotenoid-rich extract and a gum arabic solution (20% *w*/*v*), using an Ultra-Turrax® IKA T25 (Labotechnic, Staufen, Germany) at 11,200× *g* for 3 min. The emulsion was atomized according to Rocha, Fávaro-Trindade, and Grosso [28] with slight modifications in the drying temperature. A spray dryer (Model MSD 1.0, Labmaq do Brasil, Ribeirão Preto, Brazil) equipped with a spray nozzle of 1.2 mm and inlet air temperature of 140 °C, air speed of 2.5 m/s, feed flow of 10 mL/min, and air pressure of 823,759 N/m$^2$ was used. During the drying procedure, the emulsion fed to the dryer was kept under magnetic stirring.

2.3.2. Microparticles Obtained by Spray Chilling (SC)

Solid lipid microparticles were prepared by spray chilling [29]. Dispersions with different proportions of carotenoid-rich extract and high-melting-point vegetable fat were prepared using an Ultra-Turrax® IKA T25 (Labotechnic, Staufen, Germany) at 11,200× *g* for 3 min and 64 °C. The mixture was atomized using the same spray dryer equipment (Model MSD 1.0, Labmaq do Brasil, Ribeirão Preto, Brazil) coupled to a spray nozzle of 1.2 mm and the following conditions: 1.0 kgf/cm$^2$ air pressure, a feed flow of 40 mL/min, and a temperature of t 13 °C.

2.3.3. Microparticles Obtained by the Combination of Spray Drying and Chilling (SDC)

The SD33 formulation was firstly prepared by spray drying as described previously (Section 2.3.1). SDC microparticles were then prepared by dispersing the SD33 microparticles into vegetable fat (*w*/*w*) at 11,200× *g* for 3 min and 64 °C. The atomization conditions used were the same as described previously for the spray chilling process.

*2.4. Characterization of Microparticles and Storage Stability*

The microparticles were placed in vials and kept in desiccators containing saturated magnesium chloride solution to create a storage environment of 33% ± 5% relative humidity (RH), at 25 ± 5 °C for 90 days. Water activity, color, mean diameter, and particle size were determined at the beginning (time 0) and the end (after 90 days) of storage. The total carotenoid content of the samples was evaluated every 15 days [30].

2.4.1. Water Activity

The water activity ($a_w$) of the microparticles was measured utilizing an Aqualab instrument (Series 3 TE—Decagon Devices, Pullman, WA, USA) at room temperature. Measurements were conducted in triplicate.

2.4.2. Instrumental Color Analysis

The color parameters L (luminosity), a* (red–green), and b* (yellow–blue) of the samples were evaluated using a portable colorimeter (Mini Scan XE Plus—Hunterlab, Reston, VA, USA). Measurements were executed in triplicate.

2.4.3. Particle Size and Distribution

Particle size distribution and the mean diameters of the particles were determined using laser diffraction (SaldI-201-V, Shimadzu, Kyoto, Japan). Pure ethanol was used as the dispersant of microparticles obtained by spray drying, while distilled water was used for the microparticles prepared by spray chilling and the combined process. Measurements were performed in triplicate.

2.4.4. Determination of Total Carotenoid Content in the Nonencapsulated and Encapsulated Extract

The total carotenoid content was determined according to a spectrophotometric method [11], using a UV/Vis spectrophotometer (Genesys 10S Thermo Scientific, São Paulo/SP, Brasil) at 450 nm.

Carotenoids were extracted following Pelissari et al. [29] with slight modifications of the solvents used. Carotenoid-rich extracts and SD microparticles were dispersed in hexane, while SC and SDC microparticles were dispersed in petroleum ether. The mixtures were agitated for 1 min and kept in an ultrasound bath (USC-1400, Unique, Indaiatuba, Brazil) for 20 min. Distilled water was then added and agitated (Multi Reax, Heidolph Instruments, Schwabach, Germany) for 2 min. They were then centrifuged at 4930× *g* for

10 min, and hexane or petroleum ether-rich phases were transferred to cuvettes. The total carotenoid content was determined according to the following equation:

$$C = A \times V \times 10^4 / Abs_{1cm}^{1\%} \times m, \quad (1)$$

where $C$ (μg/g) is the total carotenoid content, $A$ (nm) is the absorbance of the sample at 450 nm, $V$ (Ml) is the final volume, and $Abs_{1cm}^{1\%}$ the extinction coefficient of $\beta$-carotene, which has a value of 2592 $cm^{-1}$ in petroleum ether [11] and 2560 $cm^{-1}$ in hexane [31]; $m$ is the sample mass (g). The determinations were performed in triplicate.

#### 2.4.5. Encapsulation Efficiency (EE)

Encapsulation efficiency ($EE$) was calculated as the ratio between the total carotenoid content present in the microparticles ($C_0$) and the total carotenoid content in the feed material ($C_{FM}$) before drying, as indicated by Equation (2) [12,32]. Microparticle production was conducted in triplicate.

$$EE(\%) = (C_0 / C_{FM}) \times 100. \quad (2)$$

#### 2.4.6. Kinetics of Carotenoid Degradation

The degradation of carotenoids during storage was described using kinetic models to quantitatively describe the reactions that occur in the system. The kinetics and the corresponding reaction rate constants $k_0$ and $k_1$ were evaluated following zero- and first-order kinetics described by Equations (3) and (4), for nonencapsulated and encapsulated carotenoid-rich extract. The half-life of the first-order reaction was calculated according to Equation (5) [33].

$$C_t - C_0 = -k_0 \times t, \quad (3)$$

$$-\ln C_t / C_0 = k_1 \cdot t, \quad (4)$$

$$t_{1/2} = \ln 2 / k_1, \quad (5)$$

where $C_0$ (μg/g) is the initial carotenoid content (time 0), $C_t$ (μg/g) is the content at time $t$, and $t$ is the time (days). The reaction rate constants for the zero-order kinetics $k_0$ (μg/(g·s)) and first-order kinetics $k_1$ (1/s) were obtained from the slopes of linear plots of ($C_t - C_{t,0}$) vs. $t$ and ln ($C_t/C_{t,0}$) vs. $t$, respectively. The kinetic analysis was carried out using the program Origin Pro 8.5.

#### 2.4.7. Carotenoid Retention (CR)

Carotenoid retention (%) was determined according to the following equation:

$$CR = (C_{90} / C_0) \times 100, \quad (6)$$

where $C_{90}$ (μg/g) is the carotenoid concentration after 90 days of storage, whereas $C_0$ (μg/g) is the initial carotenoid concentration in the microparticles.

### 2.5. *Analysis of Selected Microparticles*

To further analyze the characteristics of the microparticles, three formulations were selected. The samples SD33, SC40, and SDC20 were chosen mainly on the basis of the results of the total carotenoid content and the reasonable stability observed during storage. They were evaluated in terms of their morphology, oxidative stability, thermal properties, and water vapor sorption, as described below.

#### 2.5.1. Scanning Electron Microscopy (SEM)

SD, SC, and SDC microparticles were placed on a double-sided carbon adhesive tape (Ted Pella Inc., Redding, CA, USA); they were coated with gold and analyzed using SEM (Benchtop Microscope Hitachi TM 300, Tokyo, Japan). SEM images of the microparticles were captured at an accelerating voltage of 15 Kv.

2.5.2. Oxidative Stability

The oxidative stability of powders was conducted according to De Leonardis and Macciola [34]. Microparticles (2.0 g) were heated from 25 °C to 120 °C under a 20 L/h airflow, using a Rancimat instrument (model 873, Metrohm, Herisau, Switzerland). The analyses were carried out in duplicate, and results were expressed as the oxidation induction time.

2.5.3. Differential Scanning Calorimetry (DSC)

Thermal properties of microparticles were evaluated using differential scanning calorimetry (DSC 2500, TA Instruments, New Castle, Delaware, DE, USA). Samples were weighed (~5 mg) into aluminum pans, sealed, and placed in the DSC instrument. An empty pan was used as a reference. Samples were equilibrated at 25 °C for 2 min, ramped to 300 °C at a rate of 10 °C/min, kept at an isothermal condition (T = 300 °C) for 2 min, and then cooled to 25 °C at a rate of 30 °C/min [35]. Data collection and analysis were conducted using Trios software (TA Instruments). Triplicates were run for each sample, and the average results were shown.

2.5.4. Dynamic Vapor Sorption (DVS)

The water vapor sorption of the microparticles was conducted using a Dynamic Vapor Sorption instrument (Surface Measurement Systems Ltd., Allentown, PA, USA). The airflow in the DVS was compressed nitrogen flowing at 200 Ml/min. The changes in sample mass at various relative humidities (between 0% and 90%) were recorded continuously at 25 °C using the DVS Analysis Macro V6.1 software. Each sample was run in duplicate.

*2.6. Statistical Analysis*

Data were examined by analysis of variance (ANOVA) and Tukey's test, at the 5% level significance, using the statistical program SAS (Statistic Analysis System) version 9.2.

**3. Results and Discussion**

*3.1. Characterization of Microparticles and Their Changes during Storage*

3.1.1. Water Activity ($a_w$)

The water activities ($a_w$) of microparticles produced by spray drying, chilling, and the combined process combination loaded with carotenoid-rich extract are shown in Table 2. SD microparticles had the lowest $a_w$, probably due to larger water evaporation at the higher inlet temperature (140 °C) of the spray drying process. After storage, $a_w$ increased due to the particle hygroscopicity (water uptake) as a consequence of the presence of gum arabic, which was used as the carrier material.

In this system, the molecular mobility increased, and the microparticles absorbed water until reaching an equilibrium condition, in an intermediate relative humidity range of 35–40% at 25 °C. This condition creates a favorable environment for binding water, which is influenced by the highly branched structure of gum arabic and its high water affinity [36]. The water activity of SD microparticles was below the minimum value of 0.60, which corresponds to a satisfactory condition to avoid microbial growth. It is worth mentioning that, even during storage, these samples exhibited great stability.

SC microparticles had a higher $a_w$ (0.80–0.95) than SD and SDC microparticles (0.41–0.52), and they varied with formulations. These $a_w$ values measured are in agreement with the $a_w$ range reported by Labuza [37], from 0.3 to 1.0, which is associated with the susceptibility of high lipid oxidation at lower water activity conditions. Silva et al. [38] presented similar results, with water activities ranging from 0.91 to 0.98, for solid lipid microparticles loaded with probiotics produced by spray chilling.

**Table 2.** Characterization of microparticles produced by spray drying (SD), chilling (SC), and their combination (SDC) before and after storage.

| Response | | Time (Days) | SD20 | SD25 | SD33 | SC20 | SC30 | SC40 | SDC10 | SDC20 |
|---|---|---|---|---|---|---|---|---|---|---|
| $a_w$ | | 0 | 0.132 ± 0.004 $^{h,B}$ | 0.210 ± 0.001 $^{f,B}$ | 0.161 ± 0.005 $^{g,B}$ | 0.913 ± 0.001 $^{a,B}$ | 0.872 ± 0.002 $^{b,B}$ | 0.801 ± 0.002 $^{c,B}$ | 0.47 ± 0.01 $^{d,B}$ | 0.412 ± 0.003 $^{e,B}$ |
| | | 90 | 0.456 ± 0.002 $^{e,A}$ | 0.449 ± 0.001 $^{e,A}$ | 0.418 ± 0.003 $^{e,A}$ | 0.953 ± 0.004 $^{a,A}$ | 0.897 ± 0.006 $^{b,A}$ | 0.833 ± 0.008 $^{c,A}$ | 0.52 ± 0.01 $^{d,A}$ | 0.468 ± 0.008 $^{e,A}$ |
| Color parameters | L* | 0 | 80.61 ± 0.01 $^{d,B}$ | 76.44 ± 0.01 $^{f,B}$ | 74.11 ± 0.01 $^{h,B}$ | 84.15 ± 0.01 $^{a,A}$ | 82.51 ± 0.24 $^{b,A}$ | 79.20 ± 1.82 $^{e,A}$ | 81.26 ± 0.02 $^{c,B}$ | 74.39 ± 0.01 $^{g,B}$ |
| | | 90 | 81.46 ± 0.01 $^{b,A}$ | 78.80 ± 0.01 $^{c,A}$ | 75.18 ± 0.01 $^{f,A}$ | 76.31 ± 0.01 $^{d,B}$ | 75.63 ± 1.42 $^{e,B}$ | 71.01 ± 2.29 $^{h,B}$ | 81.75 ± 0.01 $^{a,A}$ | 75.12 ± 0.01 $^{g,A}$ |
| | a* | 0 | 1.08 ± 0.04 $^{d,A}$ | 2.06 ± 0.03 $^{b,A}$ | 4.05 ± 0.02 $^{a,A}$ | 0.79 ± 0.02 $^{e,B}$ | 0.84 ± 0.26 $^{e,B}$ | 1.97 ± 016 $^{b,B}$ | 0.14 ± 0.02 $^{f,A}$ | 1.81 ± 0.05 $^{c,A}$ |
| | | 90 | 0.51 ± 0.03 $^{f,B}$ | 1.04 ± 0.02 $^{e,B}$ | 3.17 ± 0.04 $^{a,B}$ | 1.08 ± 0.03 $^{e,A}$ | 1.42 ± 0.02 $^{c,A}$ | 2.29 ± 0.07 $^{b,A}$ | 0.06 ± 0.01 $^{g,B}$ | 1.60 ± 0.06 $^{d,B}$ |
| | b* | 0 | 23.07 ± 0.02 $^{g,A}$ | 33.39 ± 0.05 $^{b,A}$ | 39.68 ± 0.02 $^{a,A}$ | 27.29 ± 0.05 $^{e,A}$ | 24.12 ± 0.59 $^{f,A}$ | 30.76 ± 1.47 $^{d,A}$ | 20.35 ± 0.05 $^{h,A}$ | 31.04 ± 0.08 $^{c,A}$ |
| | | 90 | 20.36 ± 0.05 $^{g,B}$ | 29.62 ± 0.02 $^{b,B}$ | 35.12 ± 0.07 $^{a,B}$ | 24.07 ± 0.09 $^{e,B}$ | 23.39 ± 0.03 $^{f,B}$ | 25.27 ± 0.07 $^{d,B}$ | 19.89 ± 0.06 $^{h,B}$ | 29.02 ± 0.10 $^{c,B}$ |
| Mean diameter (μm) | | 0 | 16.2 ± 0.3 $^{d,B}$ | 10.8 ± 0.2 $^{e,A}$ | 10.9 ± 0.6 $^{e,B}$ | 59 ± 2 $^{b,B}$ | 56.6 ± 0.4 $^{b,B}$ | 52 ± 1 $^{c,B}$ | 73 ± 1 $^{a,A}$ | 60 ± 1 $^{b,B}$ |
| | | 90 | 20 ± 1 $^{c,A}$ | 13 ± 2 $^{c,A}$ | 15 ± 2 $^{c,A}$ | 74 ± 2 $^{b,A}$ | 71.3 ± 0.8 $^{b,A}$ | 71.7 ± 0.8 $^{b,A}$ | 75 ± 2 $^{b,A}$ | 96 ± 2 $^{a,A}$ |

Values are the mean ± standard error (SE) ($n$ = 3–4 analytical replicates). Different uppercase letters in a column represent a significant difference between 0 and 90 days with the response for formulations. The different lowercase letters in a row indicate a significant difference among formulations by the Tukey test at the 5% level of significance. SD: microparticles obtained by spray drying; SC: microparticles obtained by spray chilling; SDC: microparticles obtained by spray drying and chilling combination. The numerical suffix denotes the proportion of core/wall material of each formulation.

#### 3.1.2. Color Parameters

Color coordinates L*, a*, and b* were examined during storage time, as parameters associated with the chemical stability of carotenoids encapsulated within SD, SC, and SDC microparticles (Table 2). Color parameter analyses were conducted considering that the carotenoid-rich microparticles could be used as colorants with bioactivity. SD20, SC20, and SDC10 microparticles showed higher lightness (L*), compared to the other treatments obtained through the same process. This was due to the higher content of gum arabic and/or vegetable fat, as well as the lower ratio between the concentration of extract and SD microparticles in their formulations. During the stability test, L* of the SD microparticles increased. This can be associated with carotenoid degradation, which leads to changes in the color intensity of the microparticles. McClements [39] reported that the increase in particle dimensions caused by aggregation and/or its morphology is likely to influence the light of the samples, affecting their L* value.

Regarding the parameter a* (redness), all the microparticles had positive values, which implies the subtle redness of all samples. A relatively superior magnitude of a* was observed for the SD33 microparticles, indicating more redness compared to others. The formulation was composed of a high proportion of extract, and the evaporation of the carrier solution during the spray drying led to a higher concentration of carotenoid within the microparticles.

Concerning the parameter b* (yellowness), all formulations displayed a noticeable decrease in this value during storage. Indeed, the intensity change of the yellow/orange color (Δb*) of the microparticles during storage followed the order: SC30 = SDC10 < SC20 = SDC20 < SD20 = SC40 < SD25 < SD33. These findings suggest that encapsulated carotenoids were more stable in lipid microparticles (SC and SDC) obtained by spray chilling and the combined process.

#### 3.1.3. Mean Diameter and Particle Size Distribution

The volume-weighted mean diameter ($D_{4,3}$) and particle size distribution are shown in Table 2 and Figure 1, respectively. SD20 microparticles had a significantly larger mean diameter than SD25 and SD33 microparticles. This was mainly attributed to the sample formulation. SD20 feed material had the highest concentration of the carrier agent and increased the viscosity, resulting in larger spray-dried particles compared to the others from the same process.

During storage, the size of SD microparticles had a slight increase, which was likely due to agglomeration promoted by adhesion forces. The high water content likely led to the microparticles remaining bonded after collision, promoting the formation of aggregates. Microparticle agglomeration may promote powder application by reducing dust formation [29].

Samples obtained by spray chilling had average sizes that changed over 90 days. In the spray chilling process, operating conditions such as temperature, pressure, cooling air speed, feed flow, and spray nozzle diameter may affect particle size measured by the mean diameter. In addition, intrinsic parameters, such as the composition of lipid carrier, and the ratio of the feed matrix (i.e., bioactive ingredient:carrier) can affect the size of the microparticles [40].

SDC10 microparticles were larger than the SDC20 ones after being freshly prepared. The mean diameter of the microparticles increased continuously throughout storage ($p < 0.05$), especially the SDC20 formulation. Microparticles produced by spray chilling are classified as matrices, where the compound to be protected is entrapped by the volume of the particle. In the SDC microparticle case, the microparticle structure is composed of vegetable fat (carrier) and SD microparticles (core). The observed broad range of the diameter after storage of the SDC20 microparticles could be attributed to the composition of the microparticles, which may promote aggregation. Pelissari et al. [29] reported that the agglomeration of particles formed by hydrogenated and interesterified vegetable oils as carrier material may be attributed to the presence of melted triacylglycerols, which favors the adherence among lipid particles.

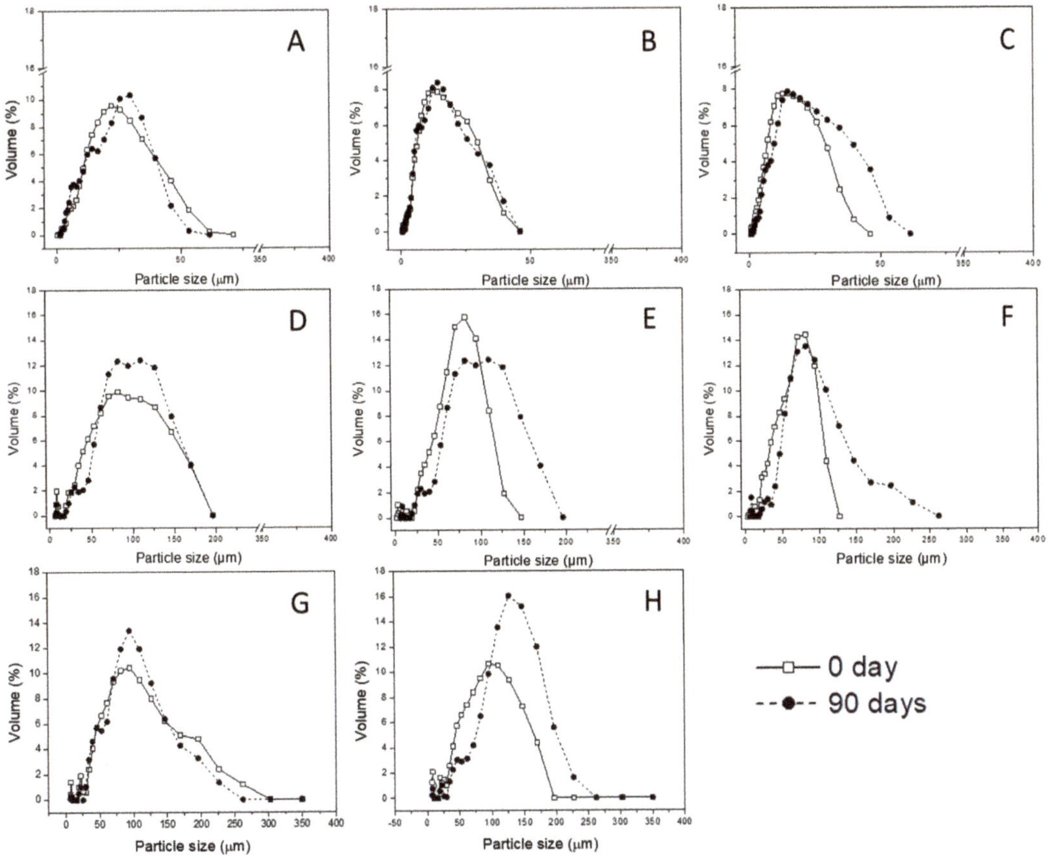


**Figure 1.** Particle size distribution of microparticles prepared by spray drying, chilling, and the combined process: (**A–C**) microparticles obtained by spray drying with a 20%, 25%, and 33% ratio core/carrier material, respectively; (**D–F**) microparticles obtained by spray chilling with a 20%, 30%, and 40% ratio core/carrier material, respectively; (**G,H**) microparticles obtained by spray drying and chilling combination with a 10% and 20% ratio core/carrier material, respectively.

Overall, the size distribution of all formulations exhibited a unimodal distribution before and after storage at 25 °C. Microparticles prepared by SD had smaller sizes compared to those of the SC and SDC microparticles, even though the same drying equipment, nozzle type, and diameter were used for atomization of the feed. The smaller droplet size of the emulsion in the feed material and the operational parameters, such as high temperature and pressure, might facilitate solvent evaporation and shorten the particle coat formation during the atomization by spray drying. Conversely, in the spray chilling process, the heat transfer between the cold air and the molten feed material is able to quickly solidify the coating material, maintaining the original size of the formed microparticle.

Particle size is a physical property relevant to many food applications, and it is also related to the sensory attributes of foods in which powders are incorporated [41]. Hansen, Allan-Wojtas, Jin, and Paulson [42] suggested that powders with particle sizes less than 100 µm would not affect the sensory quality of the food. Thus, microparticles produced by spray drying, chilling, and their combination prepared in this study had suitable sizes for practical applications. However, from an applied perspective, SD microparticles had a more suitable size for food supplementation. In addition to the particle size, the composition of microparticles produced by the different methods would affect their applications. For example, SD microparticles could be used to disperse and protect carotenoids during the processing of aqueous products, while microparticles obtained by SC and SDC would be preferable for applications that require nonaqueous media.

### 3.2. Encapsulation Efficiency (EE)

Obtaining microparticles with the best application properties is one of the primary goals of encapsulation. Factors related to the feed flow rates, inlet air temperature, carrier material type, and formulation are essential because they affect the microparticles' characteristics and encapsulation efficiency (EE). The EE of microparticles produced through spray drying, chilling, and the combined processes is shown in Table 3. The samples had high EE, ranging from 90% to 100% for SD, 90% to 97% for SC, and 82% to 94% for the SDC techniques, which demonstrates the greater entrapment of the carotenoid content in the microparticles for all treatments used.

**Table 3.** Encapsulation efficiency (%) of microparticles obtained by spray drying, chilling, and their combination.

| Formulation | Encapsulation Efficiency (%) |
|---|---|
| SD20 | $96 \pm 7$ ab |
| SD25 | $100 \pm 2$ a |
| SD33 | $90 \pm 2$ bc |
| SC20 | $96 \pm 2$ ab |
| SC30 | $97 \pm 1$ ab |
| SC40 | $90 \pm 1$ bc |
| SDC10 | $82 \pm 7$ c |
| SDC20 | $94.4 \pm 0.5$ ab |

SD: microparticles obtained by spray drying; SC: microparticles obtained by spray chilling; SDC: microparticles obtained by spray drying and chilling combination. The numerical suffix denotes the proportion of core/carrier material of each formulation. Different superscript letters indicate significant differences among the data in the same column ($p < 0.05$).

A slight carotenoid loss was observed in SD samples, which might have occurred during the atomization step, in which these components possibly adhered to the hot drying chamber walls during processing. On the other hand, in SC treatments, the degradation of carotenoids may be attributed to the fact that the vegetable fat mixed with the extract was kept at 64 °C for a long time during the processing, to improve the mixture homogenization before atomization. Although spray chilling is a technique that produces microparticles at low temperatures, due to the fact that the fat used as the carrier needs to be in a molten state, a relatively high temperature must be used, which might affect the stability of carotenoids. Additionally, exposure of the product to environmental conditions, such as light and oxygen, may have triggered some oxidation of the pigments.

Concerning the SDC microparticles, they had lower EE compared to the others. Lipids exhibit three types of crystal polymorphic phases, i.e., alpha ($\alpha$), beta-prime ($\beta'$), and beta ($\beta$), from the least to the most stable [43,44]. Thus, the carrier material can solidify following different crystallization patterns during the spray chilling process. Through atomization at low temperatures via rapid cooling, the solidification of microparticles composed of gum arabic and vegetable fat may generate an arrangement of lipid crystals that are unable to trap some of the bioactive compounds. As a result, the achieved arrangement probably led to the expulsion of carotenoids and their loss during production. A similar finding was reported by Navarro-Guajardo et al. [45].

### 3.3. Chemical Stability of Nonencapsulated and Encapsulated Carotenoid-Rich Extract

Carotenoid retentions of nonencapsulated extracts, as well as SD, SC, and SDC particles, upon storage were calculated using Equation (6) and are shown in Figure 2. It can be seen that carotenoids were degraded to a large extent during storage, despite reasonable storage conditions and the exclusion of light. Samples prepared by SD, SC, and SDC had retentions ranging from 97% to 68%, from 99% to 84%, and from 99% to 75%, respectively.

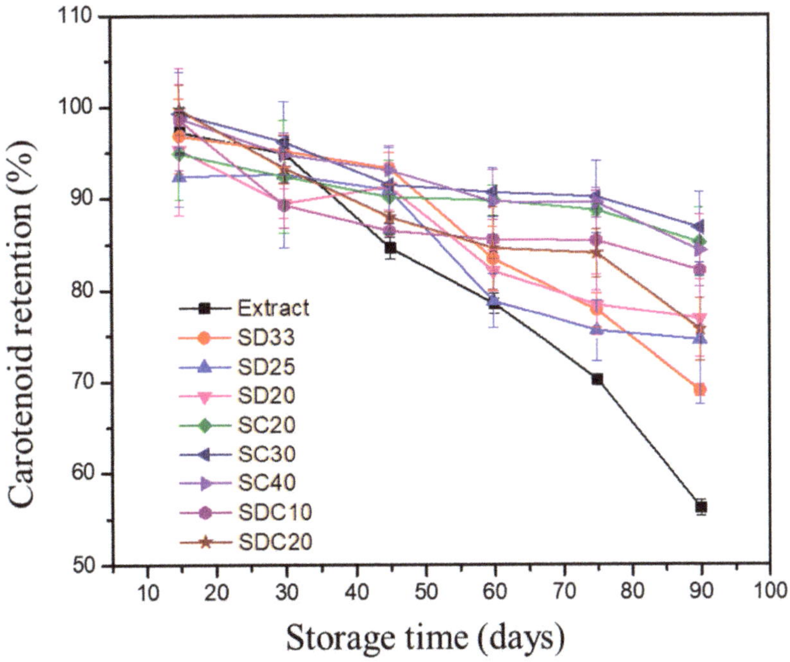


**Figure 2.** Carotenoid retention (CR, %) in nonencapsulated extracts and microparticles obtained by spray drying, chilling, and their combination, during 90 days of storage at 25 °C. SD: microparticles obtained by spray drying; SC: microparticles obtained by spray chilling; SDC: microparticles obtained by spray drying and chilling combination. The numerical suffix denotes the proportion of core/carrier material of each formulation.

Throughout the 90 days of storage time, carotenoid losses increased, despite the encapsulation processes providing protection of total carotenoids when compared to nonencapsulated extracts. However, microparticles obtained by SC and SDC displayed higher effectiveness in preserving the carotenoids, which could indicate that the composition of the carriers impacted the stability of active components. This result was in agreement with the color change observed for the different samples. According to previous studies on guaraná byproducts, the major carotenoids found in guaraná peels are all-*trans*-β-carotene, followed by *cis*-β-carotenes and lutein. However, environmental factors such as soil condition, weather, and ripening process may influence the composition of guaraná peels [4].

The potential entrapment of oxygen within the carrier matrix in the process might contribute to the oxidation of carotenoids during storage. Furthermore, isomerization may also affect the stability of these compounds. Khoo, Prasad, Kong, Jiang, and Ismail [46] and Provesi, Dias, and Amante [47] claimed that light and temperature are the major causes of carotenoids isomerization. However, Rodriguez-Amaya [48] reported that degradation of carotenoids during storage is mainly caused by enzymatic and nonenzymatic oxidation, which correlates with the oxygen level in the microparticles and the carotenoid molecular structure.

The kinetics of the carotenoid degradation was investigated, and the results are presented in Table 4. Data were fitted to the zero- and first-order kinetics according to the carotenoid content during storage. Total carotenoid retention decreased remarkably, and the fitting of the degradation and its rate were analyzed by the correlation coefficient ($R^2$) and the reaction k values, respectively. For the zero-order model, $R^2$ ranged from 0.610 to 0.957, and the $k_0$ value ranged from 0.021 to 1.558 μg/(g·s), which was associated with suitable fitting for most samples. Reasonable values of $R^2$ (0.755–0.940) combined with lower $k_1$ values (0.001–0.006 1/s) were achieved using first-order kinetics.

**Table 4.** Rate constants ($k_0$ and $k_1$), coefficient of determination ($R^2$), and half-life periods ($t_{1/2}$) for degradation of nonencapsulated and encapsulated carotenoid-rich extracts obtained by different processes, during storage in the dark at 25 °C.

| Sample | Zero-Order | | First-Order | | $t_{1/2}$ (Days) |
|---|---|---|---|---|---|
| | $k_0$ (μg/(g·s)) | $R^2$ | $k_1$ (1/s) | $R^2$ | |
| Nonencapsulated extract | 1.558 | 0.957 | 0.006 | 0.857 | 108.08 |
| SD20 | 0.089 | 0.911 | 0.003 | 0.901 | 236.93 |
| SD25 | 0.176 | 0.844 | 0.003 | 0.847 | 212.20 |
| SD33 | 0.284 | 0.805 | 0.004 | 0.783 | 168.07 |
| SC20 | 0.025 | 0.925 | 0.001 | 0.919 | 606.36 |
| SC30 | 0.035 | 0.909 | 0.001 | 0.792 | 581.55 |
| SC40 | 0.075 | 0.949 | 0.001 | 0.859 | 546.95 |
| SDC10 | 0.021 | 0.610 | 0.003 | 0.755 | 316.59 |
| SDC20 | 0.058 | 0.942 | 0.003 | 0.940 | 223.45 |

SD: microparticles obtained by spray drying; SC: microparticles obtained by spray chilling; SDC: microparticles obtained by spray drying and chilling combination. The numerical suffix denotes the proportion of core/carrier material for each formulation.

The structural characteristics of the microparticles led to different results for the kinetic parameters obtained from the formulations and processes (spray drying, chilling, and their combination). Therefore, according to these aspects, microparticles produced by spray chilling showed superior protective capacity, ensuring greater stability of the carotenoids over 90 days, as demonstrated by the half-life data.

It is worth noting that, when changing the carotenoid proportion in SD, SC, and SDC microparticles, the half-life was shorter in formulations with higher carotenoid content. This suggests that the bioactive components are less protected in such samples since more carotenoid molecules were likely located on the microparticles' surface.

Taken together, the total carotenoid concentration present in the microparticles was considered as an indicator to assess the stability of microparticles during storage, how it was affected by the formation process, and the feasibility of using these samples for food applications. These results were used to design the next steps of the present research, in which further investigation may unravel more details about the design of suitable formulations for the different processes.

### 3.4. *Characterization of Selected Microparticles by Morphology, Thermal Properties, and Moisture Sorption*

#### 3.4.1. Scanning Electron Microscopy (SEM)

Scanning electron micrographs of SD33, SC40, and SDC20 microparticles loaded with carotenoids are illustrated in Figure 3. The SD33 microparticles exhibited different sizes with irregular spherical shapes, and no cracks were found at the outer surfaces, which is important to prevent gas permeability that may affect carotenoids protection. Similar behavior was observed in microparticles of eggplant peel extract [49] and pumpkin peel extract [50] encapsulated with gum arabic.

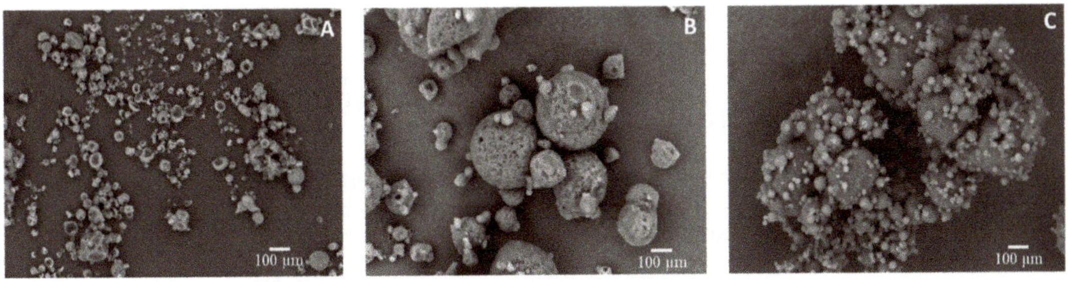


**Figure 3.** Scanning electron micrographs (1000× magnification) of microparticles obtained by (**A**) spray drying (SD33), (**B**) spray chilling (SC40), and (**C**) spray drying and chilling combination (SDC20). The numerical suffix denotes the core/wall material ratio of the selected formulation.

The cavities and irregularities on the microparticle surface obtained by spray drying are associated with rapid solvent evaporation and higher pressure inside the microparticles at high processing temperatures [51]. According to Elik, Yanık, and Göğüş [52], the temperature range between drying inlet air and the atomized droplets promotes wrinkling on the particle surface. This could be attributed to the initial expansion due to the incorporation of air in the particle, followed by its contraction in the colder chamber of the equipment. Bhusari, Muzaffar, and Kumar [53] proposed that the cavities in the microparticles containing gum arabic may be linked to the protein fraction in the gum that may promote the formation of small air pockets.

SC40 and SDC20 microparticles had spherical shapes of different sizes. The differences demonstrated that the addition of SD microparticles into the vegetable fat carrier may have affected their morphologies, mainly the surface of the microparticles. The SC40 powders showed some agglomeration and an irregular surface with some pores. Similarly, the solid lipid microparticles loaded with carotenoid [29] and guaraná seed extract [54] obtained by spray chilling showed a spherical shape with agglomeration. From a technological standpoint, these physical properties could reduce the flowability of the powders. Furthermore, holes on the microparticle surface can increase the exposure of carotenoids to oxygen, thus reducing the powder functionality in terms of its nutritional properties.

Regarding the SDC powders, the morphological analysis showed a smooth surface with a high level of agglomeration. This may be related to the complex carrier matrix formed by the combination of gum arabic and vegetable fat, as well as its water holding capacity. The collision of microparticles at moderate–high moisture contents might induce agglomeration; moreover, incomplete solidification of the carrier material could also lead to the formation of partially melted microparticles favoring agglomeration. The microparticle agglomeration can produce structures that no longer have a spherical shape. These structures may favor additional protection to the microencapsulated compounds, as the outer particles can shield the inner ones from environmental conditions. Indeed, microparticle morphology can be influenced by operational parameters, which include feed material composition, temperature, solvents used, and drying rate [55,56].

### 3.4.2. Oxidative Stability

The Rancimat method is widely used to investigate the oxidative stability of oil/lipid-containing samples [57]. The technique is based on the measurements of an induction period (IP) linked to the formation of certain components when the lipid samples are oxidized under heating (at temperatures higher than 100 °C) and under a constant aeration flow (20 L/h). In the present study, a higher IP of the encapsulated extract compared to the nonencapsulated one led to a better stability of the carotenoids within the microparticles. Data for the nonencapsulated extract, which contained oil in its formulation, as well as for selected microparticles, are presented in Table 5. Results indicated that, under the conditions applied in the test, carotenoid-rich extracts were highly sensitive to oxidation.

The oxidative stability improved significantly ($p < 0.05$) when microparticles were prepared using different carriers, which were assessed using a comparative approach. It was observed that spray drying encapsulation increased the thermal stability of the extract by at least 52-fold, in contrast spray chilling by threefold, and the combined process by 545-fold, confirming the ability of the encapsulated microparticles to protect sensitive compounds against degradation. SD encapsulation produced microparticles composed of a resistant shell, which had limited gas transfer. Regarding SC40 microparticles, although they had a lower IP compared to the other microparticles, they showed a high level of carotenoid retention after 90 days of storage (Figure 2). However, when the Rancimat test was used on the SC40 microparticles, the drastic treatment at high temperatures (120 °C) and high flow oxygen exposure caused morphological changes, increased surface area, and promoted the exposition of the lipid matrix and carotenoids, which led to rapid carotenoid degradation. When evaluating the effect of the microencapsulation combined techniques on oxidative stability, it was observed that the lack of porosity of the wall on the surfaces of SDC20 microparticles indicating a complete coverage was associated with the measured stability. This structure could have created a sturdier and dense shell matrix that reduced oxygen permeability into the microparticle core.

**Table 5.** Oxidation induction period of nonencapsulated and encapsulated carotenoid-rich extract, evaluated by Rancimat, at 120 °C.

| Sample | Induction Time (h) |
| --- | --- |
| Nonencapsulated extract | $0.04 \pm 0.00$ [d] |
| SD33 | $2.10 \pm 0.01$ [b] |
| SC40 | $0.11 \pm 0.02$ [c] |
| SDC20 | $21.8 \pm 0.3$ [a] |

SD: microparticles obtained by spray drying; SC: microparticles obtained by spray chilling; SDC: microparticles obtained by spray drying and chilling combination. The numerical suffix denotes the core/wall material ratio of the selected formulation. Values with the same lowercase letter are not statistically different ($p > 0.05$).

### 3.4.3. Thermophysical Properties

DSC is a technique commonly used to evaluate the thermal properties of bioactive compounds inside matrices, detect melting points, and record variations in the crystal structure via the displacement or disappearance of the endothermic peaks in the analyzed material [58]. The DSC curves shown in Figure 4 represent the thermal profile of the different microparticles prepared by the different utilized encapsulation processes.

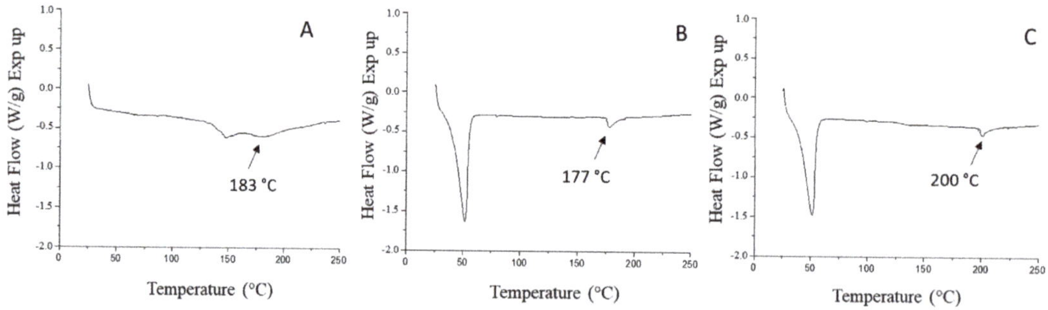


**Figure 4.** Differential scanning calorimetry (DSC) thermograms of microparticles prepared by (**A**) spray drying (SD), (**B**) spray chilling (SC), and (**C**) Spray drying and chilling combination (SDC). The numerical suffix denotes the core:wall material ratio of the selected formulation.

Regarding the first endothermic peak, for all formulations, the SD microparticles exhibited a transition that began at 141 °C and ended at 160 °C, with a melting temperature of 147 °C and an enthalpy of $8.3 \pm 0.6$ J/g powder. The broad curve was assumed to be related to the melting point of gum arabic, which agrees with the results of Mothé and Rao [59]. For the SC40 microparticles, the melting peak started at 39 °C and ended at 61 °C, having a $T_m$ of 52 °C and enthalpy of 86 ± 2 J/g powder. For the SDC20 microparticles, the melting began at 38 °C and ended at 59 °C, with a $T_m$ of 51 °C and enthalpy of 80 ± 2 J/g of. For both samples, SC40 and SDC20, the peaks described the melting range of the vegetable fat that was used as a carrier agent. Furthermore, among the three encapsulating processes used in this work, SD33, SC40, and SDC20 microparticles presented a second peak (highlighted in Figure 4), ranging from 177 to 200 °C. These peaks were probably associated with carotenoid melting at high temperatures. Similarly, in DSC curves of $\beta$-carotene, a large endothermic process was reported at about 180 °C [60]. Furthermore, Sy, Gleize, Dangles, Landrier, Veyrat, and Borel [61] found that the melting point of carotenoids ranged from 175 to 195 °C. According to the thermograms, the occurrence of two peaks suggested the formation of a heterogeneous system, and the changes in the thermal profile indicated interactions between the carrier and the core materials. The crystalline structure of microparticles suggested higher stability at room temperature and contributed to the protection of components against environmental factors [62]. In this system, the melting enthalpy was related to the energy absorbed to disrupt and breakdown the crystalline lipid structure of the carrier agent. Therefore, a more homogeneous crystalline structure necessitated a higher energy absorbed to carry out the melting transition. The melting points of the microparticles presented particular values, as different processes and carriers were used to produce the powders.

### 3.4.4. Sorption Isotherms

Sorption isotherms usually express the relationship between the relative humidity (RH) surrounding the food and the equilibrium moisture content of the sample, at a certain temperature. These data are important in food processes and quality such as packaging and storage of materials to preserve the products. Figure 5 illustrates that SD33 samples exhibited significant changes throughout the whole range of RH used when compared to the other two samples. These microparticles were formulated with gum arabic as the carrier material, which has a composition of ~98 wt.% polysaccharides and ~2 wt.% proteinaceous material [63]. When exposed to atmospheric conditions of increasing RH, the hydrophilic domain of the polysaccharide tends to reach fast equilibrium moisture with the surrounding environment. Typically, atomized particles obtained by spray drying are hygroscopic and can easily absorb water [55].

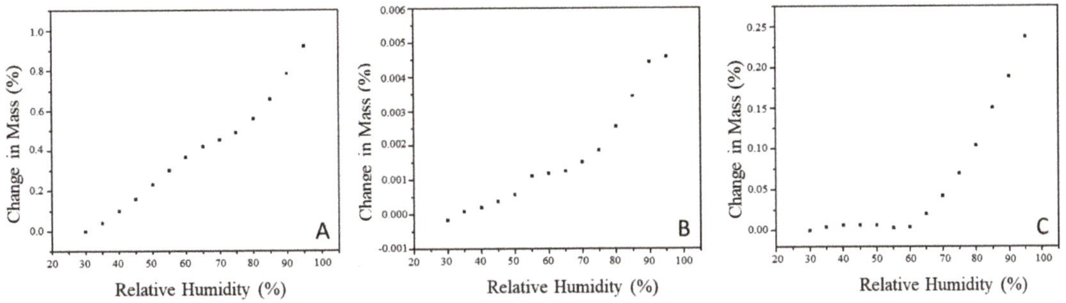


**Figure 5.** Sorption isotherms of microparticles obtained by (**A**) spray drying (SD33), (**B**) spray chilling (SC40), and (**C**) spray drying and chilling combination (SDC20). The numerical suffix denotes the core/carrier material ratio of the selected formulation.

The SC40 sample showed a lower increase in moisture content, even at a relative 95% RH. As expected, this sample presented minimum fluctuations in moisture, as the vegetable fat used for microencapsulation created a hydrophobic coating. SDC20, on the other hand, showed a slight increase in the moisture content up to 60% RH, and a large increase thereafter. The presence of gum arabic and vegetable fat in the microparticles possibly altered the balance of hydrophilic/hydrophobic interaction, favoring the two observed stages of water sorption at 25 °C.

By analyzing the sorption isotherms, it was possible to identify two zones: a region of slower adsorption at low and intermediate RH, and a region of capillary condensation, in which water absorption raised faster with increasing RH. Lourenço, Moldão-Martins, and Alves [56] reported similar results when studying the sorption isotherms of microparticles loaded with pineapple peel extract produced by spray drying using gum arabic as a carrier agent.

Generally, it can be seen that the behaviors of DVS curves were of type III (J shape). It is worth noting that, in the SD33 microparticles, this behavior was not that obvious. The isotherm profile by DVS analysis was a consequence of the physical and/or chemical transformations that occurred during the atomization, in addition to the carrier materials in the composition of the microparticles.

## 4. Conclusions

The production of microparticles loaded with carotenoids by spray drying, chilling, and a combined process was shown as an effective approach to provide pigment stability with a high encapsulation efficiency of 82% to 100%. The mean diameter and size distribution of microparticles exhibited an increase after 90 days; however, even after storage, they were within the suitable scale for industrial applications. In addition, all formulations exhibited variation in color over storage. The degradation kinetics of the carotenoids followed the zero-order kinetics for most samples, presumably driven by permeation of oxygen into the core of the microparticles. By using selected formulations for each process, micrographs demonstrated the physical details of the microparticles such as their shape, e.g., with an irregular surface for the SC microparticles and spherical shape for the microparticles prepared with the other two treatments. Micrographs also showed SD and SDC samples with a smooth surface and heterogeneous sizes. Although losses in the carotenoid content were noted in the SDC particles, they displayed good stability concerning their thermal properties, oxidative stability, and water sorption. Overall, the microencapsulation processes used in the present study not only provide practical options to improve carotenoid stability, but also explain the phenomenon behind pigment protection. Further research should be conducted to evaluate the carotenoid bioavailability within the different matrices obtained from the encapsulation techniques developed in this research for application in functional foods as colorants with bioactive potential.

**Author Contributions:** Conceptualization, L.S.P., C.E.d.C.R. and C.S.F.-T.; methodology, L.S.P., P.M.d.L. and D.C.; software, L.S.P.; validation, L.S.P.; formal analysis, L.S.P.; D.C.; investigation, L.S.P., P.M.d.L. and S.H.G.d.S.; resources, L.S.P.; data curation, L.S.P.; writing—original draft preparation, L.S.P.; writing—review and editing, L.S.P., D.C., O.H.C., C.E.d.C.R., C.S.F.-T.; visualization, L.S.P. and C.E.d.C.R.; supervision, C.S.F.-T.; project administration, L.S.P.; funding acquisition, L.S.P. All authors have read and agreed to the published version of the manuscript.

**Funding:** This research was funded by Fundação de Amparo à Pesquisa do Estado de São Paulo (FAPESP) grant number 2016/24916-2 and 2019/11113-7 and Coordenação de Aperfeiçoamento de Pessoal de Nível Superior—Brasil (CAPES) Finance Code 001.

**Institutional Review Board Statement:** Not applicable.

**Informed Consent Statement:** Not applicable.

**Data Availability Statement:** Data is contained within the article.

**Acknowledgments:** The authors thank the Fundação de Amparo à Pesquisa do Estado de São Paulo (FAPESP) for the fellowship granted to L.S. Pinho (#2016/24916-2 and #2019/11113-7) and Coordenação de Aperfeiçoamento de Pessoal de Nível Superior—Brasil (CAPES) for the fellowship granted to P.M. Lima and S.G. Sá (Finance Code 001). The authors also thank Conselho Nacional de Desenvolvimento Científico e Tecnológico (CNPq) for the research fellowship granted to C.S. Favaro-Trindade (#305115/2018-9), as well as partial financial support from the Ohio State University.

**Conflicts of Interest:** The authors declare no conflict of interest.

## References

1. da Silva, L.M.R.; de Sousa, P.H.M.; de Sousa Sabino, L.B.; do Prado, G.M.; Torres, L.B.V.; Maia, G.A.; de Figueiredo, R.W.; Ricardo, N.M.P.S. Brazilian (North and Northeast) Fruit By-Products. *Food Wastes By-Prod. Nutraceutical Health Potential* **2020**, *45*, 127–158. [CrossRef]
2. Schimpl, F.C.; da Silva, J.F.; de Carvalho Gonçalves, J.F.; Mazzafera, P. Guarana: Revisiting a highly caffeinated plant from the Amazon. *J. Ethnopharmacol.* **2013**, *150*, 14–31. [CrossRef] [PubMed]
3. Santana, Á.L.; Zanini, J.A.; Macedo, G.A. Dispersion-assisted extraction of guarana processing wastes on the obtaining of polyphenols and alkaloids. *J. Food Process. Eng.* **2020**, *43*, e13381. [CrossRef]
4. Pinho, L.S.; da Silva, M.P.; Thomazini, M.; Cooperstone, J.L.; Campanella, O.H.; da Costa Rodrigues, C.E.; Favaro-Trindade, C.S. Guaraná (*Paullinia cupana*) by-product as a source of bioactive compounds and as a natural antioxidant for food applications. *J. Food Process. Preserv.* **2021**, *45*, e15854. [CrossRef]
5. During, A.; Harrison, E.H. Mechanisms of provitamin A (carotenoid) and vitamin A (retinol) transport into and out of intestinal Caco-2 cells. *J. Lipid Res.* **2007**, *48*, 2283–2294. [CrossRef]
6. Liu, X.H.; Yu, R.B.; Liu, R.; Hao, Z.X.; Han, C.C.; Zhu, Z.H.; Ma, L. Association between lutein and zeaxanthin status and the risk of cataract: A meta-analysis. *Nutrients* **2014**, *6*, 452–465. [CrossRef]
7. Chuyen, H.; Van Eun, J.B. Marine carotenoids: Bioactivities and potential benefits to human health. *Crit. Rev. Food Sci. Nutr.* **2017**, *57*, 2600–2610. [CrossRef]
8. Milani, A.; Basirnejad, M.; Shahbazi, S.; Bolhassani, A. Carotenoids: Biochemistry, pharmacology and treatment. *Br. J. Pharmacol.* **2017**, *174*, 1290–1324. [CrossRef]
9. Cano, M.P.; Gómez-Maqueo, A.; García-Cayuela, T.; Welti-Chanes, J. Characterization of carotenoid profile of Spanish Sanguinos and Verdal prickly pear (*Opuntia ficus-indica*, spp.) tissues. *Food Chem.* **2017**, *237*, 612–622. [CrossRef]
10. Fraser, P.D.; Bramley, P.M. The biosynthesis and nutritional uses of carotenoids. *Prog. Lipid Res.* **2004**, *43*, 228–265. [CrossRef] [PubMed]
11. Rodriguez-Amaya, D.B. *A Guide to Carotenoid Analysis in Foods*; ILSI Human Nutrition Institute: Washington DC, USA, 2001. Available online: https://pdf.usaid.gov/pdf_docs/Pnacq929.pdf (accessed on 13 January 2022).
12. Sartori, T.; Consoli, L.; Hubinger, M.D.; Menegalli, F.C. Ascorbic acid microencapsulation by spray chilling: Production and characterization. *LWT-Food Sci. Technol.* **2015**, *63*, 353–360. [CrossRef]
13. Troya, D.; Tupuna-Yerovi, D.S.; Ruales, J. Effects of wall materials and operating parameters on physicochemical properties, process efficiency, and total carotenoid content of microencapsulated banana passionfruit pulp (*Passiflora tripartita var. mollissima*) by spray-drying. *Food Bioprocess Technol.* **2018**, *11*, 1828–1839. [CrossRef]
14. Anthero, A.G.D.S.; Bezerra, E.O.; Comunian, T.A.; Procópio, F.R.; Hubinger, M.D. Effect of modified starches and gum arabic on the stability of carotenoids in paprika oleoresin microparticles. *Dry. Technol.* **2020**, *39*, 1927–1940. [CrossRef]
15. Santos, P.D.F.; Rubio, F.T.V.; de Carvalho Balieiro, J.C.; Thomazini, M.; Favaro-Trindade, C.S. Application of spray drying for production of microparticles containing the carotenoid-rich tucumã oil (*Astrocaryum vulgare Mart.*). *LWT* **2021**, *143*, 111106. [CrossRef]
16. Islam, M.Z.; Kitamura, Y.; Yamano, Y.; Kitamura, M. Effect of vacuum spray drying on the physicochemical properties, water sorption and glass transition phenomenon of orange juice powder. *J. Food Eng.* **2016**, *169*, 131–140. [CrossRef]
17. Selim, K.; Tsimidou, M.; Biliaderis, C.G. Kinetic studies of degradation of saffron carotenoids encapsulated in amorphous polymer matrices. *Food Chem.* **2000**, *71*, 199–206. [CrossRef]
18. Souza, V.B.; Thomazini, M.; Barrientos, M.A.E.; Nalin, C.M.; Ferro-Furtado, R.; Genovese, M.I.; Favaro-Trindade, C.S. Functional properties and encapsulation of a proanthocyanidin-rich cinnamon extract (*Cinnamomum zeylanicum*) by complex coacervation using gelatin and different polysaccharides. *Food Hydrocoll.* **2018**, *77*, 297–306. [CrossRef]
19. Carneiro, H.C.; Tonon, R.V.; Grosso, C.R.; Hubinger, M.D. Encapsulation efficiency and oxidative stability of flaxseed oil microencapsulated by spray drying using different combinations of wall materials. *J. Food Eng.* **2013**, *115*, 443–451. [CrossRef]
20. Okuro, P.K.; de Matos Junior, F.E.; Favaro-Trindade, C.S. Technological challenges for spray chilling encapsulation of functional food ingredients. *Food Technol. Biotechnol.* **2013**, *51*, 171.
21. Chambi, H.N.M.; Alvim, I.D.; Barrera-Arellano, D.; Grosso, C.R.F. Solid lipid microparticles containing water-soluble compounds of different molecular mass: Production, characterisation and release profiles. *Food Res. Int.* **2008**, *41*, 229–236. [CrossRef]
22. Sillick, M.; Gregson, C.M. Spray chill encapsulation of flavors within anhydrous erythritol crystals. *LWT-Food Sci. Technol.* **2012**, *48*, 107–113. [CrossRef]

23. Zuidam, N.J.; Shimoni, E. Overview of microencapsulates for use in food products or processes and methods to make them. In *Encapsulation Technologies for Active Food Ingredients and Food Processing*; Springer: New York, NY, USA, 2010; pp. 3–29.
24. Consoli, L.; Grimaldi, R.; Sartori, T.; Menegalli, F.C.; Hubinger, M.D. Gallic acid microparticles produced by spray chilling technique: Production and characterization. *LWT-Food Sci. Technol.* **2016**, *65*, 79–87. [CrossRef]
25. Arslan-Tontul, S.; Erbas, M. Single and double layered microencapsulation of probiotics by spray drying and spray chilling. *LWT* **2017**, *81*, 160–169. [CrossRef]
26. Fadini, A.L.; Alvim, I.D.; Ribeiro, I.P.; Ruzene, L.G.; da Silva, L.B.; Queiroz, M.B.; de OliveiraMiguel, A.M.R.; Chaves, F.C.M.; Rodrigues, R.A.F. Innovative strategy based on combined microencapsulation technologies for food application and the influence of wall material composition. *LWT* **2018**, *91*, 345–352. [CrossRef]
27. Cuevas, M.S.; Rodrigues, C.E.; Gomes, G.B.; Meirelles, A.J. Vegetable Oils Deacidification by Solvent Extraction: Liquid– Liquid Equilibrium Data for Systems Containing Sunflower Seed Oil at 298.2 K. *J. Chem. Eng. Data* **2021**, *55*, 3859–3862. [CrossRef]
28. Rocha, G.A.; Fávaro-Trindade, C.S.; Grosso, C.R.F. Microencapsulation of lycopene by spray drying: Characterization, stability and application of microcapsules. *Food Bioprod. Process.* **2012**, *90*, 37–42. [CrossRef]
29. Pelissari, J.R.; Souza, V.B.; Pigoso, A.A.; Tulini, F.L.; Thomazini, M.; Rodrigues, C.E.; Urbano, A.; Favaro-Trindade, C.S. Production of solid lipid microparticles loaded with lycopene by spray chilling: Structural characteristics of particles and lycopene stability. *Food Bioprod. Process.* **2016**, *98*, 86–94. [CrossRef]
30. Tonon, R.V.; Brabet, C.; Hubinger, M.D. Anthocyanin stability and antioxidant activity of spray-dried açai (Euterpe oleracea Mart.) juice produced with different carrier agents. *Food Res. Int.* **2010**, *43*, 907–914. [CrossRef]
31. Hart, D.J.; Scott, K.J. Development and evaluation of an HPLC method for the analysis of carotenoids in foods, and the measurement of the carotenoid content of vegetables and fruits commonly consumed in the UK. *Food Chem.* **1995**, *54*, 101–111. [CrossRef]
32. Alvim, I.D.; Stein, M.A.; Koury, I.P.; Dantas, F.B.H.; Cruz, C.L.D.C.V. Comparison between the spray drying and spray chilling microparticles contain ascorbic acid in a baked product application. *LWT-Food Sci. Technol.* **2016**, *65*, 689–694. [CrossRef]
33. Xiao, Y.D.; Huang, W.Y.; Li, D.J.; Song, J.F.; Liu, C.Q.; Wei, Q.Y.; Zhang, M.; Yang, Q.M. Thermal degradation kinetics of all-trans and cis-carotenoids in a light-induced model system. *Food Chem.* **2018**, *239*, 360–368. [CrossRef] [PubMed]
34. De Leonardis, A.; Macciola, V. Heat-oxidation stability of palm oil blended with extra virgin olive oil. *Food Chem.* **2012**, *135*, 1769–1776. [CrossRef] [PubMed]
35. Xu, P.X.; Dai, Z.Q.; Li, D.J.; Liu, C.Q.; Wu, C.E.; Song, J.F. Preparation, optimization, characterization, and in vitro bioaccessibility of a lutein microparticle using spray drying with β-cyclodextrin and stevioside. *J. Food Process. Preserv.* **2021**, *45*, e15032. [CrossRef]
36. Sanchez, C.; Nigen, M.; Tamayo, V.M.; Doco, T.; Williams, P.; Amine, C.; Renard, D. Acacia gum: History of the future. *Food Hydrocolloids* **2018**, *78*, 140–160. [CrossRef]
37. Labuza, T.P. Properties of water as related to the keeping quality of foods. In Proceedings of the International Congress of Food Science and Technology, Washington, DC, USA, 9–14 August 1971.
38. Silva, M.P.; Tulini, F.L.; Matos-Jr, F.E.; Oliveira, M.G.; Thomazini, M.; Fávaro-Trindade, C.S. Application of spray chilling and electrostatic interaction to produce lipid microparticles loaded with probiotics as an alternative to improve resistance under stress conditions. *Food Hydrocoll.* **2018**, *83*, 109–117. [CrossRef]
39. McClements, D.J. Theoretical prediction of emulsion color. *Adv. Colloid Interface Sci.* **2002**, *97*, 63–89. [CrossRef]
40. Favaro-Trindade, C.S.; Okuro, P.K.; de Matos, F.E., Jr. 5 Encapsulation via Spray. In *Handbook of Encapsulation and Controlled release*; CRC Press: Boca Raton, FL, USA, 2015; Volume 71.
41. De Lara Pedroso, D.; Thomazini, M.; Heinemann, R.J.B.; Favaro-Trindade, C.S. Protection of Bifidobacterium lactis and Lactobacillus acidophilus by microencapsulation using spray-chilling. *Int. Dairy J.* **2012**, *26*, 127–132. [CrossRef]
42. Hansen, L.T.; Allan-Wojtas, P.M.; Jin, Y.L.; Paulson, A.T. Survival of Ca-alginate microencapsulated Bifidobacterium spp. in milk and simulated gastrointestinal conditions. *Food Microbiol.* **2002**, *19*, 35–45. [CrossRef]
43. Müller, R.H.; Radtke, M.; Wissing, S.A. Nanostructured lipid matrices for improved microencapsulation of drugs. *Int. J. Pharm.* **2002**, *242*, 121–128. [CrossRef]
44. Ghotra, B.S.; Dyal, S.D.; Narine, S.S. Lipid shortenings: A review. *Food Res. Int.* **2002**, *35*, 1015–1048. [CrossRef]
45. Navarro-Guajardo, N.; García-Carrillo, E.M.; Espinoza-González, C.; Téllez-Zablah, R.; Dávila-Hernández, F.; Romero-García, J.; Ledezma-Pérez, A.; Mercado-Silva, J.A.; Torres, C.A.P.; Pariona, N. Candelilla wax as natural slow-release matrix for fertilizers encapsulated by spray chilling. *J. Renew. Mater.* **2018**, *6*, 226–236. [CrossRef]
46. Khoo, H.E.; Prasad, K.N.; Kong, K.W.; Jiang, Y.; Ismail, A. Carotenoids and their isomers: Color pigments in fruits and vegetables. *Molecules* **2011**, *16*, 1710–1738. [CrossRef] [PubMed]
47. Provesi, J.G.; Dias, C.O.; Amante, E.R. Changes in carotenoids during processing and storage of pumpkin puree. *Food Chem.* **2011**, *128*, 195–202. [CrossRef] [PubMed]
48. Rodriguez-Amaya, D.B. Effects of processing and storage on food carotenoids. *Sight Life Newsl.* **2002**, *3*, 25–35.
49. Sarabandi, K.; Jafari, S.M.; Mahoonak, A.S.; Mohammadi, A. Application of gum Arabic and maltodextrin for encapsulation of eggplant peel extract as a natural antioxidant and color source. *Int. J. Biol. Macromol.* **2019**, *140*, 59–68. [CrossRef] [PubMed]

50. Lima, P.M.; Dacanal, G.C.; Pinho, L.S.; Pérez-Córdoba, L.J.; Thomazini, M.; Moraes, I.C.F.; Favaro-Trindade, C.S. Production of a rich-carotenoid colorant from pumpkin peels using oil-in-water emulsion followed by spray drying. *Food Res. Int.* **2021**, *148*, 110627. [CrossRef]
51. Medina-Torres, L.; Santiago-Adame, R.; Calderas, F.; Gallegos-Infante, J.A.; González-Laredo, R.F.; Rocha-Guzmán, N.E.; Núñez-Ramírez, D.M.; Manero, O. Microencapsulation by spray drying of laurel infusions (*Litsea glaucescens*) with maltodextrin. *Ind. Crops Prod.* **2016**, *90*, 1–8. [CrossRef]
52. Elik, A.; Yanık, D.K.; Göğüş, F. A comparative study of encapsulation of carotenoid enriched-flaxseed oil and flaxseed oil by spray freeze-drying and spray drying techniques. *LWT* **2021**, *143*, 111153. [CrossRef]
53. Bhusari, S.N.; Muzaffar, K.; Kumar, P. Effect of carrier agents on physical and microstructural properties of spray dried tamarind pulp powder. *Powder Technol.* **2014**, *266*, 354–364. [CrossRef]
54. Silva, M.P.; Thomazini, M.; Holkem, A.T.; Pinho, L.S.; Genovese, M.I.; Fávaro-Trindade, C.S. Production and characterization of solid lipid microparticles loaded with guaraná (*Paullinia cupana*) seed extract. *Food Res. Int.* **2019**, *123*, 144–152. [CrossRef]
55. Alves, A.I.; Rodrigues, M.Z.; Ribeiro Pinto, M.R.M.; Lago Vanzela, E.S.; Stringheta, P.C.; Perrone, Í.T.; Ramos, A.M. Morphological characterization of pequi extract microencapsulated through spray drying. *Int. J. Food Prop.* **2017**, *20* (Suppl. 2), 1298–1305. [CrossRef]
56. Lourenço, S.C.; Moldão-Martins, M.; Alves, V.D. Microencapsulation of pineapple peel extract by spray drying using maltodextrin, inulin, and arabic gum as wall matrices. *Foods* **2020**, *9*, 718. [CrossRef] [PubMed]
57. Choe, E.; Min, D.B. Mechanisms and factors for edible oil oxidation. *Compr. Rev. Food Sci. Food Saf.* **2006**, *5*, 169–186. [CrossRef]
58. Rutz, J.K.; Borges, C.D.; Zambiazi, R.C.; da Rosa, C.G.; da Silva, M.M. Elaboration of microparticles of carotenoids from natural and synthetic sources for applications in food. *Food Chem.* **2016**, *202*, 324–333. [CrossRef] [PubMed]
59. Mothé, C.G.; Rao, M.A. Thermal behavior of gum arabic in comparison with cashew gum. *Thermochim. Acta* **2000**, *357*, 9–13. [CrossRef]
60. Peinado, I.; Mason, M.; Romano, A.; Biasioli, F.; Scampicchio, M. Stability of β-carotene in polyethylene oxide electrospun nanofibers. *Appl. Surf. Sci.* **2016**, *370*, 111–116. [CrossRef]
61. Sy, C.; Gleize, B.; Dangles, O.; Landrier, J.F.; Veyrat, C.C.; Borel, P. Effects of physicochemical properties of carotenoids on their bioaccessibility, intestinal cell uptake, and blood and tissue concentrations. *Mol. Nutr. Food Res.* **2012**, *56*, 1385–1397. [CrossRef]
62. Yan, C.; Zhang, W. Coacervation processes. In *Microencapsulation in the Food Industry*; Academic Press: New York, NY, USA, 2014; pp. 125–137. [CrossRef]
63. Isobe, N.; Sagawa, N.; Ono, Y.; Fujisawa, S.; Kimura, S.; Kinoshita, K.; Miuchi, T.; Iwata, T.; Isogai, A.; Nishino, M.; et al. Primary structure of gum arabic and its dynamics at oil/water interface. *Carbohydr. Polym.* **2020**, *249*, 116843. [CrossRef]

foods

MDPI

*Article*

# Temperature and pH Stability of Anthraquinones from Native *Aloe vera* Gel, Spray-Dried and Freeze-Dried *Aloe vera* Powders during Storage

**Uzma Sadiq, Harsharn Gill and Jayani Chandrapala** *

School of Science, RMIT University, Bundoora, Melbourne, VIC 3083, Australia;
uzma.sadiq@student.rmit.edu.au (U.S.); harsharn.gill@rmit.edu.au (H.G.)
* Correspondence: jayani.chandrapala@rmit.edu.au

**Abstract:** The present study explored the stability of extracted anthraquinones (aloin, aloe-emodin and rhein) from whole-leaf *Aloe vera* gel (WLAG), its freeze-dried powder (FDP) and spray-dried powder (SDP) under varying pH and temperature conditions during storage. Each anthraquinone behaved differently under different processing parameters. The amount of anthraquinones present in the gel was higher than in FDP and SDP. The aloin contents decreased by more than 50% at 50 °C and 70 °C, while at 25 °C and 4 °C, the decrease was moderate. A substantial reduction in aloin concentration was noticed at pH 6.7, whereas it remained unaffected at pH 3.5. The temperature and pH had no significant effect on the stability of aloe-emodin. Interestingly, a small quantity of rhein was detected during storage due to the oxidative degradation of aloin into aloe-emodin and rhein. These findings can provide significant insight into retaining anthraquinones during processing while developing functional foods and nutraceuticals to obtain maximum health benefits.

**Keywords:** *Aloe vera*; anthraquinones; stability; aloin; aloe-emodin; rhein; spray drying; freeze drying; processing temperature; pH

**Citation:** Sadiq, U.; Gill, H.; Chandrapala, J. Temperature and pH Stability of Anthraquinones from Native *Aloe vera* Gel, Spray-Dried and Freeze-Dried *Aloe vera* Powders during Storage. *Foods* **2022**, *11*, 1613. https://doi.org/10.3390/foods11111613

Academic Editor: Guowen Zhang

Received: 5 May 2022
Accepted: 26 May 2022
Published: 30 May 2022



## 1. Introduction

*Aloe vera* (*Aloe barbadensis*) is a perennial, drought-resistant succulent plant that belongs to the Liliaceae family. Out of 360 known species of Aloe vera, *Aloe arborescens* Miller; *A. perryi* Baker; *A. ferox* Miller o' *A. capensis*; and *Aloe bar-badensis* Miller, also known as *Aloe vera* Linne o' *A. vulgaris* Lamarck have medicinal properties [1,2]. The Aloe vera juicy material can be divided into two parts: the yellow exudate which contains a high percentage of aloin and other anthraquinones (such as aloe-emodin, barbaloin isobarbaloin, emodin, anthracene, anthranol) and the gel which contains 99.5% water along with polysaccharides, vitamins, enzymes, amino acids, phenolic compounds, phytosterols, and anthraquinones [3–5].

Anthraquinone derivatives, including aloin (A and B), barbaloin, isobarbaloin, emodin, aloe-emodin, rhein, chrysophanol and physcion, are known to possess a wide range of biological activities. These include laxative, antibacterial, antiviral [6], anti-inflammatory, antifungal, anticancer, diuretic, hepatoprotective, vasorelaxant and antioxidant effects [7–13]. Kang et al. [14] demonstrated the inhibitory activities of anthraquinones against advanced glycation end-products (AGEs) that are formed during food processing, and Huang et al. [15] reported enhanced anticancer activity of low-dose cisplatin when used in combination with anthraquinones against human gastric cancer cells in vitro.

Despite numerous studies reporting the beneficial effects of anthraquinones, their use in the food industry has been limited because of their poor water solubility, instability and rapid degradability during secondary processing conditions [16,17]. The five free anthraquinones (chrysophanol, emodin, physcion, aloe-emodin and rhein) extracted from rhubarb showed different thermal properties [18]. It was also found that aloe-emodin and emodin are highly prone to acidic and hydrolytic degradation, moderately prone to oxidative and thermal degradation (105 °C) and least prone to basic degradation [19]. Anthraquinone concentrations vary greatly depending upon climatic conditions, species, and even within various parts of the same plant [11,20,21]. Chang and colleagues [22] reported instability of polysaccharides and barbaloin upon heating and in various solvents, respectively, though exact mechanisms of hydrolysis remain unclear. However, Pellizzoni et al. [23] further reported that this instability could not be improved by the addition of ascorbates as well as antimicrobial agents. Ding et al. [16] investigated the stability of pure aloin in phosphate buffers by differing pH, temperature and light, and reported its decomposition at 70 °C and pH 8.0. Machado and colleagues developed and validated an HPLC method to detect aloin in fresh and dried leaves of *Aloe vera* by using phosphate-buffered saline (pH 3) as an extraction solvent [21].

The use of fresh *Aloe vera* gel in various processed foods is unsustainable due to its perishability and quick ability to relinquish its functional properties. However, various drying technologies that can dramatically improve the storage stability of *Aloe vera* gel have been explored to enhance its widespread use in food products. Spray drying is the most often employed method, followed by freeze drying, although the latter is a mild process to produce a powder. Dehydrated *Aloe vera* samples produced by spray drying, industrial freeze drying, reactance window drying and radiant zone drying led to lower water activities, higher solubilities, and showed high hygroscopic behaviours compared to fresh *Aloe vera* gels [24].

Other plants, especially *Rheum*, *Rumex*, *Rhamnus* and *Cassia*, contain anthraquinones that are sensitive to various processing conditions. McDougall et al. [25] evaluated the effect of various cooking methods on *Rheum rhapontigen* and found a dramatic reduction in anthraquinone content (aglycones and glycones) within 10 min of baking. However, Yen and Chung [26] reported the thermal degradation of anthraquinones extracted from *Cassia tora* seeds during roasting.

To date, no study has examined the effect of storage temperature and pH on the stability of anthraquinones extracted from *Aloe vera* leaves. The stability of anthraquinones in spray-dried and freeze-dried *Aloe vera* powders also remains unknown. Therefore, the present work was conducted to examine the stability of three anthraquinone compounds (aloin, aloe-emodin, rhein) present in fresh and stored whole-leaf *Aloe vera* gel, spray-dried powder and freeze-dried powder under various temperature and pH conditions. FTIR spectroscopy was also used to analyse the effect of heat, pH and drying procedures on the structure of anthraquinones.

## 2. Materials and Methods

### 2.1. Materials

*Aloe vera* (*Aloe barbadensis*) leaves were obtained from Aloe vera Australia (Goodman international, Brisbane, Australia). Standards of aloin (Mikromol, item code: MM1318.01, CAS# 1415-73-2), emodin (LGC, item code: DRE-C13117900, CAS #518-82-1), aloe-emodin (TRC, item code: A575400-10MG, CAS #481-72-1), and rhein CRS (EDQM, item code: EPY0002159, CAS# 478-43-3) were purchased from Novachem Pty Ltd. (Melbourne, Australia). HPLC-grade methanol and formic acid and analytical-grade HCL and NaOH were purchased from Sigma Aldrich Pty Ltd. (Castle Hill, NSW, Australia). MilliQ water was used at all times.

### 2.2. Preparation of Whole-Leaf Aloe vera Gel

*Aloe vera* leaves along with rinds were washed with MilliQ water and cut into small pieces. These cut slices were blended using an electric blender (Nutri Ninja) to prepare the mucilaginous gel, followed by centrifugation and vacuum filtration to eliminate the rind portions. The liquid gel was then stored at −20 °C overnight until further analysis.

### 2.3. Preparation of Whole-Leaf *Aloe vera* Spray-Dried and Freeze-Dried Powders

Spray-dried whole-leaf *Aloe vera* powder was prepared using a mini spray dryer B-290 (BÜCHI Labor Technik AG, Meierseggstrasse 40 Postfach, Flawil, Switzerland) which operated at the inlet and outlet temperatures of 175 °C and 75 °C, respectively.

Freeze-dried whole-leaf *Aloe vera* powder was prepared by pre-freezing the whole-leaf *Aloe vera* gel (WLAG) at −80 °C for 24 h and then lyophilising in a freeze dryer (VirTis Model EL-SP Scientific, Hardning Highway, Buena, NJ, USA) for ten days at temperature and pressures of −40 °C and 0.25 atm, respectively.

### 2.4. Evaluation for Temperature and pH Stability

The temperature and pH stability of anthraquinones in whole-leaf *Aloe Vera* gel (WLAG), spray-dried whole-leaf *Aloe vera* powder (SDP) and freeze-dried whole-leaf *Aloe vera* powder (FDP) were studied at four different temperatures (4 °C, 25 °C, 50 °C and 70 °C) and five different pH values (3.5, 4.6, 5.6, 6.4, 6.7) at room temperature.

An accurately weighed sample portion (10 g each) was transferred into four test tubes to test the stability towards temperature. One tube was left in the refrigerator at 4 °C, while another tube was left at room temperature (25 °C). The other two tubes with WLAG were heated at 50 °C and 70 °C by using a water bath TW-20 (JULABO, Marcon Boulevard, Allentown, PA, USA) for 15 min. The same procedure was also repeated for 3- and 7-day samples.

For testing the pH stability, the sample pH was adjusted by adding 1M HCL or 1M NaOH, using the pH SevenCompact$^{TM}$ S220 (Mettler-Toledo GmbH, Im Langacher, Greifensee, Switzerland). HPLC analysis was performed on days 0, 3 and 7. All tubes were stored at room temperature and covered with aluminium foil to protect them from light.

### 2.5. Quantification of Anthraquinones by HPLC

The WLAG, pH-adjusted WLAGs and heat-treated (4 °C, 25 °C, 50 °C and 70 °C) WLAGs, SDP and FDP, were centrifuged at 6000 rpm for 20 min (Allegra 64R-Beckman Coulter, NSW, Australia). The supernatant was collected, and ten millilitre aliquots of methanol were added and vortex mixed, followed by sonication for 10 min using an ultrasonic bath (FXP12, Unisonics, NSW, Australia) at ~4 ± 1 °C. The supernatant was taken carefully and filtered by a 0.45 µm membrane filter for all samples prior to HPLC analysis.

Stock solutions (1000 µg/mL) of aloin, aloe-emodin and rhein were prepared by dissolving in HPLC grade methanol and used as standards. Working solutions of mixed standards were prepared at concentrations of 5 ppm, 10 ppm, 20 ppm and 40 ppm by diluting the stock solution in a volumetric flask with methanol. Both stock and working solutions were prepared on the day of analysis and injected directly into the HPLC system. HPLC measurements were performed on an Agilent series1250 infinity gradient HPLC (Agilent Technologies, Santa Clara, CA, USA) equipped with a 600 solvent pump and a $C_{18}$ reversed-phase packing column (Phenomenex XB-C18, 250 mm × 4.60 mm, 3.6 µm Aeris). Gradient elution was performed at 0.7 mL/min, usinga binary mobile phase consisting of water with 1% formic acid (A) and methanol (B). The elution was monitored at 254 nm and the injection volume was 20 µL. The HPLC run started with 20% methanol and increased to 30%, 45% and 60% at 3, 10 and 15 min, respectively. The methanol percentage was increased to 70% at 18 min and maintained for 8 min, followed by a gradual decrease to 60%, 40% and 20% at 30, 32 and 35 min, respectively.

A calibration curve was established for each standard as a function of their peak areas. The contents of anthraquinones present in the samples were quantified against the standards. The amount of anthraquinones was presented as micrograms per gram of dry powder in the case of spray-dried and freeze-dried powders of *Aloe vera*, while for fresh whole-leaf *Aloe vera* gel, it was the microgram per gram of solid, obtained after drying a measured quantity of fresh whole-leaf *Aloe vera* gel (WLAG) in the oven at 102 °C until a constant weight was obtained.

### *2.6. Fourier Transform Infrared (FTIR) Spectroscopy Analysis*

FTIR spectra of extracted anthraquinones from all samples were obtained using the FTIR (Spectrum two, Perkin Elmer, Australia) furnished with IRWinLab FTIR software. Measurements were taken at the wavelength range of 4000–400 $cm^{-1}$ using Milli Q water as the background. Sixteen scans were performed, and the resolution of 4 $cm^{-1}$ was used.

### *2.7. Statistical Analysis*

All measurements were performed in triplicates, and the results were expressed in means ± standard deviation. Statistical differences for quantification of anthraquinones in the controls (Table 1) were analysed by one-way analysis of variance (ANOVA) using Tukey's post hoc HSD test to establish a significant difference between samples studied (WLAG, SDP and FDP) with 95% confidence interval by SPSS 11.0 software (SPSS Inc., Chicago, IL, USA). For performing temperature and pH stability comparisons compared to the controls, the data were analysed using a two-way analysis of variance (ANOVA) followed by Dunnett's multiple comparisons. Values significantly different from controls to various temperature and pH treatments in a two-way ANOVA with Dunnett's test are indicated with asterisks (* $p < 0.01$, *** $p < 0.0003$, **** $p < 0.0001$). Data analysis was performed using Graph pad Prism 9.1.0 (GraphPad Software, Inc., San Diego, CA, USA). Here, $p$-values of less than 0.05 were considered significant for all statistical tests.

**Table 1.** Contents of aloin, aloe-emodin and rhein in whole-leaf *Aloe vera* gel (WLAG), freeze-dried powder (FDP) and spray-dried powder (SDP) of *Aloe vera*. (Values are mean and standard deviations from triplicate data. The data were analysed by one-way ANOVA and Tukey's post hoc HSD tests. The different letters in superscript (a, b, c) within rows indicate statistically significant differences ($p < 0.05$)).

| Sample (μg/g of Dry Powder) | Aloin | Aloe-Emodin | Rhein |
|---|---|---|---|
| WLAG | 6134.0 ± 15.0 [a] | 69.0 ± 0.0 [a] | 172.7 ± 0.4 [a] |
| FDP | 223.1 ± 3.9 [b] | 0.00 | 0.00 |
| SDP | 220.0 ± 0.2 [b] | 18.2 ± 0.0 [c] | 0.00 |

## 3. Results and Discussion

### *3.1. Quantification of Anthraquinones through HPLC*

Figure 1A shows the baseline separation of anthraquinones in 35 min with retention times of 16.74, 21.21, 23.40 and 28.26 for aloin, aloe-emodin, rhein, and emodin, respectively. Figure 1B is a representative chromatogram of the whole-leaf *Aloe vera* gel sample and shows aloin, aloe-emodin and rhein compounds with the same retention times as standards. The aloin and aloe-emodin contents of whole-leaf *Aloe vera* gel (WLAG on day zero) were significantly different from those of the spray-dried (SDP) and freeze-dried (FDP) powders of *Aloe vera* (Table 1).

The freshly prepared WLAG had the highest amount of aloin with an average of ~6134 μg/g and this were consistent with the concentrations (4040 to 13,410 μg/g of aloin in fresh gel and 96,270 to 247,000 μg/g of aloin in fresh latex) reported by Sánchez-Machado et al. [21]. Some of these differences were due to the usage of different extraction solvents (methanol), as Sánchez-Machado et al. used phosphate-buffered saline of pH 3 to extract aloin (lower pH conserves aloin). In addition, this variability might be due to the growing conditions of *Aloe Vera* as well [27]. Aloin contents in freeze-dried powders were ~223 μg/g, which is lower than previously reported (1140 μg/g of aloin in freeze-dried ethanol extracts; [28]). The lower concentration of aloin in freeze-dried powders might be due to its failure to preserve phenolic glycoside, as previously reported [29]. It might be due to the failure of the methanol solvent to extract aloin efficiently from the freeze-dried powders.

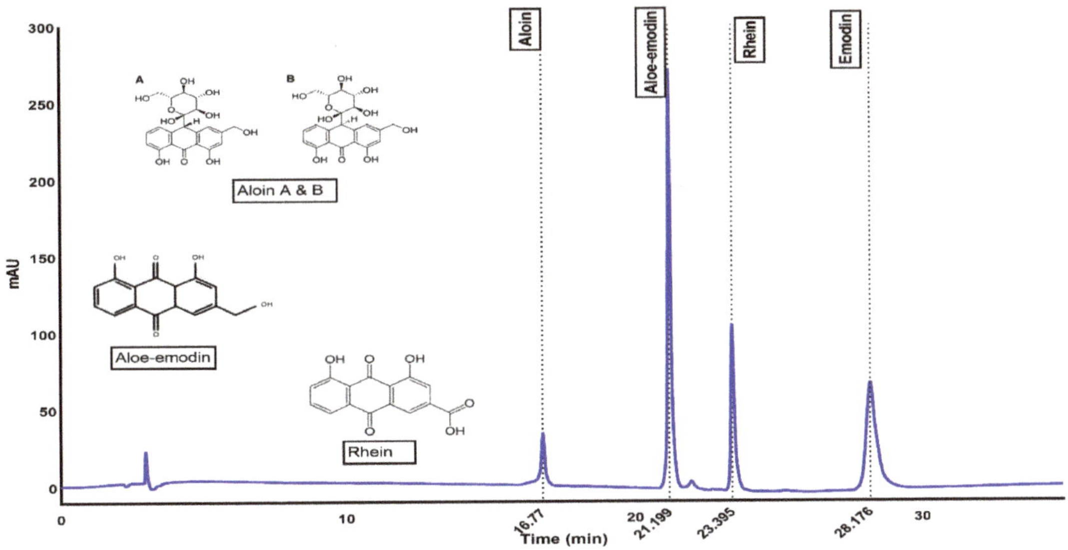


(A)

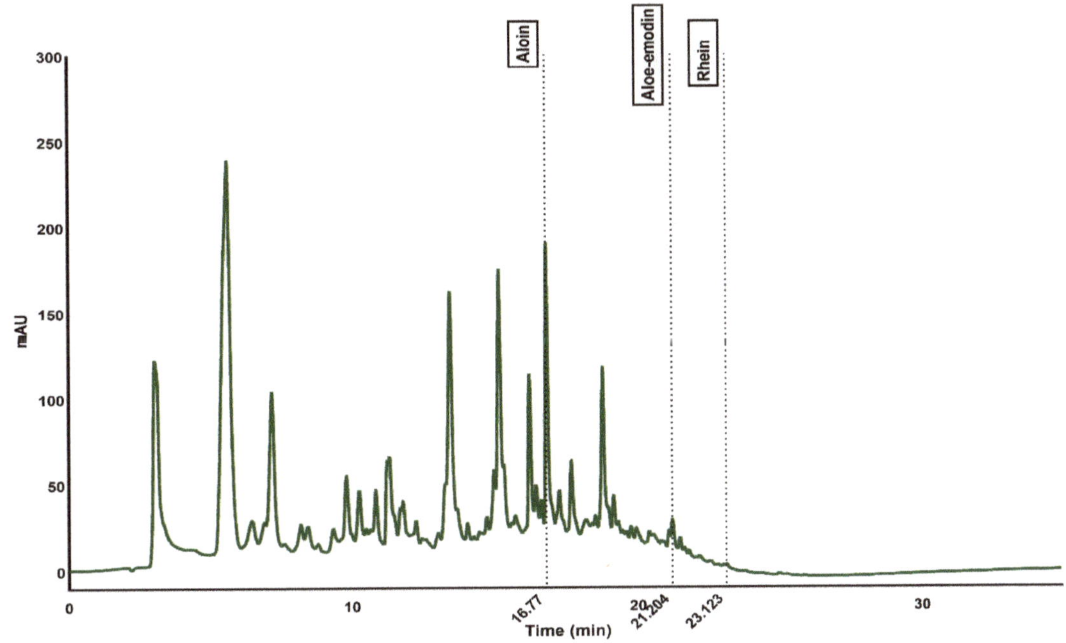


(B)

**Figure 1.** HPLC chromatographs of (**A**) reference standards of aloin, aloe-emodin and rhein and (**B**) WLAG extract containing aloin, aloe-emodin and rhein.

Aloin contents in spray-dried powder were ~220 μg/g, and these were lower in concentration in comparison to the freeze-dried powders because of the exposure of polyphenols to high temperature during spray drying. Although there are no previous data available on the quantification of aloin in spray-dried powders of *Aloe vera*, the values from the present study fall between the observed ranges (150 to 450 μg/g) of aloin in *Aloe vera* leaf powders prepared using sun-drying followed by oven drying (50 °C) methods [11]. Overall, dehydration-process-induced stresses were the reason for low aloin contents in both (FDP and SDP) powders [30]. During spray drying, thermal, automisation, mechanical and dehydration stresses caused the degradation of bioactive components. However, during lyophilisation, dehydration, pH changes, crystallisation and interfacial stresses take place [31]. Moreover, the higher percentages of aloin in WLAG compared to FDP and SDP are due to the size of the solvent (methanol) molecules and the diffusion mechanism through which it can penetrate the solid matrix and increase the polarity of the solvent [32].

The amount of aloe-emodin in WLAG on day zero was ~69 μg/g. This value was one-third of the values reported previously (270 μg/g; [11]). This may be due to the differences in climatic regions from where the products were sourced. *Aloe vera* from warmer climatic regions is reported to have a lower amount of aloe-emodin [11]. Aloe-emodin was not detected in the freeze-dried powders, although this is against the extracted 5.52 ± 0.3 μg/g of aloe-emodin from freeze-dried *Aloe vera* leaves previously reported by Lee et al. [33]. This might be due to the inefficient extraction of aglycones in methanol. Spray-dried powder contained ~18 μg/g of aloe-emodin on day zero of analysis. In the case of spray-dried powders, heating may have liberated the aglycones (aloe-emodin) from their glycosides (aloin). Similar results have previously been reported by Fu [34].

In addition, large variability in the amount of aloin and aloe-emodin was observed in WLAG on day zero (Table 1). A higher quantity of aloin of ~6134.0 μg/g was found compared to aloe-emodin at ~69.0 μg/g, and these findings agree with previous studies [11,35]. The cause for the lesser quantity of aloe-emodin was that it could not be synthesised directly in plants but instead resulted from the oxidative decomposition of its glycosides [36,37].

### 3.2. FTIR Analysis

FTIR profiles of anthraquinones extracted from WLAG exposed to different temperatures (4 °C, 25 °C, 50 °C, 70 °C) and pHs (3.5, 4.6, 5.5, 6.0, 6.7), freeze-dried powders (FDP) and spray-dried powders (SDP) of *Aloe vera* are shown in Figure 2A–C. Anthraquinones are assembled from an anthracene ring through a keto group located at 9,10 carbon as a basic core with diverse functional groups, such as -OH, -$CH_3$-COOH, CHO, $CH_2OH$ and $OCH_3$. Figure 1A illustrates the structure of aloin, aloe-emodin, and rhein present in *Aloe vera* plants. In general, all samples exhibited a broad and intense absorption band at 3400 $cm^{-1}$ which can be attributed to the -OH stretching of mannose and uronic acids and bands at 3200 $cm^{-1}$ assigned to O-H stretching of anthraquinones (phenolic groups) such as aloin and aloe-emodin. These compounds contain hydroxyl groups directly attached to an aromatic hydrocarbon. Two bands at 2930 $cm^{-1}$ and 2850 $cm^{-1}$ are related to C-H symmetric stretching and asymmetric stretching of methylene ($CH_2$), respectively, indicating long aliphatic chains (-CH) of anthraquinones. Peaks at 1725 $cm^{-1}$ of (C=O) stretching suggest the presence of carbonyl groups. Aromatic double bonds (C=C) of benzene rings could be seen at 1546 $cm^{-1}$ and 1641 $cm^{-1}$, indicating vinyl ether and the aloin compounds.

The minor peaks at 1402 $cm^{-1}$ are from the -COO of carboxylate compounds. Peaks between 1250 $cm^{-1}$ and 1000 are the C-O-C stretching of ($COCH_3$), probably related to methyl acyl groups, indicating the presence of *O*-acetyl ester. Peaks at 1070 $cm^{-1}$ and 1126 $cm^{-1}$ are associated with the C–O and C–H bending vibration associated with rhamnogalacturonan, a side-chain constituent of pectins [38–41].

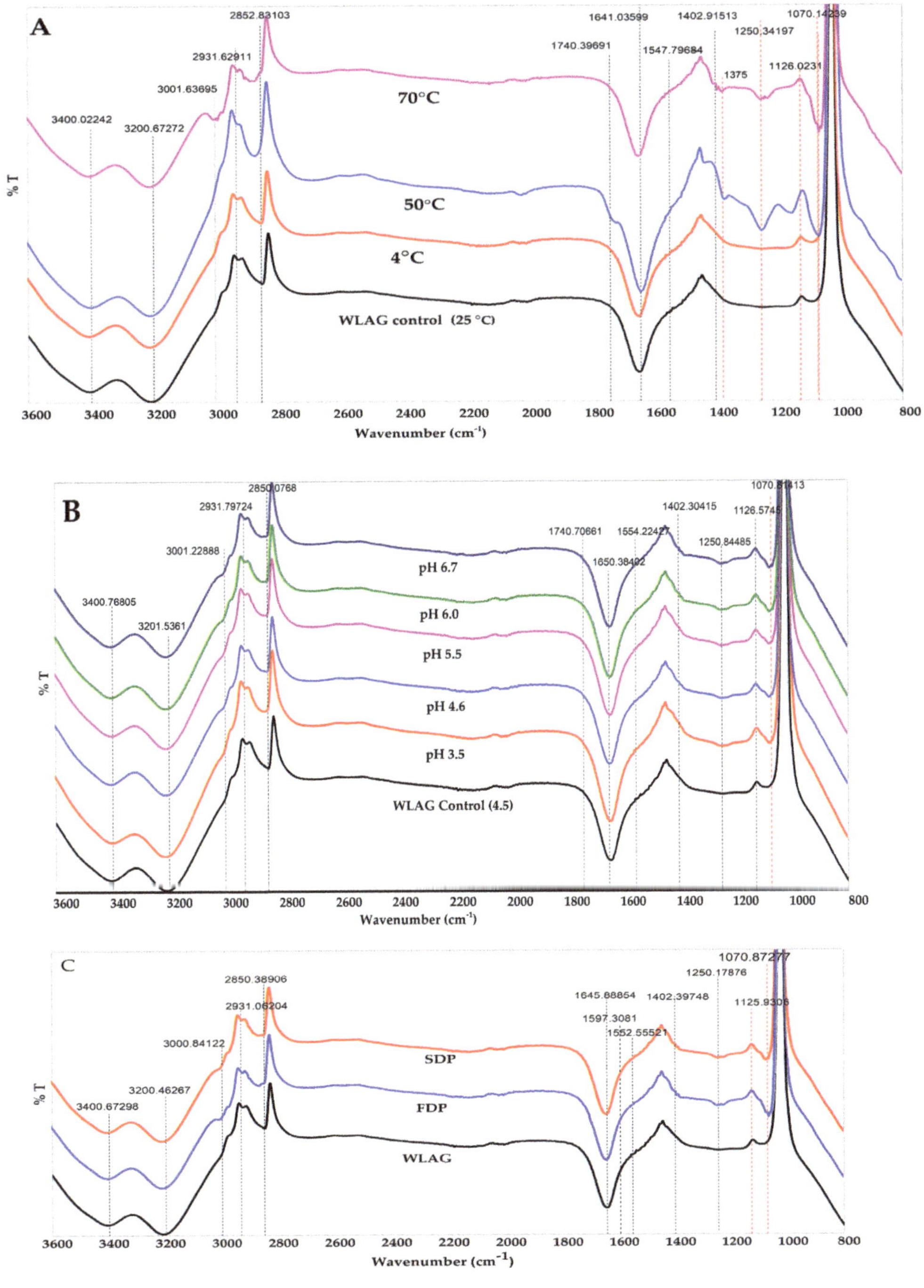


**Figure 2.** FTIR spectra of (**A**) anthraquinone extracted from whole leaf *Aloe vera* gel (WLAG) at various temperatures, (**B**) anthraquinone extracted from whole leaf *Aloe vera* gel at various pH conditions and (**C**) anthraquinone extracted from the whole leaf *Aloe vera* gel (WLAG), freeze-dried powder (FDP) and spray-dried powder (SDP) of *Aloe vera*.

The peak patterns obtained from FTIR (Figure 2A) for samples heated at 50 °C and 70 °C indicated a change in the chemical structure of aloin at wavenumbers 1740 $cm^{-1}$ and 1250 $cm^{-1}$ corresponding to the C = O and C-O-C stretches of carbonyl groups of *O*-acetyl ester. The peak labelled 50 °C (Figure 2A) has two extra sharp edges at 1250 $cm^{-1}$ and 1740 $cm^{-1}$, and in the case of 70 °C, these are not prominent. WLAG samples at 25 °C and at 4 °C) showed no peaks at 1250 $cm^{-1}$ and 1740 $cm^{-1}$, indicating the absence of any structural modifications. However, in Figure 2A, strong peaks were identified at 1070 $cm^{-1}$ and 1126 $cm^{-1}$ in heated WLAG (50 °C and 70 °C) which correspond to C-O and C-H bonds as a result of glycosidic breakage and release of sugars (mannose or glucose) in aloin, as previously reported by Nejatzadeh-Barandozi and Enferadi and Patel et al. [42,43]. The high-intensity peaks at 1402 $cm^{-1}$ are assigned to symmetrical -COO stretching of carboxylate due to de-esterification exaggerated after heating [41]. This was previously reported by Lim and Cheong, who analysed the de-esterification of pectin by mixing pure *Aloe vera* gel with ethanol and heating it at various temperatures [39]. A change in peak intensity at 1641 $cm^{-1}$ in samples heated at 50 °C and 70 °C relates to the hydrolysis of aloin by an enzyme polygalacturonase (PG), which exists endogenously in pure *Aloe vera* gel [44].

In Figure 2B, there is no change in the position of 1740 $cm^{-1}$ and 1250 $cm^{-1}$ peaks, which corresponds to no change in C=O and C-O-C stretches of carbonyl groups of *O*-acetyl ester. This suggested that the sugar moiety in aloin A and aloin B is D-glucose and the $C_1$ position of D-glucose is connected straightforwardly to the $C_{10}$ position of the anthrone ring in a β-configuration. The β-(1–10) C–C bond remains unaffected by acidic and basic conditions [4]. Thus, there is no change in the peaks of anthraquinones in samples exposed to different pHs. The appearance of peaks at 1070 $cm^{-1}$ and 1126 $cm^{-1}$ in all samples (excluding control WLAG)) corresponded to reduction and oxidation to form stable anions [42,43,45]. At acidic pH, the reduction is a single-step two-electron two-proton process, while in alkaline pH the reduction does not involve protons and is only a case of two-electron reduction [46]. This can be verified by HPLC data indicating aloin's maximum stability at acidic pH (3.5). The appearance of peaks at 1402 $cm^{-1}$ (pH 6.0, pH 6.7) can be assigned to symmetrical -COO stretching of carboxylate, indicating the oxidation of aloin into 10-hydroxyaloin A and B [41].

Femenia and colleagues reported that deacetylation of bioactive components (polysaccharides) and loss of glycosyl residues occurs while drying above 60 °C [47], and according to Minjares-Fuentes et al. mannose units undergo deacetylation (~65%) during industrial drying (spray drying and freeze drying) [24]. As per Figure 2C, peaks at 1250 $cm^{-1}$, 1645 $cm^{-1}$ and 1402 $cm^{-1}$ can be attributed to C-O-C stretches of $-COCH_3$ groups: asymmetrical and symmetrical -COO stretching carboxylate compounds, respectively. Thus, the appearance of peaks at 1250 $cm^{-1}$ in FDP and SDP might be due to the loss of galactosyl residues and formation of hydrogen bonds between mannose (sugar) chains, which might be the cause of the loss of aloin, as it was previously reported [48] that acemannan decreased (~40–45%) by the application of hot air (70 °C) and a high-voltage electric field. The peaks at 1402 $cm^{-1}$ can be assigned to symmetrical -COO stretching of carboxylate by oxidation degradation, which intensified after mixing with the organic solvent (methanol) [41], which was not evident in the WLAG. Our results agree with the previous studies of Chokboribal et al. [49]. In the case of FDP and SDP, new band formations at 1070 $cm^{-1}$ and intensified peaks at 1126 $cm^{-1}$ correspond to the glycosidic breakage or liberation of sugars from aloin. However, these bands were absent in WLAG.

### 3.3. Stability of Anthraquinones at Different Temperatures

Aloin contents in WLAG were ~6134 μg/g at room temperature (25 °C) and the remaining contents of aloin after various thermal treatments were plotted as per Figure 3a. It is notable from Figure 3a that the aloin degradation occurs gradually while heating WLAG (at 50 °C and 70 °C). WLAG at 70 °C exhibited the highest loss of aloin with an average of 3737.61 μg/g (40% decline) remaining, and 3955.07 μg/g (35% decline) remaining at 50 °C. Aloin A and B are the two diastereoisomers of an anthron C-glycoside, aloin, also referred as barbaloin and isobarbaloin. It has been reported that aloin B (10, C1′:R, S diastereomer of aloin) is formed naturally in plants and the non-enzymatic conversion of aloin B into aloin takes place (C10, C1′:S, S diastereomer of aloin), which might describe their common anthranol form [50]. They differ only in optical rotation and circular dichroism [36]. Therefore, upon heating, aloin A is converted to 10-hydroxyaloins A and B, similar to that previously reported on pure aloin [16]. However, ~4382 μg/g of aloin remained (Figure 3a) at the refrigeration temperature (4 °C) compared to aloin contents (~6134 μg/g) originally present in the control sample at room temperature. The lower amount of aloin at 4 °C might be due to the generation of elgonica dimers A and B, which are major degradation products of aloin A [51].

Figure 3b shows the quantity of aloin in FDP at 4 °C and 25 °C with an average amount of 204.4761 μg/g and 223.15 μg/g, respectively. However, in the case of spray drying, ~193.86 μg/g of aloin at 4 °C and ~220.0 μg/g of aloin at 25 °C were found. Hence, aloin is more stable at room temperature than at refrigeration temperatures in the case of powders as well.

Storage time significantly affected the extent of aloin degradation. It can be elucidated from Figure 3d that degradation of aloin was accelerated while storing at room temperature (from day zero to day three) and boosted from day three to day seven. Decreasing gradually from days (zero to three), the highest loss of aloin was observed on the seventh day in WLAG, FDP and SDP. Remaining aloin contents were 2428.72 μg/g (39%), 113.37 μg/g (50%) and 88.47 μg/g (40%) for WLAG, FDP and SDP, respectively, at room temperature (25 °C). Remarkably, heating samples at 70 °C resulted in ~ 70% reduction of the aloin content. Similar findings were reported previously by Chang et al. [22], that during subsequent heating (60 °C, 70 °C, 80 °C, 90 °C) and storage (12 h) a gradual decrease in aloin contents (also named barbaloin) takes place. However, the similar trend of degradation in FDP was because the freeze-dried powder is very hydrophilic, so there might be a possibility that oxidation caused rapid degradation while exposed to oxygen and extracting solvent [52]. In the case of SDP, the oxidative degradation of aloin into aloe-emodin and various unidentified components occurs, similar to that previously reported on the indispensability of methanol to dissolve barbaloin (aloin) from aloe powder [22]. As per Figure 3a, at 70 °C, more than 60% of aloin degradation was observed on the seventh day, but in powders the degradation was almost 40%.

These findings agreed with the previous study on the instability of barbaloin (also called aloin A), an isomer of aloin, over one month of storage [53]. It was hypothesised that aloin has been converted into its dimers and then trimers during storage. Thermal decomposition of anthrone C-glycoside occurs after heating, resulting in its degradation products including aloe-emodin, a trihydroxyanthraquinone [54]. Figure 3d shows an exciting trend concerning aloe-emodin at various temperatures, showing 69.01 μg/g (25 °C) and 74.15 μg/g of aloe-emodin in the samples heated at 50 °C on day zero. The highest amount in samples at 50 °C vs. 25 °C on day zero is due to the production of aloe-emodin through oxidative degradation of aloin. The aloe-emodin amount on the seventh day was 104.23 02 μg/g, which is almost double as compared to its original amount at 50 °C. This increase is due to metabolite production of aloe-emodin by oxidative cleavage of aloin into its degradation products over time. These results agreed with a previous study [22] that reported the production of aloe-emodin versus the disappearance of aloin over time. In addition, the originally present aloe-emodin in samples was found to be less susceptible to thermal degradation as compared to aloin.

Interestingly, on the seventh day of storage, samples at 25 °C and 70 °C showed a less rapid increase in aloe-emodin contents from their original amount on day zero. The reason is that the disappearance of aloin and the production of aloe-emodin processes are not in equilibrium (more aloin disappeared than in the production of aloe-emodin). Another reason is that racemisation occurs faster at the beginning of degradation. At the start (on day zero), a chemical reaction enhanced aloin's conversion into aloe-emodin and the thermal decomposition of aloin. However, racemisation was less remarkable than the degradation, until racemic mixtures reached a balance on the third day and finally, racemisation was hardly detected. These findings agree with previous studies [16] that aloin A degradation was proceeded both by degradation and the racemisation mechanism.

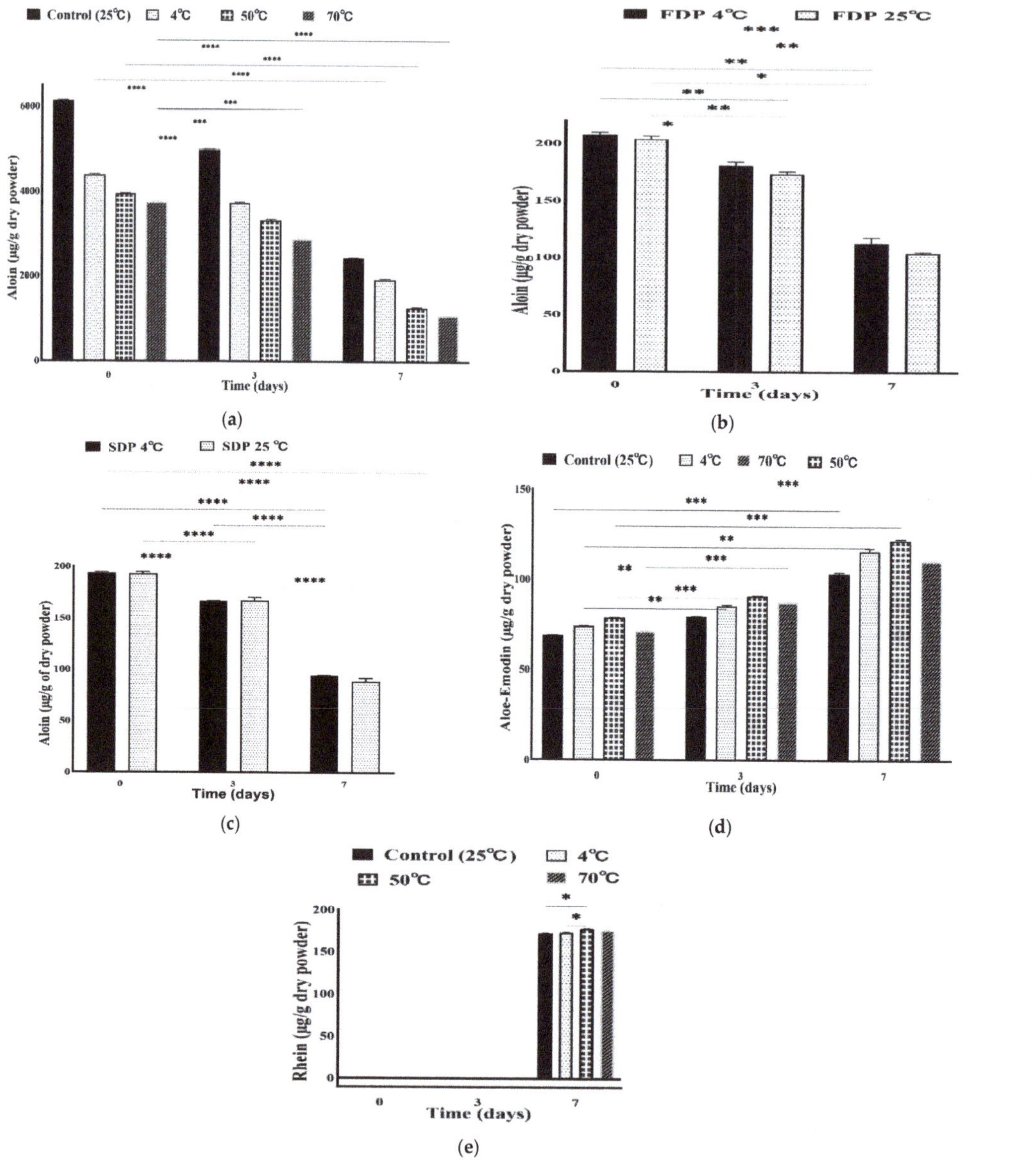


**Figure 3.** Plots of aloin in whole-leaf *Aloe vera* gel (**a**), aloin in freeze-dried powder of *Aloe vera* (**b**), aloin in spray-dried powder of *Aloe vera* (**c**) and plots of aloe-emodin (**d**) and rhein (**e**) in whole-leaf *Aloe vera* gel, vs. time under various temperature conditions. (Values are from triplicates; the mean and standard deviation are shown. Values significantly different from control to various temperature treatments in a two-way ANOVA with Dunnett's test analysis indicated with asterisks: * $p< 0.01$, ** $p < 0.001$, *** $p < 0.0003$, **** $p < 0.0001$).

Rhein was the last component detected in anthraquinone's extract from WLAG on the seventh day of storage. In contrast, no peaks of rhein were detected in WLAG on day zero or day three. The highest quantity of rhein detected was 178.25 μg/g at 50 °C, while the lowest was an average of 172.75 μg/g at 25 °C. Apparently, there was no significant difference within all samples. This can be due to the inefficient extraction of rhein with methanol [54]. The production of rhein might be due to the oxidation of aloe-emodin to rhein by a chemical reaction with the organic solvent (methanol). This has been reported previously by [55] and [56], that aloe-emodin is oxidised by chromic acid to rhein. However, no peaks of rhein were observed in FDP or SDP. This might be due to the low content of aloin, which was not enough to be oxidised into aloe-emodin, which produces rhein on further oxidation.

### 3.4. Stability of Anthraquinones at Different pH Adjustments

Additionally, pH also plays a crucial role in the stability of anthraquinones. A series of pHs (3.5, 4.6, 5.5, 6.0, 6.7) was selected depending on functional foods and nutraceutical requirements. A change in aloin contents in WLAG, FDP and SDP were plotted, as shown in Figure 4a–c, respectively. Aloin contents at natural pH (4.52) of *Aloe vera* gel were ~6134 μg/g, decreasing gradually with an increase in pH from 4.6 (5412 μg/g) to 5.5 (5452 μg/g) and finally to 6.0 (4784.75 μg/g), as shown in Figure 4a. These findings align with [50], who reported that aloin A and B undergo rapid decomposition at basic pH values.

Aloin contents in WLAG exhibited good stability at pH 3.5 with an average amount of 5890.61 μg/g, and lowest stability at pH 6.7 with an average of 4032.70 μg/g compared to the control (6134.0 μg/g) on day zero. Acid hydrolysis might be the reason for the higher amount of aloin at pH 3.5. Sánchez-Viesca and Gómez [57] reported that acid hydrolysis is the starting point of C-glycosidic cleavage, followed by the protonation of $O_2$ with the elimination of $H_2O$. At a high pH of 6.7, sodium ions combine with carbocation. C-C fission takes place, as well as oxidation into aloe-emodin-nine-anthrone. Then, a second oxido-reduction step necessarily occurs with the formation aloe-emodin, as shown in Figure 5. Finally, the generation of carbocation and ring contraction occurs with oxygen assistance. Therefore, in the presence of hydrochloric acid, a substituted exomethylene derivative must be formed, and no splitting is possible by simple hydrolysis, hence aloin remains stable [58].

Figure 4b shows the quantity of aloin in FDP at pH 3.5 and 6.7 with an average amount of 185.19 μg/g and 124.65 μg/g, respectively, as compared to the control samples (223.15 μg/g). However, in the case of spray drying, 181.45 μg/g of aloin at pH 3.5 and 109.69 μg/g of aloin at pH 6.7 were found, respectively, concerning the control (220.06 μg/g), as depicted in Figure 4c. The identical trend of decreasing the amount of aloin (~18%) at acidic pH in both SDP and FDP was due to the instability of diastereomeric 10-C-glucosylanthrones in solution, where they are supposed to interconvert via tautomeric anthrol [59]. However, in the case of pH 6.7, both powders (FDP, SDP) exhibited parallel aloin reduction (~50%) compared to the control sample. This was due to the oxidation of aloin to its oxanthrone derivatives, 10-hydroxy aloin A and B [60]. By comparing the aloin's behaviour in all samples (WLAG, FDP, SDP), it can be noted that a less rapid decline (3.9%) can be observed in WLAG as compared to its powders (17%) at acidic pH (3.5). In the aqueous phase in acidic pH (WLAG), the keto form of aloin is more stable than its powders, where the organic solvent favours the enolic form. Therefore, aloin's stability was higher in WLAG at acidic pH (3.5) than powders.

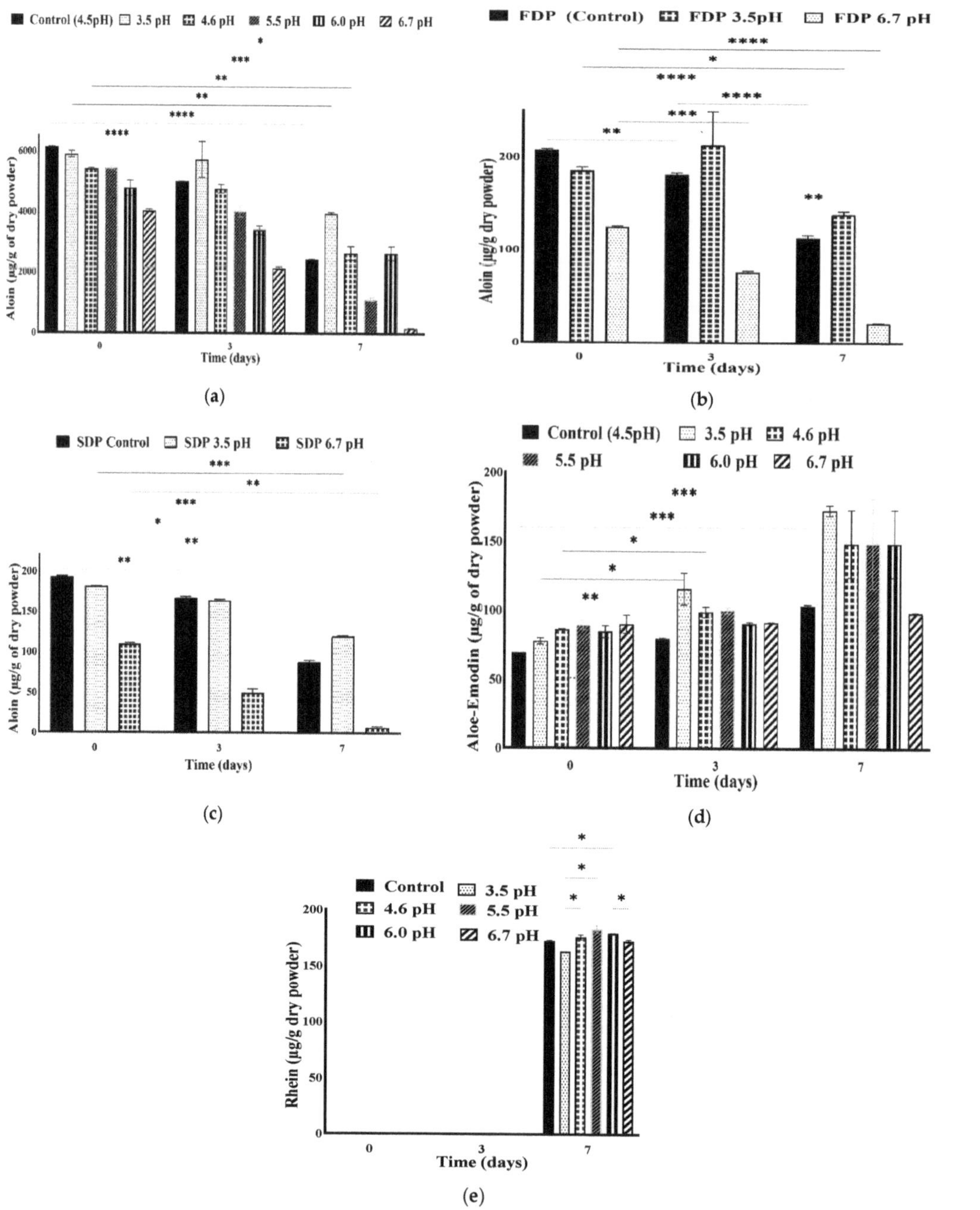


**Figure 4.** Plots of aloin in whole-leaf *Aloe vera* gel (**a**), aloin in freeze-dried powder of *Aloe vera* (**b**), aloin in spray-dried powder (**c**) of *Aloe vera* and plots of aloe-emodin (**d**) and rhein (**e**) in whole-leaf *Aloe vera* gel, vs. time under various pH conditions. (Values are from triplicates; the mean and standard deviation are shown. Values significantly different from control to various pH treatments in a two-way ANOVA with Dunnett's test analysis are indicated with asterisks: * $p < 0.01$, ** $p < 0.001$, *** $p < 0.0003$, **** $p < 0.0001$).

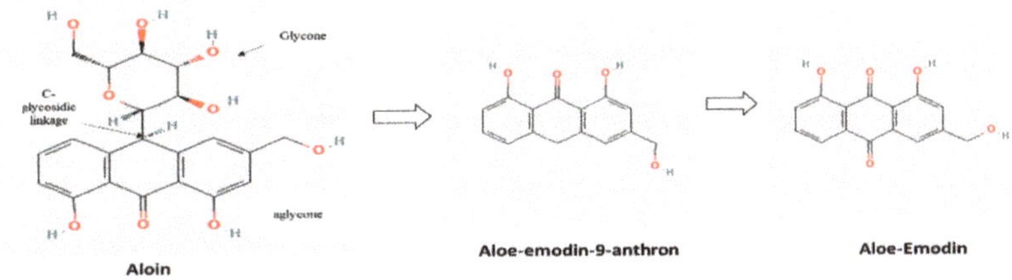


**Figure 5.** Hydrolysis of aloin into aloe-emodin-9-anthrone and subsequent oxidation to aloe-emodin, a free anthraquinone.

Regarding storage time (Figure 4a), aloin showed variable trends depending on acidic and basic conditions. WLAG samples (Figure 4a) at natural pH (4.52) with 6134 μg/g of aloin possessed an 18% decrease while moving from days zero to three and a 60% reduction from day zero to day seven. This decline was due to the conversion of aloin to 10 hydroxyaloins A and B while stored at its natural pH (i.e., 4.5). At pH 3.5, aloin exhibited good stability up to day seven, with ~70% of aloin remaining. The stability (90%) from day zero to day three is owed to the keto carbonyl function on the central ring of aloin, which caused aglycones' elevation after acid hydrolysis. Conversely, the decrease in the aloin contents at acidic pH (20%) was much less than the decrease in aloin from the control WLAG (40%) with seven days of storage. This smaller decline was due to the degradation of aloin into aloe-emodin and elgonica dimers, which occurs quickly at acidic values. These results are aligned with [61], who reported aloin's inclination after 2 h of acid hydrolysis and then a gradual decrease up to 6 h. In the case of pH 6.7, there was a sharp decline (50%) of aloin from day zero to day three, and less than 2% of aloin was detected on day seven. At alkaline conditions, a more rapid decomposition of aloin into its derivatives (10-hydroxyaloins A and B) took place, as previously reported [16,19].

Regarding Figure 4b,c, FDP and SDP also showed similar trends upon storing. Both powders exhibited good stability of aloin at pH 3.5 for three days, and just 20% degradation was observed up to seven days as compared to control. However, in the case of pH 6.7 samples, 50% of aloin degraded within three days and more than 90% decreased on day seven. Therefore, it was demonstrated that aloin undergoes oxidative degradation at higher pH values and over storage intervals in all samples (WLAG, FDP, SDP). Aloe-emodin was the second anthraquinone component detected in whole-leaf Aloe vera gel with the opposite trend from aloin, and with a slight decline while moving to acidic pH. An average of 69.01 μg/g of aloe-emodin in control WLAG (pH 4.52) and 77.59 μg/g of aloe-emodin at pH 3.5 was present (Figure 4d). However, at pH 4.6, 85.99 μg/g, at pH 5.5, 89.24 μg/g and relatively constant at pH 6.7 with 89.91 μg/g of aloe-emodin were detected. The reason for aloe-emodin concentration being more than the control is due to the production of aloe-emodin at pH 5 or below (one of the degradation products of aloin) that causes the concentration of aloe-emodin to rise while moving from acidic to basic pH (3.5–6.7) values, as previously reported by [19]. Interestingly, at pH 3.5, aloe-emodin contents increased two-thirds from zero to three days and doubled at the end of the seventh day. This was due to the keto-enol tautomerisation of aloe-emodin, whereas, in the case of pH 4.6, 5.5 and 6.0, a slight incline was observed from days zero to three and a 50% increase from its original amount at day zero. This was due to the reason that at higher pH, aloin was converted to aloe-emodin along with other degradation products: 10-hydroxyaloin A and B. Aloe-emodin at pH 6.7 remained steady from day zero to the seventh day due to its stability in basic environment and that there was not sufficient aloin (5% remained at 6.7 pH) that could be converted into its derivative, i.e., aloe-emodin. These results are in alignment with previous studies [19].

Rhein (Figure 4e) was the third anthraquinone found in WLAG samples on the seventh day of storage with an average of 172.75 μg/g. The lowest amount was at pH 3.5 (163.73 μg/g), while the highest amount at (183.50 μg/g) was found at pH 5.5. Apparently, there was no significant difference in all samples. The production of rhein is due to the oxidation of aloe-emodin to rhein by a chemical reaction with an organic solvent. This was previously reported: aloe-emodin is oxidised by chromic acid to rhein [55]. However, no peaks of rhein were observed in freeze-dried and spray-dried powders of whole-leaf *Aloe vera*. This is due to the low contents of aloin, which was not enough to oxidise it into aloe-emodin, which produces rhein on further oxidation.

## 4. Conclusions

The current study indicated that aloin decomposed quickly at higher temperatures (50 °C, 70 °C) and high pH conditions (pH 6.0, 6.7). However, WLAG retains most of its aloin at room temperature and acidic pH. The present study also revealed that aloe-emodin could be retained at high temperatures and at pH 5.0 during processing. However, aloin and aloe-emodin can be converted into degraded products over storage. Therefore, it is recommended that after extraction, the organic solvent must be evaporated to minimise its decomposition and degradation. Rhein was not present in detectable amounts in *Aloe vera* gel. However, on storage for up to seven days, little amounts of rhein were detected due to the degradation of aloe-emodin to rhein. Extra FTIR peaks arose due to the conversion of glycons to aglycon by thermal degradation and oxidation–reduction of anthraquinones. In addition, the extraction efficacy of anthraquinones was more in WLAG than FDP and SDP due to the impact of dehydration stresses on anthraquinones. It is suggested that low temperature and acidic pH are required to preserve aloin contents for functional foods and nutraceuticals to obtain the maximum health benefits.

**Author Contributions:** U.S.; Conceptualization, Methodology, Investigaton, Formal analysis, Software, Data curation, Validation, Writing-original draft., H.G.; Writing—review and editing, Supervision, Methadolgy, Validation, Funding acquisition, Resources J.C.; Writing—review and editing, Supervision, Methadology, Softwar, Data Curation, Validation. All authors have read and agreed to the published version of the manuscript.

**Funding:** This research received no external funding.

**Institutional Review Board Statement:** Not applicable.

**Informed Consent Statement:** Not applicable.

**Data Availability Statement:** The data presented in this study are available on request from the corresponding author.

**Acknowledgments:** We are thankful to the Higher Education Commission (HEC) of Pakistan and RMIT University, Australia, for providing scholarship support.

**Conflicts of Interest:** The authors declare no conflict of interest.

## References

1. Atherton, P. Aloe vera revisited. *Br. J. Phytother.* **1997**, *4*, 176–183.
2. Urch, D. *Aloe Vera-Nature's Gift*; Blackdown Publications: Bristol, UK, 1999; pp. 7–13.
3. Eshun, K.; He, Q. Aloe vera: A valuable ingredient for the food, pharmaceutical and cosmetic industries—A review. *Crit. Rev. Food Sci. Nutr.* **2004**, *44*, 91–96. [CrossRef] [PubMed]
4. Boudreau, M.D.; Olson, G.R.; Tryndyak, V.P.; Bryant, M.S.; Felton, R.P.; Beland, F.A. From the cover: Aloin, a component of the aloe vera plant leaf, induces pathological changes and modulates the composition of microbiota in the large intestines of f344/N male rats. *Toxicol. Sci.* **2017**, *158*, 302–318. [CrossRef] [PubMed]
5. Drudi, D.; Tinto, D.; Ferranti, D.; Fiorelli, F.; Pozzo, M.D.; Capitani, O. *Aloe barbadensis* miller versus silver sulfadiazine creams for wound healing by secondary intention in dogs and cats: A randomized controlled study. *Res. Vet. Sci.* **2018**, *117*, 1–9. [CrossRef]
6. Borges-Argáez, R.; Chan-Balan, R.; Cetina-Montejo, L.; Ayora-Talavera, G.; Sansores-Peraza, P.; Gómez-Carballo, J.; Cáceres-Farfán, M. In vitro evaluation of anthraquinones from Aloe vera (*Aloe barbadensis* Miller) roots and several derivatives against strains of influenza virus. *Ind. Crops Prod.* **2019**, *132*, 468–475. [CrossRef]
7. Jeremić, S.; Amić, A.; Stanojević-Pirković, M.; Marković, Z. Selected Anthraquinones as Potential Free Radical Scavengers and P-Glycoprotein Inhibitors. *Org. Biomol. Chem.* **2018**, *16*, 1890–1902. [CrossRef]
8. Fox, L.T.; Mazumder, A.; Dwivedi, A.; Gerber, M.; du Plessis, J.; Hamman, J.H. In vitro wound healing and cytotoxic activity of the gel and whole-leaf materials from selected aloe species. *J. Ethnopharmacol.* **2017**, *200*, 1–7. [CrossRef]

9. García-Sosa, K.; Villarreal-Alvarez, N.; Lübben, P.; Peña-Rodríguez, L.M. Chrysophanol, an antimicrobial anthraquinone from the root extract of Colubrina greggii. *J. Mex. Chem. Soc.* **2006**, *50*, 76–78.
10. Hu, F.; Xing, F.; Zhu, G.; Xu, G.; Li, C.; Qu, J.; Lee, I.; Pan, L. Rhein antagonizes P2X7 receptor in rat peritoneal macrophages. *Sci. Rep.* **2015**, *5*, 14012. [CrossRef]
11. Kumar, S.; Yadav, M.; Yadav, A.; Rohilla, P.; Yadav, J.P. Antiplasmodial potential and quantification of aloin and aloe-emodin in Aloe vera collected from different climatic regions of India. *BMC Complement. Altern. Med.* **2017**, *17*, 310–369. [CrossRef]
12. Lachenmeier, K.; Kuepper, U.; Musshoff, F.; Madea, B.; Reusch, H.; Lachenmeier, D.W. Quality control of Aloe vera beverages. *Electron. J. Environ. Agric. Food Chem.* **2005**, *4*, 1033–1042.
13. Salah, F.; Ghoul, Y.E.; Mahdhi, A.; Majdoub, H.; Jarroux, N.; Sakli, F. Effect of the deacetylation degree on the antibacterial and antibiofilm activity of acemannan from Aloe vera. *Ind. Crops Prod.* **2017**, *103*, 13–18. [CrossRef]
14. Kang, S.; Zhao, X.; Yue, L.; Liu, L. Main anthraquinone components in Aloe vera and their inhibitory effects on the formation of advanced glycation end-products. *J. Food Process. Preserv.* **2017**, *41*, e13160. [CrossRef]
15. Huang, L.; Wang, X.B.; Yu, Q.M.; Luo, Q.Y.; Zhang, Z.Z. Synergistic cancer growth-inhibitory effect of emodin and low-dose cisplatin on gastric cancer cells In vitro. *Trop. J. Pharm. Res.* **2015**, *14*, 1427. [CrossRef]
16. Ding, W.J.; Wu, X.F.; Zhong, J.S.; Wan, J.Z. Effects of temperature, pH and light on the stability of aloin A and characterisation of its major degradation products. *Int. J. Food Sci. Technol.* **2014**, *49*, 1773–1779. [CrossRef]
17. Yang, M.; Wei, Y.; Ashokkumar, M.; Qin, J.; Han, N.; Wang, Y. Effect of ultrasound on binding interaction between emodin and micellar casein and its microencapsulation at various temperatures. *Ultrason. Sonochem.* **2020**, *62*, 104861. [CrossRef]
18. Zhang, L.-M.; Xie, W.-G.; Wen, T.-T.; Zhao, X. Thermal behavior of five free anthraquinones from rhubarb. *J. Therm. Anal. Calorim.* **2010**, *100*, 215–218. [CrossRef]
19. Narayanan, S.; Jadhav, A.; Kadam, V. Forced degradation studies of aloe emodin and emodin by HPTLC. *Indian J. Pharm. Sci.* **2015**, *77*, 795–798. [CrossRef]
20. Kaparakou, E.H.; Kanakis, C.D.; Gerogianni, M.; Maniati, M.; Vekrellis, K.; Skotti, E.; Tarantilis, P.A. Quantitative determination of aloin, antioxidant activity, and toxicity of Aloe vera leaf gel products from Greece. *J. Sci. Food Agric.* **2021**, *101*, 414–423. [CrossRef]
21. Sánchez-Machado, D.I.; López-Cervantes, J.; Mariscal-Domínguez, M.F.; Cruz-Flores, P.; Campas-Baypoli, O.N.; Cantú-Soto, E.U.; Sanches-Silva, A. An HPLC Procedure for the Quantification of Aloin in Latex and Gel from *Aloe barbadensis* Leaves. *J. Chromatogr. Sci.* **2017**, *55*, 251–257. [CrossRef]
22. Chang, X.L.; Wang, C.; Feng, Y.; Liu, Z. Effects of heat treatments on the stabilities of polysaccharides substances and barbaloin in gel juice from Aloe vera Miller. *J. Food Eng.* **2006**, *75*, 245–251. [CrossRef]
23. Pellizzoni, M.; Molinari, G.P.; Lucini, L. Stability of the main Aloe fractions and Aloe-based commercial products under different storage conditions. *Agrochimica* **2011**, *55*, 288–296.
24. Minjares-Fuentes, R.; Femenia, A.; Comas-Serra, F.; Rosselló, C.; Rodríguez-González, V.; González-Laredo, R.; Gallegos-Infante, J.; Medina-Torres, L. Effect of different drying procedures on physicochemical properties and flow behavior of Aloe vera (*Aloe barbadensis* Miller) gel. *LWT* **2016**, *74*, 378–386. [CrossRef]
25. McDougall, G.J.; Dobson, P.; Jordan-Mahy, N. Effect of different cooking regimes on rhubarb polyphenols. *Food Chem.* **2010**, *119*, 758–764. [CrossRef]
26. Yen, G.-C.; Chung, D.-Y. Antioxidant Effects of Extracts from Cassia t ora L. Prepared under Different Degrees of Roasting on the Oxidative Damage to Biomolecules. *J. Agric. Food Chem.* **1999**, *47*, 1326–1332. [CrossRef] [PubMed]
27. Wamer, W.G.; Vath, P.; Falvey, D.E. In vitro studies on the photobiological properties of aloe emodin and aloin A. *Free Radic. Biol. Med.* **2003**, *34*, 233–242. [CrossRef]
28. Park, M.; Park, J.; Kim, N.Y.; Shin, Y.G.; Choi, Y.S.; Lee, J.G.; Kim, K.H.; Lee, S.K. Analysis of 13 phenolic compounds in Aloe species by high performance liquid chromatography. *Phytochem. Anal. Int. J. Plant Chem. Biochem. Tech.* **1998**, *9*, 186–191. [CrossRef]
29. Lindroth, R.; Pajutee, M. Chemical analysis of phenolic glycosides: Art, facts, and artifacts. *Oecologia* **1987**, *74*, 144–148. [CrossRef]
30. Jovanović, A.A.; Lević, S.M.; Pavlović, V.B.; Marković, S.B.; Pjanović, R.V.; Đorđević, V.B.; Nedović, V.; Bugarski, B.M. Freeze vs. Spray drying for dry wild thyme (*Thymus serpyllum* L.) extract formulations: The impact of gelatin as a coating material. *Molecules* **2021**, *26*, 3933. [CrossRef]
31. Emami, F.; Vatanara, A.; Park, E.J.; Na, D.H. Drying technologies for the stability and bioavailability of biopharmaceuticals. *Pharmaceutics* **2018**, *10*, 131. [CrossRef]
32. Jawade, N.R.; Chavan, A.R. Ultrasonic-assisted extraction of aloin from aloe vera gel. *Procedia Eng.* **2013**, *51*, 487–493. [CrossRef]
33. Lee, Y.S.; Yang, T.-J.; Park, S.U.; Baek, J.H.; Wu, S.; Lim, K.-B. Induction and Proliferation of Adventitious Roots from'Aloe vera'Leaf Tissues for'in vitro'Production of Aloe-emodin. *Plant Omics* **2011**, *4*, 190–194.
34. Fu, H. Free radical scavenging and leukemia cell growth inhibitory properties of onion powders treated by different heating processes. *J. Food Sci.* **2004**, *69*, SNQ50–SNQ54. [CrossRef]
35. Dagne, E.; Bisrat, D.; Viljoen, A.; Van Wyk, B. Chemistry of Aloe species. *Curr. Org. Chem.* **2000**, *4*, 1055–1078. [CrossRef]
36. Grun, M.; Franz, G. Studies on the biosynthesis of aloins in aloe-arborescens mill. *Arch. Der. Pharm.* **1982**, *315*, 231–241. [CrossRef]
37. Han, Y.-S.; Van der Heijden, R.; Verpoorte, R. Biosynthesis of anthraquinones in cell cultures of the Rubiaceae. *Plant Cell Tissue Organ Cult.* **2001**, *67*, 201–220. [CrossRef]

38. Fardsadegh, B.; Jafarizadeh-Malmiri, H. Aloe vera leaf extract mediated green synthesis of selenium nanoparticles and assessment of their in vitro antimicrobial activity against spoilage fungi and pathogenic bacteria strains. *Green Process. Synth.* **2019**, *8*, 399–407. [CrossRef]
39. Lim, Z.X.; Cheong, K.Y. Effects of drying temperature and ethanol concentration on bipolar switching characteristics of natural Aloe vera-based memory devices. *Phys. Chem. Chem. Phys.* **2015**, *17*, 26833–26853. [CrossRef]
40. Niamlang, S.; Buranut, T.; Niansiri, A.; Sirivat, A. Controlled aloin release from crosslinked polyacrylamide hydrogels: Effects of mesh size, electric field strength and a conductive polymer. *Materials* **2013**, *6*, 4787–4800. [CrossRef]
41. Torres-Giner, S.; Wilkanowicz, S.; Melendez-Rodriguez, B.; Lagaron, J.M. Nanoencapsulation of Aloe vera in Synthetic and Naturally Occurring Polymers by Electrohydrodynamic Processing of Interest in Food Technology and Bioactive Packaging. *J. Agric. Food Chem* **2017**, *65*, 4439–4448. [CrossRef]
42. Patel, C.M.; Barot, A.A.; Kumar Sinha, V. Sequential liquefaction of Nicotiana tabacum stems biomass by crude polyhydric alcohols for the production of polyols and rigid polyurethane foams. *J. Appl. Polym. Sci.* **2016**, *133*, 43974. [CrossRef]
43. Nejatzadeh-Barandozi, F.; Enferadi, S.T. FT-IR study of the polysaccharides isolated from the skin juice, gel juice, and flower of Aloe vera tissues affected by fertilizer treatment. *Org. Med. Chem. Lett.* **2012**, *2*, 33. [CrossRef] [PubMed]
44. Valverde, J.M.; Valero, D.; Martínez-Romero, D.; Guillén, F.; Castillo, S.; Serrano, M. Novel edible coating based on Aloe vera gel to maintain table grape quality and safety. *J. Agric. Food Chem.* **2005**, *53*, 7807–7813. [CrossRef] [PubMed]
45. Bierhalz, A.C.K.; Lopes, S.A.; Pires, A.L.R.; Moraes, Â.M. Development of polysaccharide-based membranes incorporating the bioactive compound aloin. *Int. J. Polym. Mater. Polym. Biomater.* **2017**, *66*, 193–202. [CrossRef]
46. Guin, P.S.; Das, S.; Mandal, P. Electrochemical reduction of quinones in different media: A review. *Int. J. Electrochem.* **2011**, *2011*, 1–22. [CrossRef]
47. Femenia, A.; García-Pascual, P.; Simal, S.; Rosselló, C. Effects of heat treatment and dehydration on bioactive polysaccharide acemannan and cell wall polymers from *Aloe barbadensis* Miller. *Carbohydr. Polym.* **2003**, *51*, 397–405. [CrossRef]
48. Sriariyakul, W.; Swasdisevi, T.; Devahastin, S.; Soponronnarit, S. Drying of aloe vera puree using hot air in combination with far-infrared radiation and high-voltage electric field: Drying kinetics, energy consumption and product quality evaluation. *Food Bioprod. Process.* **2016**, *100*, 391–400. [CrossRef]
49. Chokboribal, J.; Tachaboonyakiat, W.; Sangvanich, P.; Ruangpornvisuti, V.; Jettanacheawchankit, S.; Thunyakitpisal, P. Deacetylation affects the physical properties and bioactivity of acemannan, an extracted polysaccharide from Aloe vera. *Carbohydr. Polym.* **2015**, *133*, 556–566. [CrossRef]
50. Zonta, F.; Bogoni, P.; Masotti, P.; Micali, G. High-performance liquid chromatographic profiles of aloe constituents and determination of aloin in beverages, with reference to the EEC regulation for flavouring substances. *J. Chromatogr. A* **1995**, *718*, 99–106. [CrossRef]
51. Sousa, E.T.; da Silva, M.M.; de Andrade, S.J.; Cardoso, M.P.; Silva, L.A.; de Andrade, J.B. Evaluation of thermal stability of quinones by thermal analysis techniques. *Thermochim. Acta* **2012**, *529*, 1–5. [CrossRef]
52. Abascal, K.; Ganora, L.; Yarnell, E. The effect of freeze-drying and its implications for botanical medicine: A review. *Phytother.Res. Int. J. Devoted Pharmacol. Toxicol. Eval. Nat. Prod. Deriv.* **2005**, *19*, 655–660. [CrossRef] [PubMed]
53. Yasuda, K.; Uehara, S.-i.; Takano, I.; Shindo, T.; Nishijima, M. Stability of barbaloin in aqueous solution. *Food Preserv. Sci.* **2000**, *26*, 85–90. [CrossRef]
54. Sehgal, I.; Winters, W.D.; Scott, M.; Kousoulas, K. An in vitro and in vivo toxicologic evaluation of a stabilized aloe vera gel supplement drink in mice. *Food Chem. Toxicol.* **2013**, *55*, 363–370. [CrossRef] [PubMed]
55. Bloomer, J.L.; Stagliano, K.W.; Gazzillo, J.A. Preparation of functionalized juglone acetates and juglones via 1,4-dimethoxynaphthalene derivatives: Synthesis of anthraquinones related to rhein and aloe-emodin. *J. Org. Chem.* **1993**, *58*, 7906–7912. [CrossRef]
56. Cahn, R.; Simonsen, J. Experiments on the constitution of the aloins. Part III. *J. Chem. Soc.* **1932**, *382*, 2573–2582. [CrossRef]
57. Sánchez-Viesca, F.; Gómez, R. The mechanism of nitric acid degradation of the C-glycoside Aloin to Aloe-emodin. *Am. J. Chem.* **2020**, *10*, 1–5.
58. Assary, R.S.; Kim, T.; Low, J.J.; Greeley, J.; Curtiss, L.A. Glucose and fructose to platform chemicals: Understanding the thermodynamic landscapes of acid-catalysed reactions using high-level ab initio methods. *Phys. Chem. Chem. Phys.* **2012**, *14*, 16603–16611. [CrossRef]
59. Korth, H.-G.; Mulder, P. Anthrone and related hydroxyarenes: Tautomerization and hydrogen bonding. *J. Org. Chem.* **2013**, *78*, 7674–7682. [CrossRef]
60. Rauwald, H.W. Naturally occurring quinones and their related reduction forms: Analysis and analytical methods. *PZ Wiss* **1990**, *3*, 169–181.
61. Chiang, H.-M.; Lin, Y.-T.; Hsiao, P.-L.; Su, Y.-H.; Tsao, H.-T.; Wen, K.-C. Determination of marked components—aloin and aloe-emodin—in Aloe vera before and after hydrolysis. *J. Food Drug Anal.* **2012**, *20*, 646–652.

foods

MDPI

Article

# Colloidal and Acid Gelling Properties of Mixed Milk and Pea Protein Suspensions

Isabelle Carolina Oliveira [1], Iuri Emmanuel de Paula Ferreira [1], Federico Casanova [2,*], Angelo Luiz Fazani Cavallieri [1], Luis Gustavo Lima Nascimento [3], Antônio Fernandes de Carvalho [3] and Naaman Francisco Nogueira Silva [1,*]

1 Center of Natural Sciences, Federal University of São Carlos (UFSCar), Buri 18290-000, SP, Brazil; isabelle.4862@gmail.com (I.C.O.); ferreira.iep@gmail.com (I.E.d.P.F.); angelo.lf.cavallieri@gmail.com (A.L.F.C.)

2 Food Production Engineering Group, DTU Food, Technical University of Denmark, Søltofts Plads 227, DK-2800 Lyngby, Denmark

3 Department of Food Technology, Federal University of Viçosa (UFV), Viçosa 36570-900, MG, Brazil; luis.g.nascimento@ufv.br (L.G.L.N.); antoniofernandes@ufv.br (A.F.d.C.)

* Correspondence: feca@food.dtu.dk (F.C.); naaman.nogueira@ufscar.br (N.F.N.S.); Tel.: +55-159-9676-0076 (F.C.); +45-4525-2716 (N.F.N.S.)

**Citation:** Oliveira, I.C.; de Paula Ferreira, I.E.; Casanova, F.; Cavallieri, A.L.F.; Lima Nascimento, L.G.; de Carvalho, A.F.; Nogueira Silva, N.F. Colloidal and Acid Gelling Properties of Mixed Milk and Pea Protein Suspensions. *Foods* **2022**, *11*, 1383. https://doi.org/10.3390/foods11101383

Academic Editor: Soottawat Benjakul

Received: 9 March 2022
Accepted: 4 May 2022
Published: 11 May 2022



**Abstract:** The present study aims to describe colloidal and acid gelling properties of mixed suspensions of pea and milk proteins. Mixed protein suspensions were prepared by adding pea protein isolate to rehydrated skimmed milk (3% *w/w* protein) to generate four mixed samples at 5, 7, 9, and 11% *w/w* total protein. Skimmed milk powder was also used to prepare four pure milk samples at the same protein concentrations. The samples were analyzed in regard to their pH, viscosity, color, percentage of sedimentable material, heat and ethanol stabilities, and acid gelling properties. Mixed suspensions were darker and presented higher pH, viscosity, and percentage of sedimentable material than milk samples. Heat and ethanol stabilities were similar for both systems and were reduced as a function of total protein concentration. Small oscillation rheology and induced syneresis data showed that the presence of pea proteins accelerated acid gel formation but weakened the final structure of the gels. In this context, the results found in the present work contributed to a better understanding of mixed dairy/plant protein functionalities and the development of new food products.

**Keywords:** dairy proteins; caseins; plant proteins; protein beverages; acid gel; colloidal stability

## 1. Introduction

According to the World Resources Report [1], the worldwide consumption of animal-source foods may increase 68% by 2050 due to the population growth. For this reason, the mixture of plant and animal proteins has been seen as an answer to the global demand for dairy and meat products in the upcoming decades. Field pea (*Pisum sativum*, L) is one of the plants that can be used as an alternative protein source. In general, pea seeds contain 10–20% (*w/w*) fiber, 40–50% (*w/w*) starch, and 18–30% (*w/w*) protein, which can vary according to environmental and genotypic factors. When compared to soy or other plant proteins, pea protein presents high digestibility, less allergenicity, low production cost, and high nutritional value [2,3]. However, when used in food systems, the poor water solubility and the pronounced taste, described as "beany", limits its application [4,5]. Therefore, the implementation of pea protein into a mixed system with an animal protein source, such as milk, has the potential to produce nutritious foods with acceptable sensorial characteristics and distinctive functional properties.

In this sense, milk and dairy products are consumed by billions of people due to their industrial versatility, high nutritional value, and good taste. When used as an ingredient, milk is a major component of diverse dairy products, such as yogurt, butter, cheese, creams, and high-protein beverages. This diversity of application is possible because of milk proteins (mainly caseins and whey proteins), which are responsible for its functional properties. These include the ability to form gels, emulsions, and foams [6–8]. Mixed systems of milk and different plant proteins have been the focus of recent studies to evaluate their gelling and emulsifying properties [9–12]. The results demonstrate the possibility of creating innovative food products that can fulfill an emerging market of consumers who worry increasingly about sustainability and want to partially replace animal-based proteins [13].

Although the literature regarding the functional properties of food matrices containing plant and dairy proteins is growing [9,11,12,14], there is a disparity in the research between academic purposes and evaluations for industrial application. In this context, the present work aims to describe colloidal and acid gelling properties of mixed milk/pea protein suspensions, with total protein concentrations ranging from 3.0% (*w*/*w*) to 11.0% (*w*/*w*). Here, the main strategy was to carry out the experiments in a manner that can be replicated by the dairy industry. Regarding colloidal aspects, the suspensions were analyzed in relation to pH, viscosity, color attributes, amount of sedimentable material, and heat and ethanol stabilities. Concurrently, the acid gelling properties were evaluated using small oscillation rheology and induced syneresis.

## 2. Materials and Methods

### 2.1. Materials

Pea protein isolate (PPI) (NUTRALYS S85F) was provided by Roquette (Lestrem, France), and low heat skimmed milk powder (SMP) was provided by courtesy of Itambé (Sete Lagoas, Brazil). The PPI was composed of 83% *w*/*w* proteins, 3.8% *w*/*w* minerals and 0.1% *w*/*w* moisture; the composition of SMP was 35% *w*/*w* proteins, 50% carbohydrates, 7.5% minerals, and 4% *w*/*w* moisture. This information was provided by the manufacturers. Sodium azide was purchased from Sigma Aldrich (São Paulo, Brazil).

### 2.2. Methods

#### 2.2.1. Preparation of Suspensions

The SMP was rehydrated in distillated water with 3% *w*/*w* proteins. Sodium azide was added at 0.03% *w*/*w*, to prevent microbial growth, and the sample was stirred at room temperature for 1 h. To generate the mixed protein suspensions, the PPI was added to the rehydrated skimmed milk with 3% (*w*/*w*) proteins to achieve total protein concentrations of 5, 7, 9, and 11% (*w*/*w*). The same procedure was applied to prepare pure skimmed milk samples with the same total protein concentrations. In total, nine samples were produced, and the skimmed milk with 3% (*w*/*w*) proteins was used as the control (Table 1). All the protein suspensions were stirred overnight at room temperature and then stored at 4 °C prior to analysis.

**Table 1.** Protein composition of the samples used in the present work.

| Samples | Milk Protein (%) | Pea Protein (%) | Total Protein (%) |
|---|---|---|---|
| Control | 3 | - | 3 |
| Skimmed milk | 5 | - | 5 |
| | 7 | - | 7 |
| | 9 | - | 9 |
| | 11 | - | 11 |
| Mixed protein suspensions | 3 | 2 | 5 |
| | 3 | 4 | 7 |
| | 3 | 6 | 9 |
| | 3 | 8 | 11 |

#### 2.2.2. Colloidal Properties

##### pH of the Suspensions

The pH of the samples was measured at room temperature using a calibrated pH-meter, PHS-3E (Satra, São Paulo, Brazil).

##### Viscosity

Viscosity analyses were carried out using a controlled stress rheometer, MCR 102 (Anton Paar, Ostfildern, Germany). The rheological properties of protein suspensions were obtained under steady state shear at 20 °C with a stainless-steel cone-plane geometry (diameter 50 mm and cone angle 1°). The flow curves were determined using a shear rate within the range of 0 and 300 $s^{-1}$. Three shear stress sweeps (up–down–up steps) were performed to verify the presence of shear time effects (thixotropy) [15,16]. Data from the third flow curve (steady state conditions) were adjusted to classical rheological models; the Newtonian Equation (1) or power law model Equation (2) was used according to the best coefficient of determination ($R^2$).

$$\sigma = \eta.\gamma \quad (1)$$

$$\sigma = k.\gamma^n \quad (2)$$

where $\sigma$ is the shear stress (Pa), $\gamma$ is the shear rate ($s^{-1}$), $\eta$ is the Newtonian viscosity, k is the consistency index (Pa.$s^n$), and $n$ (dimensionless) is the behavior index. The test was carried out in duplicate. For Newtonian fluids, the viscosity value was considered the Newtonian viscosity, while for shear thinning fluids, apparent viscosity was obtained at 60 $s^{-1}$ of the third sweep.

##### Color

The color was measured using a Colorium7 Chromameter (Delta Color, São Leopoldo, Brazil) [17]. The equipment was calibrated with a blacklight trap and white calibration ceramic. The parameters, L* (luminosity), a* (redness/greenness), and b* (yellowness/blueness) were obtained through i7 software (Delta Color, São Leopoldo, Brazil). The test was conducted in duplicate.

##### Sedimentable Material (SM)

The amount of SM was determined by centrifugation (Centrifuge NT 820—Novatecnica, Piracicaba, Brazil) of 10 g of each sample in separate centrifuge tubes (15 mL), at room temperature for 30 min at 4856× *g*. To express the amount of SM as a percentage, the mass of the obtained pellets was divided by the mass of the sample submitted to centrifugation, and the result was multiplied by 100. The test was performed in duplicate.

##### Heat Stability

Thermal stability was determined by adding 2 mL of the protein suspensions in hermetically sealed glass tubes, followed by their immersion in an oil bath at 140 °C. The thermal stability was defined as the time (in seconds) until visual observation of protein coagulation commencement [18,19]. The test was made in triplicate.

##### Ethanol Stability

Ethanol stability was determined by mixing equal volumes of protein suspensions with ethanol solutions (Labsynth—São Paulo, Brazil) in Petri dishes with increasing ethanol concentrations. Ethanol concentrations ranged from 60 to 100% *v*/*v* and ascended at 2.5% intervals. The highest ethanol concentration without visual signs of coagulation was defined as the ethanol stability of the samples [18,19]. The test was carried out in triplicate.

#### 2.2.3. Gelling Properties

##### Gel Formation and Rheological Analyses

Before acidification, all milk samples and mixed protein suspensions were heated at 85 °C for 5 min and subsequently cooled to 42 °C. This step was intended to induce protein denaturation and consequent aggregation, which is a classic strategy employed by the dairy industry to improve the firmness of fermented milks. Then, gel formation was monitored in triplicate under isothermal conditions (42 °C) by oscillatory measurements in the same controlled stress rheometer equipped with the same cone-plate geometry used for steady state flow measurements (Section 2.2.2 Viscosity). Glucono delta-lactone (GDL) (Sigma Aldrich—São Paulo, Brazil) was added to the protein suspensions at 2% (*w*/*w*), followed by agitation for 1 min, and then transferred to the rheometer. Time sweeps were conducted at an oscillation frequency of 0.1 Hz within the linear viscoelastic domain (0.1% of oscillatory strain). The gel point (tg) was considered as the time when the crossover between elastic (G′) and viscous (G″) moduli took place, in accordance with Cavallieri and da Cunha [20].

##### pH Kinetics

During the acidification of the samples, the pH was measured at 42 °C after 20 min intervals using a calibrated pH meter PHS-3E (Satra, Brazil).

##### Induced Syneresis

GDL was added to 15 g of protein suspensions at 2% (*w*/*w*) and at 42 °C in cylindrical plastic flasks (−1.2 cm in diameter), followed by agitation for 1 min. The flasks were kept under rest at 42 °C for 4 h. Immediately after this gelling process, the induced syneresis was determined as the percentage of the liquid drained after centrifugation (NT820—Novatecnica, Brazil) of the gels at 1800× *g* for 30 min [21].

#### 2.2.4. Statistical Analysis

The experiments were replicated three independent times. In each run, the batch of skimmed milk at 3% (*w*/*w*) proteins was split into nine samples and treatments were assigned to the samples at random. Firstly, the data were explored through descriptive statistics (means, standard deviations, and plots). Then, second-degree polynomials were fitted to describe the effect of protein concentration on the colloidal characteristics of the samples. Non-significant terms were removed from the model during a stepwise procedure, and the best model was selected according to the lowest AIC (Akaike's Information Criterion). For each fitted model, the significance was assessed using Wald's test, the explanatory power was given by the adjusted coefficient of determination, and the F-test was carried out to evaluate the lack of fit. The adequacy of the fitted models was examined through residual analyses, and the normality and homogeneity of variances were evaluated by Shapiro–Wilk's and Levene's tests, respectively. The Box-Cox's optimal transformation was applied to the values of viscosity to normalize the data and stabilize the variance. The fittings were plotted with the original data and error bars (SE, $n = 3$ per treatment). The fitted models, the adjusted coefficients of determination, the regression mean-squared errors, and F-tests for the lack of fit are provided as in Supplementary Material—Table S1.

There was no suitable polynomial model for describing the relation between heat stability and protein concentrations for milk samples and mixed protein suspensions. Thus, the heat stabilities of these samples were compared within each protein dosage by a Student's *t*-test. A power analysis was conducted considering similar scenarios to those verified in our data; in all the simulations, the *t*-test ensured 85% of power or more on differentiating the means of experimental groups.

The statistical hypotheses were tested considering a significance level of 5%. All the statistical analyses were conducted using R software [22]. The research data are available at an open access repository [23].

## 3. Results and Discussion

### 3.1. Colloidal Characterization of the Suspensions

#### 3.1.1. pH

The pH data are shown in Figure 1. The pH of milk samples decreased as the total protein concentration increased. Compared to mixed protein suspensions, the lower pH values found for milk samples were due to the presence of acidic compounds naturally found in milk, such as caseins, whey proteins, and phosphate salts [6]. In contrast to the samples containing only milk proteins, the mixed suspensions exhibited stable pH values despite the increase in protein concentration. This has not been commonly reported in other studies evaluating pH levels of pea protein suspensions as a function of protein concentration. In general, this is explained by the fact that the pH values of isolated pea protein suspensions are corrected before performing further experiments. However, Jiang et al. [24] used the same PPI applied in the present work (NUTRALYS S85F) and described the pH as being close to 7.0 of PPI suspensions at 25–30 g protein per liter, which is consistent with the results presented in Figure 1.

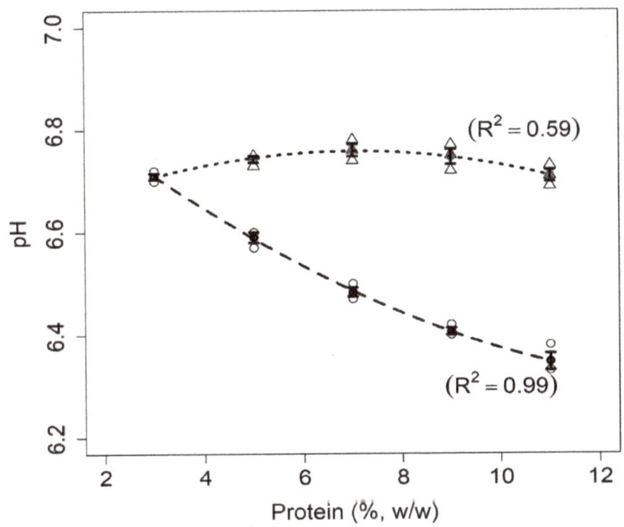


**Figure 1.** Relationship between pH and protein concentrations for milk (dashed line) and mixed protein (dotted line) samples. The figure shows the data points for milk (circles) and mixed samples (triangles), error bars (SE), fitted curves, and adjusted coefficients of determination ($R^2$).

#### 3.1.2. Viscosity

Figure 2 highlights the viscosity values of the suspensions as a function of the total protein concentration. Milk samples and mixed protein suspensions exhibited Newtonian behavior up to 9% (*w/w*) protein (Supplementary Materials—Figure S1); therefore, the Newtonian viscosity of these samples was used in Figure 2. For the samples at 11% (*w/w*) protein, the apparent viscosity presented in Figure 2 was obtained at a shear rate of 60.0 $s^{-1}$. In fact, a pseudoplastic behavior was observed for these two highly concentrated samples (Supplementary Materials—Figure S2), with a behavior index of 0.86 ± 0.02 for mixed protein suspensions and 0.87 ± 0.06 for milk. The consistency index was 0.11 ± 0.02 $Pa.s^n$ for the mixture and 0.05 ± 0.03 $Pa.s^n$ for milk. Similar results were observed for skimmed milk [25] and equal mixtures of pea and milk protein [26] at similar protein concentrations.

As shown in Figure 2, the viscosity increased with total protein concentrations for both systems. Nevertheless, the increase in viscosity was higher in the presence of pea protein levels above 7% (*w/w*). The viscosity results for skimmed milk samples are in accordance with the literature, as the viscosity of milk increases as a function of the total solids content [27,28]. For the mixed protein suspensions, it is plausible that the low solubility of pea proteins [29] was responsible for the higher viscosity values. Indeed, the presence of aggregates or insoluble material increases the volume fraction of protein suspensions, which consequently increases the apparent viscosity for a given temperature and the shear rate [30]. This result is in line with the determination of the amount of sedimentable material (Section 3.1.4).

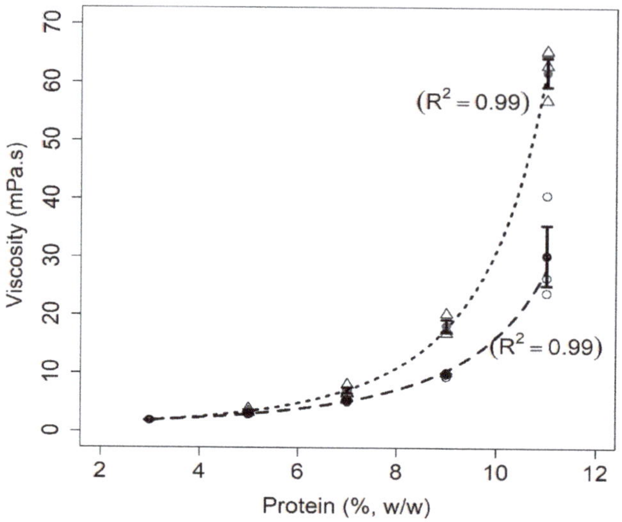


**Figure 2.** Relationship between viscosity and protein concentrations for milk (dashed line) and mixed protein (dotted line) samples. The figure shows the data points for milk (circles) and mixed samples (triangles), error bars (SE), fitted curves, and adjusted coefficients of determination ($R^2$).

### 3.1.3. Color

Figure 3 shows the parameters L*, a*, and b* of the samples. An increase in the lightness (L*—Figure 3A) of the samples containing only milk proteins (Figure 3A) as a function of the protein concentration was observed. This increase in L* is due to the increment of particles scattering light, mainly casein micelles [31]. The parameter a* (Figure 3B) is correlated with the presence of absorbing compounds that reflects red (positive values) and green (negative values) light. In milk samples, the main molecules responsible for parameter a* are lactoferrin for redness and riboflavin for greenness [27]. In general, the values of a* are negative for milk and dairy protein suspensions, regardless of the protein/fat concentrations and the physical treatments applied to milk [27,28]; which is consistent with the results found in the present work. The parameter b* (Figure 3C) is associated with the presence or absence of molecules that reflect yellow or blue. In the case of milk, the carotenoids associated with the fat portion are responsible for its yellow tendency [27]. As a matter of fact, all industrially produced skimmed milk using centrifuges present a residual fat portion. Therefore, the increase in the b* parameter found in this work for skimmed milk can be attributed to an increase in the fat concentration of the samples.

The L* values for mixed suspensions decreased as a function of the protein concentration, which can be attributed to the increase in the concentration of dark pigments, e.g., phenolic compounds naturally present in pulses [32]. The increase in pea protein concentration in the mixed suspensions also caused the augmentation of a* and b* parameters (Figure 3A,B). According to Saldanha do Carmo et al. [33], independent of the dehulling method applied to the production of protein-rich fractions from pea beans, the color compounds are naturally present in the resulting powder and are related to pea variety used to produce the PPI. These data are important since color is a sensorial criterion that influences consumer's preferences. Thus, when developing dairy products with pea proteins, technologists and/or engineers must take this aspect into account because the products will present a darker color than usually observed for dairy products.

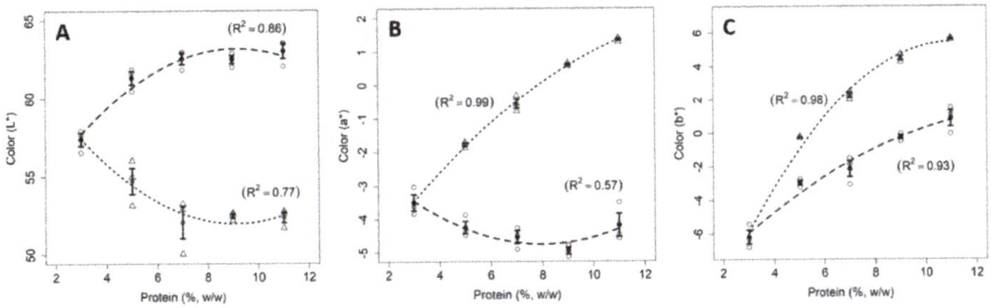


**Figure 3.** Relationship between color parameters and protein concentrations for skimmed milk (dashed line) and mixed protein (dotted line) samples. The figure shows the data points for milk (circles) and mixed samples (triangles), error bars (SE), fitted curves, and adjusted coefficients of determination ($R^2$). The parameters L* represents lightness (**A**), a* red (+), and green (−) components (**B**), and b* yellow (+) and blue (−) components (**C**).

### 3.1.4. Sedimentable Material (SM)

The percentages for SM are shown in Figure 4. The results show that the SM percentages were quite stable for all milk samples. However, the SM percentages were linearly correlated with the total protein content of the mixed protein suspensions and were significantly higher than those of the milk samples. This result was somewhat expected since only the albumin fraction of the PPI is water soluble, and it accounted for approximately 20% of the total protein content of commercial PPI [2]. In fact, some research groups work only with the soluble fraction of pea proteins when searching for new functionalities in association with dairy proteins, which can be considered as a valid scientific approach [12,34]. In the present study, however, we decided to use PPI powder as a whole because this strategy is closer to an industrial application of PPI in mixed protein suspensions. The separation of the soluble fraction from PPI, or pea flour, is a process that requires time, resources, and chemicals [32]. In addition, the disposal of the insoluble protein fraction should also be considered. On one hand, the insoluble protein fraction can be undesirable if the industrial goal is to produce protein beverages. On the other, this insoluble fraction is not necessarily a problem if the food product is more solid-like (elastic), such as some types of fermented milk products, such as cheese analogues, ice cream, and desserts.

### 3.1.5. Heat Stability

The results for heat stability are presented in Table 2. For both systems, the heat stability decreased insofar as the total protein content increased. The results for the milk samples are comparable to those reported by Dumpler and Kulozik [35] and Huppertz et al. [19] for similar pH values, protein concentrations, and temperatures. In effect, caseins are considered rheomorphic proteins that are naturally resistant to heating, without presenting a well-defined tertiary structure [36]. Moreover, caseins are present in milk as supramolecular aggregates, known as casein micelles, which are electrostatically stabilized by an external layer of κ-casein that prevents their aggregation. At high temperatures (above 140 °C) for extended periods (several minutes), milk coagulates as a result of the collapse of the external layer caused by a reduction in pH originating from lactose degradation [37]. As can be inferred from Figure 1 and Table 2, the milk sample at 11% (*w*/*w*) protein is less heat stable and more acidic (lower pH) than the milk sample at 3% (*w*/*w*) protein. Nevertheless, pH reduction is not the only factor contributing to decreased heat stability. It is also necessary to consider that the ionic calcium concentration of the milk samples naturally increased as a function of the total protein concentration. This was due to SMP being employed to prepare these samples. Indeed, the increase in ionic calcium concentration induced a reduction of casein micelle stability [38] and contributed to the reduction of heat stability of the milk samples (Table 2).

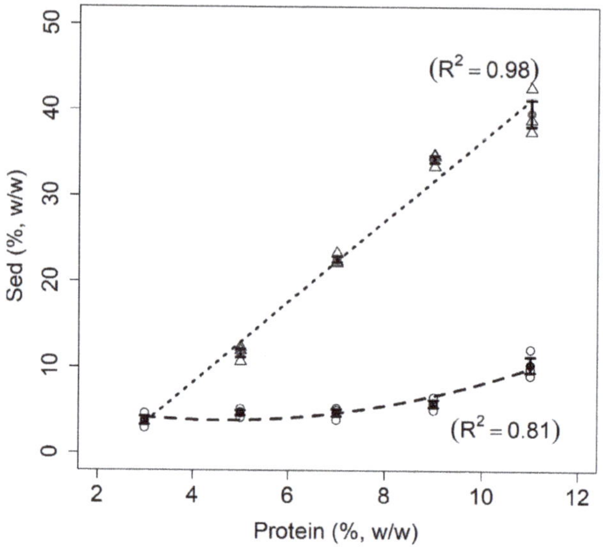


**Figure 4.** Relationship between sedimentable material (Sed % *w*/*w*) and protein concentrations for milk (dashed line) and mixed protein (dotted line) samples. The figure shows the data points for milk (circles) and mixed samples (triangles), error bars (SE), fitted curves, and adjusted coefficients of determination ($R^2$).

The mixed protein suspensions presented significantly lower heat stability than the skimmed milk samples ($p < 0.05$). Pea proteins are mainly composed of globulin and albumin fractions, which exhibit well-defined secondary and tertiary structures that are naturally more sensitive to heating [2,36]. High temperatures induce the denaturation of pea proteins, which leads to hydrophobic and covalent inter-protein interactions, resulting in aggregation and precipitation [32,39]. However, there are no comparative results in the literature, since this is the first work evaluating heat stability of mixed protein suspensions. It is noteworthy that if the mixed suspensions were acidified (pH ~6.0 or lower), they would tend to form heat-induced gels during heating instead of precipitating, as observed by Schmitt et al. [40].

**Table 2.** Heat stability (in sec.) of milk samples and mixed protein suspensions represented as mean (±standard error), and comparisons between milk and mixed protein suspensions within each protein concentration according to a Student's *t*-test.

| Protein % (*w*/*w*) | Milk (sec.) | Mixed Suspensions (sec.) | *t*-Statistics | *p*-Values |
|---|---|---|---|---|
| 3 | 1314 ± 61 | – | – | |
| 5 | 506 ± 27 | 380 ± 25 | 3.44 | 0.026 * |
| 7 | 591 ± 5 | 284 ± 35 | 8.63 | 0.000 *** |
| 9 | 414 ± 5 | 178 ± 5 | 35.48 | 0.000 *** |
| 11 | 176 ± 6 | 144 ± 8 | 3.16 | 0.034 * |

Signif. Codes: * $p < 0.05$; *** $p < 0.001$.

3.1.6. Ethanol Stability

The data for ethanol stability are shown in Figure 5. The ethanol stability test is globally employed by the dairy industry to indirectly evaluate the heat stability of milk prior to pasteurization or commercial sterilization. In fact, as an organic solvent, ethanol reduces the dielectric constant of the aqueous medium, which in turn reduces the hydration of the external layer of casein micelles and diminishes their stability [31]. This effect is more pronounced when the pH is low and the concentration of ionic calcium is high [36,37]. For the milk sample at 3% (*w*/*w*) protein, ethanol stability was 92.5%, in accordance with Gulati et al. [41]. For other milk samples, the ethanol stability linearly decreased as a function of total protein concentration, as was observed for heat stability (Section 3.1.5).

In the case of the mixed protein suspensions, they exhibited higher ethanol stability than milk samples at intermediate protein concentrations (from 5 to 9% *w*/*w*). The increase in ethanol stability caused by the addition of pea protein might be due to the capacity for pea proteins to bind calcium. Silva et al. [12] showed that the concentration of ionic calcium diminished from 6 mM in casein micelles suspensions to approximately 2 mM when 6% (*w*/*w*) pea protein was added at pH 5.8. The authors concluded that the pea proteins acted as chelating agent for calcium ions. However, at 11% *w*/*w* protein, the stability of the mixed protein suspension was lower than that of milk samples at the same protein concentration. In effect, another factor that might have contributed to the reduction in ethanol stability for the more concentrated milk samples and mixed protein suspensions was the high concentration per se. It is well established that the increase in the concentration of colloidal particles enhances the probability of collisions due to Brownian motion. Therefore, the particles in concentrated protein suspensions, under destabilizing conditions, such as high temperature or high ethanol concentration, are more likely to interact with each other, increasing the aggregation and precipitation rates [31].

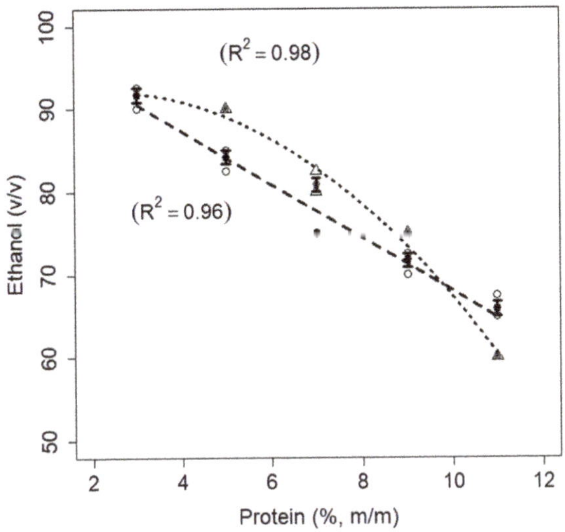


**Figure 5.** Relationship between ethanol stability and protein concentrations for skimmed milk (dashed line) and mixed protein (dotted line) samples. The figure shows the data points for milk (circles) and mixed samples (triangles), error bars (SE), fitted curves, and adjusted coefficients of determination ($R^2$).

### *3.2. Rheological Analysis of the Gels*

The experimental conditions for studying the acid gelling properties of milk and mixed suspensions were set in order to simulate industrial yogurt production taking place at 42 °C for 4 h. Therefore, Table 3 presents the overall characteristics of the gels, including the gel point (tg), the complex shear modulus (G*) at tg, the final G* and pH (obtained after 240 min of acidification), and induced syneresis. Figure 6 presents the evolution of G* against the reduced time (t/tg). Figure 7 assembles the evolution of G* and pH as a function of time after GDL addition for 240 min.

**Table 3.** Gel point ($t_g$), complex modulus (G*), and corresponding pH for mixed protein suspensions and milk samples.

| Protein Concentration % (*w/w*) | $t_g$ (sec.) | | G* at $t_g$ (Pa) | | Final G* (Pa) | | Final pH | | Induced Syneresis % (*w/w*) | |
|---|---|---|---|---|---|---|---|---|---|---|
| | Milk | Mixed | Milk | Mixed | Milk | Mixed | Milk | Mixed | Milk | Mixed |
| 3 | 240 ± 138 | - | 0.16 ± 0.18 | - | 227.6 ± 14.7 | - | 3.76 ± 0.04 | | 72.1 ± 1.9 | - |
| 5 | 642 ± 84 | 90 ± 48 | 0.27 ± 0.07 | 0.11 ± 0.03 | 688.5 ± 24.3 | 535.9 ± 23.7 | 4.19 ± 0.05 | 3.88 ± 0.04 | 62.0 ± 1.5 | 49.6 ± 6.7 |
| 7 | 834 ± 216 | 48 ± 6 | 0.32 ± 0.14 | 0.90 ± 0.12 | 954.6 ± 162.4 | 545.3 ± 34.9 | 4.54 ± 0.03 | 4.13 ± 0.05 | 32.9 ± 8.1 | 46.5 ± 0.45 |
| 9 | 1152 ± 126 | 54 ± 12 | 0.45 ± 0.16 | 2.77 ± 3.77 | 1054.8 ± 160.1 | 900.7 ± 298.4 | 4.73 ± 0.01 | 4.28 ± 0.04 | 2.3 ± 0.2 | 46.5 ± 8.7 |
| 11 | 1140 ± 288 | - | 0.60 ± 0.35 | - | 2076.9 ± 144.2 | 278.0 ± 68.4 | 4.83 ± 0.01 | 4.46 ± 0.03 | 1.5 ± 0.5 | 25.9 ± 1.1 |

### 3.2.1. Initial Gelation Stage

As shown in Table 3, the start of gelation (tg) for milk samples was delayed as protein content increased. Although the pH was not measured at tg, the increase in buffering capacity of the more concentrated milk samples can explain the longer periods required to achieve tg. In the present study, SMP was used to increase the protein content of milk samples. Thus, in addition to the buffering capacity of caseins and whey proteins, there is a significant contribution of milk salts, such as colloidal calcium phosphate and soluble phosphate, carbonate, and citrate [42]. Concomitantly, the values for G* at tg slightly increased in milk samples, which can be related to the higher amount of protein per unit volume, allowing for more protein–protein interactions at the beginning of the gelation process. When comparing milk and mixed suspensions, it was observed that the presence of pea proteins reduced tg. Similar results were noticed during the GDL-induced acidification of 14.8% *w/w* mixed (1:1 ratio) milk and pea protein suspensions at 30 °C for 20 h in the presence of 1.0% (*w/w*) NaCl [9]. The authors attributed the decrease in tg to a lower buffering capacity of suspensions containing pea proteins [9]. It is noteworthy that, here, the viscosity of mixed suspensions was higher than that of milk samples at comparable protein contents (Figure 2) and could partially contribute to the reduction in tg in the presence of pea proteins. Regarding the mixed suspension at 11% (*w/w*) protein (Table 3), the elastic modulus (G′) was higher than the viscous modulus (G″) during the initial moments of rheological analysis. Afterwards, it was not possible to determine tg; i.e., this sample was already gelled when the rheometer started to measure its rheological properties.

The variation in the complex shear modulus (G*) during gelation is shown against t/tg in Figure 6. Indeed, shifting the data along the horizontal axis eliminates the effect of the GDL hydrolysis kinetic, and the representation of G* against the reduced time (t/tg) makes it possible to differentiate the gels with respect to their structure [20]. Therefore, the results demonstrated that the structure formation of the gels containing pea proteins are distinctive of those formed by only milk proteins, and that the protein concentration had a significant effect on the gel structure (Figure 6). In general, for both systems, the increase in the protein concentration leads to more elastic gels. However, the rates at which the structure of pure milk gels evolve are much faster than those of the mixed samples (Figure 6) despite similar initial G* values at tg (Table 3). This kind of antagonistic effect between milk and plant proteins on the acid gelling properties was also reported by Ben-Harb et al. and Roesch et al. [9,43]. They suggested that the formation of different-sized protein aggregates during heating and acidification was correlated with the formation of large pores in mixed protein matrices, leading to more brittle acid gels. Moreover, according to Mession et al. [34], legumin aggregates formed during heating and acidification of pea proteins have a low potential for producing non-covalent interactions, which disturb the formation of more structured and elastic gels.

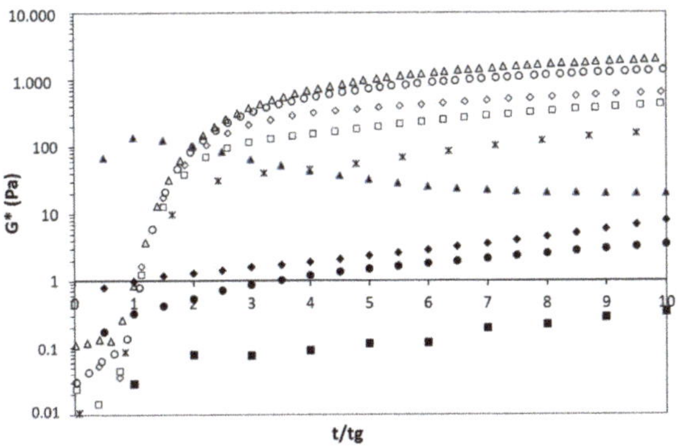


**Figure 6.** The evolution of the complex shear modulus (G*) as a function of reduced time (t/tg) for milk (open symbols) and mixed suspensions (closed symbols). The square, diamond, circle, and triangle markers represent the 5, 7, 9, and 11% (*w*/*w*) protein concentrations, respectively. The control milk sample is represented by the "x" symbol.

### 3.2.2. Evolution of Gelling Properties

The viscoelastic properties of milk and mixed gels obtained under 0.1% of oscillatory strain and at 0.1 Hz after 240 min following GDL addition are shown in Figure 7. The results for acid milk gels (Figure 7A,C) were in accordance with the literature for experiments performed under similar protein concentrations, pH, and temperature [44–46]. It should be highlighted that, during the production of fermented milks, the heat treatment applied to milk prior to acidification induces the formation of whey protein/casein aggregates that increases the firmness of the acid gels compared to those obtained from non-thermally treated milks [44,45]. In the present work, the acid milk gels became more elastic as the protein content increased (Figure 7A and Table 3), which is a direct result of a denser protein matrix in association with the heating step performed before acidification.

Concerning the mixed gels, the viscoelastic behavior during acidification (Figure 7B,D) depended on the pea protein concentrations, which corroborates previous studies [12,43]. For the same total protein content, the mixed gels were less elastic than pure milk gels (Figure 7A,C; Table 3). Although a certain degree of co-gelation between whey and pea proteins in the mixed gels can be assumed [47], this result can be attributed to the fact that plant and milk proteins present a competitive dynamic, forming independent networks that do not strongly interact in acidic conditions [40]. Furthermore, increasing the proportion of plant protein in acid mixed gels leads to the formation of large pores in the protein matrix, reducing the strength of the gels [9,43]. Thus, the formation of large pores can be related to the reduction in G′ for mixed gels at 11% (*w*/*w*) protein (Figure 7B and Table 3).

Another important aspect of protein gels is pH. Here, the evolution of pH as a function of acidification time is represented in Figure 7E,F for milk and mixed gels, respectively; final pH values are given in Table 3. For both samples, the higher the protein concentration, the higher the final pH. However, for equal protein concentrations, the milk gels presented higher values than the mixed ones. This effect can be attributed to a greater buffering capacity of milk proteins and salts (as discussed in Section 3.2.1). Additionally, except for the milk sample at 3% *w*/*w* and the mixed sample at 5% *w*/*w* protein, the pH values of the other samples were close to the average isoelectric point of milk and pea proteins (~4.5–4.6) [6,48]. This information is relevant insofar as the differences in the gelling properties between milk and mixed samples observed in the present work did not have a major influence on pH.

Lastly, the induced syneresis results are presented in Table 3. Syneresis is an important sensorial aspect of fermented milks that is related to the rearrangement and shrinking of the protein matrix under acidic conditions. With respect to the milk gels, the total protein concentration greatly influenced syneresis, varying from 72.1% at 3% (*w*/*w*) protein to 1.5%

at 11% ($w/w$) protein concentration. In this case, a highly hydrated and denser protein network, with a consequently low porosity, can explain the improvement in water holding capacity of the milk gels as a function of the total protein content [49]. Regarding the mixed gels, even though the protein concentration had a significant effect on induced syneresis, the results were more limited compared to the milk samples. Table 3 shows that the mixed gel at 5% ($w/w$) protein exhibited lower induced syneresis than milk gels at 3 and 5% ($w/w$) protein concentration. For mixed gels from 7 to 11% ($w/w$), the induced syneresis was higher than those of milk gels. As observed by Youssef et al. [50], a partial replacement of milk by pea proteins can intensify the syneresis in fermented milks obtained from diverse lactic acid bacteria. Overall, syneresis is not necessarily correlated with gel strength [49]. Additionally, its reduction as a function of the total protein content for both types of gels in the present study can be attributed to the increase in structural (nonsolvent) and immobilized water molecules that are present in open pores of the gelled matrix and between chains of pea and milk proteins [31].

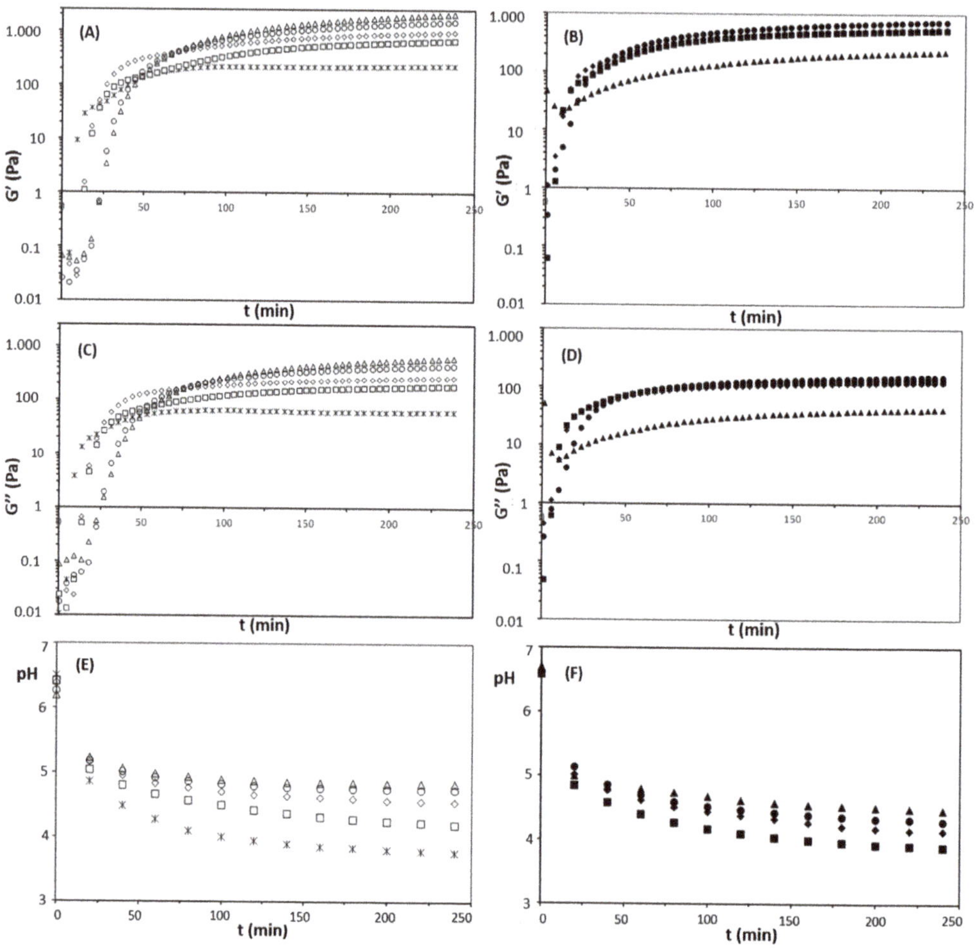


**Figure 7.** Evolution of elastic (G′) (**A**,**B**) and viscous shear moduli (G″) (**C**,**D**), and pH (**E**,**F**) as a function of acidification time (t) for milk (open symbols) and mixed suspensions (closed symbols). The square, diamond, circle, and triangle markers represent 5, 7, 9, and 11% ($w/w$) protein, respectively. The control milk sample is represented by the "x" symbol.

## 4. Conclusions

The experimental strategy adopted in the present work was established in a manner that can be replicated by the food industry to develop new products that combine high-quality plant and dairy proteins. The data analysis produced models that accurately describe the physicochemical behavior of milk and mixed protein suspensions, including food attributes that are important for industry. These include characteristics such as color, viscosity, and heat and ethanol stabilities. The equations that predict these characteristics are available in the Supplementary Material. In this regard, mixed suspensions were darker and presented higher pH, viscosity, and percentages of sedimentable material than milk samples. At the same time, heat and ethanol stabilities were similar for both systems and were reduced as a function of total protein concentration. The rheological analysis of the acid-induced gels made from milk and pea protein suspensions showed that the presence of pea proteins accelerates gel formation but weakens the structure of mixed gels compared to pure milk gels. Taken together, the results presented in this study can be useful for the current transition towards more plant-based foods and contributes to the comprehension of the functional properties of mixed dairy/plant protein systems, such as gel, emulsion, and foam formation. Notably, the insoluble fraction of PPI can be a sensorial problem to the development of a non-viscous mixed milk/pea protein beverage. Notwithstanding, for the production of gelled foods, this insoluble fraction can be integrated into the gel matrix without negative consequences to the sensorial acceptance of the final product.

**Supplementary Materials:** The following are available online at https://www.mdpi.com/article/10.3390/foods11101383/s1, Figure S1: Flow curves of milk (open symbols) and pea suspensions (closed symbols). The square, diamond, and circle markers represent the 5, 7 and 9% (*w*/*w*) protein, respectively. The control milk sample is represented by the "x" symbol. The Dashed Line represents the best fit Newtonian model, Figure S2: Flow curves of milk (open symbols) and pea suspensions (closed symbols). The triangle marker represents the 11% (*w*/*w*) protein. The dashed line represents the best fit power-law model, Table S1: Milk responses (y) in relation to the protein concentrations (x) for skim milk and mixed milk systems. Table shows the fitted models, adjusted coefficients of determination (R_adj^2), regression mean squared errors (σ̂^2), and F-tests for the lack of fit.

**Author Contributions:** I.C.O. conceived and designed the experiment, collected data, performed data analysis, and contributed to writing. I.E.d.P.F. conceived and designed the experiment, performed data analysis, and contributed to writing. F.C. conceived and designed the experiment, contributed to writing, and critically revised the manuscript. A.L.F.C. conceived and designed the experiment, collected data, performed data analysis, and contributed to writing. L.G.L.N. performed the analysis, contributed to writing, and critically revised the manuscript. A.F.d.C. conceived and designed the experiment, contributed to writing, and critically revised the manuscript. N.F.N.S. conceived and designed the experiment, performed the analysis, contributed to writing, critically revised the manuscript, and participated in funding acquisition. All authors have read and agreed to the published version of the manuscript.

**Funding:** This work was supported by Sao Paulo Research Foundation (Grant Number: 2020/08255-1) and the Brazilian National Council for Scientific and Technological Development (Grant Number: 121204/2019-7).

**Informed Consent Statement:** Not applicable.

**Data Availability Statement:** The research data are available at an open access repository: https://data.mendeley.com/datasets/bw4sxsyg37/1, accessed on 9 May 2022.

**Conflicts of Interest:** The authors declare no conflict of interest.

## References

1. Searchinger, T.; Waite, R.; Hanson, C.; Ranganathan, J. Creating a sustainable food future a creating a sustainable food future world resources report. *World Resour. Rep.* **2019**, *1*, 558.
2. Lu, Z.X.; He, J.F.; Zhang, Y.C.; Bing, D.J. Composition, physicochemical properties of pea protein and its application in functional foods. *Crit. Rev. Food Sci. Nutr.* **2020**, *60*, 2593–2605. [CrossRef] [PubMed]

3. Tulbek, M.C.; Lam RS, H.; Wang, Y.C.; Asavajaru, P.; Lam, A. Pea: A Sustainable Vegetable Protein Crop. In *Sustainable Protein Sources*; Nadathur, S.R., Wanasundara, J.P.D., Scanlin, L., Eds.; Academic Press: Cambridge, MA, USA, 2016.
4. Roland WS, U.; Pouvreau, L.; Curran, J.; Van De Velde, F.; De Kok, P.M.T. Flavor aspects of pulse ingredients. *Cereal Chem.* **2017**, *94*, 58–65. [CrossRef]
5. Wang, Y.; Guldiken, B.; Tulbek, M.; House, J.D.; Nickerson, M. Impact of alcohol washing on the flavour profiles, functionality and protein quality of air classified pea protein enriched flour. *Food Res. Int.* **2020**, *132*, 109085. [CrossRef] [PubMed]
6. Walstra, P.; Wouters, J.T.M.; Geurts, T.J. *Dairy Science and Technology*, 2nd ed.; Taylor & Francis: Boca Raton, FL, USA, 2006.
7. Minj, S.; Anand, S. Whey proteins and its derivatives: Bioactivity, functionality, and current applications. *Dairy* **2020**, *1*, 233–258. [CrossRef]
8. Ghanimah, M.A. Functional and technological aspects of whey powder and whey protein products. *Int. J. Dairy Technol.* **2018**, *71*, 454–459. [CrossRef]
9. Ben-Harb, S.; Panouillé, M.; Huc-Mathis, D.; Moulin, G.; Saint-Eve, A.; Irlinger, F.; Bonnarme, P.; Michon, C.; Souchon, I. The rheological and microstructural properties of pea, milk, mixed pea/milk gels and gelled emulsions designed by thermal, acid, and enzyme treatments. *Food Hydrocoll.* **2018**, *77*, 75–84. [CrossRef]
10. Kristensen, H.T.; Denon, Q.; Tavernier, I.; Gregersen, S.B.; Hammershøj, M.; Van Der Meeren, P.; Dewettinck, K.; Dalsgaard, T.K. Improved food functional properties of pea protein isolate in blends and co-precipitates with whey protein isolate. *Food Hydrocoll.* **2021**, *113*, 106556. [CrossRef]
11. Grasberger, K.F.; Gregersen, S.B.; Jensen, H.B.; Sanggaard, K.W.; Corredig, M. Plant-dairy protein blends: Gelation behaviour in a filled particle matrix. *Food Struct.* **2021**, *29*, 100198. [CrossRef]
12. Silva JV, C.; Cochereau, R.; Schmitt, C.; Chassenieux, C.; Nicolai, T. Heat-induced gelation of mixtures of micellar caseins and plant proteins in aqueous solution. *Food Res. Int.* **2019**, *116*, 1135–1143. [CrossRef]
13. Mintel. Global Food and Drink Trends 2030. 2020. Available online: https://downloads.mintel.com/private/vKd7N/files/817369/ (accessed on 9 May 2022).
14. Mession, J.L.; Roustel, S.; Saurel, R. Interactions in casein micelle—Pea protein system (Part II): Mixture acid gelation with glucono-δ-lactone. *Food Hydrocoll.* **2017**, *73*, 344–357. [CrossRef]
15. Castro, W.F.; Cruz, A.G.; Bisinotto, M.S.; Guerreiro, L.M.R.; Faria, J.A.F.; Bolini, H.M.A.; Cunha, R.L. Development of probiotic dairy beverages: Rheological properties and application of mathematical models in sensory evaluation. *J. Dairy Sci.* **2013**, *96*, 16–25. [CrossRef]
16. Silva, N.; Saint-Jalmes, A.; de Carvalho, A.F.; Gaucheron, F. Development of casein microgels from cross-linking of casein micelles by genipin. *Langmuir* **2014**, *30*, 10167–10175. [CrossRef]
17. Rossini, K.; Noreña, C.P.; Brandelli, A. Changes in the color of white chocolate during storage: Potential roles of lipid oxidation and non-enzymatic browning reactions. *J. Food Sci. Technol.* **2011**, *48*, 305–311. [CrossRef]
18. Casanova, F.; Nogueira Silva, N.F.; Gaucheron, F.; Nogueira, M.H.; Teixeira AV, N.C.; Perrone, I.T.; Alves, M.P.; Fidelis, P.C.; Carvalho, A.F.d. Stability of casein micelles cross-linked with genipin: A physicochemical study as a function of pH. *Int. Dairy J.* **2017**, *68*, 70–74. [CrossRef]
19. Huppertz, T.; Grosman, S.; Fox, P.F.; Kelly, A.L. Heat and ethanol stabilities of high-pressure-treated bovine milk. *Int. Dairy J.* **2004**, *14*, 125–133. [CrossRef]
20. Cavallieri AL, F.; da Cunha, R.L. The effects of acidification rate, pH and ageing time on the acidic cold set gelation of whey proteins. *Food Hydrocoll.* **2008**, *22*, 439–448. [CrossRef]
21. Gilbert, A.; Rioux, L.E.; St-Gelais, D.; Turgeon, S.L. Studying stirred yogurt microstructure using optical microscopy: How smoothing temperature and storage time affect microgel size related to syneresis. *J. Dairy Sci.* **2020**, *103*, 2139–2152. [CrossRef]
22. R Core Team. R: A Language and Environment for Statistical Computing (No. 2018). R Foundation for Statistical Computing. 2018. Available online: https://www.r-project.org/ (accessed on 9 May 2022).
23. Oliveira, I.; Ferreira, I.; Silva, N. Colloidal Stability of Mixed Protein Systems—Milk and Isolated Pea Protein Suspensions. Version 1. 2020. Available online: https://data.mendeley.com/datasets/bw4sxsyg37/1 (accessed on 9 May 2022).
24. Jiang, S.; Ding, J.; Andrade, J.; Rababah, T.M.; Almajwal, A.; Abulmeaty, M.M.; Feng, H. Modifying the physicochemical properties of pea protein by pH-shifting and ultrasound combined treatments. *Ultrason. Sonochem.* **2017**, *38*, 835–842. [CrossRef]
25. Christiansen, M.V.; Dave, A.; Skibsted, L.H.; Ahrné, L. Functional properties of skim milk concentrates produced by reverse osmosis filtration and reconstituted commercial powders. *Int. Dairy J.* **2022**, *126*, 105225. [CrossRef]
26. Štreimikytė, P.; Keršienė, M.; Eisinaitė, V.; Jasutienė, I.; Lesauskaitė, V.; Damulevičienė, G.; Knašienė, J.; Leskauskaitė, D. Formulating protein-based beverages for the dysphagia diets of the elderly: Viscosity, protein quality, in vitro digestion, and consumers acceptability. *J. Sci. Food Agric.* **2020**, *100*, 3895–3901. [CrossRef] [PubMed]
27. Cheng, N.; Barbano, D.M.; Drake, M.A. Effect of pasteurization and fat, protein, casein to serum protein ratio, and milk temperature on milk beverage color and viscosity. *J. Dairy Sci.* **2019**, *102*, 2022–2043. [CrossRef] [PubMed]
28. Mercan, E.; Sert, D.; Akın, N. Effect of high-pressure homogenisation on viscosity, particle size and microbiological characteristics of skim and whole milk concentrates. *Int. Dairy J.* **2018**, *87*, 93–99. [CrossRef]
29. Bogahawaththa, D.; Chau, N.H.B.; Trivedi, J.; Dissanayake, M.; Vasiljevic, T. Impact of controlled shearing on solubility and heat stability of pea protein isolate dispersed in solutions with adjusted ionic strength. *Food Res. Int.* **2019**, *125*, 108522. [CrossRef]

30. McPhie, M.G.; Daivis, P.J.; Snook, I.K. Viscosity of a binary mixture: Approach to the hydrodynamic limit. *Phys. Rev. E Stat. Nonlinear Soft Matter Phys.* **2006**, *74*, 031201. [CrossRef]
31. Walstra, P. *Physical Chemistry of Foods*; CRC Press: New York, NY, USA, 2002.
32. Lam, A.C.Y.; Can Karaca, A.; Tyler, R.T.; Nickerson, M.T. Pea protein isolates: Structure, extraction, and functionality. *Food Rev. Int.* **2018**, *34*, 126–147. [CrossRef]
33. Saldanha do Carmo, C.; Silventoinen, P.; Nordgård, C.T.; Poudroux, C.; Dessev, T.; Zobel, H.; Holtekjølen, A.K.; Draget, K.I.; Holopainen-Mantila, U.; Knutsen, S.H.; et al. Is dehulling of peas and faba beans necessary prior to dry fractionation for the production of protein- and starch-rich fractions? Impact on physical properties, chemical composition and techno-functional properties. *J. Food Eng.* **2020**, *278*, 109937. [CrossRef]
34. Mession, J.L.; Roustel, S.; Saurel, R. Interactions in casein micelle—Pea protein system (part I): Heat-induced denaturation and aggregation. *Food Hydrocoll.* **2017**, *67*, 229–242. [CrossRef]
35. Dumpler, J.; Kulozik, U. Heat stability of concentrated skim milk as a function of heating time and temperature on a laboratory scale–Improved methodology and kinetic relationship. *Int. Dairy J.* **2015**, *49*, 111–117. [CrossRef]
36. Holt, C.; Carver, J.A.; Ecroyd, H.; Thorn, D.C. Invited review: Caseins and the casein micelle: Their biological functions, structures, and behavior in foods. *J. Dairy Sci.* **2013**, *96*, 6127–6146. [CrossRef]
37. De Kruif, C.G.; Holt, C. Casein micelle structure, functions and interactions. In *Advanced Dairy Chemistry Volume 1 Proteins*, 3rd ed.; Fox, P.F., McSweeney, P.L.H., Eds.; Springer: New York, NY, USA, 2003; pp. 233–276.
38. Horne, D.S.; Muir, D.D. Alcohol and Heat Stability of Milk Protein. *J. Dairy Sci.* **1990**, *73*, 3613–3626. [CrossRef]
39. Bogahawaththa, D.; Bao Chau, N.H.; Trivedi, J.; Dissanayake, M.; Vasiljevic, T. Impact of selected process parameters on solubility and heat stability of pea protein isolate. *LWT* **2019**, *102*, 246–253. [CrossRef]
40. Schmitt, C.; Silva, J.V.; Amagliani, L.; Chassenieux, C.; Nicolai, T. Heat-induced and acid-induced gelation of dairy/plant protein dispersions and emulsions. *Curr. Opin. Food Sci.* **2019**, *27*, 43–48. [CrossRef]
41. Gulati, A.; Hennessy, D.; O'Donovan, M.; McManus, J.J.; Fenelon, M.A.; Guinee, T.P. Dairy cow feeding system alters the characteristics of low-heat skim milk powder and processability of reconstituted skim milk. *J. Dairy Sci.* **2019**, *102*, 8630–8647. [CrossRef]
42. Salaün, F.; Mietton, B.; Gaucheron, F. Buffering capacity of dairy products. *Int. Dairy J.* **2005**, *15*, 95–109. [CrossRef]
43. Roesch, R.; Juneja, M.; Monagle, C.; Corredig, M. Aggregation of soy/milk mixes during acidification. *Food Res. Int.* **2004**, *37*, 209–215. [CrossRef]
44. Anema, S.G. Effect of milk solids concentration on whey protein denaturation, particle size changes and solubilization of casein in high-pressure-treated skim milk. *Int. Dairy J.* **2008**, *18*, 228–235. [CrossRef]
45. Mahomud, M.S.; Katsuno, N.; Zhang, L.; Nishizu, T. Physical, rheological, and microstructural properties of whey protein enriched yogurt influenced by heating the milk at different pH values. *J. Food Process. Preserv.* **2017**, *41*, e13236. [CrossRef]
46. Li, Y.; Corredig, M. Acid induced gelation behavior of skim milk concentrated by membrane filtration. *J. Texture Stud.* **2020**, *51*, 101–110. [CrossRef]
47. Wong, D.; Vasanthan, T.; Ozimek, L. Synergistic enhancement in the co-gelation of salt-soluble pea proteins and whey proteins. *Food Chem.* **2013**, *141*, 3913–3919. [CrossRef]
48. Boye, J.I.; Aksay, S.; Roufik, S.; Ribéreau, S.; Mondor, M.; Farnworth, E.; Rajamohamed, S.H. Comparison of the functional properties of pea, chickpea and lentil protein concentrates processed using ultrafiltration and isoelectric precipitation techniques. *Food Res. Int.* **2010**, *43*, 537–546. [CrossRef]
49. Meletharayil, G.H.; Patel, H.A.; Huppertz, T. Rheological properties and microstructure of high protein acid gels prepared from reconstituted milk protein concentrate powders of different protein contents. *Int. Dairy J.* **2015**, *47*, 64–71. [CrossRef]
50. Yousseef, M.; Lafarge, C.; Valentin, D.; Lubbers, S.; Husson, F. Fermentation of cow milk and/or pea milk mixtures by different starter cultures: Physico-chemical and sensorial properties. *LWT Food Sci. Technol.* **2016**, *69*, 430–437. [CrossRef]

foods

MDPI

*Article*

# Preparation and Characterization of Solid Acid Catalysts for the Conversion of Sesamin into Asarinin in Sesame Oil

**Qiong Yu, Xue-De Wang *, Hua-Min Liu and Yu-Xiang Ma**

Institute of Special Oilseed Processing and Technology, College of Food Science and Engineering, Henan University of Technology, Zhengzhou 450001, China; zqymmh@163.com (Q.Y.); liuhuamin5108@163.com (H.-M.L.); myx366@163.com (Y.-X.M.)

* Correspondence: 13903865584@126.com or wangxuede1962@126.com; Tel.: +86-037-167-758-025

**Abstract:** Asarinin, an isomer of sesamin, has attracted attention because it has stronger biological properties than sesamin. The research on the conversion of sesamin into asarinin is limited. In this study, solid acid catalysts were screened and applied to promote the conversion of sesamin into asarinin in sesame oil. The results showed that citric acid loaded on zeolite beta (CTAH) was the optimal catalyst for asarinin production among the prepared catalysts. Characterization showed that CTAH had the greatest pore volume, largest surface area and strongest acid content. Response surface methodology (RSM) was applied to optimize the reaction conditions for asarinin yield using CTAH. The optimal reaction conditions were as follows: temperature, 85 °C; time, 2.7 h; catalyst amount, 1.6%. The predicted and experimental values of asarinin yield were 50.79 and 51.80 mg/100 g, respectively. The peroxide value and color in sesame oil samples treated with CTAH were clearly improved. In short, CTAH is a solid acid catalyst with potential application in the industrial conversion of sesamin into asarinin and in the improvement of sesame oil.

**Keywords:** asarinin; solid acid catalyst; characterization; optimization; physicochemical properties

**Citation:** Yu, Q.; Wang, X.-D.; Liu, H.-M.; Ma, Y.-X. Preparation and Characterization of Solid Acid Catalysts for the Conversion of Sesamin into Asarinin in Sesame Oil. *Foods* **2022**, *11*, 1225. https://doi.org/10.3390/foods11091225

Academic Editors: Federico Casanova and Michela Verni

Received: 12 April 2022
Accepted: 22 April 2022
Published: 24 April 2022



## 1. Introduction

There is growing interest in lignans because they have many important physiological properties, including anti-inflammatory, antineoplastic and hypocholesterolemic effects [1–3]. Among the various lignans found in *Sesamum indicum* L., asarinin, an isomer of sesamin and a fat-soluble furanofuran type lignan, has particularly attracted attention due to its plentiful biological properties [4–6]. Asarinin has been shown to reduce basal and TNF-α-induced vascular smooth muscle cell migration and proliferation, thus preventing atherosclerosis [5]. It has been reported that asarinin significantly inhibits the decrease in SOD activity caused by 6-hydroxydopamine in PC12 cells [7]. Asarinin also has been shown to inhibit mast cell activation by preventing the phosphorylation of Src family kinases [6].

Asarinin has stronger biological activity than sesamin, particularly in promoting fatty acid oxidizing liver enzyme gene expression and decreasing cholesterol. The hepatic concentrations of phospholipid were significantly higher in rats fed asarinin (42.2 ± 0.9 μmol/g) than in animals fed sesamin (37.9 ± 1.2 μmol/g) [8]. Kuo et al. found that, at the concentration of 32.25 μM, asarinin had a superoxide anion scavenging activity rate of 20.4%, whereas the scavenging activity rate of sesamin was 12.1% [9]. It was reported that asarinin showed remarkably stronger antioxidant activity in DPPH ($IC_{50}$ = 0.21 mg/mL) radical scavenging activities than sesamin ($IC_{50}$ = 3.05 × $10^{10}$ mg/mL) [10]. During the acid bleaching and deodorization of sesame oil processing, one of the major lignans, sesamin, is converted to asarinin [11,12]. The conversion of sesamin to asarinin is reversible; the ratio of sesamin and asarinin remains more or less the same as the reaction proceeds [11,13,14].

The mechanism of conversion of sesamin into asarinin is displayed in Figure 1 [11,13–15]. Different catalysts have been used for the conversion of sesamin to asarinin. Tsai et al. reported that no asarinin was observed, regardless of the ratio, in a mixture of sesamin and sesamolin treated with sulfuric acid or formic acid in anhydrous toluene [11]. Wang et al. reported that the highest percentage of asarinin (53.0%) was achieved using hydrochloric acid as a catalyst in an ethanol–sesamin system. They also demonstrated that, in an ethanol–sesamin system, nearly 60.0% asarinin was produced when hydrochloric acid and ferric chloride were added in a certain proportion [16]. The content of asarinin reached 2.06 μg/mL when sesame samples were treated with hydrochloric acid [14]. Although these homogenous catalysts exhibit good activity, they have several limitations, such as the difficulty of catalyst removal and corrosion [17]. In contrast, solid acid catalysts have remarkable advantages including reusability, separability, and efficiency in the conversion of sesamin to asarinin [18]. Zeolite is a widely reported solid acid catalyst. Zeolite beta, having high acidity and pore structures, exhibits a high catalytic activity among various zeolites. Many recent studies have reported using modified zeolite for acid-catalyzed reactions such as esterification of free fatty acids and isomerization of alkanes [18,19].

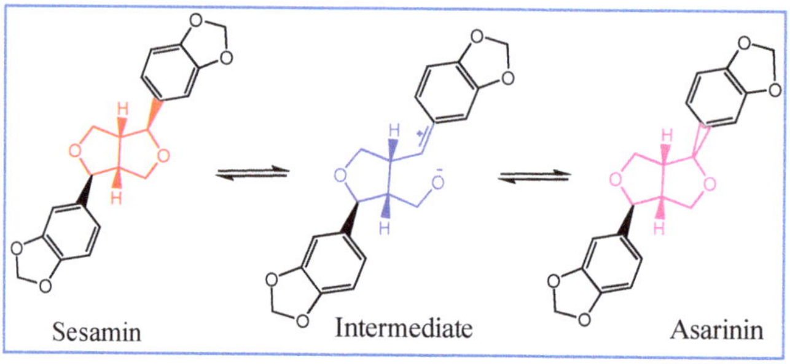


**Figure 1.** The mechanism of conversion of sesamin into asarinin.

At present, the research on the solid acid-catalyzed conversion of sesamin into asarinin is limited. As a result, this study represents exploration of a new catalyst which may be expected to be active in converting sesamin to asarinin in cold-pressed sesame oil, thereby enhancing the nutritional value of sesame oil. The best of the catalysts, namely citric acid loaded on zeolite beta (CTAH), was then investigated in depth. The parameters of temperature, time, catalyst amount and the concentration of citric acid loaded on zeolite beta were evaluated, and their effects on asarinin yield and the physicochemical properties of sesame oil were also assessed.

## 2. Materials and Methods

### *2.1. Materials and Chemicals*

Sesamin with a purity exceeding 98.0% was purchased from Shanghai Macklin Biochemical Co., Ltd. (Shanghai, China). Asarinin with a purity exceeding 98.0% was obtained from Solarbio® Life Sciences Co., Ltd. (Beijing, China). All standards were stored at 4 °C in darkness. The hydrogen type of zeolite beta (Hβ) with an Si/Al ratio of 25 was obtained from the Nankai University catalyst plant (Tianjin, China). Citric acid (CTA) was obtained from Shanghai Yuanye Biological Technology Co., Ltd. (Shanghai, China). The analytical grade chemicals including phosphotungstic acid (PTA), phosphomolybdic acid (PMA), ferric sulfate (FCS), ferrous sulfate (FRS), and ferric chloride hexahydrate (FCH) were obtained from Tianjin Kermel Chemical Reagents Co., Ltd. (Tianjin, China). Cold-pressed sesame oil was produced in a hydraulic press (Bafang Ltd., model XL-600, Suzhou, China).

### *2.2. Synthesis and Characterizaiton of Catalysts*

#### 2.2.1. Catalyst Synthesis

Various solid acid catalysts were synthesized according to the impregnation method using Hβ as carrier material; the prepared catalysts were denoted as FCSH, FRSH, FCHH, PMAH, PTAH, and CTAH [20,21]. The impregnation procedure was carried out by soaking about 10.0 g of Hβ in 100 mL 0.1 mol/L reagent solution and continuously stirring at room temperature for 16.0 h. After the impregnation process was finished, the mixture was centrifuged. The samples were then dried, except CTA, which had to be first washed with distilled water to neutral and then dried. All the solids were dried at 65 °C, then calcined at 500 °C for 4.0 h. BHβ was prepared by calcinating Hβ directly in a muffle furnace. DHβ was obtained by modifying Hβ with 0.1 mol/L NaOH based on the literature [22]. DBHβ was prepared by modifying BHβ in the same way [19]. The prepared catalysts together with their abbreviations are shown in Table 1.

**Table 1.** The various solid acid catalysts prepared in this study are listed in the table with their corresponding serial number and abbreviations.

| Serial Number | Abbreviations | Sample Name |
|---|---|---|
| 1 | Hβ | Hydrogen type of zeolite beta |
| 2 | PMA | Phosphomolybdic acid |
| 3 | FCH | Ferric chloride hexahydrate |
| 4 | PTA | Phosphotungstic acid |
| 5 | DHβ | Alkali modified Hβ |
| 6 | BHβ | Calcined Hβ |
| 7 | DBHβ | Alkali modified BHβ |
| 8 | PMAH | Phosphomolybdic acid loaded on Hβ |
| 9 | CTAH | Citric acid loaded on Hβ |
| 10 | FCHH | Ferric chloride hexahydrate loaded on Hβ |
| 11 | PMAB | Phosphomolybdic acid loaded on BHβ |
| 12 | CTAB | Citric acid loaded on BHβ |
| 13 | FCHB | Ferric chloride hexahydrate loaded on BHβ |
| 14 | PTAH | Phosphotungstic acid loaded on Hβ |
| 15 | FRSH | Ferrous sulfate loaded on Hβ |
| 16 | FCSH | Ferric sulfate loaded on Hβ |

2.2.2. Catalyst Characterization

The Brunauer–Emmett–Teller (BET) surface area measurements of various catalysts were performed using a 3-Flex automatic specific surface area analyzer (Micromeritics, USA) after removing $N_2$. The pore size and pore volume were calculated by the BET method using experimental points, and pore size was calculated by Barrett–Joyner–Halenda (BJH) analysis [23]. $NH_3$-TPD profiles ($NH_3$ temperature programmed desorption) of different catalysts samples were recorded on a chemisorption analyzer (Xian Quan, TP 5080, Tianjin, China) [24]. Samples were pre-treated at 500 °C with a heating ramp: 10 °C/min by passing high purity helium for 1 h, and then cooled to 120 °C. Then, purity anhydrous ammonia was introduced to the saturation of the sample and the TPD analysis was performed from room temperature to 600 °C at a heating rate of 10 °C/min.

### 2.3. *Effect of Catalyst on Asarinin Yield*

Prior to response surface methodology (RSM), a serious of experiments were used to select the ranges for independent variables. Various reaction temperatures (30, 40, 60, 80, 100, 120 and 140 °C) were used when reaction time, catalyst amount ($w_{catalyst}/w_{oil}$) and the concentration of citric acid loaded on Hβ were fixed at 2.0 h, 1.0% and 0.4 mol/L, respectively. Reaction times of 0.5, 1.0, 1.5, 2.0, 2.5, 3.0 and 4.0 h were tested, when reaction temperature, catalyst amount ($w_{catalyst}/w_{oil}$) and the concentration of citric acid loaded on Hβ were fixed at 80 °C, 1.0% and 0.4 mol/L, respectively. As for catalyst amount ($w_{catalyst}/w_{oil}$), 0.2, 0.4, 0.6, 0.8, 1.0, 1.2, 1.4 and 1.6% were used when reaction time, temperature and the concentration of citric acid loaded on Hβ were fixed at 2.0 h, 80 °C and 0.4 mol/L, respectively. Finally, the concentrations of citric acid loaded on Hβ of 0.1, 0.2, 0.4, 0.6, 0.8, 1.0 and 1.5 mol/L were used when reaction time, temperature and catalyst amount ($w_{catalyst}/w_{oil}$) were fixed at 2.0 h, 80 °C and 1.0%, respectively.

The statistical analysis and optimization of data were performed by the Design-expert 8.0 statistical software package [25–27]. The optimal reaction conditions of asarinin yield were obtained through the Box–Behnken design.

### 2.4. Preparation of the Tested Sesame Oil Samples

All the tested sesame oil samples were prepared under the optimal reaction conditions as indicated by RSM. The oil samples were cooled at room temperature naturally, and then centrifuged (4000 r/min, 5 min), filtered and stored at −20 °C. Cold-pressed sesame oil without any treatment was recorded as CSO, and cold-pressed sesame oil samples treated with citric acid and with citric acid loaded on Hβ were denoted as sesame oil sample CTA and sesame oil sample CTAH, respectively.

### 2.5. Effect of Catalyst on the Physicochemical Properties of Sesame Oil

#### 2.5.1. Acid Value (AV), Peroxide Value (POV) and Color

Peroxide value and acid value were measured following previously reported methods [28]. The color of sesame oil was detected by a PFX-I series spectro-colorimeter (Lovibond Co., Ltd., Amesbury, UK). The optical path was 25.4 mm.

#### 2.5.2. Determination of Lignans

Lignans were measured based on the method published in the literature [14,29]. The determinations of sesamin and asarinin were performed using the high-performance liquid chromatography equipped with ultraviolet-visible detector (HPLC-UV) method; HPLC-UV chromatograms of cold-pressed sesame oil (CSO) and CSO treated with citric acid loaded on Hβ (CTAH) are displayed in Figure 2.

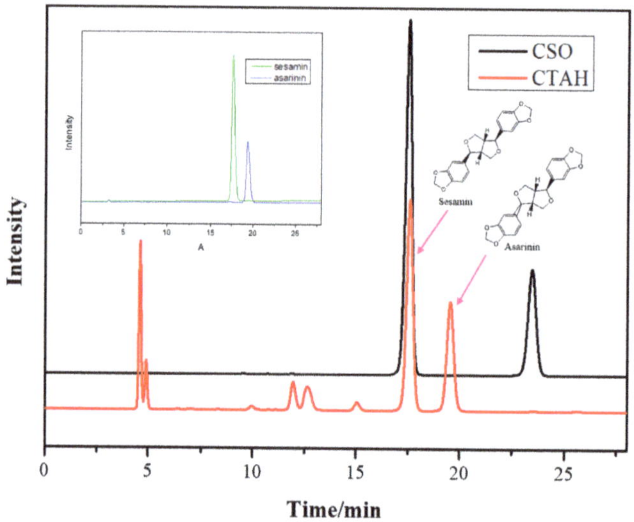


**Figure 2.** HPLC-UV chromatogram of the standard substance (A); cold-pressed sesame oil (CSO) and CSO were treated with citric acid loaded on Hβ (CTAH). Detection wavelength, 287 nm. Retention times and chemical structures of sesamin and asarinin in sesame oil are displayed.

### 2.6. Statistical Analysis

All measurements were conducted in triplicate. Probability values were considered statistically significant with one-way ANOVA using the SPSS package version 19.0. All significant differences were assessed by Duncan's test. Values are denoted as means ± standard deviations.

## 3. Results and Discussion

### 3.1. Catalyst Screening

The sample names of the various types of catalysts prepared are shown in Table 1. The effects of different types of catalysts on the conversion of sesamin into asarinin are displayed in Figure 3. The reaction conditions were as follows: temperature, 80 °C; time, 2 h; catalyst amount, 1.0%. As illustrated in Figure 3, there was no significant difference among the sesame oils treated with Hβ (No. 1 in Figure 3), DHβ (No. 5 in Figure 3) and BHβ (No. 6 in Figure 3), indicating that the modification of Hβ had no effect on the production of asarinin. Compared with the pure reagents PMA (No. 2 in Figure 3) and PTA (No. 4 in Figure 3), the influences of PMAH (No. 8 in Figure 3) and PTAH (No. 14 in Figure 3) on the conversion of sesamin in CSO were significantly reduced. Asarinin was not detected in the sesame oil samples treated with FRS and FCS, which are not shown in this figure. Figure 1 shows that the maximum amount of asarinin was produced when the sesame oil sample treated with FCH (No. 3 in Figure 3), and the next was the sesame oil sample treated with CTAH (No. 9 in Figure 3), demonstrating that FCH and CTAH could catalyze the production of asarinin. Apparently, in the CSO samples treated with FCSH (No. 16 in Figure 3), FCHH (No. 10 in Figure 3), FRSH (No. 15 in Figure 3) and PMAH (No. 8 in Figure 3), very little asarinin was produced at the reaction temperature of 80 °C. The reason for this may be that the active sites of the Hβ were occupied by these chemical regents [30,31]. In other words, the catalytic effect could not be enhanced with FCH, PMA or PTA loaded on Hβ. The color of the sesame oil sample treated with FCH was dark brown. Although this sample showed excellent catalysis, this color of sesame oil in food products would not be accepted by consumers. Thus, CTAH was chosen as the best catalyst for further study.

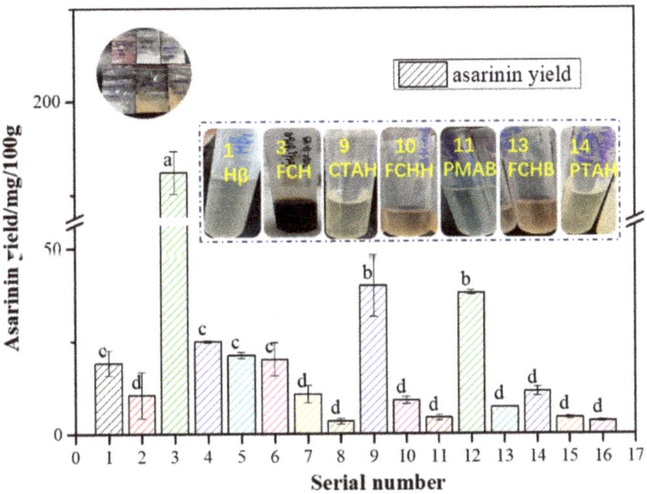


**Figure 3.** Effects of various catalysts on the yield of asarinin under the following reaction conditions: reaction temperature, 80 °C; reaction time, 2 h; catalyst amount, 1.0%. Different letters above the bars show significant differences among various treatments. The various solid acid catalysts represented by the serial number in the Figure are shown in Table 1.

### 3.2. Catalyst Characterization

#### 3.2.1. Nitrogen Adsorption and Desorption Analysis

The structure properties including surface area, pore size and pore volume of different catalysts were obtained from nitrogen ($N_2$) adsorption and desorption isotherms (Figure 4a–d and Table 2). The catalysts Hβ, CTAH, PTAH and FCHH exhibited a combination of type I and type IV isotherms, on the basis of the classification of international union of pure and applied chemistry (IUPAC). It is clear that, at lower relative pressures ($P/P_0$ = 0.0–0.1), the catalysts exhibited isotherm type I, which corresponds to microporous material. A hysteresis loop was observed at the upper section of isotherm over the relative pressure range from 0.6 to 0.9, indicating that there were mesopores in the solid. This is further confirmed in the BET analysis results in Table 2. The surface area, pore size, micropore volume and mesopore volume of CTAH were the biggest among the selected catalysts because the solid contains both micropores and mesopores. Similar phenomena have been reported previously [32,33].

**Table 2.** Surface and acidic properties of various solid acid catalysts.

| | | Hβ | CTAH | PTAH | FCHH |
|---|---|---|---|---|---|
| Surface area ($m^2/g$) | | 489.39 | 537.99 | 360.61 | 435.70 |
| External surface area ($m^2/g$) | | 134.90 | 178.27 | 96.76 | 129.29 |
| Pore size (nm) | | 4.27 | 4.57 | 4.22 | 4.04 |
| Pore volume ($cm^3/g$) | | 0.52 | 0.61 | 0.38 | 0.44 |
| Micropore volume ($cm^3/g$) | | 0.17 | 0.18 | 0.13 | 0.15 |
| Mesopore volume ($cm^3/g$) | | 0.35 | 0.44 | 0.25 | 0.29 |
| $NH_3$ acidity (mmol/g) | weak | 0.21 | 0.17 | 0.20 | 0.09 |
| | strong | 0.06 | 0.10 | 0.10 | 0.00 |
| | strong/weak | 0.29 | 0.60 | 0.49 | 0.04 |
| Acid strength (°C) | weak | 244 | 251 | 256 | 261 |
| | strong | 351 | 351 | 351 | 600 |

Notes: Hβ: Hydrogen type of zeolite beta; CTAH: Citric acid loaded on Hβ; PTAH: Phosphotungstic acid loaded on Hβ; FCHH: Ferric chloride hexahydrate loaded on Hβ.

### 3.2.2. Acidic Properties

Figure 4e shows that two $NH_3$ desorption peaks are present in the $NH_3$-TPD profiles of Hβ, CTAH, PTAH and FCHH catalysts. According to the literature [19,34], the strong peaks in the 150–300 °C range are assigned to the weak acid sites in the catalysts, and peaks in the 300–500 °C range can be ascribed to the strong acid sites. The acid amounts and strengths of different catalysts are displayed in Table 2. The number of acid sites of the samples was determined by the amount of desorbed ammonia, and acidic strength was determined by desorption temperature [21,34]. Compared with Hβ, the number of weak acid sites of CTAH, PTAH and FCHH was less by 19.43%, 4.27% and 56.87%, respectively; the number of strong acid sites of CTAH and PTAH was greater by 67.21% and 60.66%, respectively. The order of the ratio of strong/weak was: CTAH (0.60) > PTAH (0.49) > Hβ (0.29) > FCHH (0.04). The results indicate that CTA and PTA modification favors the formation of strong acid sites [20,21]. In general, the acid strength of CTAH was strongest among the prepared catalysts.

## 3.3. *Effect of Catalyst on Asarinin Yield*

According to the characterization and these preliminary tests of various solid acid catalysts, CTAH showed the best compromise between activity and structural properties; it had: (1) the highest pore volume (0.61 $cm^3/g$) and surface area (537.99 $m^2/g$); (2) the highest strong acidity (0.10 mmol/g); and (3) high asarinin yield (39.79 mg/100 g). The reason for the good catalytic effect may be that the change in catalyst structure makes it more suitable for the conversion of sesamin into asarinin. Consequently, the effects of CTAH on the reaction conditions for asarinin yield were further investigated.

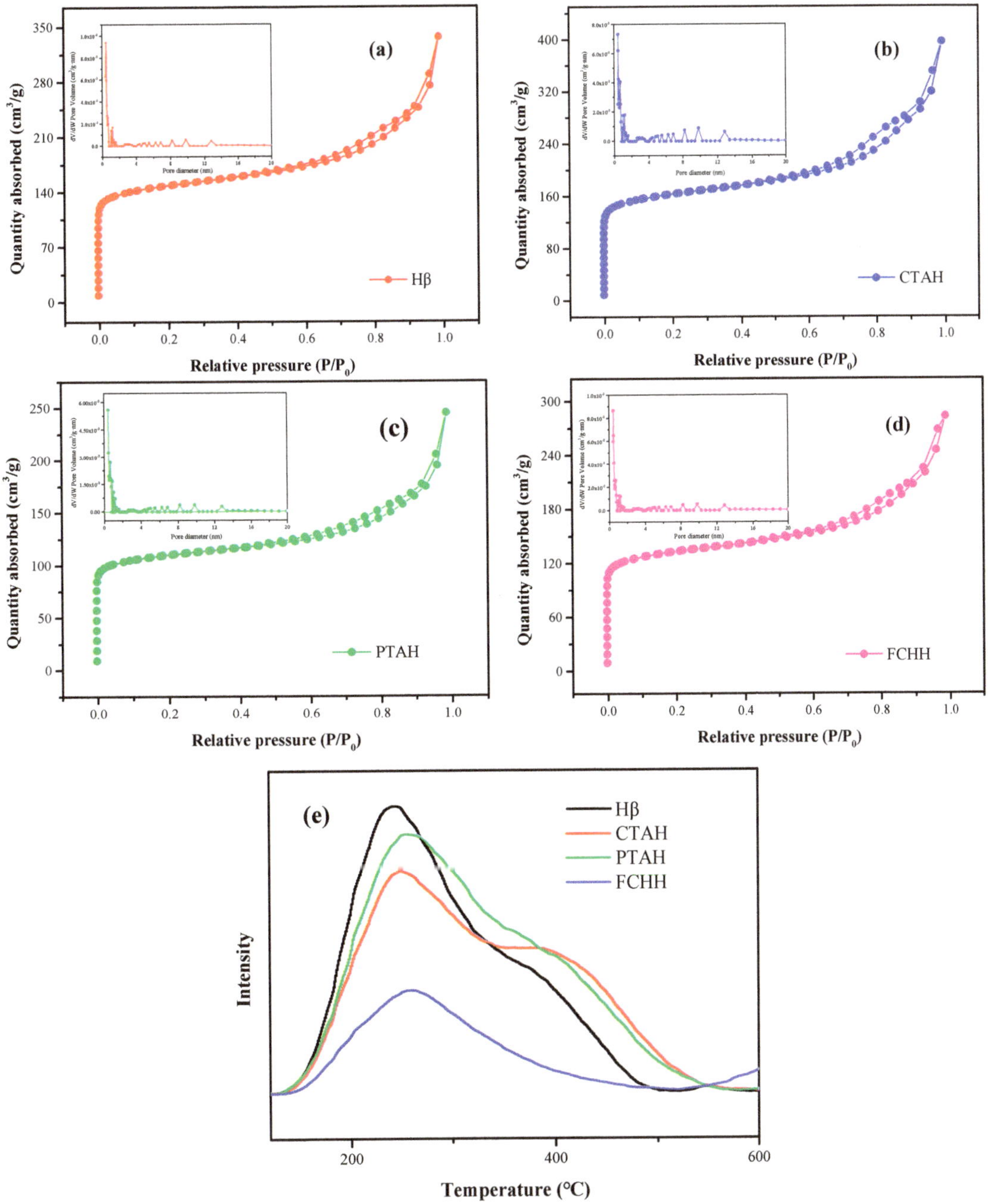


**Figure 4.** (**a**–**d**): $N_2$ adsorption and desorption isotherms and pore size distribution of various catalysts; (**e**) $NH_3$-TPD curves of various catalysts. Hβ: Hydrogen type of zeolite beta; CTAH: Citric acid loaded on Hβ; PTAH: Phosphotungstic acid loaded on Hβ; FCHH: Ferric chloride hexahydrate loaded on Hβ.

### 3.3.1. Single Factors Analysis

The effects of reaction temperature, time, catalyst amount and the concentration of CTA loaded on Hβ on asarinin production are shown in Figure 5a–d. It can be seen that the content of asarinin increased gradually with increases in time, temperature and catalyst amount. Different concentrations of CTA loaded on Hβ had little effect on asarinin yield.

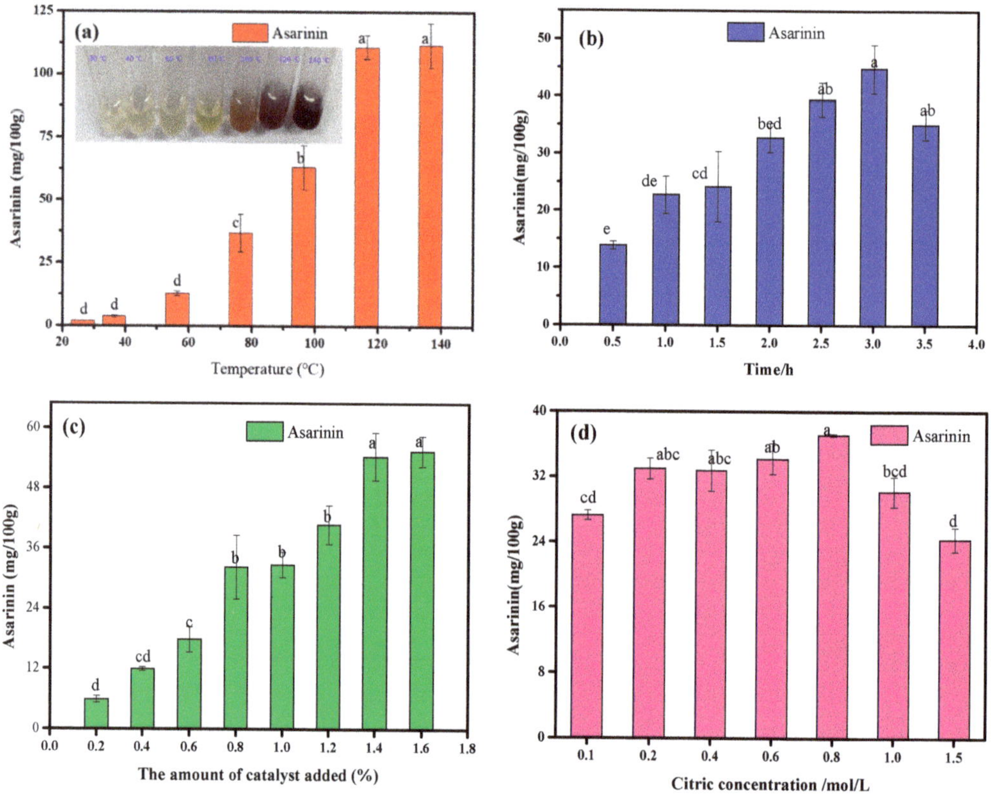


**Figure 5.** Effects of various reaction conditions on asarinin production: (**a**) catalyst amount, 1.0%; reaction time, 2.0 h; concentration of citric acid loaded on Hβ, 0.4 mol/L; (**b**) catalyst amount, 1.0%; reaction temperature, 80 °C; concentration of citric acid loaded on Hβ, 0.4 mol/L; (**c**) reaction temperature, 80 °C; reaction time, 2.0 h; concentration of citric acid loaded on Hβ, 0.4 mol/L; (**d**) catalyst amount, 1.0%; reaction temperature, 80 °C; reaction time, 2.0 h. Different letters above the bars show significant differences among various conditions.

As shown in Figure 5a, the yield of asarinin increased within the reaction temperature range from 30 to 140 °C. In terms of the sensory qualities of sesame oil, flavor and color underwent significant deterioration at the temperatures above 100 °C (see the picture in Figure 5a). Therefore, 80 °C was selected as the appropriate temperature for further study. The production of asarinin increased with reaction time from 0.5 to 3.0 h, and then decreased after 3.0 h (Figure 5b). The yields of asarinin were 39.44 and 44.87 mg/100 g at 2.5 and 3.0 h, respectively, and there was no significant difference between them. Considering the cost of time and energy, 2.5 h was chosen as the time for further tests. The variation in asarinin yield according to catalyst amount is displayed in Figure 5c. There was no significant difference in asarinin yield when catalyst amount varied between 1.4% and 1.6%. Considering the yield of asarinin, adding 1.6% catalyst was considered the better choice.

The effects of different concentrations of CTA loaded on Hβ on the content of asarinin yield are displayed in Figure 4d. There was no significant difference in the production of asarinin when the concentration of citric acid was between 0.2 and 0.8 mol/L. As the concentration of citric acid increased, the yield of asarinin decreased. The reason for this was related to the activity of catalyst. According to the literature, high concentration of citric acid will lead to the collapse of the Hβ framework [20]. There was no significant difference in asarinin yield among different concentrations of citric acid loaded on Hβ (Figure 5d). Therefore, this factor was not taken into account in the following selection of response surface factor.

### 3.3.2. Optimized Experiment

The yield of asarinin depends on reaction parameters such as reaction temperature, time and type of catalyst. The ranges of variables were decided on the basis of single factor experiments as reported above. In the present study, three factors (catalyst amount, reaction temperature and reaction time) were selected as the most significant in determining asarinin yield for RSM. These variables are presented in Table 3.

**Table 3.** The levels of independent variables used for RSM.

| Level/Variables | Catalyst Amount ($X_1$)/% | Reaction Temperature ($X_2$)/°C | Reaction Time ($X_3$)/h |
|---|---|---|---|
| −1 | 1.2 | 75 | 2.0 |
| 0 | 1.4 | 80 | 2.5 |
| 1 | 1.6 | 85 | 3.0 |

### 3.3.3. Determination of Optimal Reaction Conditions

Table 4 exhibits the results of ANOVA analysis of asarinin yield. The model F-value of 16.97 and $p$-value of 0.0006 demonstrated statistical significance of the quadratic model. Coefficient of determination ($R^2$) was 0.9562, which indicates that the model represented the data accurately. This demonstrated that the relationship between the independent variables (catalyst amount, temperature and time) and the response (asarinin yield) can be explained based on the regression model. Thus, this model was fit for the further analysis of the influence of process parameters.

**Table 4.** ANOVA analysis on asarinin yield.

| Source | Sum of Squares | Degree of Freedom | Mean Square | F Value | $p$ Value Prob > F | |
|---|---|---|---|---|---|---|
| Model | 844.84 | 9 | 93.87 | 16.97 | 0.0006 * | significant |
| $X_1$ | 270.12 | 1 | 270.12 | 48.84 | 0.0002 * | |
| $X_2$ | 285.78 | 1 | 285.78 | 51.67 | 0.0002 * | |
| $X_3$ | 175.17 | 1 | 175.17 | 31.67 | 0.0008 * | |
| $X_1X_2$ | 20.46 | 1 | 20.46 | 3.70 | 0.0959 | |
| $X_1X_3$ | 27.01 | 1 | 27.01 | 4.88 | 0.0628 | |
| $X_2X_3$ | 1.04 | 1 | 1.04 | 0.19 | 0.6775 | |
| $X_1^2$ | 0.00 | 1 | 0.00 | 0.00 | 0.9870 | |
| $X_2^2$ | 10.56 | 1 | 10.56 | 1.91 | 0.2095 | |
| $X_3^2$ | 51.67 | 1 | 51.67 | 9.34 | 0.0184 | |
| Residual | 38.72 | 7 | 5.53 | | | |
| Lack of Fit | 31.50 | 3 | 10.50 | 5.82 | 0.0610 | not significant |
| Pure Error | 7.22 | 4 | 1.80 | | | |
| Cor Total | 883.56 | 16 | | | | |
| Std. Dev. | 2.35 | | R-Squared | 0.9562 | | |
| Mean | 39.99 | | Adj R-Squared | 0.8998 | | |
| C.V.% | 5.88 | | Pred R-Squared | 0.4169 | | |
| Press | 515.23 | | Adeq Precision | 13.6019 | | |

* Values are significant at 95% confidence level.

The model equation was significant at 95% confidence level with non-significant lack of fit [25,27,35]. The goodness of fit of this model was verified by the coefficient of $R^2$ (0.9456) and the adjusted $R^2$ (0.8756). Nonsignificant lack-of-fit (0.06) in comparison with the pure error substantiated the fitness of the quadratic model. The empirical relationship between asarinin yield and the investigated variables in coded units is shown in the equation:

$$A = 42.39 + 5.81\,X_1 + 5.98\,X_2 + 4.68\,X_3 - 2.26\,X_1X_2 - 2.60\,X_1X_3 + 0.51\,X_2X_3 - 0.02\,X_1^2 - 1.58\,X_2^2 - 3.50\,X_3^2 \quad (1)$$

where A is the yield of asarinin (response), and $X_1$, $X_2$, and $X_3$ are the coded values of the experiment variables, namely catalyst amount ($X_1$), temperature ($X_2$) and time ($X_3$). The influence of each factor was significant for the first-order linear effect ($X_1$, $X_2$, $X_3$) ($p < 0.05$). All the quadratic terms $X_1^2$, $X_2^2$, $X_3^2$ and interactive effects ($X_1X_2$, $X_1X_3$, $X_2X_3$) were non-significant ($p > 0.05$). The results showed that the three main factors showed significant linear effects on the yield of asarinin. The 3D response surface and the response contour plots of the three independent variables are exhibited in Figure 6. The predicted asarinin yield was 50.79 mg/100 g under the optimal conditions: reaction temperature, 85 °C; time, 2.7 h; and catalyst amount, 1.6%. The experimental value was 51.72 mg/100 g.

### 3.4. Determination of Physicochemical Properties and Lignans of Tested Sesame Oils

Under the optimal reaction conditions, the effects of CTAH on the basic physicochemical properties of sesame oil samples were further evaluated. In this section, the acid value, peroxide value, color (red value, yellow value) and lignans of sesame oil samples treated with different methods were compared and analyzed. AV (acid value) and POV (peroxide value), the classic indicators applied to evaluate the oxidation degree of vegetable oils, were also detected here [36–38].

The acid values and peroxide values of the sesame oil samples CSO, CTA and CTAH are exhibited in Table 5. Acid value is a direct index of the percentage of free fatty acids in a certain amount of sesame oil. It is a measure of the degree to which the triglycerides in the oil have been decomposed into free fatty acids by lipase action or high temperature [37,39]. It can be seen from Table 5 that there was no significant difference among the tested sesame oil samples CSO, CTA and CTAH in terms of acid value. The peroxide value of sesame oil is commonly used as a quality parameter [37,40]. The peroxide values of the sesame oil samples are exhibited in Table 5. Obviously, the sesame oil sample treated with citric acid loaded on Hβ (CTAH) had the lowest peroxide value (0.14 meq/kg), followed by CSO (0.80 meq/kg). The peroxide values of the different treated sesame oil samples showed significant differences ($p < 0.001$). Compared with the treated sesame oil sample CSO, the peroxide value of the treated sesame oil sample CTAH was 82.5% less. This might be due to the lignans in the treated sesame oil samples [41].

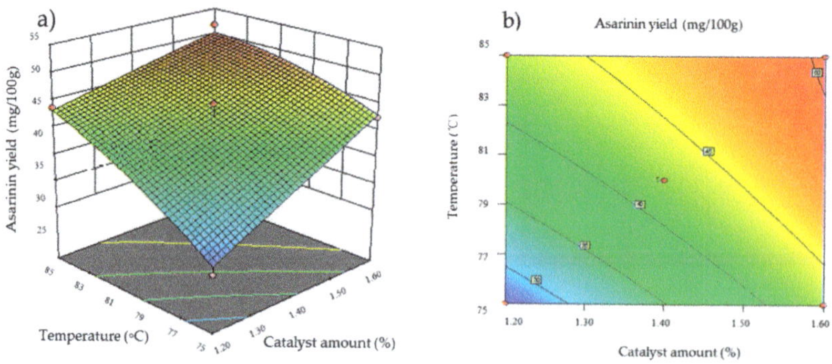


**Figure 6.** *Cont.*

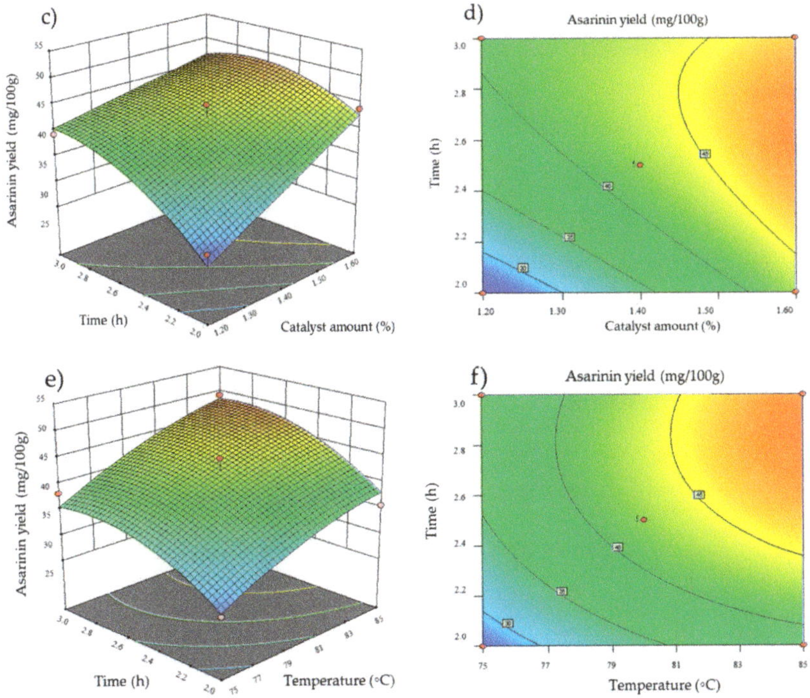

**Figure 6.** Response surface plots and corresponding contour plots of the effects of different factors on asarinin yield. Interaction of temperature and catalyst amount (**a**,**b**); interaction of time and catalyst amount (**c**,**d**); interaction of time and temperature (**e**,**f**).

**Table 5.** Acid value, peroxide value and lignans of different treated sesame oil samples.

| Samples | Acid Value (mg KOH/g) | Peroxide Value (meq/kg) | Sesamin (mg/100 g) | Asarinin (mg/100 g) | Red Value | Yellow Value |
|---|---|---|---|---|---|---|
| CSO | 1.95 ± 0.05 [a] | 0.80 ± 0.02 [b] | 489.74 ± 2.89 [a] | Nd | 1.25 ± 0.07 [a] | 4.60 ± 0.14 [a] |
| CTA | 2.06 ± 0.16 [a] | 1.82 ± 0.03 [a] | 497.20 ± 2.24 [b] | Nd | 0.90 ± 0.14 [b] | 3.75 ± 0.21 [b] |
| CTAH | 1.92 ± 0.00 [a] | 0.14 ± 0.00 [c] | 358.99 ± 0.83 [c] | 83.52 ± 0.52 [a] | 0.55 ± 0.07 [c] | 1.45 ± 0.07 [c] |

Notes: Values are means ± SD (standard deviation), mean value of different superscript letters in the same column are significantly different at $p < 0.05$. nd represents not detected. CSO: cold-pressed sesame oil; CTA: CSO treated with citric acid; CTAH: CSO treated with citric acid loaded on Hβ.

Many published studies have demonstrated that sesamin can be converted into asarinin under the conditions of heat and acid [13–15]. The contents of sesamin and asarinin are exhibited in Table 5. There was no asarinin in the sesame oil samples CSO and CTA. In other words, CTA could not promote the conversion of sesamin. Compared with the sesame oil sample CTA, the sesamin content of sesame oil sample CTAH was significantly decreased by 27.8%, indicating that CTAH could markedly promote the production of asarinin. Compared with CSO, the red value and yellow value of the sesame oil sample CTAH were significantly reduced by 56.0% and 68.5%, respectively. The results showed that CTAH has a decolorization effect [42]. CTAH, an eco-friendly and sustainable solid acid catalyst, is expected to decolorize vegetable oils. The catalyst is used in the decolorization process in the refining of leaching sesame oil, which can not only have a decolorization effect, but also promote the conversion of sesamin into asarinin, thereby improving the added-value of leaching sesame oil.

## 4. Conclusions

In this study, various solid acid catalysts were prepared, characterized and screened according to their ability to convert sesamin into asarinin. The selected catalyst, citric acid loaded on zeolite beta (CTAH), was investigated for its ability to convert sesamin into asarinin. The effects of reaction conditions on asarinin yield and the physicochemical properties of sesame oil samples were assessed. The results showed that the catalyst CTAH was the best among the catalysts tested. CTAH had the greatest surface area, largest pores and strong acid sites. The optimal reaction conditions were: reaction temperature of 85 °C, reaction time of 2.7 h and catalyst amount of 1.6%. The predicted and experimental values of asarinin yield were 50.79 and 51.80 mg/100 g, respectively. Under the optimal reaction conditions, there was no significant difference among the acid values of sesame oil samples. Compared with the treated sesame oil sample CSO, the peroxide value of the treated sesame oil sample CTAH was 82.5% less. In summary, CTAH is a solid acid catalyst with great potential for industrial use in converting sesamin into asarinin, thereby not only improving the basic physicochemical properties of sesame oil, but also increasing its biological activity. This study provides a reference for the process conditions needed in using a solid acid catalyst to produce asarinin and improve cold-pressed sesame oil.

**Author Contributions:** Q.Y.: investigation, writing—original draft preparation, methodology, software, formal analysis; X.-D.W.: resources, supervision, visualization, funding acquisition; H.-M.L.: Conceptualization, supervision, writing—review and editing, project administration; Y.-X.M.: resources, funding acquisition, methodology, formal analysis. All authors have read and agreed to the published version of the manuscript.

**Funding:** This work was supported by an earmarked fund for National Natural Science Foundation of China (31972004), Doctor Research Fund of Henan University of Technology (2020BS007) and the earmarked fund for Modern Agro-industry Technology Research System (CARS14-1-29).

**Data Availability Statement:** The data presented in this study are available in this article.

**Acknowledgments:** The authors would like to acknowledge support from a range of funds, such as namely, the National Natural Science Foundation of China (31972004), Doctor Research Fund of Henan University of Technology (2020BS007) and the earmarked fund for Modern Agro-industry Technology Research System (CARS14-1-29).

**Conflicts of Interest:** The authors declare no conflict of interest.

## References

1. Bodoira, R.; Velez, A.; Andreatta, A.E.; Martínez, M.; Maestri, D. Extraction of bioactive compounds from sesame (*Sesamum indicum* L.) defatted seeds using water and ethanol under sub-critical conditions. *Food Chem.* **2017**, *237*, 114–120. [CrossRef] [PubMed]
2. Kancharla, P.K.; Arumugam, N. Variation of oil, sesamin, and sesamolin content in the germplasm of the ancient oilseed crop *Sesamum indicum* L. *J. Am. Oil Chem. Soc.* **2020**, *97*, 475–583. [CrossRef]
3. Majdalawieh, A.F.; Dalibalta, S.; Yousef, S.M. Effects of sesamin on fatty acid and cholesterol metabolism, macrophage cholesterol homeostasis and serum lipid profile: A comprehensive review. *Eur. J. Pharmacol.* **2020**, *885*, 173417. [CrossRef] [PubMed]
4. Jeong, S.H.; Jang, J.H.; Cho, H.Y.; Lee, Y.B. Simultaneous determination of asarinin, β-eudesmol, and wogonin in rats using ultraperformance liquid chromatography–tandem mass spectrometry and its application to pharmacokinetic studies following administration of standards and Gumiganghwal-tang. *Biomed. Chromatogr.* **2020**, *35*, e5021. [CrossRef] [PubMed]
5. Freise, C.; Querfeld, U. The lignan (+)-episesamin interferes with TNF-α-induced activation of VSMC via diminished activation of NF-kB, ERK1/2 and AKT and decreased activity of gelatinases. *Acta Physiol.* **2015**, *213*, 642–652. [CrossRef] [PubMed]
6. Hou, Y.J.; Hu, T.; Wei, D.; Gao, J.P.; Che, D.L.; Wang, C.; He, H.Z. (−)-Asarinin inhibits mast cells activation as a Src family kinase inhibitor. *Int. J. Biochem. Cell Biol.* **2020**, *121*, 105701. [CrossRef] [PubMed]
7. Park, H.J.; Lee, K.S.; Zhao, T.T.; Lee, K.E.; Lee, M.L. Effects of asarinin on dopamine biosynthesis and 6-hydroxydopamine-induced cytotoxicity in PC12 cells. *Arch. Pharm. Res.* **2017**, *40*, 631–639. [CrossRef] [PubMed]
8. Ide, T.; Lim, J.S.; Odbayar, T.; Nakashima, Y. Comparative study of sesame lignans (sesamin, episesamin and sesamolin) affecting gene expression profile and fatty acid oxidation in rat liver. *J. Nutr. Sci. Vitaminol.* **2009**, *55*, 31–43. [CrossRef] [PubMed]
9. Kuo, P.; Lin, M.; Chen, G.; Yiu, T.; Tzen, J.T.C. Identification of methanol-soluble compounds in sesame and evaluation of antioxidant potential of its lignans. *J. Agric. Food Chem.* **2011**, *59*, 3214–3219. [CrossRef] [PubMed]

10. Gao, J.H.; Wang, R.D.; Lu, X.; Jia, C.; Sun, Q.; Huang, J.N.; Wei, S.L.; Ma, L. Enzymatic preparation and structure-activity relationship of sesaminol. *J. Oleo Sci.* **2021**, *70*, 1261–1274. [CrossRef] [PubMed]
11. Tsai, H.Y.; Lee, W.J.; Chu, I.H.; Hung, W.C.; Su, N.W. Formation of samin diastereomers by acid-catalyzed transformation of sesamolin with hydrogen peroxide. *J. Agric. Food Chem.* **2020**, *68*, 6430–6438. [CrossRef] [PubMed]
12. Fukuda, Y.; Nagata, M.; Osawa, T.; Namiki, M. Contribution of lignan analogues to antioxidative activity of refined unroasted sesame seed oil. *J. Am. Oil Chem. Soc.* **1986**, *63*, 1027–1031. [CrossRef]
13. Li, C.Y.; Chow, T.J.; Wu, T.S. The epimerization of sesamin and asarinin. *J. Nat. Prod.* **2005**, *68*, 1622–1624. [CrossRef] [PubMed]
14. Chen, J.C.; Chen, Y.Z.; Tian, J.J.; Ge, H.F.; Liang, X.F.; Xiao, J.B.; Lin, H.T. Simultaneous determination of four sesame lignans and conversion in *Monascus* aged vinegar using HPLC method. *Food Chem.* **2018**, *256*, 133–139. [CrossRef]
15. Yu, Q.; Ma, Y.X.; Qin, Z.; Luo, X.R.; Liu, H.M.; Wang, X.D. Using solid acid catalysts to improve the oxidative stability of cold-pressed sesame oil. *LWT* **2021**, *141*, 110928. [CrossRef]
16. Wang, X.D.; Huang, X.; Liu, S.; Cui, Y.D. Catalytic epimerization of hydrochloric acid-ferric chloride of sesamin to synthesis of asarinin and its mechanism. *CIESC J.* **2015**, *66*, 455–460.
17. Liu, M.; Yin, Y.X.; Guo, X.W.; Song, C.S. Oxalic acid modification of β zeolite for dehydration of 2-(4′-ethylbenzoyl) benzoic acid. *Ind. Eng. Chem. Res.* **2017**, *56*, 8850–8856. [CrossRef]
18. Mansir, N.; Taufiq-Yap, Y.H.; Rashid, U.; Lokman, I.M. Investigation of heterogeneous solid acid catalyst performance on low grade feedstocks for biodiesel production: A review. *Energy Convers. Manag.* **2017**, *141*, 171–182. [CrossRef]
19. Hoff, T.C.; Gardner, D.W.; Thilakaratne, R.; Proano-Aviles, J.; Brown, R.C.; Tessonnier, J.P. Elucidating the effect of desilication on aluminum-rich ZSM-5 zeolite and its consequences on biomass catalytic fast pyrolysis. *Appl. Catal. A-Gen.* **2017**, *529*, 68–78. [CrossRef]
20. Bai, G.Y.; Ma, Z.; Shi, L.J.; Lan, X.W.; Wang, Y.L.; Han, J.; Qiu, M.D.; Fu, H.L.; Liu, P.D. Continuous synthesis of bis(indolyl)phenylmethane over acid modified Hβ zeolite. *Appl. Catal. A-Gen.* **2012**, *427–428*, 114–118. [CrossRef]
21. Patel, A.; Narkhede, N. 12-Tungstophosphoric acid anchored to zeolite Hβ: Synthesis, characterization, and biodiesel production by esterification of oleic acid with methanol. *Energy Fuel* **2012**, *26*, 6025–6032. [CrossRef]
22. Wu, Y.H.; Tian, F.P.; Liu, J.; Song, D.; Jia, C.Y.; Chen, Y.Y. Enhanced catalytic isomerization of α-pinene over mesoporous zeolite beta of low Si/Al ratio by NaOH treatment. *Microporous Mesoporous Mater.* **2012**, *162*, 168–174. [CrossRef]
23. Yang, Q.L.; Sun, Y.X.; Sun, W.H.; Qin, Z.; Liu, H.M.; Ma, Y.X.; Wang, X.D. Cellulose derived biochar: Preparation, characterization and Benzo[a]pyrene adsorption capacity. *Grain Oil Sci. Technol.* **2021**, *4*, 182–190. [CrossRef]
24. Viswanadham, B.; Nekkala, N.; Rohitha, C.N.; Vishwanathan, V.; Chary, K.V.R. Synthesis, characterization and catalytic dehydration of glycerol to acrolein over phosphotungstic acid supported Y-zeolite catalysts. *Catal. Lett.* **2018**, *148*, 397–406. [CrossRef]
25. Fasuan, T.O.; Omobuwajo, T.O.; Gbadamosi, S.O. Optimization of simultaneous recovery of oil and protein from sesame (*Sesamum indicum*) seed. *J. Food Process. Preserv.* **2018**, *42*, e13341. [CrossRef]
26. Sivakanthan, S.; Jayasooriya, A.P.; Madhujith, T. Optimization of the production of structured lipid by enzymatic interesterification from coconut (*Cocos nucifera*) oil and sesame (*Sesamum indicum*) oil using response surface methodology. *LWT* **2019**, *101*, 723–730. [CrossRef]
27. Cao, J.; Xiong, N.; Zhang, Y.; Dai, Y.W.; Wang, Y.Y.; Lu, L.Y.; Jiang, L. Using RSM for optimum of optimum production of peptides from edible bird's nest by-product and characterization of its antioxidant's properties. *Foods* **2022**, *11*, 1071. [CrossRef]
28. Wu, S.M.; Wang, L.; Shu, F.Y.; Cao, W.M.; Chen, F.X.; Wang, X.G. Effect of refining on the lignan content and oxidative stability of oil pressed from roasted sesame seed. *Int. J. Food Sci. Technol.* **2013**, *48*, 1187–1192. [CrossRef]
29. Shi, L.K.; Zheng, L.; Jin, Q.Z.; Wang, X.G. Effects of adsorption on polycyclic aromatic hydrocarbon, lipid characteristic, oxidative stability, and free radical scavenging capacity of sesame oil. *Eur. J. Lipid Sci. Technol.* **2017**, *119*, 1700150. [CrossRef]
30. MacQueen, B.; Royko, M.; Crandall, B.S.; Heyden, A.; Pagán-Torres, Y.J.; Lauterbach, J. Kinetics study of the hydrodeoxygenation of xylitol over a $ReO_x$-Pd/$CeO_2$ catalyst. *Catalysts* **2021**, *11*, 108. [CrossRef]
31. Zhang, J.W.; Uknalis, J.; Chen, L.; Moreau, R.A.; Ngo, H. Development of magnesium oxide–zeolite catalysts for isomerization of fatty acids. *Catal. Lett.* **2019**, *149*, 303–312. [CrossRef]
32. Singh, H.K.G.; Yusup, S.; Quitain, A.T.; Kida, T.; Sasaki, M.; Cheah, K.W.; Ameen, M. Production of gasoline range hydrocarbons from catalytic cracking of linoleic acid over various acidic zeolite catalysts. *Environ. Sci. Pollut. Res.* **2019**, *26*, 34039–34046. [CrossRef] [PubMed]
33. Oginni, O.; Singh, K.; Oporto, G.; Dawson-Andoh, B.; McDonald, L.; Sabolsky, E. Effect of one-step and two-step $H_3PO_4$ activation on activated carbon characteristics. *Bioresour. Technol. Rep.* **2019**, *8*, 100307. [CrossRef]
34. Li, F.; Min, R.; Li, J.; Gao, L.Y.; Xue, W.; Wang, Y.J.; Zhao, X.Q. Condensation reaction of methyl *N*-phenylcarbamate with formaldehyde over Hβ catalyst. *Ind. Eng. Chem. Res.* **2014**, *53*, 5406–5412. [CrossRef]
35. Iglesias-Carres, L.; Mas-Capdevila, A.; Bravo, F.I.; Bladé, C.; Arola-Arnal, A.; Muguerza, B. Optimization of extraction methods for characterization of phenolic compounds in apricot fruit (*Prunus armeniaca*). *Food Funct.* **2019**, *10*, 6492–6502. [CrossRef] [PubMed]
36. Zhu, M.T.; Shi, T.; Guo, Z.Y.; Liao, H.X.; Chen, Y. Comparative study of the oxidation of cold-pressed and commercial refined camellia oil during storage with $^1H$ and $^{31}P$ NMR spectroscopy. *Food Chem.* **2020**, *321*, 126640. [CrossRef] [PubMed]

37. Tenyang, N.; Ponka, R.; Tiencheu, B.; Djikeng, F.T.; Azmeera, T.; Karuna, M.S.L.; Prasad, R.B.N.; Womeni, H.M. Effects of boiling and roasting on proximate composition, lipid oxidation, fatty acid profile and mineral content of two sesame varieties commercialized and consumed in far-north region of cameroon. *Food Chem.* **2017**, *221*, 1308–1316. [CrossRef] [PubMed]
38. Moghtadaei, M.; Soltanizadeh, N.; Goli, S.A.H. Production of sesame oil oleogels based on beeswax and application as partial substitutes of animal fat in beef burger. *Food Res. Int.* **2018**, *108*, 368–377. [CrossRef]
39. Hama, J.R. Comparison of fatty acid profile changes between unroasted and roasted brown sesame (*Sesamum indicum* L.) seeds oil. *Int. J. Food Prop.* **2017**, *20*, 957–967. [CrossRef]
40. Elhussein, E.; Bilgin, M.; Şahin, S. Oxidative stability of sesame oil extracted from the seeds with different origins: Kinetic and thermodynamic studies under accelerated conditions. *J. Food Process Eng.* **2018**, *41*, e12878. [CrossRef]
41. Mishra, S.K.; Belur, P.D.; Iyyaswami, R. Use of antioxidants for enhancing oxidative stability of bulk edible oils: A review. *Int. J. Food Sci. Technol.* **2021**, *56*, 1–12. [CrossRef]
42. Dadfar, T.; Sahari, M.A.; Barzegar, M. Bleaching of olive oil by membrane filtration. *Eur. J. Lipid Sci. Technol.* **2020**, *122*, 1900151. [CrossRef]

*foods*

MDPI

*Article*

# Traditional Cooking Methods Affect Color, Texture and Bioactive Nutrients of *Undaria pinnatifida*

**Shan Jiang [1,†], Meiqi Yu [1,†], Yingzhen Wang [1], Wei Yin [2], Pengfei Jiang [1], Bixiang Qiu [3] and Hang Qi [1,*]**

1 National Engineering Research Center of Seafood, Liaoning Provincial Aquatic Products Deep Processing Technology Research Center, School of Food Science and Technology, Dalian Polytechnic University, Dalian 116034, China; jiangshan_dlpu@163.com (S.J.); yumeiqi_dlpu@163.com (M.Y.); wangyz2036@163.com (Y.W.); dpb@gaishi.cn (P.J.)
2 Dalian Gaishi Food Co., Ltd., Dalian 116047, China; 13840940070@163.com
3 Fujian Yida Food Co., Ltd., Fuzhou 350500, China; 13905028186@163.com
* Correspondence: qihang@dlpu.edu.cn; Tel.: +86-411-86318785
† These authors contributed equally to this work.

**Abstract:** *Undaria pinnatifida* (*U. pinnatifida*) is an edible brown seaweed with high health value. The objective of this study was to evaluate the effect of traditional cooking methods (i.e., blanching, steaming, boiling and baking) on the color, texture and bioactive nutrients of *U. pinnatifida*, so as to screen out the traditional cooking methods more suitable for *U. pinnatifida*. In this study, methods of blanching and boiling resulted in better reduction in total color difference (0.91 ± 0.58 and 0.79 ± 0.34, respectively) and retention of chlorophyll A (62.99 ± 1.27 µg/g FW and 51.35 ± 1.69 µg/g FW), along with better elevation of fucoxanthin content (increased by 11.05% and 18.32%, respectively). Baking method got the best retention of total phenol content (1.62 ± 0.11 mg GAE/g DW), followed by methods of boiling and blanching (1.51 ± 0.07 mg GAE/g DW and 1.43 ± 0.05 mg GAE/g DW). Among these cooking methods, blanching and boiling seemed to be the more suitable for *U. pinnatifida* compared to other methods. These results could help to determine the better cooking methods for *U. pinnatifida* products and provide a scientific and theoretical basis for improving human dietary health.

**Keywords:** *U. pinnatifida*; traditional cooking methods; quality characteristics; color; texture; bioactive nutrients

**Citation:** Jiang, S.; Yu, M.; Wang, Y.; Yin, W.; Jiang, P.; Qiu, B.; Qi, H. Traditional Cooking Methods Affect Color, Texture and Bioactive Nutrients of *Undaria pinnatifida*. *Foods* **2022**, *11*, 1078. https://doi.org/10.3390/foods11081078

Academic Editors: Federico Casanova and Michela Verni

Received: 7 March 2022
Accepted: 7 April 2022
Published: 8 April 2022



## 1. Introduction

A healthy diet has been a hot topic in recent years. As the public becomes more health conscious, more and more vegetables are being considered as a part of a healthy lifestyles. Therefore, there is a growing demand for high value-added vegetables. *Undaria pinnatifida* (*U. pinnatifida*) is a large annual temperate seaweed found mainly in the coastal areas of China, Japan and South Korea [1]. *U. pinnatifida* has a high economic and medicinal value and is rich in polysaccharides, protein, vitamins, minerals, polyphenols and phytosterols [2]. The components of *U. pinnatifida* showed various biological activities, such as hypolipidemic, antihypertensive, immunoregulation and antitumor activity [3]. In addition, the application of polyphenols and pigments in seaweeds has also become a research hotspot in recent years. Utilizing their unique physicochemical properties and health functions, they could be used as preservatives, antioxidants and colorants. In a variety of foods, they could achieve the purpose of improving quality, extending storage period and developing new health foods, and have broad application prospects in food production.

In order to provide reasonable guidance for cooking food and improve the sensory and nutritional characteristics of each meal, it is necessary to determine the optimal cooking methods [4]. The color, texture, structural state and bioactive nutrients are the main factors to measure the value of vegetables, and cooking methods greatly affect these important indicators. The cooking and processing methods of vegetables can preserve/improve the nutrition and quality of vegetables to a certain extent [5]. Many studies have evaluated the effects of different cooking methods on the qualities of vegetables [6,7]. In a study on the effect of different baking conditions on the sensory quality of barley, it was found that baking in the low to medium temperature range improved the sensory quality [8]. Red cabbage cooked by the sous-vide method was more purple and more flavorful than traditionally cooked cabbage [9]. Akdaş and Bakkalbaşı [10] found that microwaving and stir-frying increased the value of the greenness in kale. Additionally, cooking methods have different effects on the bioactive ingredients in vegetables. They may increase the bioavailability of polyphenols, fucoxanthin and chlorophyll; however, it may also lead to a loss of their contents [11]. Moreover, the extent of these changes depends on a number of factors, such as the type and quality of raw materials, the used methods, the degree of heating, immersion in the processing medium, the extraction solvent, as well as the pH [12–14]. Giusti et al. revealed that the contents of free and bound phenolic compounds were greatly reduced during cooking and most of the phenolic compounds were leached into water [15]. Compared with sous-vide and traditional processing, cook-vide had less harm to the total phenol content of green bean. Moreover, sous-vide had better protective effect on the total phenolic content of carotenes [16]. According to the study by Gu et al., the steaming process resulted in a significant loss of phenolic/flavonoid components while the baking process had minimal impact on the active components [5].

At present, there are few studies within our knowledge comparing the effects of different traditional cooking methods on the quality and bioactive nutrients of *U. pinnatifida* and analyzing structural changes from a microscopic perspective. Therefore, in this study, four traditional cooking methods (blanching, steaming, boiling and baking) were used to compare the changes in characteristics (color, texture, moisture and microstructure) and bioactive nutrients (total phenols, chlorophyll A and fucoxanthin) of *U. pinnatifida*. The aims of this study were to: (1) compare the effects of four traditional cooking methods on the color, texture and bioactive nutrients of *U. pinnatifida*; and (2) screen for better cooking methods to preserve bioactive nutrients, texture and color in cooking. This study provides useful guidance on cooking selection of *U. pinnatifida*, as well as a theoretical basis and basic reference for cooking *U. pinnatifida* to improve human dietary health.

## 2. Materials and Methods

### 2.1. Materials and Reagents

The dried *U. pinnatifida* was supplied by Dalian LiaoHai Aquatic Food Trading Co., Ltd. (Dalian, China). It was stored in a sealing bag before analysis. Folin–Ciocalteu's phenol reagent was obtained from Sangon Biotech (Shanghai) Co., Ltd. (Shanghai, China). Fucoxanthin (≥98%, HPLC grade), chlorophyll A (≥95%, HPLC grade) and gallic acid (≥98%, HPLC grade) were provided by Sigma-Aldrich Co., Ltd. (Shanghai, China). Methanol (HPLC grade) was purchased from Sibiquan Chemical Co., Ltd. (Shanghai, China). Acetone (HPLC grade) was supplied by Sinopharm Chemical Reagents Co., Ltd. (Shanghai, China). All other chemicals used for the analysis were analytical grade.

### 2.2. Preparation of *U. pinnatifida* Samples

The dried *U. pinnatifida* was immersed in deionized water until fully extended and drained to remove excess water. The rehydrated *U. pinnatifida* was then cut into equal sizes and used for subsequent experiments at 100 g per aliquot.

### 2.3. Conditions of Cooking Methods

Four traditional cooking methods were used in this study, including blanching, steaming, boiling and baking. Initial tests were carried out on each sample to optimize the cooking conditions. The shortest cooking times were used for all methods to achieve adequate palatability and taste according to the Chinese eating habits.

Blanching: 100 g of prepared *U. pinnatifida* sample was immersed in 1 L of boiling water (100 °C) for 3 min, and the blanched sample was then wiped dry and cooled on ice.

Steaming: after the water in the steamer began to boil, the *U. pinnatifida* sample (100 g) was placed in the steamer for 10 min and then quickly cooled on ice to prevent further heating.

Boiling: the *U. pinnatifida* sample (100 g) was boiled in 1 L boiling water (100 °C) for 10 min in a covered stainless-steel pot, imitating traditional cooking methods. The boiled samples were dried with kitchen papers and cooled on ice immediately.

Baking: the *U. pinnatifida* sample was baked in the oven (SCC WE 101, RATIONAL, Landsberg a. Lech, Germany) for 10 min at 160 °C.

Partially cooled samples (approximately 10 g for each sample) were used for immediate analysis of moisture, color, texture and chlorophyll A. Another portion (approximately 90 g for each sample) was placed in a drying oven (PH-070A, Shanghai YiHeng Technology Co., Ltd., Shanghai, China) at 50 °C for 16 h. The dried samples were crushed and passed through a 200-mesh sieve. The dried powder was sealed in the vacuum bag and stored at 4 °C for further analysis. The raw *U. pinnatifida* was analyzed as the control.

### 2.4. *Quality Characteristics*

#### 2.4.1. Color Analysis

The colors of the raw and cooked *U. pinnatifida* samples were measured using a chromameter (UltraScan Pro, HunterLab, Reston, VA, USA). In the CIE color system, the negative coordinates of $a^*$ and the positive coordinates of $b^*$ represent the green and yellow intensities, respectively [17]. The calculation of the total color difference ($\Delta E$) was performed according to the formula previously described by Armesto et al. [18]:

$$\Delta E_{\mathrm{ab}} = \sqrt{({L_2}^* - {L_1}^*)^2 + ({a_2}^* - {a_1}^*)^2 + ({b_2}^* - {b_1}^*)^2}$$

where ${L_1}^*$, ${a_1}^*$, ${b_1}^*$ are the values of the raw sample. ${L_2}^*$, ${a_2}^*$, and ${b_2}^*$ are the values of the samples subjected to different cooking methods.

#### 2.4.2. Texture Profile Analysis

For texture profile, the *U. pinnatifida* samples were analyzed by texture analyzer (TA-XT plus, Stable Micro Systems Ltd., Vienna, UK) equipped with a 5 mm diameter (P/5) probe. The parameters were set as follows: compression variable, 30%; pre-test speed, 2.0 mm/s; test speed: 1.0 mm/s; post test speed, 2.0 mm/s, which was according to the procedure described by Peng et al. [19].

#### 2.4.3. Low-Field Nuclear Magnetic Resonance (LF-NMR) Measurement

The water state of raw and cooked samples was determined using a LF-NMR analyzer (MesoQMR23-060H, Suzhou (Shanghai) Niumag Electronic Technology Co., Ltd., Shanghai, China). The decay signal was acquired using the Carr–Purcell–Meiboom–Gill (CPMG) pulse sequence. The main parameters were according to Jiang et al., as follows [20]: TE (time echo) = 0.5 ms, TW (time waiting) = 4000 ms, NS (number of scan) = 8, NECH (number of echo) = 8000.

#### 2.4.4. Scanning Electron Microscope (SEM)

The microstructure of raw and cooked *U. pinnatifida* was observed using a SEM (JSM-7800F, Tokyo, Japan). According to the method of Jiang et al., *U. pinnatifida* was freeze-dried in a freeze-dryer at −80 °C for 48 h. The samples obtained were fractured in liquid nitrogen and structural changes in the fracture surface were observed under SEM with ×500 magnification [20].

### 2.5. Bioactive Nutrients

#### 2.5.1. Total Phenol

The total phenolic compounds were extracted with modifications as described by Ummat et al. [21]. Briefly, *U. pinnatifida* sample powder was mixed with 50% ethanol (1:15, $w/v$) and sonicated for 30 min at 25 °C. The mixture was incubated in a thermostatic shaker (THZ-82, Zhibri Instruments Ltd., Changzhou, China) at 50 °C for 7.5 h at 120 rpm, then filtered through Whatman #1 filter paper (Whatman International Co., Ltd., Maidstone, UK) and the supernatant was collected. The extracts were then stored at 4 °C in the dark until analysis.

Total phenol content (TPC) was determined using the modified Folin–Ciocalteau method [20]. First 600 μL of deionized water and 50 μL of Folin–Ciocalteu phenol reagent were added sequentially to 50 μL of the extract. Next, 20% sodium carbonate solution (150 μL) was added after 1 min, then deionized water was added to the mixture to 1 mL and incubated in the dark at 25 °C for 2 h. The microplate reader (Infinite 200, Tecan Austria Co., Ltd., Grodig, Austria) was used to measure the absorbance at 760 nm. A standard curve with continuous gallic acid solution was used for calibration. TPC was expressed as gallic acid equivalent (GAE) per gram of dry weight (DW).

#### 2.5.2. Fucoxanthin

The *U. pinnatifida* sample powder was mixed with 80% methanol (1:10, $w/v$) and sonicated at room temperature for 30 min. The mixture was centrifuged at 3040× $g$ for 10 min and the supernatant was evaporated in the rotary evaporator at 40 °C until dryness. The same volume of methanol (HPLC grade) was used to redissolve and the resulting solution was filtered through a 0.22 μm syringe filter to obtain the fucoxanthin extract.

The fucoxanthin was determined by HPLC according to the procedure described by Sui et al. [22]. Methanol was used as the mobile phase in a C18 column (5 μm, 4.6 × 250 mm, Shimadzu, Kyoto, Japan) at 40 °C with a flow rate of 0.5 mL/min. The injection volume was 10 μL, and the data acquisition time was 10 min. Detection of fucoxanthin was performed at 450 nm and the results were expressed as μg per gram of DW.

#### 2.5.3. Chlorophyll A

Chlorophyll A was extracted with modification as described by Havlíková et al. [23]. First, 5 mL acetone and 1 mL deionized water were mixed with 1 g of *U. pinnatifida* sample, sonicated at room temperature for 2 h. Then the mixture was centrifuged at 3040× $g$ for 10 min and the supernatant was collected. The supernatant (5 mL) was evaporated to dryness in a rotary evaporator at 40 °C and redissolved with acetone (5 mL, HPLC grade). The solution was filtered through 0.45 μm nylon microporous filter membrane to obtain the chlorophyll A extract.

The chlorophyll A extract was also analyzed by Shimadzu HPLC series system. The C18 column (5 μm, 4.6 × 150 mm, Shimadzu, Japan) was used to achieve the separation at 30 °C. The mobile phase was 100% methanol (HPLC grade) with a flow rate of 0.8 mL/min and the total run time was 20 min. Detection was performed by a DAD at 430 nm and the injection volume was 10 μL. The results were expressed as μg per gram of fresh weight (FW).

### 2.6. Statistical Analysis

All experiments were conducted at least in triplicate. Data are reported as means ± standard deviation. Statistical comparisons were performed by analysis of variance (ANOVA) with Duncan's multiple range tests, and the significant differences were identified at a level of $p < 0.05$ using SPSS software (ver. 20.0; SPSS Inc., Chicago, IL, USA).

## 3. Results and Discussion

### *3.1. Effects of Cooking Methods on Quality Characteristics of U. pinnatifida*

#### 3.1.1. Color of *U. pinnatifida* in Different Cooking Methods

Color plays an important role in the acceptability of vegetables after cooking [24]. The effects of different cooking methods on color coordinates and total color difference ($\Delta E$) were observed in Table 1. Cooking samples other than blanching significantly increased the $a^*$ value compared to the raw sample, which meant the greenness was reduced. The greenness of the blanching sample was increased, this result was consistent with the previous reported results. Pellegrini et al. evaluated the effect of different cooking methods on the color changes of three Brassica vegetables and the results showed that blanching can prevent vegetables from discoloring after they were cooked to a certain extent [25]. Additionally, the result of the reduction in greenness by steaming was consistent with the results reported by Zhong et al., who found that steaming broccoli florets lost the greenness (increased $a^*$ values) [26]. For $b^*$ value and $L^*$ value, except the $b^*$ value of the baking was significantly decreased, other cooking methods had no significant change compared with the control group. The total color difference ($\Delta E$) of the samples with different cooking methods was shown in Table 1, with the total color difference ranging between 0.79 ± 0.34 and 7.65 ± 0.49. The raw *U. pinnatifida* was considered as the control, baking had a greater effect on the $\Delta E$ than steaming (1.76 ± 0.85), blanching (0.91 ± 0.58) and boiling (0.79 ± 0.34), as the baking had a greater effect on the a* and $b^*$ values. Mashiane et al., found that steaming reduced the $\Delta E$ in both African pumpkin and pumpkin leaves, whilst boiling resulted in greater $\Delta E$ compared to raw leaves, which was contrary to our results and may be caused by differences in cooking conditions and raw materials [24].

**Table 1.** Influence of different cooking methods on the color of *Undaria pinnatifida* (*U. pinnatifida*).

| Color Properties | Raw | Blanching | Steaming | Boiling | Baking |
|---|---|---|---|---|---|
| $L^*$ | 23.60 ± 0.56 $^{ab}$ | 23.48 ± 0.90 $^{ab}$ | 24.44 ± 0.98 $^{a}$ | 24.04 ± 0.37 $^{ab}$ | 23.07 ± 0.98 $^{b}$ |
| $a^*$ | −4.27 ± 0.22 $^{a}$ | −4.35 ± 0.25 $^{a}$ | −3.52 ± 0.30 $^{b}$ | −3.69 ± 0.19 $^{b}$ | −1.04 ± 0.13 $^{c}$ |
| $b^*$ | 8.58 ± 0.52 $^{a}$ | 8.67 ± 0.57 $^{a}$ | 8.95 ± 1.26 $^{a}$ | 8.51 ± 0.21 $^{a}$ | 1.73 ± 0.47 $^{b}$ |
| $\Delta E$ | — | 0.91 ± 0.58 $^{a}$ | 1.76 ± 0.85 $^{b}$ | 0.79 ± 0.34 $^{a}$ | 7.65 ± 0.49 $^{c}$ |
| Sample color |  |  |  |  |  |

Data are expressed as mean ± standard deviation ($n$ = 9). Different letters in the same row indicate significant differences at ($p < 0.05$).

#### 3.1.2. Effects of Cooking Methods on Texture of *U. pinnatifida*

Texture is an important sensory attribute affecting food acceptability [27]. The textural parameters of *U. pinnatifida* samples by different cooking methods were shown in Table 2, including the hardness, cohesiveness, chewiness and resilience. The results showed that the samples with different cooking methods underwent significant changes compared to the raw sample. Among them, the texture of the baking sample changed so much that the parameters could not be detected in the same way as the others. From Table 2, almost all texture parameters followed the same general trend, the degree of change caused by cooking compared to the control group was blanching < boiling < steaming, with the lowest change occurring in the blanching. Hardness, cohesiveness, chewiness and resilience were significantly reduced after various cooking methods ($p < 0.05$). This phenomenon may be due to impaired cell wall integrity, which in turn affects the textural properties [28]. Hardness was probably the most relevant texture feature for solid foods and played a key role in consumer acceptance and market value. Compared with the raw *U. pinnatifida* (1353.11 ± 50.78), the hardness values of cooked *U. pinnatifida* decreased significantly, with the steaming sample (1041.11 ± 96.87) the most obviously decreased in all samples, followed by boiling (1172.61 ± 37.69) and blanching (1182.27 ± 110.77) samples. Cohesiveness represents the internal force that holds the sample together until compression reaches the sample fracture point and depends on the properties of the sample and external factors such as humidity [29]. Chewiness referred to the energy required for chewing, which was correlated with hardness and cohesiveness [20]. Therefore, the trend of chewiness was consistent with the trends in hardness, i.e., raw (16,711.54 ± 3544.40) > blanching (10,514.54 ± 2108.38) > boiling (800.02 ± 31.39) > steaming (621.74 ± 133.73).

**Table 2.** Influence of different cooking methods on the texture of *U. pinnatifida*.

| Texture Parameters | Raw | Blanching | Steaming | Boiling | Baking |
|---|---|---|---|---|---|
| Hardness | 1353.11 ± 50.78 [a] | 1182.27 ± 110.77 [b] | 1041.11 ± 96.87 [c] | 1172.61 ± 37.69 [b] | nd |
| Cohesiveness | 16.86 ± 2.98 [a] | 12.35 ± 1.54 [b] | 1.03 ± 0.10 [c] | 1.01 ± 0.02 [c] | nd |
| Chewiness | 16,711.54 ± 3544.40 [a] | 10,514.54 ± 2108.38 [b] | 621.74 ± 133.73 [c] | 800.02 ± 31.39 [c] | nd |
| Resilience | 0.63 ± 0.04 [a] | 0.53 ± 0.03 [b] | 0.45 ± 0.05 [c] | 0.43 ± 0.03 [c] | nd |

Data are expressed as mean ± standard deviation ($n = 9$). Different letters in the same row indicate significant differences at ($p < 0.05$). nd: not detected.

### 3.1.3. LF-NMR Analysis

Different cooking methods can result in different moisture distribution changes, thus affecting their physicochemical properties. The transverse relaxation time of LF-NMR can reflect the water state and distribution. As shown in Figure 1, the $T_2$ relaxation spectrum was obtained by multi-exponential fitting of the CPMG raw data. The smaller the $T_2$, the tighter the bond between water and matter [30]. There were mainly three peaks in the $T_2$ relaxation curves. The details of three peaks were $T_{21}$ (0.1–10 ms), $T_{22}$ (10–100 ms) and $T_{23}$ (100–1000 ms) corresponding to the bound, immobilized and free water [31]. The results showed that the moisture changes in the *U. pinnatifida* were the most significant after baking, which showed a complete loss of free water from the baking sample and a significant reduction in $T_{22}$ and $T_{21}$ compared to the control group, implying a decrease in the mobility of immobilized and bound water. The fact was probably related to the higher temperature promoting the internal water losses during baking [6]. While $T_{23}$ shifted to the right in the three cooking samples (i.e., blanching, steaming and boiling), indicating that the free water mobility of samples was stronger than raw samples. Additionally, the area of free water of the steaming sample was significantly reduced, which meant that the way of steaming caused serious loss of free water in the *U. pinnatifida*.

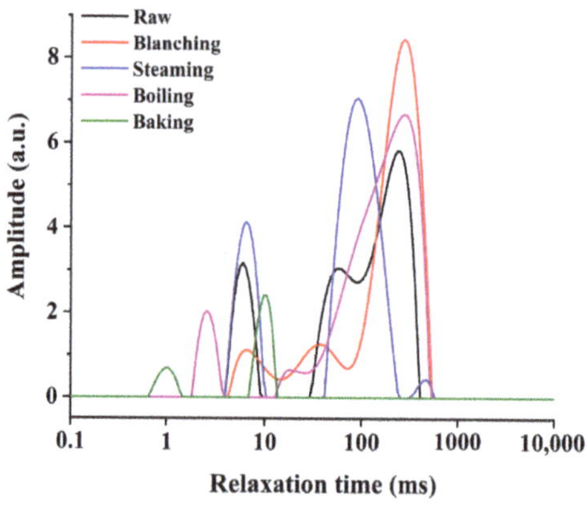


**Figure 1.** The transverse relaxation time ($T_2$) curves of the raw and cooked *U. pinnatifida* under different cooking methods.

#### 3.1.4. Scanning Electron Microscopy (SEM)

To investigate the effect of different cooking methods on the microstructure of raw and cooked *U. pinnatifida*, freeze-dried samples were observed with scanning electron microscopy (SEM). As shown in Figure 2, the structure of the fracture surfaces of raw (Figure 2a), blanching (Figure 2b), steaming (Figure 2c), boiling (Figure 2d) and baking (Figure 2e,f) samples were shown. The microstructure of *U. pinnatifida* was significantly altered by the four different cooking methods. It could be seen that the pores of blanching samples were uneven and loose, with collapsed pore walls and disorganized arrangements, compared with raw samples. The structure of the steaming sample stuck together. After boiling, the structure of the *U. pinnatifida* changed significantly, with a significant increase in pore size and the formation of a larger cavity structure. The most pronounced changes in the samples were observed after baking treatment, where the overall sample became extremely thin and the structure was difficult to see even under multiple observations under the same conditions (×500) as the other cooking conditions, therefore needed to be observed at higher magnifications (×2000, Figure 2f). The porous structure of the baking sample was completely transformed into a uniform dense scale-like structure. This may be related to the severe loss of water from the sample during baking treatment. Yang et al. had confirmed that the structure of food changes after thermal treatment, their results showed that the surface of fried potato sticks changed under heating [32].

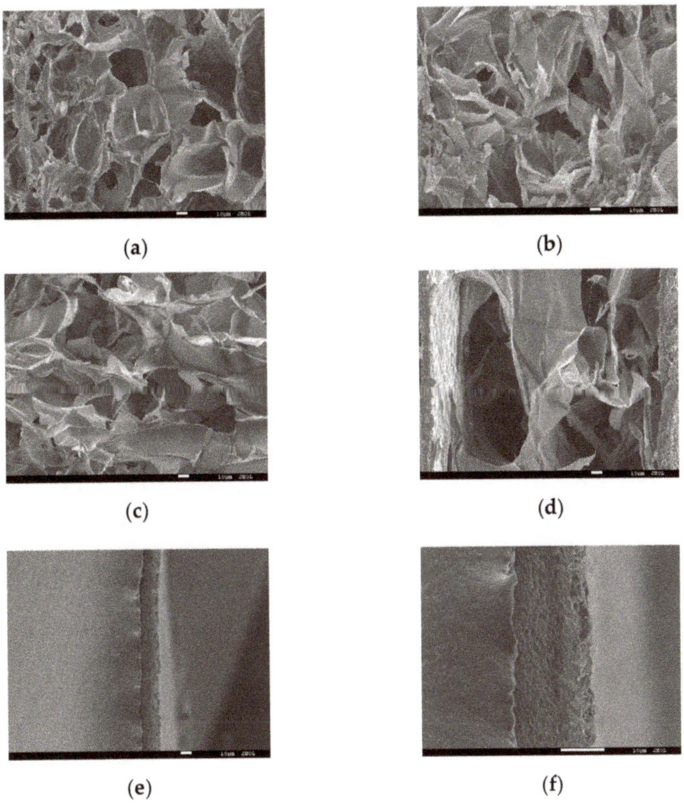

**Figure 2.** Representative scanning electron microscope (SEM) images of *U. pinnatifida* under different cooking methods (10 μm length scale bar): (**a**) raw (×500); (**b**) blanching (×500); (**c**) steaming (×500); (**d**) boiling (×500); (**e**) baking (×500); (**f**) baking (×2000).

### 3.2. Effects of Different Cooking Methods on TPC of U. pinnatifida

The results of the evaluation of TPC in *U. pinnatifida* samples were shown in Table 3. TPC of raw *U. pinnatifida* was 1.91 ± 0.08 mg GAE/g DW, which decreased significantly after cooking ($p < 0.05$). This result indicated the breaking down of polyphenols in the *U. pinnatifida* samples during cooking. Among the four cooking methods, it was observed that the TPC in the steamed *U. pinnatifida* lost the most (32.46%), followed by the blanching samples (25.13%), boiling samples (20.94%) and baking samples (15.18%). High temperature oxidation, leaching during cooking and dissolution of phenolic compounds in hot water may lead to the loss of polyphenols during the cooking [20]. Previous studies have also reported the deleterious effects of different cooking methods on phenolic compounds in vegetables. Armesto et al. investigated some of the effects of cooking methods on the total phenol content of Galega kale [18]. It was found that all cooking methods (boiling, microwaving steaming and pressure cooking) reduced the total phenol content of kale. Sun et al. also found that the TPC in the sweet potato leaves after cooking (microwave, boiling and frying) was significantly reduced compared to untreated sweet potato leaves [33]. Lemos et al. reported that baking, steaming and boiling resulted in the loss of the TPC [34]. Jiang et al. also found a reduction in the TPC of *U. pinnatifida* after air frying and microwave cooking [20].

**Table 3.** Effects of different cooking methods on the contents of total phenols, fucoxanthin and chlorophyll A in *U. pinnatifida*.

| Cooking Methods | Total Phenols mg GAE/g DW | Fucoxanthin µg/g DW | Chlorophyll A µg/g FW |
|---|---|---|---|
| Raw | 1.91 ± 0.08 $^{a}$ | 206.99 ± 7.43 $^{c}$ | 83.43 ± 9.63 $^{a}$ |
| Blanching | 1.43 ± 0.05 $^{c}$ | 229.86 ± 2.24 $^{ab}$ | 62.99 ± 1.27 $^{b}$ |
| Steaming | 1.29 ± 0.07 $^{d}$ | 211.85 ± 2.04 $^{bc}$ | 42.28 ± 2.13 $^{c}$ |
| Boiling | 1.51 ± 0.07 $^{bc}$ | 244.91 ± 7.67 $^{a}$ | 51.35 ± 1.69 $^{bc}$ |
| Baking | 1.62 ± 0.11 $^{b}$ | 216.85 ± 12.27 $^{bc}$ | 27.04 ± 0.98 $^{d}$ |

Data are expressed as mean ± standard deviation ($n = 3$). Different letters in the same column indicate significant differences at ($p < 0.05$).

### 3.3. Effects of Different Cooking Methods on Fucoxanthin Content of U. pinnatifida

The fucoxanthin content was shown in Table 3. Different cooking methods had different effects on the fucoxanthin content in *U. pinnatifida*. The fucoxanthin of raw *U. pinnatifida* was 206.99 ± 7.43 µg/g DW, which increased after the four cooking treatments. Among them, the fucoxanthin content after boiling (244.91 ± 7.67 µg/g DW) and blanching (229.86 ± 2.24 µg/g DW) treatment increased significantly. The results were consistent with the study by Susanto et al., who found higher levels of fucoxanthin in brown algae Sargassum ilicifolium under different blanching treatments than in untreated ones [35]. The increase in content may be because the heat treatment may increase free fucoxanthin in the protein-bound state, such as fucoxanthin chlorophyll a/c binding protein [36]. In addition, the heat treatment may increase the diffusivity of the solvent by affecting the microstructure of the *U. pinnatifida*, thus increasing the permeability of the cell wall, which made the extraction of the fucoxanthin easier [37]. Heating inactivated fucoxanthin-destroying endogenous oxidases, such as polyphenol oxidase and peroxidase, which could prevent further fucoxanthin degradation [35].

### 3.4. Effects of Different Cooking Methods on Chlorophyll A Content of U. pinnatifida

The source of greenness in *U. pinnatifida* was mainly chlorophyll, and its content was related to $a^*$ value to a certain extent. The contents of chlorophyll A in blanching, steaming, boiling and baking samples were 62.99 ± 1.27 µg/g FW, 42.28 ± 2.13 µg/g FW, 51.35 ± 1.69 µg/g FW, and 27.04 ± 0.98 µg/g FW, respectively. These contents were significantly lower than that in the control group (83.43 ± 9.63 µg/g FW). Compared with the raw sample, the degree of chlorophyll A loss was blanching (24.50%) < boiling (38.45%) < steaming (49.32%) < baking

(67.59%). The changes in chlorophyll A contents and greenness ($-a^*$) were almost consistent. The loss of chlorophyll A content during boiling or blanching in water may be due to degradation of chlorophyll to pheophytins and cell contents were leached in boiling water [24]. Similar to our results, Mazzeo et al. observed that steamed and boiled vegetables had lower chlorophyll A content than fresh samples [4]. Possible causes were cell rupture due to heat treatment, resulting in degradation and/or loss of chlorophyll. Additionally, Gonnella et al. observed that all cooking treatments (steaming, boiling, microwaving and sous vide) significantly reduced the content of chlorophyll A of asparagus spears [38]. The most severe loss of chlorophyll A content was observed in the baking samples, probably due to the severe thermal degradation caused by the high temperature of 160 °C. Similarly, Chen et al. confirmed that the chlorophyll A content of Nori (red algae), sea lettuce (green algae) and kombu (brown algae) decreased after boiling and microwaving [39].

## 4. Conclusions

This study showed that different traditional cooking methods had different effects on the bioactive nutrients, color, texture, etc., of *U. pinnatifida*. The influence of cooking methods on bioactive nutrients is important to enhance the health benefits of cooked seaweeds for consumers. Therefore, it is of great significance to select the suitable cooking methods to reduce the loss of bioactive nutrients and quality during cooking. Based on the results of this study, blanching and boiling appeared to be better for *U. pinnatifida* than other methods. In our previous study [20], it was concluded that microwaving was more suitable for cooking *U. pinnatifida* than high temperature and pressure and air frying. However, the mildest traditional cooking methods that preserve quality characteristics and bioactive nutrients still need to be investigated in future studies.

**Author Contributions:** S.J. and M.Y. performed the experiments and drafted the manuscript. Y.W. and P.J. participated in the experiments and performed experiments and data analysis. W.Y. and B.Q. contributed analysis and designed the experiment. H.Q. conceived, designed and supervised the study. All authors have read and agreed to the published version of the manuscript.

**Funding:** The National Key Research and Development Program of China (2019YFD-0902001), Key Science and Technology Program of Liaoning Province (2020JH1/10200001), Dalian Science and Technology Foundation for Distinguished Young Scholars (2018RJ07).

**Data Availability Statement:** Not Applicable.

**Acknowledgments:** This work was supported financially by the National Key Research and Development Program of China (2019YFD-0902001), Key Science and Technology Program of Liaoning Province (2020JH1/10200001), Dalian Science and Technology Foundation for Distinguished Young Scholars (2018RJ07).

**Conflicts of Interest:** There are no conflict to declare. Wei Yin and Bixiang Qiu participated in the experimental design and data analysis. Wei Yin is the director of R&D and Production at Dalian Gaishi Food Co., Ltd.; Bixiang Qiu is the General Manager of Fujian Yida Food Co., Ltd. Dalian Gaishi Food Co., Ltd. and Fujian Yida Food Co., Ltd. have no conflict of interest with this research.

## References

1. Lee, J.B.; Hayashi, K.; Hashimoto, M.; Nakano, T.; Hayashi, T. Novel antiviral fucoidan from sporophyll of *Undaria pinnatifida* (Mekabu). *Chem. Pharm. Bull.* **2004**, *52*, 1091–1094. [CrossRef] [PubMed]
2. Boulom, S.; Robertson, J.; Hamid, N.; Ma, Q.; Lu, J. Seasonal changes in lipid, fatty acid, α-tocopherol and phytosterol contents of seaweed, *Undaria pinnatifida*, in the Marlborough Sounds, New Zealand. *Food Chem.* **2014**, *161*, 261–269. [CrossRef] [PubMed]
3. Wang, S.; Li, Y.; White, W.; Lu, J. Extracts from New Zealand *Undaria pinnatifida* Containing Fucoxanthin as Potential Functional Biomaterials against Cancer in Vitro. *J. Funct. Biomater.* **2014**, *5*, 29–42. [CrossRef] [PubMed]
4. Mazzeo, T.; Paciulli, M.; Chiavaro, E.; Visconti, A.; Fogliano, V.; Ganino, T.; Pellegrini, N. Impact of the industrial freezing process on selected vegetables -Part II. Colour and bioactive compounds. *Food Res. Int.* **2015**, *75*, 89–97. [CrossRef]
5. Gu, R.; Chang, X.; Bai, G.; Li, X.; Di, Y.; Liu, X.; Sun, L.; Wang, Y. Effects of household cooking methods on changes of tissue structure, phenolic antioxidant capacity and active component bioaccessibility of quinoa. *Food Chem.* **2021**, *350*, 129138. [CrossRef]
6. Feng, Q.; Jiang, S.; Feng, X.; Zhou, X.; Wang, H.; Li, Y.; Wang, J.; Tang, S.; Chen, Y.; Zhao, Y. Effect of different cooking methods on sensory quality assessment and in vitro digestibility of sturgeon steak. *Food Sci. Nutr.* **2020**, *8*, 1957–1967. [CrossRef]

7. Shi, S.; Wang, X.; Wu, X.; Shi, W. Effects of four cooking methods on sensory and taste quality of *Portunus trituberculatus*. *Food Sci. Nutr.* **2020**, *8*, 1115–1124. [CrossRef]
8. Schlörmann, W.; Zetzmann, S.; Wiege, B.; Haase, N.U.; Greiling, A.; Lorkowski, S.; Dawczynski, C.; Glei, M. Impact of different roasting conditions on chemical composition, sensory quality and physicochemical properties of waxy-barley products. *Food Funct.* **2019**, *10*, 5436–5445. [CrossRef]
9. Iborra-Bernad, C.; Tárrega, A.; García-Segovia, P.; Martínez-Monzó, J. Advantages of sous-vide cooked red cabbage: Structural, nutritional and sensory aspects. *LWT—Food Sci. Technol.* **2014**, *56*, 451–460. [CrossRef]
10. Akdaş, Z.Z.; Bakkalbaşı, E. Influence of different cooking methods on color, bioactive compounds, and antioxidant activity of kale. *Int. J. Food Prop.* **2017**, *20*, 877–887. [CrossRef]
11. Barakat, H.; Rohn, S. Effect of different cooking methods on bioactive compounds in vegetarian, broccoli-based bars. *J. Funct. Foods* **2014**, *11*, 407–416. [CrossRef]
12. Murador, D.; Braga, A.R.; Da Cunha, D.; De Rosso, V. Alterations in phenolic compound levels and antioxidant activity in response to cooking technique effects: A meta-analytic investigation. *Crit. Rev. Food Sci.* **2018**, *58*, 169–177. [CrossRef]
13. Pérez-Burillo, S.; Rufián-Henares, J.Á.; Pastoriza, S. Effect of home cooking on the antioxidant capacity of vegetables: Relationship with Maillard reaction indicators. *Food Res. Int.* **2019**, *121*, 514–523. [CrossRef]
14. Qi, X.; Cheng, L.; Li, X.; Zhang, D.; Wu, G.; Zhang, H.; Wang, L.; Qian, H.; Wang, Y. Effect of cooking methods on solubility and nutrition quality of brown rice powder. *Food Chem.* **2019**, *274*, 444–451. [CrossRef]
15. Giusti, F.; Capuano, E.; Sagratini, G.; Pellegrini, N. A comprehensive investigation of the behaviour of phenolic compounds in legumes during domestic cooking and in vitro digestion. *Food Chem.* **2019**, *285*, 458–467. [CrossRef]
16. Koç, M.; Baysan, U.; Devseren, E.; Okut, D.; Atak, Z.; Karataş, H.; Kaymak-Ertekin, F. Effects of different cooking methods on the chemical and physical properties of carrots and green peas. *Innov. Food Sci. Emerg.* **2017**, *42*, 109–119. [CrossRef]
17. Managa, M.G.; Remize, F.; Garcia, C.; Sivakumar, D. Effect of moist cooking blanching on colour, phenolic metabolites and glucosinolate content in Chinese cabbage (*Brassica rapa* L. Sub sp. Chinensis). *Foods* **2019**, *8*, 399. [CrossRef]
18. Armesto, J.; Gómez-Limia, L.; Carballo, J.; Martínez, S. Effects of different cooking methods on some chemical and sensory properties of Galega kale. *Int. J. Food Sci. Technol.* **2016**, *51*, 2071–2080. [CrossRef]
19. Peng, J.; Yi, J.; Bi, J.; Chen, Q.; Wu, X.; Zhou, M.; Liu, J. Freezing as pretreatment in instant controlled pressure drop (DIC) texturing of dried carrot chips: Impact of freezing temperature. *LWT—Food Sci. Technol.* **2018**, *89*, 365–373. [CrossRef]
20. Jiang, S.; Wang, Y.; Song, H.; Ren, J.; Zhao, B.; Zhu, T.; Yu, C.; Qi, H. Influence of Domestic Cooking on Quality, Nutrients and Bioactive Substances of *Undaria pinnatifida*. *Foods* **2021**, *10*, 2786. [CrossRef]
21. Ummat, V.; Tiwari, B.K.; Jaiswal, A.K.; Condon, K.; Garcia-Vaquero, M.; O'Doherty, J.; O'Donnell, C.; Rajauria, G. Optimisation of ultrasound frequency, extraction time and solvent for the recovery of polyphenols, phlorotannins and associated antioxidant activity from brown seaweeds. *Mar. Drugs* **2020**, *18*, 250. [CrossRef]
22. Sui, Y.; Gu, Y.; Lu, Y.; Yu, C.; Zheng, J.; Qi, H. Fucoxanthin@Polyvinylpyrrolidone Nanoparticles Promoted Oxidative Stress-Induced Cell Death in Caco-2 Human Colon Cancer Cells. *Mar. Drugs* **2021**, *19*, 92. [CrossRef]
23. Havlíková, L.; Šatínský, D.; Opletal, L.; Solich, P. A Fast Determination of Chlorophylls in Barley Grass Juice Powder Using HPLC Fused-Core Column Technology and HPTLC. *Food Anal. Method* **2014**, *7*, 629–635. [CrossRef]
24. Mashiane, P.; Mashitoa, F.M.; Slabbert, R.M.; Sivakumar, D. Impact of household cooking techniques on colour, antioxidant and sensory properties of African pumpkin and pumpkin leaves. *Int. J. Gastron. Food Sci.* **2021**, *23*, 100307. [CrossRef]
25. Pellegrini, N.; Chiavaro, E.; Gardana, C.; Mazzeo, T.; Contino, D.; Gallo, M.; Riso, P.; Fogliano, V.; Porrini, M. Effect of different cooking methods on color, phytochemical concentration, and antioxidant capacity of raw and frozen brassica vegetables. *J. Agric. Food Chem.* **2010**, *58*, 4310–4321. [CrossRef]
26. Zhong, X.; Dolan, K.D.; Almenar, E. Effect of steamable bag microwaving versus traditional cooking methods on nutritional preservation and physical properties of frozen vegetables: A case study on broccoli (*Brassica oleracea*). *Innov. Food Sci. Emerg.* **2015**, *31*, 116–122. [CrossRef]
27. Ikoko, J.; Kuri, V. Osmotic pretreatment effect on fat intake reduction and eating quality of deep-fried plantain. *Food Chem.* **2007**, *102*, 523–531. [CrossRef]
28. Nguyen, T.V.; Tran, T.Y.; Lam, D.; Bach, L.; Nguyen, D.C. Effects of microwave blanching conditions on the quality of green asparagus (*Asparagus officinalis* L.) butt segment. *Food Sci. Nutr.* **2019**, *7*, 3513–3519. [CrossRef]
29. Adhikari, B.; Howes, T.; Bhandari, B.R.; Truong, V. Stickiness in Foods: A Review of Mechanisms and Test Methods. *Int. J. Food Prop.* **2001**, *4*, 1–33. [CrossRef]
30. Sun, Q.; Zhang, M.; Yang, P. Combination of LF-NMR and BP-ANN to monitor water states of typical fruits and vegetables during microwave vacuum drying. *LWT—Food Sci. Technol.* **2019**, *116*, 108548. [CrossRef]
31. Khan, M.I.H.; Wellard, R.M.; Nagy, S.A.; Joardder, M.U.H.; Karim, M.A. Investigation of bound and free water in plant-based food material using NMR T2 relaxometry. *Innov. Food Sci. Emerg.* **2016**, *38*, 252–261. [CrossRef]
32. Yang, D.; Wu, G.; Li, P.; Qi, X.; Zhang, H.; Wang, X.; Jin, Q. The effect of fatty acid composition on the oil absorption behavior and surface morphology of fried potato sticks via LF-NMR, MRI, and SEM. *Food Chem.* **2020**, *7*, 100095. [CrossRef]
33. Sun, H.; Mu, T.; Xi, L.; Song, Z. Effects of domestic cooking methods on polyphenols and antioxidant activity of sweet potato leaves. *J. Agric. Food Chem.* **2014**, *62*, 8982–8989. [CrossRef] [PubMed]

34. Lemos, M.A.; Aliyu, M.M.; Hungerford, G. Influence of cooking on the levels of bioactive compounds in Purple Majesty potato observed via chemical and spectroscopic means. *Food Chem.* **2015**, *173*, 462–467. [CrossRef]
35. Susanto, E.; Fahmi, A.S.; Agustini, T.W.; Rosyadi, S.; Wardani, A.D. Effects of different heat processing on fucoxanthin, antioxidant activity and colour of Indonesian Brown seaweeds. *IOP Conf. Ser. Earth Environ. Sci.* **2017**, *55*, 012063. [CrossRef]
36. Di Valentin, M.; Meneghin, E.; Orian, L.; Polimeno, A.; Büchel, C.; Salvadori, E. Triplet–triplet energy transfer in fucoxanthin-chlorophyll protein from diatom *Cyclotella meneghiniana*: Insights into the structure of the complex. *BBA—Bioenerg.* **2013**, *1827*, 1226–1234. [CrossRef]
37. Nie, J.; Chen, D.; Lu, Y.; Dai, Z. Effects of various blanching methods on fucoxanthin degradation kinetics, antioxidant activity, pigment composition, and sensory quality of *Sargassum fusiforme*. *LWT—Food Sci. Technol.* **2021**, *143*, 111179. [CrossRef]
38. Gonnella, M.; Durante, M.; Caretto, S.; D'Imperio, M.; Renna, M. Quality assessment of ready-to-eat asparagus spears as affected by conventional and sous-vide cooking methods. *LWT—Food Sci. Technol.* **2018**, *92*, 161–168. [CrossRef]
39. Chen, K.; Roca, M. Cooking effects on chlorophyll profile of the main edible seaweeds. *Food Chem.* **2018**, *266*, 368–374. [CrossRef]

*foods*

MDPI

*Article*

# Effect of Extraction Method on the Bioactive Composition, Antimicrobial Activity and Phytotoxicity of Pomegranate By-Products

**Lara Campos [1,2,*], Luana Seixas [3], Susana Dias [1,2], António M. Peres [4], Ana C. A. Veloso [3,5,6] and Marta Henriques [1,2]**

1 Polytechnic Institute of Coimbra, Coimbra Agriculture School, Bencanta, 3045-601 Coimbra, Portugal; sudias@esac.pt (S.D.); mhenriques@esac.pt (M.H.)
2 CERNAS, Research Centre for Natural Resources, Environment and Society, Coimbra Agriculture School, Bencanta, 3045-601 Coimbra, Portugal
3 Polytechnic Institute of Coimbra, ISEC, DEQB, Rua Pedro Nunes—Quinta da Nora, 3030-199 Coimbra, Portugal; luana.seixas00@gmail.com (L.S.); anaveloso@isec.pt (A.C.A.V.)
4 Centro de Investigação de Montanha (CIMO), ESA, Instituto Politécnico de Bragança, Campus de Santa Apolónia, 5300-253 Bragança, Portugal; peres@ipb.pt
5 CEB—Centre of Biological Engineering, University of Minho, Campus de Gualtar, 4715-057 Braga, Portugal
6 LABBELS—Associate Laboratory, Braga/Guimarães, Portugal
* Correspondence: lara.campos@esac.pt

**Abstract:** Pomegranate by-products can be an asset to the food industry due to the richness in bioactive and antimicrobial compounds. This work studied the influence of conventional solvent and sonication-assisted extraction methods on the bioactive profile, antimicrobial properties, and phytotoxicity effect of the peels and seeds extracts from Acco, Big Full, and Wonderful pomegranate cultivars. The bioactive composition of the extracts was evaluated for the content of total phenolics, total flavonoids, and antioxidant activity (expressed as the half-maximal inhibitory concentration—$IC_{50}$) by spectrophotometric methods, while the tannins were determined by titration and the anthocyanins were estimated by the pH-differential method. For the evaluation of the antimicrobial activity, the disk diffusion method of Kirby-Bauer was adapted through inhibition halos against *Escherichia coli*, *Pseudomonas aeruginosa*, *Staphylococcus aureus*, *Bacillus cereus*, and *Yarrowia lipolytica*. The extracts' phytotoxicity was evaluated in vitro on garden-cress seeds. Extracts from conventional extraction were richer in total phenolics, expressed as gallic acid equivalents (0.16–0.73 mg GAE/mg extract), while those from sonication-assisted extraction had higher contents of total flavonoids, expressed as catechin equivalents (0.019–0.068 mg CATE/mg extract); anthocyanins, expressed as cyanidin-3-glucoside (0.06–0.60 μg C3G/mg, dry basis); and antioxidant activity ($IC_{50}$, 0.01–0.20 mg/mL). All extracts were more effective against Gram-positive bacteria and yeasts than Gram-negative bacteria. In general, the sonication-assisted extracts led to higher inhibition halos (8.7 to 11.4 mm). All extracts presented phytotoxicity against garden-cress seeds in the tested concentrations. Only the lowest concentration (0.003 mg/mL) enabled the germination of seeds and root growth, and the sonication-assisted extracts showed the highest Munoo-Liisa vitality index (51.3%). Overall, sonication-assisted extraction obtained extracts with greater bioactive and antimicrobial potential and less phytotoxicity.

**Keywords:** sonication-assisted extraction; solvent extraction; *Punica granatum* L.; pomegranate peels; pomegranate seeds; antimicrobial activity; phytotoxicity

**Citation:** Campos, L.; Seixas, L.; Dias, S.; Peres, A.M.; Veloso, A.C.A.; Henriques, M. Effect of Extraction Method on the Bioactive Composition, Antimicrobial Activity and Phytotoxicity of Pomegranate By-Products. *Foods* **2022**, *11*, 992. https://doi.org/10.3390/foods11070992

Academic Editors: Theodoros Varzakas, Arun K. Bhunia, Michela Verni and Federico Casanova

Received: 20 January 2022
Accepted: 26 March 2022
Published: 29 March 2022



## 1. Introduction

In the European Union, food waste generated during food processing represents up to 39% of the total food waste [1]. A large amount of the food waste corresponds to by-products that are rich in valuable compounds. Pomegranate (*Punica granatum* L.) is a fruit grown all over the world, predominantly in western Asia and the Mediterranean region [2]. The proportions of peel:arils:seeds of the pomegranate fruit are, respectively, 50:40:10 [3,4], which means that during pomegranate processing, about 60% of the fruit is produced as a by-product that could potentially be discarded [5], representing additional costs for its disposal [6]. However, these by-products contain important amounts of phenolic compounds (such as flavonoids and tannins), sugars, organic acids, and minerals, and have antioxidant, antifungal, and antibacterial activities [2–5,7].

Phenolic compounds from plant matrices are responsible for several benefits to human health. These benefits come from the ability of these phytochemicals to alter enzymatic and chemical reactions [8]. Due to the increasing awareness of consumers about the benefits of enhancing their quality of life through the consumption of natural compounds (e.g., prebiotics, probiotics, supplements, dietary fibers, or functional foods), intense research has been carried out on their importance, mechanisms of action, and recovery processes. The recognized biological potential of pomegranate by-products makes them good candidates for reintegration into the industry chain (transformed or incorporated into other products) after the recovery of the compounds of interest [9]. Although there are several methods capable of recovering these compounds, such as supercritical fluid extraction, microwave-assisted extraction, extraction with pressurized liquids, and extraction with pressurized hot water [10], the most applied is conventional solvent extraction.

Conventional extraction methods include organic solvent extraction and distillation [11], with Soxhlet extraction, maceration, and hydrodistillation [12,13] being the classical techniques. These methods generally depend on the effect of solvent, temperature, and extraction time on the matrix. The increase in the extraction temperature promotes the mass transfer and diffusion of the compounds present in the matrix to the solvent and enhances the solubility of the extracted compounds [14].

However, conventional methods often have practical, economic, and social concerns that are difficult to overcome or fail to achieve sustainability. Some of its drawbacks are related to the matrix overheating, which can lead to loss of functionality or stability of the final product (degradation of compounds of interest during extraction) [14,15], high energy consumption (and resources in general) [10], emission of volatile organic compounds and the related polluting effect [15,16], and difficulties in complying with increasingly strict safety regulations. Thus, greener technological processes have emerged, which seem to overcome some of these problems [10,15]. Sonication-assisted extractions, which are a type of ultrasound extraction, can be used for this purpose as the extraction can be completed in less time with high repeatability [15,17].

Ultrasound has been recognized as a potential method of extracting oils, proteins, and bioactive compounds from plants [18] because the propagation of the pressure waves and the resulting cavitation forces disrupts the cell walls and improves the release of substances into the solvent [12,15,19]. This extraction method has the main advantages of not being destructive of active ingredients in plant matrices and intensifying the extraction of bioactive compounds (such as phenolics) [10].

This work aims to compare the efficiency of two extraction methods (conventional and sonication-assisted extraction) on the bioactive quality of the recovered extracts. For this, the phytochemical and antimicrobial potential of the extracts of the peels and seeds of three pomegranate cultivars (Acco, Big Full, and Wonderful) grown in the Alentejo region (southeast Portugal) were assessed through the evaluation of the content of total phenolic compounds (TPC), total flavonoids (TF), tannins (TAN), anthocyanins (ANT), antioxidant activity (AA), inhibition halos against various microorganisms, and, finally, its phytotoxicity towards garden-cress seeds.

On the other hand, to date, few research studies have provided information on the bioactive and antimicrobial potential of pomegranate peels and seeds, namely for fruits grown under Portuguese agroclimatic conditions [20], and no studies have been found for Big Full cultivar worldwide. In this context, it was also intended to contribute with new knowledge in order to strengthen the hypothesis that these by-products can be valorized and used as raw materials for food, pharmaceutical, or cosmetic industries, and, even, for agricultural purposes.

## 2. Materials and Methods

Peels and seeds from Acco, Big Full, and Wonderful cultivars were submitted to two extraction methods: conventional and sonication-assisted extraction, using an ethanol:water mixture (50%, $v/v$). The obtained extracts were then analyzed to quantify the TPC, TF, TAN, ANT, and AA. The antimicrobial potential was evaluated through inhibition halos against two Gram-positive bacteria (*Bacillus cereus* and *Staphylococcus aureus*), two Gram-negative bacteria (*Escherichia coli* and *Pseudomonas aeruginosa*), and one yeast (*Yarrowia lipolytica*). The phytotoxicity of the extracts that presented the best bioactive and antimicrobial potential was evaluated against garden-cress seeds.

### 2.1. Pomegranate Cultivars, Peels, and Seeds Recovery and Preparation

Pomegranate fruits from Acco, Big Full, and Wonderful cultivars were supplied by POM Portugal Lda and harvested in the Alentejo region, in the southeast of Portugal (GPS coordinates 37.81717, −8.19534). All pomegranate cultivars were grown under the same climatic conditions and underwent the same agronomic practices (fertilization, irrigation, harvesting, storage, and post-harvest treatments). Although Acco (Akko) and Wonderful are already known and studied cultivars, Big Full is a new and improved pomegranate cultivar (from Acco) and, thus, is less studied.

Several factors can influence the amount of bioactive compounds in a matrix, and the fruit maturation stage is one of them. It was observed that the antioxidant activity decreases as the fruit grows, since the development of the fruit leads to a decrease in the phenolic acids content [21]. In addition, fruit development changes the composition of flavonoids [22]. To limit this possible influence, in this work, all fruits were harvested fully mature, by evaluating the soluble solids content (°Brix) of the juice, which was 15° Brix for Acco, 16° Brix for Big Full, and 15° Brix for Wonderful.

After harvesting, the ripe fruits were transported in appropriate boxes to the Agriculture School pilot plant (Coimbra). Fruits were washed and sorted to remove the rotten ones. Proper fruits were cut in 4 pieces, and the arils were separated from the peels using a grape stem removing machine (COSVAL, Mizar 60, Cosvalinox, Oliveirinha, Portugal). The arils were crushed and squeezed using a pressing machine (Aguinox, Marmorier 30 × 40, Águeda, Portugal), and the pomegranate juice was discarded. The peels and seeds were then coarsely ground and dried in a forced hot air dryer (Conterm Drying Oven 2000210, J.P. Selecta, Spain) at 70 °C for 24 h. The final moisture contents of the peels and seeds were, respectively, 17.8% and 21.2% for Acco; 10.1% and 14.1% for Big Full; and 19.1% and 15.7% for Wonderful. Before extraction, the dry material was finely ground into a powder using a mix grinder (Classic 123, 700 W, Moulinex, Écully, France), packed in sealed plastic bags, and stored at room temperature (20–25 °C) protected from the light.

### 2.2. Chemicals

Methanol (Ceamed, Lda., Funchal), Folin and Ciocalteu's phenol reagent (Biochem Chemopharma, Cosne-Cours-sur-Loire, France), sodium hydroxide p.a. (Eka, Netherlands), sodium acetate (Honeywell, Charlotte, CA, USA), sodium nitrite p.a. (Merck, Darmstadt, Germany), indigo carmine (labkem, Spain), potassium permanganate (Acofarma, Madrid, Spain), sodium carbonate anhydrous p.a., potassium chloride (Panreac, Spain), absolute ethanol and sulfuric acid 95–97% (Chem-Lab, Zedelgem, Belgium), gallic acid (GA), catechin (CAT), aluminum chloride (99%), and 2,2-diphenyl-1-picrylhydrazyl (DPPH), purchased from Sigma-Aldrich (St. Louis, MO, USA), were used as chemicals for the extractions and bioactive characterization of the extracts.

Agar Powder (VWR, Lutterworth, UK); amphotericin B solution (Sigma, USA); blank, kanamycin (K) 30UG, and penicillin G (P) 2IU discs (LIOFILCHEM, Roseto degli Abruzzi, Italy); D(+)-glucose monohydrate and yeast extract (Scharlau, Barcelona, Spain); dimethyl sulfoxide (DMSO) and Mueller-Hinton Agar (Merck, Germany); Nutrient Agar (Biolab, Budapest, Hungary); and Nutrient Broth and Peptone (Cultimed, Spain) were used as chemicals for antimicrobial characterization of the extracts.

### 2.3. *Pomegranate Peels and Seeds Extraction Methods*

Ethanol:water mixture (EtOH 50%, *v*/*v*) was prepared and used as solvent for the extraction of the bioactive compounds from pomegranate peels and seeds powders, and a solid:solvent ratio of 0.02 g/mL was applied. The conventional solid:liquid extraction was conducted for 4 h, in sealed glass flasks under continuous stirring (200 rpm) andimmersed in a water bath at 50 °C. For sonication-assisted extraction, the Q700 sonicator (QSonica, Newtown, CT, USA) equipped with a probe (CL-334, Qsonica, USA) was used and the pomegranate peels and seeds were sonicated at room temperature, for 20 min under a frequency of 20 kHz, which was selected based on the literature data regarding different plant matrices, aiming to avoid the formation of free radicals that is promoted by frequencies greater than 20 kHz [10,23–27]. After extraction, the solid:liquid mixtures were filtered under vacuum. The obtained liquid extracts were used for tannin and anthocyanin content determination. Then, the solvent was evaporated at 90 °C, until ca. 20 mL, using a rotary evaporator (Rotavapor R-210, Buchi, Switzerland) under vacuum. The concentrated extracts were frozen at −18 °C overnight, freeze-dried (UNICRYO MC 4l −60 °C, Uniequip, Planegg, Germany) and stored at −18 °C until further analysis. Four independent extractions were made for each cultivar, type of by-product, and extraction. The extraction yields (EY) were calculated as the mass of the extract recovered from the mass of the dry pomegranate material used for the extraction (mg of extract per 100 mg of dry by-product) and expressed as percentage.

### 2.4. *Bioactive Compounds Quantification*

#### 2.4.1. Total Phenolic Compounds Determination

The TPC in the pomegranate by-product extracts was determined spectrophotometrically following the method proposed by Singleton and Rossi [28]. The peels and seeds extracts were dissolved in methanol:water (70:30, *v*/*v*) at a concentration of 0.2 mg/mL. Gallic acid was used as standard for the calibration curve, within a concentration range of 0.07–0.70 mg/mL ($Abs_{750nm} = 9.729 \times TPC - 0.019$; $R^2 = 0.995$). A 200 µL of Folin-Ciocalteu reagent was added to 200 µL of the dissolved extract and placed in a water bath at 40 °C. After 4 min, 1600 µL of 5% $Na_2CO_3$ (*w*/*v*) were added. After 20 min of adding the Folin–Ciocalteu reagent, the extracts were removed from the water bath, and the absorbance was measured at 750 nm on a UV/VIS Spectrometer T80+ (PG Instruments Ltd., Lutterworth, UK). A methanol:water solution (70:30, *v*/*v*) was used as a blank sample. TPC were expressed as mg of gallic acid equivalent per mg of extract (mg GAE/mg extract).

#### 2.4.2. Total Flavonoids Determination

The TF were assessed spectrophotometrically according to the method proposed by Kim et al. [29]. The extracts were dissolved in a methanol:water solution (50:50, *v*/*v*) at a concentration of 0.6 mg/mL. Catechin was used as standard for the calibration curve, within a concentration range of 0.08–0.40 mg/mL ($Abs_{510nm} = 2.985 \times TF + 0.049$; $R^2 = 0.990$). First, 4 mL of distilled water were added to 1 mL of the extract solution, followed by 0.3 mL of 5% $NaNO_2$ (*w*/*v*). After 5 min, 0.3 mL of 10% $AlCl_3$ (*w*/*v*) were added to the mixture. Then, 6 min later, 2 mL of NaOH (1 M) and 2.4 mL of distilled water were also added, and the absorbance of the final mixture was measured at 510 nm in a UV/VIS Spectrometer (T80+, PG Instruments Ltd.). A methanol:water solution (50:50, *v*/*v*) was used as blank sample. TF were expressed as mg of catechin equivalent per mg of extract (mg CATE/mg extract).

2.4.3. Tannins Determination

The TAN content was analyzed according to the method described by Atassanova and Christova-Bagdassarian [30], with some modifications. After extraction and vacuum filtration of the solid part, 5 mL of each extract solution were mixed with 5 mL of indigo solution and 150 mL of distilled deionized water. An aqueous solution of $KmnO_4$ (0.1 N) was used for titration, until the blue color turns to a golden yellow. A blank test was performed using 5 mL of the extraction solvent EtOH 50%, instead of an extract solution. The tannin content was determined using Equation (1) and expressed as percentage of tannin mass per total sample mass on a dry basis (%, $w/w$, db).

$$\%TAN = \frac{(V - V_0) \times 0.004157 \times 100}{W} \quad (1)$$

where, $V$ is the volume of $KmnO_4$ solution spent for titration of the sample, $V_0$ is the volume of $KmnO_4$ solution spent for titration of the blank, 0.004157 is tannin equivalent in 1 mL of 0.1 N of $KmnO_4$ solution, and $W$ is the weight of the peels or seeds powder used.

2.4.4. Anthocyanins Determination

Anthocyanin content (ANT) was estimated by the pH-differential method of Sellappan et al. [31] with modifications, using two buffer systems: potassium chloride buffer, pH 1.0 (0.025 M), and sodium acetate buffer, pH 4.5 (0.4 M). An amount of 0.4 mL of each extract solution was mixed separately with 1.6 mL of each buffer, and the absorbance was read at 510 and 700 nm in a UV/VIS Spectrometer (T80+, PG Instruments Ltd.). Distilled water was used as the blank solution to calibrate the spectrometer.

Monomeric anthocyanin pigment concentration in the extract solution was calculated by Equation (2) and expressed as µg of cyanidin-3-glucoside (C3G) per mg of sample (µg C3G/mg sample, db).

$$\text{ANT}\ (\mu\text{g C3G/mg sample}) = \frac{Abs \times MW \times DF \times 1000}{\in \times 1} \quad (2)$$

where, $Abs = (Abs_{510nm} - Abs_{700nm})_{pH1.0} - (Abs_{510nm} - Abs_{700nm})_{pH4.5}$, $MW$ is the molecular weight of C3G (449.2 g/mol), $DF$ is the dilution factor; and $\in$ is the molar absorptivity of C3G (26,900).

2.4.5. Antioxidant Activity (DPPH Radical Scavenging Assay)

The AA of the extracts was measured spectrophotometrically [32–34]. DPPH was dissolved in EtOH to obtain a concentration of 0.3 mM. The extracts were dissolved, at least in 3 different concentrations, in EtOH 50%. An amount of 2.5 mL of EtOH 50% was used as control and 2.5 mL of the extract solutions were used as samples. Then, 1 mL of DPPH solution was added to the control and samples and left for 30 min protected from light. After this period, the absorbance was measured at 517 nm (UV/VIS Spectrometer T80+, PG Instruments Ltd.). For each sample, a blank with 2.5 mL of the extract solution and 1 mL of EtOH was used. The AA was calculated using Equation (3).

$$\%AA = \left(1 - \frac{Abs_s - Abs_b}{Abs_c}\right) \times 100 \quad (3)$$

where, $Abs_s$ is the absorbance of the sample, $Abs_b$ is the absorbance of the blank, and $Abs_c$ is the absorbance of the control.

The $IC_{50}$ (half-maximal inhibitory concentration, i.e., the amount of antioxidant required to decrease the initial DPPH concentration by 50%), was determined by the linear fitting of the data of %AA vs. extract concentrations (%AA must give results below and above of 50%). Finally, the AA of each extract was expressed in terms of $IC_{50}$, in mg/mL.

All the analyses of TPC, TF, TAN, ANT, and AA were performed in duplicates of two independent assays.

### 2.5. Antimicrobial Analysis

#### 2.5.1. Microorganisms' Activation and Preparation

The antimicrobial evaluation of the extracts was performed against pathogenic Gram-positive bacteria (*Bacillus cereus* ATCC 10876 and *Staphylococcus aureus* ATCC 29213), Gram-negative bacteria (*Escherichia coli* ATCC 25922 and *Pseudomonas aeruginosa* ATCC 27853), and a yeast (*Yarrowia lipolytica* ISA 1774).

Based on previous studies [35–37], 10% DMSO (dimethylsulfoxide) was selected to be used as a solvent to study the antimicrobial activity of pomegranate extracts that were diluted to a concentration of 0.30 g/mL.

Before each test, all microorganisms were activated in their respective media and incubated at specific temperature and time. Pathogenic bacteria were incubated at 37 °C for 24 h in Mueller-Hinton Agar (MHA) and yeast at 25 °C for 72 h in GYPA medium (20 g glucose, 10 g yeast extract, 20 g peptone, and 15 g agar powder, for 1000 mL of distilled water). After incubation, the microorganisms were suspended in Nutrient Broth (NB) or GYP medium (for bacteria and yeasts, respectively), in sterile test tubes, measuring their density with a Densichek Plus densitometer (BIOMERIEUX, Linda-a-Velha, Portugal), until reaching 0.5 McFarland ($1.5 \times 10^8$ CFU/mL for bacteria; $2.0 \times 10^6$ CFU/mL for yeast).

#### 2.5.2. Inhibition Halos

To determine the inhibition halos of the extracts against each microorganism was used the disk diffusion method of Kirby-Bauer with adaptations. An amount of 250 μL of the inoculum was placed in a Petri dish with MHA (for bacteria) or GYPA (for yeast). Blank discs (in triplicate) were submerged with 20 μL of each re-suspended extract and placed on top of the inoculum. One blank disc with 20 μL of distilled water was used as a negative control. Kanamycin and penicillin discs were used as positive controls for bacteria, and 20 μL of amphotericin B solution was placed in a blank disc to serve as a positive control for yeast. The Petri dishes were incubated at 37 °C for 24 h for bacteria and at 25 °C for 48 h for yeast. The diameters of the inhibition halos of the extracts and respective controls were measured in millimeters. The effectiveness of the extracts is determined by comparing their inhibition halos with those of the positive control. The results of the inhibition halos allowed to determine the Antimicrobial Activity Index (AAI), according to Equation (4), adapted from Vancheva et al. [38].

$$\%\,\text{AAI} = -1 \times \frac{A - E}{A + E} \times 100 \tag{4}$$

where, $A$ is the mean value of the inhibition halo (mm) promoted by the positive control (antibiotic/antifungal) and $E$ is the mean value of the inhibition halo (mm) promoted by the extract.

AAI ranges between [−100% to 100%]. AAI equal to −100% means the extract showed no inhibition halos; AAI values in the range ]−100 to 0%[ mean the extract showed smaller inhibition halos than the control; AAI equal to 0% means the extract and control presented the same result in relation to the inhibition halos; AAI values in the range ]0 to 100%[ mean the extract showed higher inhibition halos than the control; and AAI equal to 100% means the control showed no inhibition halos.

This analysis was performed in triplicate with three independent assays.

### 2.6. Phytotoxicity Assay

For germination tests, each extract was first prepared with distilled water to a concentration of 0.30 g/mL and diluted to 0.03, 0.01, and 0.003 g/mL. From each extract concentration, 5 mL was placed in Petri dishes (Ø 10 cm) with filter paper, and 10 seeds of garden-cress (*Lepidium sativum* L., purchased from a local market) were placed equidistantly. Distilled water was used for the control under the same conditions. All Petri dishes were placed in an incubator at 25 °C for 5 days, protected from light. Every day the number of germinated seeds and the root length were recorded.

The results were analyzed according to the guidelines of EN 16086-1 and EN 16086-2 [39,40], where the Number of Germinated Seeds, Germination Rate (%), Root Length (mm), and the Munoo–Liisa Vitality Index (%) were determined. The Munoo–Liisa Vitality Index (Table 1) allowed for classifying the extracts according to their phytotoxicity.

**Table 1.** Phytotoxicity classification (adapted from [41]).

| Munoo–Liisa Vitality Index (%) | Classification |
| --- | --- |
| >100 | Enhances germination and root growth |
| 80–100 | Nonphytotoxic |
| 60–80 | Moderately phytotoxic |
| 40–60 | Phytotoxic |
| <40 | Very phytotoxic |

### 2.7. Statistical Analysis

All data are expressed as mean values ± standard deviation. Statistical analysis was performed using GraphPad Prims Software version 8.0.2 (GraphPad Software, Inc., San Diego, CA, USA). The normality distribution of the data was evaluated by the Shapiro-Wilk test at a significance level of 5%. A three-way analysis of variance (3-way ANOVA) was applied to infer the statistical significance of the effects under study as well as the respective interactions in the bioactive and antimicrobial potentials results. If the 2-way and 3-way interaction effects were not significant, Tukey's test was further used to determine the differences among means obtained for different samples. All analyses were performed at a 5% significance level. Correlations between parameters were established using Pearson's correlation coefficient ($r$). A one-way analysis of variance (ANOVA) using Tukey's test was performed to determine the differences between the means obtained in phytotoxicity analysis at significance level of 5%.

Principal component analysis (PCA) was performed using statistical program R (version 2.15.1), at a 5% significance level. This analysis was applied as an unsupervised pattern recognition tool to evaluate the overall potential of the EY, TPC, TF, TAN, ANT, AA, and inhibition halos data, determined based on conventional analytical techniques, to classify the extracts according to the pomegranate extraction method (conventional or sonication-assisted), by-product (peels or seeds), or the pomegranate cultivar (Acco, Big Full, and Wonderful).

## 3. Results and Discussion

### 3.1. Extraction Yield and Bioactive Compounds

The extraction yields (EY), total phenolic compounds (TPC), total flavonoids (TF), tannins (TAN), anthocyanins (ANT), and the antioxidant activity (AA, expressed as $IC_{50}$) of the extracts of peels and seeds of Acco, Big Full, and Wonderful cultivars, obtained using the two studied extraction methods, are shown in Table 2. The three main effects under study (cultivar, by-product, and extraction method) significantly influenced the TPC, TF, ANT, and AA values ($p < 0.0001$). Since for each parameter under study one or more of the 2-way/3-way interactions were statistically significant ($p$-value < 0.05), the significance of the main effects could not be further interpreted based on the output of the post-hoc multicomparison tests. The EY and TAN were only affected by the cultivar and the by-product. The highest overall TPC was obtained with conventional extraction compared to the sonication-assisted extraction (0.16 to 0.73 mg GAE/mg extract vs. 0.11 to 0.50 mg GAE/mg extract, respectively). The TF, ANT, and AA were enhanced by the sonication-assisted extraction, reaching 0.019 to 0.068 mg CATE/mg extract for TF, 0.06 to 0.60 µg C3G/mg db for ANT, and 0.010 to 0.200 mg/mL for AA. Figure S1 presents a visual representation of the PCA results and the output clearly pointed out that each of the main effects considered could be effectively differentiated using an unsupervised PCA model based on the first three principal components (PCs).

**Table 2.** Extraction yield (EY, %, db), total phenolic compounds (TPC, mg GAE/mg extract), total flavonoids (TF, mg CATE/mg extract), tannins (TAN, % $w/w$, db), anthocyanins (ANT, µg C3G/mg, db), and antioxidant activity (AA expressed in terms of $IC_{50}$, mg/mL, db) of the peels and seeds of three pomegranate cultivars (Acco, Big Full, Wonderful) according to extraction method (conventional vs. sonication-assisted).

| Extraction Method | By-Product | Cultivar | EY | TPC | TF | TAN | ANT | AA ($IC_{50}$) |
|---|---|---|---|---|---|---|---|---|
| Conventional | Peels | Acco | 49.9 ± 0.9 | 0.39 ± 0.02 | 0.029 ± 0.002 | 16.7 ± 0.6 | 0.05 ± 0.02 | 0.024 ± 0.000 |
| | | Big Full | 51.0 ± 0.3 | 0.73 ± 0.18 | 0.052 ± 0.003 | 25.3 ± 0.2 | 0.12 ± 0.01 | 0.180 ± 0.005 |
| | | Wonderful | 46.3 ± 4.8 | 0.32 ± 0.01 | 0.042 ± 0.005 | 18.7 ± 0.3 | 0.00 ± 0.01 | 0.022 ± 0.001 |
| | Seeds | Acco | 59.2 ± 1.3 | 0.21 ± 0.03 | 0.008 ± 0.000 | 3.5 ± 0.5 | 0.14 ± 0.01 | 0.063 ± 0.004 |
| | | Big Full | 35.7 ± 1.2 | 0.16 ± 0.03 | 0.007 ± 0.001 | 1.8 ± 0.3 | 0.22 ± 0.02 | 0.398 ± 0.017 |
| | | Wonderful | 32.0 ± 1.7 | 0.23 ± 0.00 | 0.021 ± 0.001 | 9.6 ± 1.0 | 0.08 ± 0.01 | 0.042 ± 0.003 |
| Sonication | Peels | Acco | 49.1 ± 0.4 | 0.37 ± 0.04 | 0.038 ± 0.004 | 16.1 ± 1.0 | 0.18 ± 0.03 | 0.024 ± 0.013 |
| | | Big Full | 54.9 ± 1.2 | 0.50 ± 0.05 | 0.047 ± 0.002 | 26.7 ± 1.4 | 0.29 ± 0.03 | 0.010 ± 0.000 |
| | | Wonderful | 47.8 ± 0.5 | 0.33 ± 0.00 | 0.038 ± 0.002 | 18.3 ± 0.5 | 0.06 ± 0.05 | 0.021 ± 0.001 |
| | Seeds | Acco | 48.9 ± 1.3 | 0.14 ± 0.00 | 0.019 ± 0.002 | 3.5 ± 0.7 | 0.28 ± 0.03 | 0.067 ± 0.003 |
| | | Big Full | 36.8 ± 0.4 | 0.11 ± 0.00 | 0.032 ± 0.004 | 1.3 ± 0.3 | 0.60 ± 0.13 | 0.200 ± 0.001 |
| | | Wonderful | 30.8 ± 1.0 | 0.19 ± 0.02 | 0.068 ± 0.010 | 8.5 ± 0.2 | 0.18 ± 0.08 | 0.030 ± 0.000 |
| $p$-value | Cultivar (A) | | <0.0001 | <0.0001 | <0.0001 | <0.0001 | <0.0001 | <0.0001 |
| | By-product (B) | | <0.0001 | <0.0001 | <0.0001 | <0.0001 | <0.0001 | <0.0001 |
| | Extraction method (C) | | 0.0778 | 0.0005 | <0.0001 | 0.3244 | <0.0001 | <0.0001 |
| | A × B interaction | | <0.0001 | <0.0001 | <0.0001 | <0.0001 | 0.0038 | <0.0001 |
| | A × C interaction | | <0.0001 | 0.0083 | 0.0009 | 0.0869 | <0.0001 | <0.0001 |
| | B × C interaction | | <0.0001 | 0.4289 | <0.0001 | 0.1395 | 0.0045 | 0.0652 |
| | A × B × C interaction | | 0.0244 | 0.0139 | <0.0001 | 0.0829 | 0.0086 | 0.0777 |

Results are expressed as mean values ± standard deviation of four independent extractions for EY ($n$ = 4) and duplicates of two independent extractions for TPC, TF, TAN, ANT, and $IC_{50}$ ($n$ = 4). Three-way ANOVA ($p$-value < 0.05).

A study conducted by Passafiume et al. [42] on pomegranate juice concluded that in terms of phenolic compounds and antioxidant activity, Wonderful cultivar presented higher values than the Acco cultivar, but the anthocyanin content of Acco juice was higher. These findings in the juice are similar to those found in this study for both peels and seeds extracts from the two mentioned cultivars. In terms of seeds, extracts from Big Full cultivar had the lowest bioactive levels, however, an opposite trend was observed for peels, as the Big Full extracts are those with the highest bioactive potential, independent of the extraction method used. Many studies point out that the Wonderful cultivar has the highest antioxidant activity amongst other cultivars [43]. In this study, the peels extract of the new cultivar Big Full surpasses not only the AA (in sonication-assisted extraction) of Wonderful's extracts, but also the amounts of TPC, TF, TAN, and ANT (for both extractions methods).

It should be mentioned that the EY, TPC, and AA values found in the present study (Table 2) for peels extracted using the sonication-assisted extraction method are greater than those previously reported for pomegranate peels extracts also obtained by sonication by Tabaraki et al. [44], Sharayei et al. [45], and Ranjha et al. [46] (EY of 37.5%, TPC of 0.03–0.07 mg GAE/mg, and AA of 0.44 mg/mL $IC_{50}$, by the DPPH method), but lower than those described by Bandara et al. [47] (EY of 38%, TPC of 0.64 mg GAE/mg, and AA of 0.003 mg/mL $IC_{50}$). These findings clearly showed that different extraction conditions (frequency, time, solvent), as well as the pomegranate cultivar, greatly affects the extraction performance, which is useful to optimize the extraction conditions in each case under study. Overall, the results of Bandara et al. [47] suggest that higher frequency-time of extraction may enhance the amount of the extracted bioactive compounds.

It has been described that pomegranate fruit and its by-products have one of the highest antioxidant activities among other fruits and their by-products, being reported as a linear correlation between the phenolic content and the antioxidant activity [48–53]. Thus, for the bioactive compounds evaluated, the existence of linear correlations was assessed through the Pearson's correlation test (Table 3). According to the Shapiro–Wilk test, with the exception of AA ($p$-value < 0.05), all variables presented a normal distribution ($p$-value $\geq$ 0.05) Since the antioxidant activity is inversely proportional to the $IC_{50}$ of the extract, the correlations with AA are negative. As expected, strong correlations were found between TPC and TF ($r$ = 0.723), TAN ($r$ = 0.861), and AA ($r$ = −0.789). The strongest AA correlation was observed with TAN and TF with coefficients ($r$) −0.893 and −0.862, respectively. These correlations have already been described, since the hydrolysable tannins and flavonoids of pomegranate peels can contribute to AA [54,55]. These findings are similar to those reported by Masci et al. [50], Yan et al. [56], and Orak et al. [55] in peels. ANT did not correlate with TPC, TF, and AA, but showed a negative correlation with TAN ($r$ = −0.377). These results are in agreement with the studies of Orak et al. [55] on peels.

**Table 3.** Pearson's correlation coefficient ($r$) and the related significance between the total phenolic compounds (TPC), total flavonoids (TF), tannins (TAN), anthocyanins (ANT), and antioxidant activity (AA, $IC_{50}$).

| | TPC | TF | TAN | ANT | AA |
|---|---|---|---|---|---|
| TPC | 1.000 | | | | |
| TF | 0.723 ** | 1.000 | | | |
| TAN | 0.861 ** | 0.854 ** | 1.000 | | |
| ANT | −0.320 | −0.073 | −0.377 * | 1.000 | |
| AA | −0.789 * | −0.862 ** | −0.893 ** | 0.209 | 1.000 |

* $p$-value < 0.05, ** $p$-value < 0.0001.

### 3.2. Antimicrobial Potential

Several studies have confirmed that pomegranate's phenolics and flavonoids are related to a high antimicrobial potential against foodborne pathogens [48,57–59] and antifungal properties [22].

Table 4 shows the values of the inhibition halos obtained with the different extracts against the tested microorganisms (Figure S2 shows some examples of the inhibition halos obtained for the studied extracts). The inhibition halos obtained for the controls (penicillin, kanamycin, and amphotericin B) are given in Table S1. *E. coli* showed resistance to all extracts. Moreover, the use of Big Full seeds against *P. aeruginosa* and *B. cereus* was not effective. This may be attributed to the lowest bioactive potential of these extracts. Significant differences ($p$ < 0.05) were found regarding the extraction method, where the sonication-assisted extraction originated the extracts with the highest inhibition halos against *P. aeruginosa* (5.6 to 13.0 mm), *S. aureus* (9.3 to 14.0 mm), and *Y. lipolytica* (9.2 to 13.8 mm). It was also observed that the type of by-product resulted in significantly different inhibition halos against *P. aeruginosa*, *S. aureus*, and *B. cereus* ($p$ < 0.05).

Globally, the results found in this study regarding the pomegranate peel antimicrobial potential are in line with the literature data. McCarrell et al. [60] reported that pomegranate peel extracts (0.33 g/mL) did not inhibit *E. coli* or *P. aeruginosa* but inhibited *S. aureus* (14 mm), showing higher antimicrobial potential than some extracts obtained in the present study (7.9 to 14 mm). On the other hand, Panichayupakaranant et al. [61] showed that peels' extracts (0.20 g/mL) did not inhibit *E. coli* but were effective against *S. aureus* (15.2 to 19.4 mm of halos), which is in agreement with the results presented in this study. Alexandre et al. [20] found that pomegranate peel extract (0.50 g/mL) inhibited *E. coli* (22 mm), *P. aeruginosa* (31 mm), *S. aureus* (22 mm), and *B. cereus* (19 mm), suggesting that the use of a higher extract concentrations can promote or enhance the antimicrobial capacity. The mentioned studies confirmed the findings of Silva et al. [62] and Hanani et al. [63], which described *S. aureus* as one of the most sensitive bacteria to pomegranate extracts, and this was also confirmed in the present work (7.9 to 11.5 mm for conventional extracts and 9.3 to 14.0 mm for sonication assisted extracts, Table 4).

**Table 4.** Inhibition halos (mm) for each extract vs. microorganism tested.

| Extraction Method | By-Product | Cultivar | *E. coli* | *P. aeruginosa* | *S. aureus* | *B. cereus* | *Y. lipolytica* |
|---|---|---|---|---|---|---|---|
| Conventional | Peels | Acco | R | 12.3 ± 1.0 | 10.6 ± 0.7 | 10.8 ± 0.7 | 11.5 ± 1.0 |
| | | Big Full | R | 10.4 ± 1.9 | 11.5 ± 1.5 | 13.4 ± 1.7 | 15.7 ± 0.8 |
| | | Wonderful | R | 10.5 ± 0.5 | 11.3 ± 0.6 | 11.1 ± 1.2 | 9.2 ± 0.9 |
| | Seeds | Acco | R | 2.9 ± 4.4 | 9.7 ± 0.9 | 7.6 ± 0.7 | 9.2 ± 0.9 |
| | | Big Full | R | R | 7.9 ± 0.6 | R | 8.3 ± 1.5 |
| | | Wonderful | R | 9.7 ± 0.7 | 11.2 ± 1.4 | 10.2 ± 0.8 | 10.1 ± 1.6 |
| Sonication | Peels | Acco | R | 13.0 ± 1.2 | 11.3 ± 1.4 | 11.7 ± 1.8 | 13.8 ± 1.5 |
| | | Big Full | R | 11.6 ± 2.0 | 14.0 ± 0.9 | 15.5 ± 1.5 | 11.4 ± 2.2 |
| | | Wonderful | R | 11.4 ± 1.9 | 13.1 ± 0.6 | 11.2 ± 0.7 | 10.7 ± 1.2 |
| | Seeds | Acco | R | 5.6 ± 4.6 | 10.2 ± 1.5 | 8.4 ± 0.8 | 10.7 ± 2.0 |
| | | Big Full | R | R | 9.3 ± 0.7 | R | 9.2 ± 1.3 |
| | | Wonderful | R | 10.3 ± 1.2 | 10.7 ± 0.9 | 6.6 ± 5.1 | 12.2 ± 1.8 |
| *p*-value | Cultivar (A) | | - | <0.0001 | 0.0009 | <0.0001 | 0.0826 |
| | By-product (B) | | - | <0.0001 | <0.0001 | <0.0001 | <0.0001 |
| | Extraction method (C) | | - | 0.0111 | 0.0002 | 0.9024 | 0.0202 |
| | A × B interaction | | - | <0.0001 | <0.0001 | <0.0001 | <0.0001 |
| | A × C interaction | | - | 0.6016 | 0.1894 | 0.0020 | <0.0001 |
| | B × C interaction | | - | 0.5583 | 0.0025 | 0.0066 | 0.0049 |
| | A × B × C interaction | | - | 0.3813 | 0.2031 | 0.1033 | 0.0002 |

R: resistant. Results are expressed as mean values ± standard deviation of triplicates of three independent assays ($n$ = 9). Three-way ANOVA ($p$-value < 0.05).

With the results presented, it is also possible to expect that these extracts may possess bacteriostatic activity—it causes inhibition of bacterial growth but not death (external action required to cause death) [37,61].

Regarding antimicrobial activity against fungi, Rosa-Burgos et al. [64] and Hlima et al. [37] reported that pomegranate peels' extracts inhibited the growth of several filamentous fungi (e.g., *Alternaria alternata*, *Aspergillus flavus*, *Aspergillus niger*, *Aspergillus parasiticus*, *Botrytis cinerea*, *Fusarium culmorum*, *Fusarium graminearum Fusarium oxysporum*, and *Fusarium verticillioides*), with inhibitions halos of 8 to 15 mm. Likewise, the present study showed that peel and seed extracts, obtained from both extraction methods and from any of the three cultivars, demonstrated antimicrobial activity against *Y. lipolytica* (8.3–15.7 mm), which is a single-cell fungus.

The fact that the extracts have demonstrated inhibition against the tested microorganisms (except *E. coli*) suggests that their compositions comprise a wide range of compounds with antimicrobial properties. Many mechanisms can lead to a higher antimicrobial activity, from the chemical composition of the matrix to the active compounds' extraction method. For example, the sterilization of extracts by autoclaving appears to increase antimicrobial activity compared to sterilization by filtration [60].

The antimicrobial activity of pomegranate by-products may be indicative of the presence of metabolic toxins or a board spectrum of antibacterial compounds that act against Gram-positive and Gram-negative bacteria and yeasts [65]. In fact, Gullon et al. [66] suggest that the antimicrobial effects may be attributed to the combination of several bioactive compounds that cause microbial death through numerous mechanisms.

Thus, Pearson's correlation was applied between the results of the bioactive compounds found in by-products' extracts (TPC, TF, TAN, ANT, and AA) and the antimicrobial potential, based on the inhibition halos (Table 5). According to the Shapiro–Wilk test, with the exception of AA and the inhibition halos of *P. aeruginosa* ($p$-value < 0.05), all variables presented a normal distribution ($p$-value ≥ 0.05). Positive correlations were found for TPC, TF, and TAN with the inhibition halos for all microorganisms ($p$ < 0.05), which showed that a higher bioactive potential led to a higher antimicrobial activity, although this straightforward finding was not always verified by other researchers. Furthermore, strong correlations between AA and the inhibition halos of *P. aeruginosa* ($r = -0.902$), *S. aureus* ($r = -0.756$) and *B. cereus* ($r = -0.700$) were found. As antimicrobial activity and AA are derived from the same compounds, it is expected there is a correlation between the two.

**Table 5.** Pearson's correlation coefficient ($r$) and significance between the bioactive potential of each extract and the inhibition halos caused by pomegranate extracts against the tested microorganisms.

| | *P. aeruginosa* | *S. aureus* | *B. cereus* | *Y. lipolytica* |
|---|---|---|---|---|
| TPC | 0.621 * | 0.605 * | 0.750 * | 0.764 * |
| TF | 0.672 * | 0.740 * | 0.647 * | 0.664 * |
| TAN | 0.811 * | 0.840 ** | 0.868 ** | 0.624 * |
| ANT | −0.607 * | −0.301 | −0.557 | −0.178 |
| AA | −0.902 ** | −0.756 * | −0.700 * | −0.403 |

TPC, total phenolic compounds; TF, total flavonoids; TAN tannins; ANT, anthocyanins; AA, antioxidant activity ($IC_{50}$). * $p$-value < 0.05, ** $p$-value < 0.001.

Tables 6–8 show the antimicrobial activity index (AAI) for the extracts in relation to kanamycin, penicillin, and amphotericin B, respectively. The AAI evaluates the inhibition halos caused by the extracts relatively to a control (antibiotic/antifungal). The inhibition halos caused by kanamycin, penicillin, and amphotericin B are presented in Table S1.

**Table 6.** Antimicrobial activity index (%) for peels and seeds extracts of Acco, Big Full, and Wonderful cultivars, from both extractions, relative to kanamycin.

| Extraction Method | By-Product | Cultivar | *E. coli* | *P. aeruginosa* | *S. aureus* | *B. cereus* |
|---|---|---|---|---|---|---|
| Conventional | Peels | Acco | −100.0 ± 0.0 | −29.7 ± 2.7 | −16.3 ± 0.4 | −21.9 ± 3.1 |
| | | Big Full | −100.0 ± 0.0 | −37.1 ± 4.8 | −12.5 ± 4.4 | −11.5 ± 5.4 |
| | | Wonderful | −100.0 ± 0.0 | −36.8 ± 1.2 | −13.1 ± 0.7 | −20.6 ± 5.6 |
| | Seeds | Acco | −100.0 ± 0.0 | −81.3 ± 32.3 | −20.6 ± 1.9 | −38.0 ± 2.9 |
| | | Big Full | −100.0 ± 0.0 | −100.0 ± 0.0 | −30.3 ± 2.0 | −100.0 ± 0.0 |
| | | Wonderful | −100.0 ± 0.0 | −40.3 ± 1.2 | −13.7 ± 4.4 | −24.4 ± 3.2 |
| Sonication | Peels | Acco | −100.0 ± 0.0 | −28.5 ± 2.5 | −16.2 ± 5.8 | −23.8 ± 8.6 |
| | | Big Full | −100.0 ± 0.0 | −34.1 ± 8.9 | −5.6 ± 1.2 | −9.7 ± 2.3 |
| | | Wonderful | −100.0 ± 0.0 | −34.3 ± 4.5 | −9.1 ± 0.4 | −25.3 ± 2.2 |
| | Seeds | Acco | −100.0 ± 0.0 | −63.4 ± 31.9 | −21.6 ± 7.1 | −38.1 ± 3.6 |
| | | Big Full | −100.0 ± 0.0 | −100.0 ± 0.0 | −25.7 ± 2.6 | −100.0 ± 0.0 |
| | | Wonderful | −100.0 ± 0.0 | −38.4 ± 3.3 | −18.8 ± 3.6 | −54.4 ± 40.0 |

Results are expressed as mean values ± standard deviation of three replicates ($n$ = 3).

Regarding AAI for kanamycin (Table 6), all results are negative indicating that this antibiotic had a higher inhibition (greater inhibition halos) than the tested extracts. *S. aureus* presented the values closer to 0%, since the diameter of the inhibition halos of the extracts are closer to the halos originated by kanamycin, while *E. coli* gave values of −100% because, contrary to kanamycin, none of the extracts inhibited this Gram-negative bacterium. This means that kanamycin has a greater potential of inhibition against the studied microorganisms than any of the extracts tested.

**Table 7.** Antimicrobial activity index (%) for peels and seeds extracts of Acco, Big Full, and Wonderful cultivars, from both extractions, relative to penicillin.

| Extraction Method | By-Product | Cultivar | *E. coli* | *P. aeruginosa* | *S. aureus* | *B. cereus* |
|---|---|---|---|---|---|---|
| Conventional | Peels | Acco | −100.0 ± 0.0 | −33.4 ± 2.6 | 8.7 ± 0.5 | 100.0 ± 0.0 |
| | | Big Full | −100.0 ± 0.0 | −40.7 ± 4.7 | 12.5 ± 4.4 | 100.0 ± 0.0 |
| | | Wonderful | −100.0 ± 0.0 | −40.3 ± 1.2 | 11.9 ± 0.7 | 100.0 ± 0.0 |
| | Seeds | Acco | −100.0 ± 0.0 | −82.4 ± 30.4 | 4.3 ± 2.0 | 100.0 ± 0.0 |
| | | Big Full | −100.0 ± 0.0 | −100.0 ± 0.0 | −6.2 ± 2.2 | 0.0 ± 0.0 |
| | | Wonderful | −100.0 ± 0.0 | −43.7 ± 1.2 | 11.3 ± 4.5 | 100.0 ± 0.0 |
| Sonication | Peels | Acco | −100.0 ± 0.0 | −34.3 ± 2.4 | −6.4 ± 5.9 | 100.0 ± 0.0 |
| | | Big Full | −100.0 ± 0.0 | −39.7 ± 8.4 | 4.3 ± 1.2 | 100.0 ± 0.0 |
| | | Wonderful | −100.0 ± 0.0 | −39.9 ± 4.3 | 0.9 ± 0.3 | 100.0 ± 0.0 |
| | Seeds | Acco | −100.0 ± 0.0 | −66.5 ± 29.0 | −11.9 ± 7.3 | 100.0 ± 0.0 |
| | | Big Full | −100.0 ± 0.0 | −100.0 ± 0.0 | −16.1 ± 2.7 | 0.0 ± 0.0 |
| | | Wonderful | −100.0 ± 0.0 | −43.8 ± 3.1 | −9.0 ± 3.7 | 100.0 ± 0.0 |

Results are expressed as mean values ± standard deviation of three replicates ($n$ = 3).

**Table 8.** Antimicrobial activity index (%) for peels and seeds extracts of Acco, Big Full, and Wonderful cultivars, from both extractions, relatively to amphotericin B.

| Extraction Method | By-Product | Cultivar | *Y. lipolytica* |
|---|---|---|---|
| Conventional | Peels | Acco | 67.9 ± 1.0 |
| | | Big Full | 75.4 ± 0.5 |
| | | Wonderful | 61.4 ± 3.2 |
| | Seeds | Acco | 61.2 ± 2.8 |
| | | Big Full | 56.8 ± 6.3 |
| | | Wonderful | 64.0 ± 4.4 |
| Sonication | Peels | Acco | 42.1 ± 1.5 |
| | | Big Full | 33.4 ± 9.1 |
| | | Wonderful | 31.3 ± 0.4 |
| | Seeds | Acco | 31.0 ± 6.0 |
| | | Big Full | 23.0 ± 5.7 |
| | | Wonderful | 36.7 ± 4.4 |

Results are expressed as mean values ± standard deviation of three replicates ($n$ = 3).

Regarding AAI for penicillin (Table 7), the values ranged from −100% to 100%. Once again, *E. coli* presented an AAI of −100% because no inhibition halos were detected for the extracts. For *P. aeruginosa*, all values were negative because penicillin had a greater inhibitory effect than the evaluated extracts. Some extracts, most of which peel extracts, promoted higher inhibition halos compared to penicillin, showing a greater inhibition against *S. aureus* (positive values for AAI). Since penicillin presented no effect on *B. cereus*, the extracts reached an AAI of 100% (with the exception of the Big Full seed extract).

The highest values of AAI are reported for the extracts relatively to amphotericin B (Table 8). The positive values revealed that all extracts caused more inhibition of *Y. lipolytica* than the antifungal used as the positive control.

### 3.3. Phytotoxicity Assay

Despite the beneficial characteristics of pomegranate by-products, it is important to remark that like any other plant extract, extracts from pomegranate peel or seeds can also present toxicity [59]. The evaluation of the toxicity of the extracts is of paramount importance when applied to foods, which was evaluated by studying the phytotoxicity of the extracts obtained. Thus, in case of phytotoxicity, these results may also allow a glimpse of their possible application as natural herbicides in agriculture. To the authors' best knowledge, no studies have been conducted to evaluate the phytotoxicity of pomegranate peel extracts. For this assay, the germination test was conducted with garden-cress seeds submitted to different concentrations of Acco, Big Full, and Wonderful peel extracts, from the two studied extractions methods (conventional and sonication-assisted). Only the peels' extracts were selected for this test, since they had greater bioactive and antimicrobial potential than the seeds. Garden-cress seeds (*Lepidium sativum*) were chosen based on the work of Luo et al. [67], where it is described that these seeds are the most used for phytotoxicity studies.

To understand which concentration of the extract did not cause toxicity to the garden-cress seeds, a preliminary test was performed with different concentrations of the extracts: 0.30 g/mL (100%), 0.03 g/mL (10%), 0.01 g/mL (3.33%), and 0.003 g/mL (1%). The highest concentration was chosen based on the antimicrobial assays, since it was the concentration that inhibited different pathogenic microorganisms. For five days, the number of germinated seeds and the root length were monitored and compared to the control, for which distilled water was used (Figure S3).

It was found that the concentration of 0.30 g/mL totally inhibited seed germination; the concentration of 0.03 g/mL partially inhibited germination; and only the lowest concentrations (0.01 and 0.003 g/mL) allowed the seeds' germination. Regarding the root length, at 0.30 and 0.03 g/mL, there was no root growth for any of the seeds. At the concentration of 0.01 g/mL, only part of the roots managed to grow. Only at the lowest extract concentration (0.003 g/mL) were acceptable roots for all seeds developed. Thus, the concentration of 0.003 g/mL (1% of the initial one) was chosen for further tests.

The germination rate of 100% was observed for all peel extracts with a concentration of 0.003 g/mL, regardless of the pomegranate cultivar or extraction method applied. Only the Wonderful peel extract obtained by conventional extraction had a lower germination rate (96.7%). Regarding root length (Table 9), the extracts obtained by the sonication-assisted extraction method were the ones that least affected root growth. Even so, the control showed roots with twice of the length of the roots submitted to the extracts ($p$-value < 0.0001), which indicated that they inhibited the growth of garden-cress roots. For both extraction methods, extracts from cultivar Acco showed the lowest inhibition of root growth.

**Table 9.** Root length (mm) of garden cress seeds, in the presence of water (control) and pomegranate peel extracts (0.003 g/mL).

| Extraction Method | Cultivar | Day 1 | Day 2 | Day 3 | Day 4 | Day 5 |
|---|---|---|---|---|---|---|
| Control | | <5.0 | 15.0 ± 2.0 a | 27.2 ± 3.2 a | 36.5 ± 3.4 a | 41.0 ± 3.2 a |
| Conventional | Acco | <5.0 | 7.5 ± 1.3 b | 11.4 ± 2.6 bc | 14.6 ± 4.3 bc | 17.4 ± 5.7 bc |
| | Big Full | <5.0 | 6.5 ± 1.5 b | 9.5 ± 0.5 b | 12.5 ± 2.3 c | 14.6 ± 0.7 b |
| | Wonderful | <5.0 | 7.1 ± 0.7 b | 11.9 ± 1.7 bc | 14.0 ± 1.2 c | 15.7 ± 1.4 b |
| Sonication | Acco | <5.0 | 7.7 ± 1.1 b | 14.3 ± 3.6 c | 19.8 ± 5.9 c | 23.4 ± 4.8 c |
| | Big Full | <5.0 | 7.0 ± 1.1 b | 14.3 ± 1.2 c | 18.0 ± 2.5 bc | 20.7 ± 3.4 bc |
| | Wonderful | <5.0 | 7.6 ± 0.4 b | 15.2 ± 1.3 c | 17.6 ± 1.3 bc | 19.0 ± 1.6 bc |
| $p$-value | | - | <0.0001 | <0.0001 | <0.0001 | <0.0001 |

Results are expressed as mean values ± standard deviation of 10 replicates of three independent assays ($n$ = 30). One-way ANOVA ($p$-value < 0.0001). Different small letters in the same column represent statistical differences between extracts and the control on the same day.

The Munoo-Liisa vitality index at the fifth day of the root growth is presented in Table 10. According to the phytotoxicity classification of Trautmann and Krasny [41] (Table 1), all extracts at the concentration 0.003 g/mL can be classified as phytotoxic for garden-cress seeds, except the extracts from Big Full and Wonderful cultivars, obtained by conventional extraction, which are considered very phytotoxic.

**Table 10.** Munoo-Liisa vitality index (%) for pomegranate peel extracts (0.003 g/mL) and phytotoxicity classification.

| Extraction Method | Cultivar | Day 5 | Phytotoxicity Classification |
|---|---|---|---|
| *Conventional* | Acco | 42.4 ± 13.9 | Phytotoxic |
| | Big Full | 35.7 ± 1.7 | Very phytotoxic |
| | Wonderful | 37.1 ± 5.4 | Very phytotoxic |
| *Sonication* | Acco | 57.1 ± 11.8 | Phytotoxic |
| | Big Full | 50.5 ± 8.2 | Phytotoxic |
| | Wonderful | 46.3 ± 3.9 | Phytotoxic |

Results are expressed as mean values ± standard deviation of three replicates ($n$ = 3).

The observed phytotoxicity can be attributed to the phenolic compounds present in the extracts, which are phytotoxic and may also be responsible for plant necrosis if present in higher concentrations [68,69]. Furthermore, the phytotoxicity effect of phenolic compounds seemed to be related to their lipophilic or hydrophilic character, as it has been reported that lipophilic phenolics can cause greater phytotoxicity than hydrophilic ones [70]. A Pearson's correlation was performed between the bioactive compounds and the root length of the garden-cress seeds on the fifth day of exposure to the peel extracts, but no significant correlations were found ($p > 0.05$). However, the correlation of root length with TPC was negative, indicating that higher amounts of phenolics can inhibit root length development, as previously reported.

## 4. Conclusions

In general, for both peels and seeds, the sonication-assisted extraction method, using EtOH 50% as a solvent, proved to be more effective for the extraction of the bioactive compounds (TF and ANT) of pomegranate by-products, leading to extracts with higher AA and antimicrobial potential and lower phytotoxicity. In relation to TPC, conventional extraction was more effective, and TAN extraction is not influenced by the extraction method. Pomegranate peels showed promising bioactive and antimicrobial activities compared to seeds. Among the cultivars studied, the new cultivar Big Full showed the best bioactive potential. In terms of antimicrobial activity, extracts at a concentration of 0.30 g/mL inhibit *P. aeruginosa*, *S. aureus*, *B. cereus*, and *Y. lipolytica*, but not *E. coli*. Regarding phytotoxicity studies, all peel extracts were highly phytotoxic, even at very low concentrations (0.003 g/mL). Since it is described that antioxidant activity, antimicrobial activity, and phytotoxicity can be attributed to the phenolic compounds, it would be important to determine the phenolic profile of each extract in order to understand the cause–effect of these compounds in relation to the biological activities of the extracts.

This study contributed to increase the knowledge about the antimicrobial properties of pomegranate seed extracts and the phytotoxicity of the peel extracts, since studies are scarce in the literature. In addition, data regarding the Big Full cultivar are reported for the first time, showing the promising potential of its by-products. Moreover, the results regarding the phytotoxicity of pomegranate peels can contribute to expanding their possible applications in the agricultural sector, as possible natural substitutes for herbicides.

**Supplementary Materials:** The following are available online at https://www.mdpi.com/article/10.3390/foods11070992/s1, Table S1: Inhibition halos (mm) caused by penicillin, kanamycin (antibiotics), and amphotericin B (antifungal) on the tested microorganisms ($n$ = 18). Figure S1: 3D-PCA plots for the pomegranate extracts classification based on the experimental analysis according to: (A) the extraction method (conventional or sonication-assisted); (B) the type of by-product (peels or seeds); and (C) the cultivar (Acco, Big Full, or Wonderful). Figure S2: Inhibition halos formed due to the antimicrobial activity of the studied extracts: (A) schematic representation of a Petri dish with extracts (E), water (W), kanamycin (K), and penicillin (P); (B) inhibition halos of Acco peel extracts (sonication extraction) against *P. aeruginosa;* (C) inhibition halos of Big Full peel extracts (conventional extraction) against *S. aureus*; and (D) inhibition halos for Wonderful peel extracts (sonication extraction) against *B. cereus*. Figure S3. Preliminary phytotoxicity test against garden-cress seeds with extracts from Wonderful peel (sonication extraction) to determine (A) the number of germinated seeds and (B) the root length (mm) of the germinated seeds.

**Author Contributions:** Conceptualization: L.C., S.D., A.C.A.V. and M.H.; methodology: L.C., L.S., S.D., A.M.P., A.C.A.V. and M.H.; investigation: L.C., L.S., S.D., A.M.P., A.C.A.V. and M.H.; resources: A.C.A.V. and M.H.; writing–original draft preparation: L.C.; writing–review and editing: L.C., S.D., A.M.P., A.C.A.V. and M.H. All authors have read and agreed to the published version of the manuscript.

**Funding:** This work was supported by the Portuguese Foundation for Science and Technology (FCT) under the scope of the strategic funding of CEB (UIDB/04469/2020), CERNAS (UIDB/00681/2020), CIMO (UIDB/00690/2020), and the Associate Laboratory SusTEC (LA/P/0007/2020). The European Regional Development Fund funded MobFood operation (LISBOA-01-0247-FEDER-024524). L.C. acknowledges research grants CEB-BI-14-2019 and FCT-IPC-i2A-CERNAS/Escola de Verão/BI-01-08, and L.S. acknowledges the research grant FCT-IPC-i2A-CERNAS/Escola de Verão/BII-01-07, all provided by FCT. The Article Processing Charge (APC) was funded by the Polytechnic Institute of Coimbra.

**Institutional Review Board Statement:** Not applicable.

**Informed Consent Statement:** Not applicable.

**Data Availability Statement:** Data are contained within this article.

**Acknowledgments:** The authors would like to acknowledge the technicians Isabel Herder Costa, Rosinda Leonor Pato, and Jorge Viegas and Katarzyna Kujawa, Martyna Frąszczak, and Mateusz Majda for the laboratory assistance.

**Conflicts of Interest:** The authors declare no conflict of interest.

## References

1. Roukas, T.; Kotzekidou, P. Pomegranate peel waste: A new substrate for citric acid production by *Aspergillus niger* in solid-state fermentation under non-aseptic conditions. *Environ. Sci. Pollut. Res.* **2020**, *27*, 13105–13113. [CrossRef]
2. Kandylis, P.; Kokkinomagoulos, E. Food applications and potential health benefits of pomegranate and its derivatives. *Foods* **2020**, *9*, 122. [CrossRef] [PubMed]
3. Turrini, F.; Zunin, P.; Catena, S.; Villa, C.; Alfei, S.; Boggia, R. Traditional or hydro-diffusion and gravity microwave coupled with ultrasound as green technologies for the valorization of pomegranate external peels. *Food Bioprod. Process.* **2019**, *117*, 30–37. [CrossRef]
4. Nazeam, J.A.; AL-Shareef, W.A.; Helmy, M.W.; El-Haddad, A.E. Bioassay-guided isolation of potential bioactive constituents from pomegranate agrifood by-product. *Food Chem.* **2020**, *326*, 126993. [CrossRef]
5. Gullón, P.; Astray, G.; Gullón, B.; Tomasevic, I.; Lorenzo, J.M. Pomegranate Peel as Suitable Source of High-Added Value Bioactives: Tailored Functionalized Meat Products. *Molecules* **2020**, *25*, 2859. [CrossRef] [PubMed]
6. Natalello, A.; Hervás, G.; Toral, P.G.; Luciano, G.; Valenti, B.; Mendoza, A.G.; Pauselli, M.; Priolo, A.; Frutos, P. Bioactive compounds from pomegranate by-products increase the in vitro ruminal accumulation of potentially health promoting fatty acids. *Anim. Feed Sci. Technol.* **2020**, *259*, 114355. [CrossRef]
7. Singh, B.; Singh, J.P.; Kaur, A.; Singh, N. Antimicrobial potential of pomegranate peel: A review. *Int. J. Food Sci. Technol.* **2019**, *54*, 959–965. [CrossRef]
8. Shahidi, F.; Varatharajan, V.; Oh, W.Y.; Peng, H. Phenolic compounds in agri-food by-products, their bioavailability and health effects. *J. Food Bioact.* **2019**, *5*, 57–119. [CrossRef]

9. Verotta, L.; Panzella, L.; Antenucci, S.; Calvenzani, V.; Tomay, F.; Petroni, K.; Caneva, E.; Napolitano, A. Fermented pomegranate wastes as sustainable source of ellagic acid: Antioxidant properties, anti-inflammatory action, and controlled release under simulated digestion conditions. *Food Chem.* **2018**, *246*, 129–136. [CrossRef]
10. Ameer, K.; Shahbaz, H.M.; Kwon, J.H. Green Extraction Methods for Polyphenols from Plant Matrices and Their Byproducts: A Review. *Compr. Rev. Food Sci. Food Saf.* **2017**, *16*, 295–315. [CrossRef]
11. Pan, Z.; Qu, W.; Ma, H.; Atungulu, G.G.; McHugh, T.H. Continuous and pulsed ultrasound-assisted extractions of antioxidants from pomegranate peel. *Ultrason. Sonochem.* **2012**, *19*, 365–372. [CrossRef] [PubMed]
12. Azmir, J.; Zaidul, I.S.M.; Rahman, M.M.; Sharif, K.M.; Mohamed, A.; Sahena, F.; Jahurul, M.H.A.; Ghafoor, K.; Norulaini, N.A.N.; Omar, A.K.M. Techniques for extraction of bioactive compounds from plant materials: A review. *J. Food Eng.* **2013**, *117*, 426–436. [CrossRef]
13. Magangana, T.P.; Makunga, N.P.; Fawole, O.A.; Opara, U.L. Processing factors affecting the phytochemical and nutritional properties of pomegranate (*Punica granatum* L.) peel waste: A review. *Molecules* **2020**, *25*, 4690. [CrossRef] [PubMed]
14. Santos, M.P.; Souza, M.C.; Sumere, B.R.; da Silva, L.C.; Cunha, D.T.; Bezerra, R.M.N.; Rostagno, M.A. Extraction of bioactive compounds from pomegranate peel (*Punica granatum* L.) with pressurized liquids assisted by ultrasound combined with an expansion gas. *Ultrason. Sonochem.* **2019**, *54*, 11–17. [CrossRef]
15. Galanakis, C.M. Recovery of high added-value components from food wastes: Conventional, emerging technologies and commercialized applications. *Trends Food Sci. Technol.* **2012**, *26*, 68–87. [CrossRef]
16. Xi, J.; He, L.; Yan, L. gong. Continuous extraction of phenolic compounds from pomegranate peel using high voltage electrical discharge. *Food Chem.* **2017**, *230*, 354–361. [CrossRef] [PubMed]
17. Sasidharan, S.; Chen, Y.; Saravanan, D.; Sundram, K.M.; Yoga Latha, L. Extraction, isolation and characterization of bioactive compounds from plants' extracts. *Afr. J. Tradit. Complement. Altern. Med.* **2011**, *8*, 1–10. [CrossRef]
18. Kalamara, E.; Goula, A.M.; Adamopoulos, K.G. An integrated process for utilization of pomegranate wastes—Seeds. *Innov. Food Sci. Emerg. Technol.* **2015**, *27*, 144–153. [CrossRef]
19. Goula, A.; Lazarides, H. Integrated processes can turn industrial food waste into valuable food by-products and/or ingredients: The cases of olive mill and pomegranate wastes. *J. Food Eng.* **2015**, *167*, 45–50. [CrossRef]
20. Alexandre, E.M.C.; Silva, S.; Santos, S.A.O.; Silvestre, A.J.D.; Duarte, M.F.; Saraiva, J.A.; Pintado, M. Antimicrobial activity of pomegranate peel extracts performed by high pressure and enzymatic assisted extraction. *Food Res. Int.* **2019**, *115*, 167–176. [CrossRef] [PubMed]
21. Shaygannia, E.; Bahmani, M.; Zamanzad, B.; Rafieian-Kopaei, M. A Review Study on *Punica granatum* L. *J. Evidence-Based Complement. Altern. Med.* **2016**, *21*, 221–227. [CrossRef]
22. Smaoui, S.; Hlima, H.B.; Mtibaa, A.C.; Fourati, M.; Sellem, I.; Elhadef, K.; Ennouri, K.; Mellouli, L. Pomegranate peel as phenolic compounds source: Advanced analytical strategies and practical use in meat products. *Meat Sci.* **2019**, *158*, 107914. [CrossRef]
23. Azwanida, N.N. A Review on the Extraction Methods Use in Medicinal Plants, Principle, Strength and Limitation. *Med. Aromat. Plants* **2015**, *4*, 1000196. [CrossRef]
24. Abid, M.; Jabbar, S.; Wu, T.; Hashim, M.M.; Hu, B.; Lei, S.; Zeng, X. Sonication enhances polyphenolic compounds, sugars, carotenoids and mineral elements of apple juice. *Ultrason. Sonochem.* **2014**, *21*, 93–97. [CrossRef]
25. Fonteles, T.V.; Leite, A.K.F.; da Silva, A.R.A.; Fernandes, F.A.N.; Rodrigues, S. Sonication Effect on Bioactive Compounds of Cashew Apple Bagasse. *Food Bioprocess Technol.* **2017**, *10*, 1854–1864. [CrossRef]
26. Das, P.R.; Eun, J.B. A comparative study of ultra-sonication and agitation extraction techniques on bioactive metabolites of green tea extract. *Food Chem.* **2018**, *253*, 22–29. [CrossRef]
27. Das, S.; Ray, A.; Nasim, N.; Nayak, S.; Mohanty, S. Effect of different extraction techniques on total phenolic and flavonoid contents, and antioxidant activity of betelvine and quantification of its phenolic constituents by validated HPTLC method. *3 Biotech.* **2019**, *9*, 37. [CrossRef]
28. Singleton, V.L.; Rossi, J. Colorimetry of total phenolics with phosphomolybdic-phosphotungstic acid reagents. *Am. J. Enol. Vitic.* **1965**, *16*, 144–158. [CrossRef]
29. Kim, D.; Jeong, S.W.; Lee, C.Y. Antioxidant capacity of phenolic phytochemicals from various cultivars of plums. *Food Chem.* **2003**, *81*, 321–326. [CrossRef]
30. Atanassova, M.; Christova-bagdassarian, V. Determination of tannins content by titrimetric method for comparison of different plant species. *J. Univ. Chem. Technol. Metall.* **2009**, *44*, 413–415.
31. Sellappan, S.; Akoh, C.C.; Krewer, G. Phenolic Compounds and Antioxidant Capacity of Georgia-Grown Blueberries and Blackberries. *J. Agric. Food Chem.* **2002**, *50*, 2432–2438. [CrossRef]
32. Mensor, L.L.; Menezes, F.S.; Leitão, G.G.; Reis, A.S.; dos Santos, T.C.; Coube, C.S.; Leitão, S.G. Screening of Brazilian plant extracts for antioxidant activity by the use of DPPH free radical method. *Phyther. Res.* **2001**, *15*, 127–130. [CrossRef] [PubMed]
33. Staško, A.; Brezová, V.; Biskupic, S.; Mišík, V. The potential pitfalls of using 1,1-diphenyl-2-picrylhydrazyl to characterize antioxidants in mixed water solvents. *Free Radic. Res.* **2007**, *41*, 379–390. [CrossRef] [PubMed]
34. Rodríguez-Rojo, S.; Visentin, A.; Maestri, D.; Cocero, M.J. Assisted extraction of rosemary antioxidants with green solvents. *J. Food Eng.* **2012**, *109*, 98–103. [CrossRef]

35. Sarker, S.D.; Nahar, L.; Kumarasamy, Y. Microtitre plate-based antibacterial assay incorporating resazurin as an indicator of cell growth, and its application in the in vitro antibacterial screening of phytochemicals. *Methods* **2007**, *42*, 321–324. [CrossRef] [PubMed]
36. De Bona, E.D.A.M.; da Silva Pinto, F.G.; Fruet, T.K.; Jorge, T.C.M.; de Moura, A.C. Comparação de métodos para avaliação da atividade antimicrobiana e determinação da concentração inibitória mínima (cim) de extratos vegetais aquosos e etanólicos. *Arq. Inst. Biol.* **2014**, *81*, 218–225. [CrossRef]
37. Hlima, H.B.; Bohli, T.; Kraiem, M.; Ouederni, A.; Mellouli, L.; Michaud, P.; Abdelkafi, S.; Smaoui, S. Combined effect of *Spirulina platensis* and *Punica granatum* peel extacts: Phytochemical content and antiphytophatogenic activity. *Appl. Sci.* **2019**, *9*, 5474. [CrossRef]
38. Vancheva, T.; Encheva-malinova, M.; Tatyozova, M.; Gochev, V.; Stoyanova, M.; Moncheva, P.; Bogatzevska, N. Antimicrobial Activity of Essential Oils against Pepper Bacterial Spot Agents. In Proceedings of the First national Conference of Biotechnology, Sofia, Bulgaria, 17–18 October 2014; Volume 100, pp. 200–207.
39. *EN 16086-1*; Soil Improvers and Growing Media—Determination of Plant Response—Part 1: Pot Growth Test with Chinese Cabbage, European Standard. German Institute for Standardization: Berlin, Germany, 2011; 1–25.
40. *EN 16086-2*; Soil Improvers and Growing Media—Determination of Plant Response—Part 2: Petri Dish Test Using Cress, European Standard. German Institute for Standardization: Berlin, Germany, 2011; 1–18.
41. Trautmann, N.M.; Krasny, M.E. *Composting in the Classroom—Scientific Inquiry for High School Students*; Kendall/Hunt Publishing Company: Dubuque, IA, USA, 1998.
42. Passafiume, R.; Perrone, A.; Sortino, G.; Gianguzzi, G.; Saletta, F.; Gentile, C.; Farina, V. Chemical–physical characteristics, polyphenolic content and total antioxidant activity of three Italian-grown pomegranate cultivars. *NFS J.* **2019**, *16*, 9–14. [CrossRef]
43. Stover, E.; Mercure, E.W. The pomegranate: A new look at the fruit of paradise. *HortScience* **2007**, *42*, 1088–1092. [CrossRef]
44. Tabaraki, R.; Heidarizadi, E.; Benvidi, A. Optimization of ultrasonic-assisted extraction of pomegranate (*Punica granatum* L.) peel antioxidants by response surface methodology. *Sep. Purif. Technol.* **2012**, *98*, 16–23. [CrossRef]
45. Sharayei, P.; Azarpazhooh, E.; Zomorodi, S.; Ramaswamy, H.S. Ultrasound assisted extraction of bioactive compounds from pomegranate (*Punica granatum* L.) peel. *Lwt* **2019**, *101*, 342–350. [CrossRef]
46. Ranjha, M.M.A.N.; Amjad, S.; Ashraf, S.; Khawar, L.; Safdar, M.N.; Jabbar, S.; Nadeem, M.; Mahmood, S.; Murtaza, M.A. Extraction of Polyphenols from Apple and Pomegranate Peels Employing Different Extraction Techniques for the Development of Functional Date Bars. *Int. J. Fruit Sci.* **2020**, *20*, S1201–S1221. [CrossRef]
47. Bandara, U.Y.; Witharana, C.; Soysa, P. Extraction, total phenol content, flavonoid content, free radical scavenging capacity and phytochemical screening of the parts of sri lankan pomegranate (*Punica granatum* L.) fruit. *Curr. Trends Biotechnol. Pharm.* **2020**, *14*, 70–80. [CrossRef]
48. Miguel, M.G.; Neves, M.A.; Antunes, M.D. Pomegranate (*Punica granatum* L.): A medicinal plant with myriad biological properties—A short review. *J. Med. Plants Res.* **2010**, *4*, 2836–2847.
49. Elfalleh, W.; Hannachi, H.; Tlili, N.; Yahia, Y.; Nasri, N.; Ferchichi, A. Total phenolic contents and antioxidant activities of pomegranate peel, seed, leaf and flower. *J. Med. Plants Res.* **2012**, *6*, 4724–4730. [CrossRef]
50. Masci, A.; Coccia, A.; Lendaro, E.; Mosca, L.; Paolicelli, P.; Cesa, S. Evaluation of different extraction methods from pomegranate whole fruit or peels and the antioxidant and antiproliferative activity of the polyphenolic fraction. *Food Chem.* **2016**, *202*, 59–69. [CrossRef]
51. Kalaycıoğlu, Z.; Erim, F.B. Total phenolic contents, antioxidant activities, and bioactive ingredients of juices from pomegranate cultivars worldwide. *Food Chem.* **2017**, *221*, 496–507. [CrossRef]
52. Derakhshan, Z.; Ferrante, M.; Tadi, M.; Ansari, F.; Heydari, A.; Hosseini, M.S.; Conti, G.O.; Sadrabad, E.K. Antioxidant activity and total phenolic content of ethanolic extract of pomegranate peels, juice and seeds. *Food Chem. Toxicol.* **2018**, *114*, 108–111. [CrossRef]
53. Loizzo, M.R.; Aiello, F.; Tenuta, M.C.; Leporini, M.; Falco, T.; Tundis, R. Pomegranate (*Punica granatum* L.). In *Nonvitamin and Nonmineral Nutritional Supplements*, 2nd ed.; Nabavi, S.M., Silva, A.S., Eds.; Elsevier Inc.: Amsterdam, The Netherlands, 2019; pp. 467–472.
54. More, P.R.; Arya, S.S. Intensification of bio-actives extraction from pomegranate peel using pulsed ultrasound: Effect of factors, correlation, optimization and antioxidant bioactivities. *Ultrason. Sonochem.* **2021**, *72*, 105423. [CrossRef]
55. Orăk, H.H.; Yagar, H.; Isbilir, S.S. Comparison of antioxidant activities of juice, peel, and seed of pomegranate (*Punica granatum* L.) and inter-relationships with total phenolic, Tannin, anthocyanin, and flavonoid contents. *Food Sci. Biotechnol.* **2012**, *21*, 373–387. [CrossRef]
56. Yan, L.; Zhou, X.; Shi, L.; Shalimu, D.; Ma, C.; Liu, Y. Phenolic profiles and antioxidant activities of six Chinese pomegranate (*Punica granatum* L.) cultivars. *Int. J. Food Prop.* **2017**, *20*, S94–S107. [CrossRef]
57. Braga, L.C.; Leite, A.A.M.; Xavier, K.G.S.; Takahashi, J.A.; Bemquerer, M.P.; Chartone-Souza, E.; Nascimento, A.M.A. Synergic interaction between pomegranate extract and antibiotics against *Staphylococcus aureus*. *Can. J. Microbiol.* **2005**, *51*, 541–547. [CrossRef]
58. Duman, A.D.; Ozgen, M.; Dayisoylu, K.S.; Erbil, N.; Durgac, C. Antimicrobial activity of six pomegranate (*Punica granatum* L.) varieties and their relation to some of their pomological and phytonutrient characteristics. *Molecules* **2009**, *14*, 1808–1817. [CrossRef] [PubMed]

59. Akhtar, S.; Ismail, T.; Fraternale, D.; Sestili, P. Pomegranate peel and peel extracts: Chemistry and food features. *Food Chem.* **2015**, *174*, 417–425. [CrossRef] [PubMed]
60. McCarrell, E.M.; Gould, S.W.J.; Fielder, M.D.; Kelly, A.F.; El Sankary, W.; Naughton, D.P. Antimicrobial activities of pomegranate rind extracts: Enhancement by addition of metal salts and vitamin C. *BMC Complement. Altern. Med.* **2008**, *8*, 64. [CrossRef] [PubMed]
61. Panichayupakaranant, P.; Tewtrakul, S.; Yuenyongsawad, S. Antibacterial, anti-inflammatory and anti-allergic activities of standardised pomegranate rind extract. *Food Chem.* **2010**, *123*, 400–403. [CrossRef]
62. da Silva, J.A.T.; Rana, T.S.; Narzary, D.; Verma, N.; Meshram, D.T.; Ranade, S.A. Pomegranate Biology And Biotechnology: A Review. *Sci. Hortic.* **2013**, *160*, 85–107. [CrossRef]
63. Hanani, Z.A.N.; Yee, F.C.; Nor-Khaizura, M.A.R. Effect of pomegranate (*Punica granatum* L.) peel powder on the antioxidant and antimicrobial properties of fish gelatin films as active packaging. *Food Hydrocoll.* **2019**, *89*, 253–259. [CrossRef]
64. Rosas-Burgos, E.C.; Burgos-Hernández, A.; Noguera-Artiaga, L.; Kačániová, M.; Hernández-García, F.; Cárdenas-López, J.L.; Carbonell-Barrachina, Á.A. Antimicrobial activity of pomegranate peel extracts as affected by cultivar. *J. Sci. Food Agric.* **2017**, *97*, 802–810. [CrossRef] [PubMed]
65. Malviya, S.; Arvind; Jha, A.; Hettiarachchy, N. Antioxidant and antibacterial potential of pomegranate peel extracts. *J. Food Sci. Technol.* **2014**, *51*, 4132–4137. [CrossRef]
66. Gullon, B.; Pintado, M.E.; Pérez-Álvarez, J.A.; Viuda-Martos, M. Assessment of polyphenolic profile and antibacterial activity of pomegranate peel (*Punica granatum*) flour obtained from co-product of juice extraction. *Food Control* **2016**, *59*, 94–98. [CrossRef]
67. Luo, Y.; Liang, J.; Zeng, G.; Chen, M.; Mo, D.; Li, G.; Zhang, D. Seed germination test for toxicity evaluation of compost: Its roles, problems and prospects. *Waste Manag.* **2018**, *71*, 109–114. [CrossRef] [PubMed]
68. Ortega, M.C.; Moreno, M.T.; Ordovás, J.; Aguado, M.T. Behaviour of different horticultural species in phytotoxicity bioassays of bark substrates. *Sci. Hortic.* **1996**, *66*, 125–132. [CrossRef]
69. Guranna, P.; Hoolageri, H.C. Studies on establishment of aseptic culture in pomegranate cv. Bhagwa. *Annu. Res. Rev. Biol.* **2017**, *21*, 1–7. [CrossRef]
70. Pinho, I.A.; Lopes, D.V.; Martins, R.C.; Quina, M.J. Phytotoxicity assessment of olive mill solid wastes and the influence of phenolic compounds. *Chemosphere* **2017**, *185*, 258–267. [CrossRef] [PubMed]

foods

MDPI

Article

# Interaction between Fish Skin Gelatin and Pea Protein at Air-Water Interface after Ultrasound Treatment

Davide Odelli [1,2,3], Krystalia Sarigiannidou [1], Alberto Soliani [1], Rodolphe Marie [4], Mohammad Amin Mohammadifar [1], Flemming Jessen [1], Giorgia Spigno [2], Mar Vall-llosera [5], Antonio Fernandes de Carvalho [3], Michela Verni [6] and Federico Casanova [1,*]

1 Research Group for Food Production Engineering, National Food Institute, Technical University of Denmark, SøltoftsPlads, 2800 Kongens Lyngby, Denmark; davide.odelli@ufv.br (D.O.); krystaliasari@gmail.com (K.S.); albertosoliani97@gmail.com (A.S.); moamo@food.dtu.dk (M.A.M.); fjes@food.dtu.dk (F.J.)
2 Department for Sustainable Food Process (DiSTAS), Università Cattolica del Sacro Cuore, Via E. Parmense 84, 29122 Piacenza, Italy; giorgia.spigno@unicatt.it
3 Departamento de Tecnologia de Alimentos, Universidade Federal de Viçosa (UFV), Vicosa 36570-900, Brazil; antoniofernandes@ufv.br
4 Department of Health Technology, Technical University of Denmark, Ørsted Plads, 2800 Kongens Lyngby, Denmark; rcwm@dtu.dk
5 Department of Biology and Biological Engineering, Food and Nutrition Science, Chalmers University of Technology, SE-41296 Gothenburg, Sweden; marvallju@gmail.com
6 Department of Soil, Plant and Food Sciences, University of Bari "Aldo Moro", 70126 Bari, Italy; michela.verni@uniba.it
* Correspondence: feca@food.dtu.dk; Tel.: +45-45252716

**Abstract:** The interaction between fish skin gelatin (FG) and pea protein isolate (PPI) was investigated at the air-water interface (A-W) before and after a high intensity (275 W, 5 min) ultrasound treatment (US). We analyzed the properties of the single protein suspensions as well as an equal ratio of FG:PPI (MIX), in terms of ζ-potential, particle size, molecular weight, bulk viscosity and interfacial tension. The foaming properties were then evaluated by visual analysis and by Turbiscan Tower. Confocal laser scanning microscopy (CLSM) was employed to explore the role of the proteins on the microstructure of foams. The results showed that the ultrasound treatment slightly influenced physicochemical properties of the proteins, while in general, did not significantly affect their behavior both in bulk and at the air-water interface. In particular, PPI aggregate size was reduced (−48 nm) while their negative charges were increased (−1 mV) after the treatment. However, when the proteins were combined, higher molecular weight of aggregates, higher foam stability values (+14%) and lower interfacial tension (IFT) values (47.2 ± 0.2 mN/m) were obtained, leading us to assume that a weak interaction was developed between them.

**Keywords:** foaming properties; fish skin gelatin; pea protein; interfacial properties; Turbiscan Tower; CLSM

**Citation:** Odelli, D.; Sarigiannidou, K.; Soliani, A.; Marie, R.; Mohammadifar, M.A.; Jessen, F.; Spigno, G.; Vall-llosera, M.; de Carvalho, A.F.; Verni, M.; et al. Interaction between Fish Skin Gelatin and Pea Protein at Air-Water Interface after Ultrasound Treatment. *Foods* **2022**, *11*, 659. https://doi.org/10.3390/foods11050659

Academic Editor: Filipa V. M. Silva

Received: 17 January 2022
Accepted: 14 February 2022
Published: 23 February 2022



## 1. Introduction

Foam can be described as a two-phase system in which gas bubbles are uniformly dispersed into a continuous liquid phase. Gas bubbles are separated by a thin continuous layer of liquid called the lamellar phase [1]. In food products, this system could be very complex, containing different mixtures of gases and liquids, which contribute to the texture and the palatability of foods. In order to prevent their agglomeration and coalescence that can cause their collapse, different type of surfactant could be added to ensure foam stability during the time. Thanks to their hydrophilic and hydrophobic groups, these elements can stabilize the interface interacting with the liquid and the gas phases simultaneously. Between food surfactants, proteins are largely employed in food industry. Due to their ability of being adsorb at the air-water interface, proteins can significantly affect the foaming properties and, thanks to their different structures, vary in their behavior: ideal protein foaming agents should be able to rapidly stabilize the system at low concentrations and over different pH range. Egg white, dairy, gelatins, gluten and soy proteins are the main surfactants used in food foams due to their high efficiency [2].

However, more sustainable proteins sources have been investigated [3]. Due to its special properties, such as gelation, film forming ability and interfacial properties, gelatin is widely employed in food, pharmaceutical and cosmetic industries to improve stability, elasticity, and texture of many products [4,5]. Gelatin is a denatured and biodegradable protein derived from either alkaline or acid partial hydrolyzation of collagen [6]. Recently, the biochemical and physico-chemical properties of gelatin from saithe fish skin, has been investigated by Casanova et al. [7] to stabilize foams. The authors investigated the foaming properties of fish skin gelatin after different combinations of time and high intensity ultrasound treatment.

In the last years, pea protein isolate extracted from *Pisum sativum*, which is the European most cultivated protein source, has been employed in the food industry as a functional ingredient [8,9]. Its interfacial properties are mainly given by vicilin, legumin and convicilin, globular proteins from which it is composed. However, due to their large and compact structure, their ability to adsorb at the air-water interface can be limited [10]. Different methods to improve PPI functional properties were investigated, such as pH and/or heat treatment and polysaccharide addition. Xiong et al. [10] tried to improve their foaming properties by applying a high intensity ultrasound treatment resulting in an encouraging enhancement in foam stability. Hinderink et al. [11] tried to improve the interfacial properties of pea protein isolate by blending it with dairy proteins. The authors concluded that the interfacial properties of individual proteins are not additive, but they can react giving stronger or more elastic layers, which can affect the product stability at the interface. The aim of this study is therefore to investigate the behavior of fish skin gelatin (FG), pea protein isolate (PPI) and their mixture (MIX) in three different ratios (100:0, 50:50 and 0:100) in bulk and at the air-water interface before and after ultrasound treatment.

## 2. Materials and Methods

Fish skin gelatin (FG) was purchased from Sigma Aldrich (Sigma, St. Louis, MO, USA) while F85F Pea Protein Isolate (PPI) was kindly donated by Roquette Frères (Lestrem, France). The composition of FG and PPI powders (i.e., moisture, ash, protein content, mineral composition) and in solution (i.e., thermal properties and surface hydrophobicity) were already assessed by a previous study conducted by Vall-llosera et al. [12]. A phosphate-buffered saline solution (PBS buffer at pH 7.2–7.4) was used for the preparation of the samples. The solution was prepared in double-distilled water with a concentration of 50 mM $Na_2HPO_4/NaH_2PO_4$ and 150 mM of NaCl.

### 2.1. Samples Preparation

FG powder was dissolved into 200 mL of PBS buffer at a concentration of 18 g/L and stirred for 24 h at room temperature (20 ± 1 °C) to reach a complete solubilization of the protein. PPI solution was stirred for at least 48 h at 4 °C according to Hinderink et al. [11]. The 50:50 ratio solution (MIX), was prepared mixing 100 mL of both the protein solutions maintaining the same concentration. All the solutions were stored in blue caps glass containers (250 mL volume). Sodium azide (Sigma, St. Louis, MO, USA) at 0.02% was added to prevent microbial activity.

#### Standardization of the FG and PPI Solutions

Since PPI powder presents 80% of protein, 5 solutions at different concentrations (100, 110, 120, 150 and 200 g/L) of PPI were prepared in PBS buffer and stored at 4 °C under stirring for 48 h. The solution was then centrifuged at 12,298 g for 1 h. The supernatant was then evaluated in terms of soluble protein content by Dumas method, using a nitrogen conversion factor of 6.25. A calibration line (Figure S1, Table S1) was designed to standardize the supernatant solution at 18 g/L of soluble protein, which was obtained with a starting concentration of 110 g/L.

## 2.2. Ultrasound Treatment

An aliquot of 50 mL of FG, PPI and MIX was treated with ultrasounds (Branson, Danbury, CT, USA) into a 100 mL glass beaker. To prevent the increasing temperature due to the treatment, the samples were located in an ice bucket for the whole treatment period. The samples were sonicated in a continuous mode at a 20 kHz frequency, 275 W (50% of amplitude) for 5 min.

## 2.3. pH and Temperature Values

The pH of all the samples was measured before and after the treatment at room temperature (20 ± 1 °C), by an automated pH meter (Metrohm 780, Smedeland, Glostrup, Denmark). The pH values are reported as the average of three repeated measurements and their standard deviations.

## 2.4. Hydrodynamic Diameter (Dh)

The particle size of the untreated and sonicated solutions was measured using a Zetasizer Nano Series (Malvern Instruments, Malvern, UK) instrument. For the DLS analysis, the solutions were diluted 1:10 with Milli-Q water before the injection in capillary cells and analyzed with a wavelength of 633 nm and a scattering angle of 173°. The particle size values were calculated according to the Stokes-Einstein equation as follows:

$$D_h = \frac{K_B T}{3\pi\eta D_t} \quad (1)$$

where $D_h$ is the hydrodynamic diameter of the particles, $D_t$ is the diffusion coefficient, which was extracted from the fit of the correlation curve using the cumulative method, $K_B$ is the Boltzmann's constant, $T$ the temperature and $\eta$ the solvent viscosity (Pa $s^{-1}$). The $D_h$ is hereby reported as the average of three repeated measurements and their standard deviations.

## 2.5. ζ-Potential Measurements

ζ-Potential of the particles was determined by applying a voltage of 50 V. The values were calculated with the Henry equations as follows:

$$\zeta = \frac{3\eta\mu}{2\varepsilon f(\kappa R_h)} \quad (2)$$

where $\mu$ is the electrophoretic mobility (V $Pa^{-1}$ $s^{-1}$), $\eta$ the solvent viscosity (Pa $s^{-1}$), $\varepsilon$ is the medium dielectric constant (dimensionless), $\kappa$ is the Debye length or the thickness of the double electric layer around the molecules (nm) and $R_h$ is the hydrodynamic radius (nm). A value of 1.5 was used for $f(\kappa R_h)$, which is the Henry's constant, according to the Smoluchowski approximation [13] since the measurements were conducted in an aqueous medium.

## 2.6. SEC-MALS

Molecular weight of PP, FG and MIX, before and after ultrasound treatment was determined using size-exclusion chromatography. For that purpose, 9 mg/mL sample were prepared in phosphate buffer (pH 7.2) and filtered with 0.1 µm pore size filter. The HPLC (Agilent, Santa Clara, CA, USA) was equipped with WTC-015S5 column (300 × 7.8 mm, 150 Å maximum pore size) Wyatt Technology, Santa Barbara, CA, USA). The elute was monitored by a UV detector at 280 nm, a DAWN 8 light-scattering detector (Wyatt Technology, Santa Barbara, CA, USA) and an Optilab differential refractometer (Wyatt Technology). The flow rate was 0.8 mL/min and the injection volume 50 µL. The mobile phase was phosphate buffer, pH 7.2. 200 µL/L proClin (Sigma, St. Louis, MO, USA) was added to prevent the microbial growth. The buffer was prior filtered with a sterile single use vacuum filter (Thermo Fisher Scientific, Roskilde, Denmark) with pore size of 0.1 µm. Data analysis and molecular weight calculations were performed using the ASTRA software (7.3.2 Version, Wyatt Technology Europe, Dernbach, Germany).

### 2.7. In Bulk Rheological Analysis

The rheological properties of the solutions were evaluated with a controlled-stress rheometer (StressTech HR Cannon instruments, State College, PE, USA) equipped with a double gap geometry. A volume of 15 mL was analyzed 20 °C. Oscillatory measurements were used as a first step to find the linear viscoelastic region (LVR), which indicates a Newtonian behavior. LVR was found within a shear rate between 1–100 $s^{-1}$, which was used for the flow measurements. Flow data were fitted with the Power-law model and the apparent viscosity of the sample solutions was obtained using the following equations:

$$\tau = m\,\gamma^{\,n} \tag{3}$$

$$\mu_a = m\,\gamma^{\,n-1} \tag{4}$$

where $\tau$ is the shear stress, $m$ the consistency coefficient, $\gamma$ the shear rate, $\mu_a$ the apparent viscosity, and $n$ is the flow behavior index. Fluids can be described by three different behaviors depending on $n$ value: shear thinning ($n < 1$), shear thickening ($n > 1$) and Newtonian fluids ($n = 1$). Over the shear rate range applied, all of the sample solutions displayed Newtonian flow behavior. Frequency sweep with a frequency range between 1–10 Hz was implemented to determine viscoelastic behavior of the solutions at a 5% fixed strain value.

### 2.8. Foaming Properties

A volume of 50 mL of solution were put into a glass beaker of 100 mL and foams were produced by Ultraturrax (Colonial Scientific, DI 25 basic yellow line, Richmond, VA, USA) at 9500 rpm for 1 min. Immediately after the whipping process, the foams were poured into a 100 mL graduated cylinder and sealed with parafilm. Foam capacity (*FC*), foam stability (*FS*) and liquid fraction (*LF*), were evaluated by a visual analysis, according to the following equations:

$$FC = \frac{V_t - V_0}{V_t} \times 100\% \tag{5}$$

$$FS = \frac{V_{t\ time}}{V_0} \times 100\% \tag{6}$$

$$LF = \frac{V_0 - Vt_{liquid}}{V_0} \times 100\% \tag{7}$$

where $V_t$ represents the volume of the foam after homogenization, $V_0$ is the initial volume of the protein solution, $V_{t\ time}$ and $V_{t\ liquid}$ the volume of the foam and the volume of the drainage liquid, respectively, after 0, 5, 10, 15, 20, 30, 45, 60, 75, and 90 min. All the experiments were conducted in triplicate and conducted by the same person at room temperature (20 °C ± 1).

### 2.9. Bubble Size with Turbiscan Tower

Bubble size was evaluated with Turbiscan Tower (Formulaction, Toulouse, France). The analysis is based on Static Multiple Light Scattering principle. Briefly, a beam of light at 880 nm is sent on the sample and two different detectors acquired the value of the Backscattering (BS) and Transmission (T) all over the sample height. These values depend on the dispersed particle in a sample that are able to scatter the light source. In the case of foam, the scattering caused by bubbles is evaluated. By monitoring the sample over time, the instrument gives an evaluation of the evolution of bubbles and therefore on foam stability. According to Mie Theory [14], it is possible to calculate the size of the bubbles as follows:

$$BS = f\left(\varphi,\ d,\ n_p,\ n_f\right) \tag{8}$$

where BS is the value of backscattered light, $\varphi$ is the particle concentration (air fraction in the case of a foam), $d$ is the diameter of the particles, $n_p$ and $n_f$ are the refractive index of the dispersed and continuous phase, respectively. Immediately after the whipping process, the foams were transferred into a 55 mm high tube and loaded into the instrument. The measurements were conducted for 90 min with a scan on the sample every 1.5 min.

### 2.10. Confocal Laser Scanning Microscopy (CLSM)

Confocal laser scanning microscopy (CLSM) was utilized to confirm the presence of a protein layer at the air-water (A-W) interface. The solutions were initially diluted to 9 g/L. Rhodamine B (Sigma Aldrich, Gillingham, UK) was used to dye the proteins before following the same whipping process with Ultraturrax for 1 min at 9500 rpm. Immediately after the foam's formation, a volume of 0.5 mL of each sample was loaded into an 8-chambers microscope slide and covered with a coverslip. The dye was excited at 556 nm and a movie of each sample was taken using a 100× lens (Nikon CFI) on a confocal microscope spinning disc constituted by an inverted microscope (Nikon $Ti_2$). This was equipped with a laser source (405/488/561/640 nm), a confocal spinning disc module (Yokogawa CSU-W1, 50 um pinholes), a quad-band emission filter (440/521/607/700 nm) and a sCMOS camera (Photometrics Prime95B). However, due to the relative instability of FG foams, only PPI and MIX samples were analyzed by this technique. Indeed, FG foams drained too much liquid into the sample well, covering all of the air bubbles and hiding them from the microscope vision.

### 2.11. Interfacial Properties

The interfacial tension (IFT) between the liquid phase (pure water and protein solutions) and the gas phase (air) was measured with an optical tensiometer OCA 25 (Data-Physics Instruments, Filderstadt, Baden-Württemberg, Germany) using a static pendant drop method. IFT was recorded for 1000 s at room temperature (20 ± 1 °C), using a constant drop volume of 20 µL. The IFT values of the solutions were calculated based on the Laplace equation monitoring the shape of the droplet.

### 2.12. Statistical Analysis

The mean differences with ±standard deviation (SD) were analyzed. Statistical analysis was performed using student's test t and one-way analysis of variance test (ANOVA), with a level of significance $p < 0.05$.

## 3. Results and Discussion

### 3.1. pH Values

Table 1 enlists the pH values of all the sample and a comparison between untreated and sonicated ones. FG did not present any significant changes in pH values after the US treatment, stabilizing around 7.1, while PPI showed a slight decrease from 7.3 to 7.2. As expected, MIX sample exhibited a pH value range between FG and PPI values, which showed a slight decrease after sonication, going from 7.2 to 7.1. These results are in agreement with O'sullivan et al. [9]. The authors described a significant pH decrease of all the US treated animal and vegetable proteins considered, including bovine gelatin, fish gelatin, egg white, soy, rice and pea protein isolates. This is mainly due to the deprotonation of acidic amino acid residues contained within the agglomerated structure of the untreated proteins [15].

**Table 1.** pH values of the samples (Fish gelatin FG, pea protein isolate PPI, and their mixture MIX) as the average of three repeated measurements before and after a high intensity ultrasound treatment (275 W, 5 min).

| pH | FG | PPI | MIX |
|---|---|---|---|
| Untreated | $7.13 \pm 0.1^{a}$ | $7.34 \pm 0.2^{b}$ | $7.27 \pm 0.1^{d}$ |
| Sonicated | $7.14 \pm 0.1^{a}$ | $7.24 \pm 0.1^{c}$ | $7.15 \pm 0.3^{e}$ |

Means with different superscripts in each column indicate significant differences ($p < 0.05$).

### 3.2. Hydrodynamic Diameter ($D_h$) and ζ-Potential

Figure 1 shows the mean hydrodynamic diameter of the sample before and after US treatment. The large standard deviation associated with PPI are in accordance with the results obtained in previous studies on PPI conducted by O'Sullivan et al., and Xiong et al. [9,10]. The authors found a bimodal size distribution: one population has a similar size as the untreated protein (159.07 ± 66.78 nm) while the other one is significantly reduced by the US treatment. As expected, PPI size was reduced after the treatment (112 ± 47.49 nm). A similar behavior was observed for the MIX sample, going from 160.23 ± 69.5 to 115.67 ± 47.40 nm. This trend could be explained by the significant difference in the size between PPI and FG. The reduction of the size is probably due to the disruption of non-covalent associative forces, such as hydrogen bonds and hydrophobic interactions due to the cavitation effect of ultrasound waves [9]. However, FG exhibits a slight increase in the aggregate size after sonication, going from an average of 16.38 ± 8.64 nm to 23.03 ± 17.41 nm. This increase in the size could be due to a thermal aggregation correlated to non-covalent associative forces [9]. ζ-potential measurements present negative values for all of the samples, as presented in Figure 1. In particular, PPI and MIX have a similar tendency to decrease in value after the US treatment. Furthermore, they show similar values. ζ-potential of PPI range from −7.2 ± 0.5 mV to −8.2 ± 0.4 mV, the MIX from −6.4 ± 0.9 mV to −8.0 ± 1.2 mV while FG does not present a significant change in the values, which are stable between −1.32 ± 0.6 mV and −1.28 ± 0.5 mV.

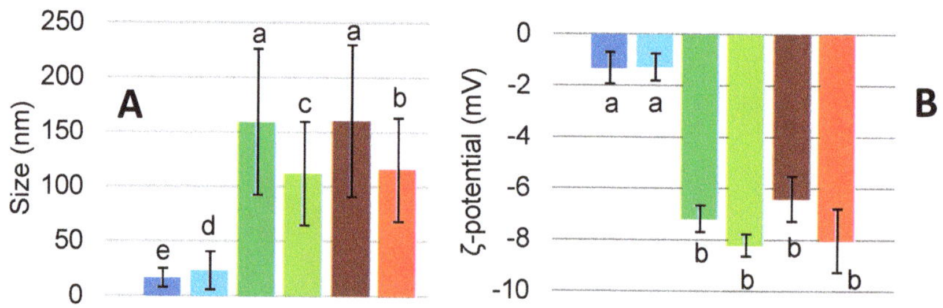


**Figure 1.** (**A**) Hydrodynamic diameter (nm) and (**B**) ζ-potential (mV) of FG (■ untreated, ■ treated), PPI (■ untreated, ■ treated) and MIX (■ untreated, ■ treated). Values are here reported as the average of three replicates and their standard deviation is shown in the error bars. a–e, different letters show significant differences ($p < 0.05$).

### 3.3. SEC-MALS

Size exclusion chromatography (SEC-MALS) was employed to determine the molecular weight (Mw) of FG, PPI and MIX prior and after sonication. The obtained results are shown on Table 2. Untreated FG exhibited fraction with Mw lower than expected, as the typical α-chain (~120 kDa) and β-chain (~200 kDa) were not observed. This might be associated to the extensive hydrolysis of collagen throughout the production process of protein [12]. Untreated PPI showed a high Mw fraction of 820 kDa, which is probably corresponded to protein aggregates, while fraction of 164.4 kDa is attributed to the trimeric form of vicilin [16]. Lower Mw fractions (61.2 and 44.1 kDa) are corresponded to vicilin and legumin subunits [16]. Vicilin monomer with Mw of 50 kDa and legumin monomer of

60 kDa had been reported in previous studies [16,17]. Untreated MIX demonstrated higher Mw fractions compared to FG and PPI (Table 2). Ultrasound treatment had no effect on Mw of FG and on lower Mw fractions of PPI. However, it was observed a decrease on Mw of high Mw aggregation of PPI. O'Sullivan et al. [9] reported no difference on molecular structure of FG and PPI after sonication (20 kHz, 95% of amplitude, for 2 min). Sonicated MIX showed higher Mw fraction compared to untreated MIX. The formation of aggregates in protein mixture caused by sonication is in accordance with the study reported by Silva et al. [18], who studied the effect of low-frequency ultrasound (20 Hz, 40% of amplitude, for 1–10 min) on whey-casein mixture.

**Table 2.** Molecular weight distribution of FG, PPI and MIX before and after sonication.

| Sample | | Fraction 1 | Fraction 2 | Fraction 3 | Fraction 4 |
|---|---|---|---|---|---|
| | | Mw (kDa) | | | |
| FG | Untreated | 33.2 | | | |
| | Sonicated | 29.8 | | | |
| PPI | Untreated | 879.4 | 164.4 | 61.2 | 44.1 |
| | Sonicated | 820.3 | 167.6 | 67.1 | 49.4 |
| MIX | Untreated | 1048.1 | 208.9 | 102.0 | 84.9 |
| | Sonicated | 2269.5 | 428.4 | 188.7 | 136.7 |

### 3.4. Rheology

Figure 2 shows the flow behavior of all the samples, untreated and sonicated, at 20 °C. The curves are presented as shear stress (τ) versus shear rate (γ). Data were evaluated with the power-law model and showed a nearly Newtonian flow behavior (n = 1) for all the samples over the shear rate range applied. Therefore, the consistency coefficient, which is equal to the apparent viscosity in Newtonian fluids, was compared for all the solutions and their values can be found in Table 3. Values were similar for all the samples, varying between 1.0 and 1.5 mPa.s. In particular, sonicated FG and PPI presented a viscosity of 1.0 mPa.s, decreasing compared to their untreated controls which showed the highest viscous behavior of 1.3 and 1.5 mPa.s, respectively. On the other hand, this decrease of viscosity with the US treatment did not occur for the MIX samples, which instead slightly increased from 1.2 to 1.3 mPa.s with the sonication. Therefore, MIX showed a viscosity between FG and PPI suggesting both protein sources contributed to its flow properties. Overall, data are in agreement with O'Sullivan et al. [9] who found a significant decrease of FG (from 1.06 ± 0.07 to 0.76 ± 0.05 dL/g) and PPI (from 0.8 ± 0.005 to 0.76 ± 0.007 dL/g) viscosity after an US treatment. In addition, MIX behavior could be explained by the complexity of the solution which is composed of a mixture of protein fractions rather than single components as FG and PPI [9].

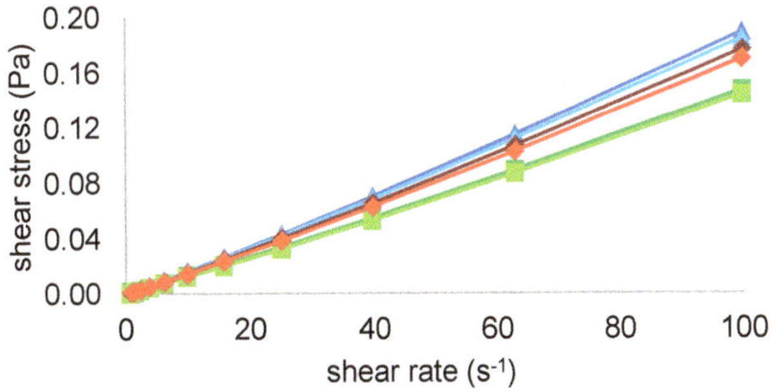


**Figure 2.** Shear stress (Pa) versus shear rate ($s^{-1}$) of FG (■ untreated, ■ treated), PPI (■ untreated, ■ treated) and MIX (■ untreated, ■ treated) at 20 °C.

**Table 3.** Power–law model parameters of all the samples at 20 °C.

| Power Law | *m* (mPa.s) * | *n* ** | $R^2$ |
|---|---|---|---|
| FG untreated | 1.3 | 1.1 | 0.99 |
| FG sonicated | 1 | 1.1 | 0.95 |
| PPI untreated | 1.5 | 1 | 0.98 |
| PPI sonicated | 1 | 1.1 | 0.98 |
| MIX untreated | 1.2 | 1.1 | 0.99 |
| MIX sonicated | 1.3 | 1 | 0.99 |

* Consistency coefficient; ** Flow behavior index.

A frequency sweep test was also employed to investigate the viscoelastic properties of the protein solutions. However, high concentrated protein solution, as the ones used for this test (18 g/L), protein molecules tend to adsorb at the A-W interface forming a viscoelastic film. Due to this effect, the measured apparent viscosity is the sum of both bulk and interfacial signals [19]. The interfacial protein layer contribution is well seen from the frequency sweep test (Figure S2). The test was conducted into a frequency range between 1–10 Hz and the viscoelastic moduli $G'$ and $G''$ were evaluated, where $G'$ represents the elastic behavior while $G''$ the viscous behavior of the fluid [20]. At low frequencies, viscous properties are predominant ($G'' > G'$) while at the end of the considered frequency range, the elastic modulus is definitely higher than the viscous one.

Protein aggregation and adsorption at the A-W interface contribute to the viscoelastic properties of the fluid giving a gel-like microstructure [21]. Finally, the crossover frequency of the moduli was analyzed. While both of FG samples exhibit frequencies over 4 Hz, PPI samples present a difference between the untreated and the sonicated groups. Untreated PPI shows a crossover around 4 Hz while after the treatment it happens around 1.5 Hz. This could be due to improved protein-protein interactions with the sonication, resulting in a previous manifestation of elastic behavior [21]. On the contrary, MIX samples exhibit the higher frequency value crossover, over 6 Hz for both untreated and sonicated, suggesting a complexity in the aggregates and interfacial film formation.

### 3.5. Foaming Properties

Foaming properties of protein solutions are generally based on the influence of many factors such as the protein nature, their isolation process, temperature, pH, concentrations, whipping method and time [22].

#### 3.5.1. Foam Capacity (*FC*)

*FC* is defined as the percentage fraction of air into the total foam volume [1]. *FC* of all the samples was determined immediately after the whipping process and is here presented as the mean of three replicates. *FC* values for FG, PPI and MIX are presented in Figure 3. We can observe that *FC* improved for sonicated FG in comparison with the untreated sample. The treatment increased the value from 46.6 ± 1.7% to 52.1 ± 1.4%, allowing FG foam to incorporate more volume of air. On the other hand, US treatment did not result in a significant difference for sonicated PPI foams compared to the control, while it slightly reduced *FC* values (−0.86%) of MIX samples. These results disagree with the study carried out by Morales et al. [23]. Indeed, the authors found a significant increase (+62%) of *FC* values for soy protein isolate foams after 5 min of a high intensity ultrasound treatment. However, these results could be explained by the different nature of soy and pea proteins. Overall, the samples reached *FC* value around 50% except for untreated FG which occurred with the lowest value (46.58 ± 1.7%) in all of the three replicates implemented.

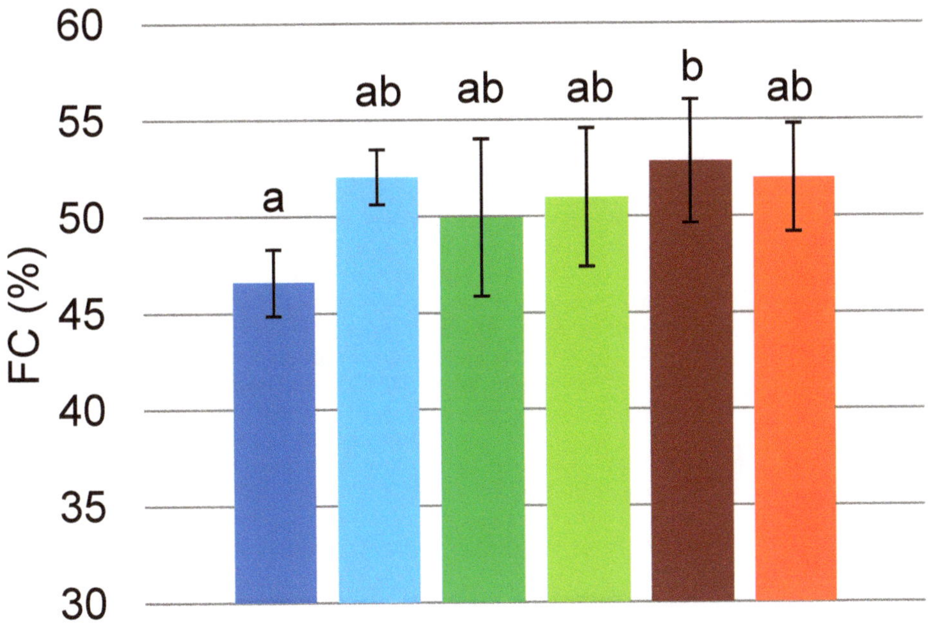


**Figure 3.** *FC* (%) of FG (■ untreated, ■ treated), PPI (■ untreated, ■ treated) and MIX (■ untreated, ■ treated). Values are here reported as the average of three replicates and their standard deviation is shown in the error bars. Different letters show significant differences ($p < 0.05$).

### 3.5.2. Foam Stability (*FS*)

*FS* can be described as the period of time in which the foam maintains its original properties when formed [1]. *FS* is mainly influenced by temperature and pH, by the nature of the proteins and their interaction, which should produce a viscous, elastic and air-impermeable film around each air bubble to stabilize the system [1]. *FS* values were visually monitored over a period of time of 90 min. *FS* values for FG, PPI and MIX are presented in Figure 4. We can observe that *FS* did not improve for all the samples after US treatment. Both FG and PPI sonicated samples showed a slight decrease in *FS* after 90 min, range from 13.33 to 12.0% and from 106.0 to 100.67%, respectively. MIX reached the same value around 114% for both untreated and sonicated samples. Furthermore, MIX foams present higher values, over the whole period considered, compared to FG and PPI foams. This may suggest a possible protein-protein interaction on the film formation around the bubbles. Globally, sonicated samples present similar values to the controls, without any statistical difference. As can be seen from the graphs, the treatment provided higher *FS* values during the firsts 25 min, while they dropped below controls values in the remaining 60 min. In particular, FG curves decreased steadily, approaching *FS* of 0% over the considered period, while both PPI and MIX stabilized above 100% during 90 min, presenting a similar behavior.

Overall, the results are in accordance with the study conducted by Xiong et al. [10]. where foaming properties of PPI were evaluated after different conditions of ultrasound treatment. In this study, *FS* values of sonicated PPI were significantly higher than those of untreated PPI, but only during the first 10 min of evaluation, whereas after 20 min, the authors found no significant difference between the two samples. This trend can be attributed to the partial unfolding of proteins induced by US, which allows them to rapidly adsorb at the A-W interface of the foam, leading to a greater stability within the first 10 min. However, in a larger period of time, protein molecules are desorbed from the interface due to the increase of air bubble sizes, a condition that leads them to interact with other desorbed molecules forming new aggregates. The aggregation phenomenon is even more pronounced within sonicated proteins because of the higher degree of exposed hydrophobic groups. This mechanism drastically reduces protein stability effect at the surface, reducing the overall *FS* values after a period of 20 min [10,24].

#### 3.5.3. Liquid Fraction (*LF*)

*LF* represents the water fraction trapped in the Plateau border between air bubbles [1]. During the time, gravitational forces cause the drainage of water from the borders resulting in a gradual collapse of the foam. Monitoring *LF* changes over time therefore can assess protein capacity of trapping water and an overall evaluation of foam stability. *LF* values were visually monitored over a period of time of 90 min for each sample and are hereby presented as the average of three replicates. *LF* values for FG, PPI and MIX are presented in Figure 5. We can observe that US treatment decreased *LF* percentages of FG and PPI samples after 90 min, going from 8.0 to 5.33% and from 8.67 to 2.0%, respectively. On the other hand, sonicated MIX sample showed a value of 9.33%, greater than 2% compared to the untreated one. Generally, all the samples tent to 0% as expected, decreasing rapidly in the firsts 20 min and steadily for the rest of the time. The obtained results are in general accordance with the study conducted by Morales et al. [23] where US were employed on soy protein isolate in an attempt of modifying their foaming functionalities. Even for soy proteins, it was not observed a significant change of liquid drainage from a foam after US treatment. However, sonicated samples showed a slight increase in the drainage velocity, and thus a faster decrease of *LF* %, as occurred in FG and PPI samples of this study.

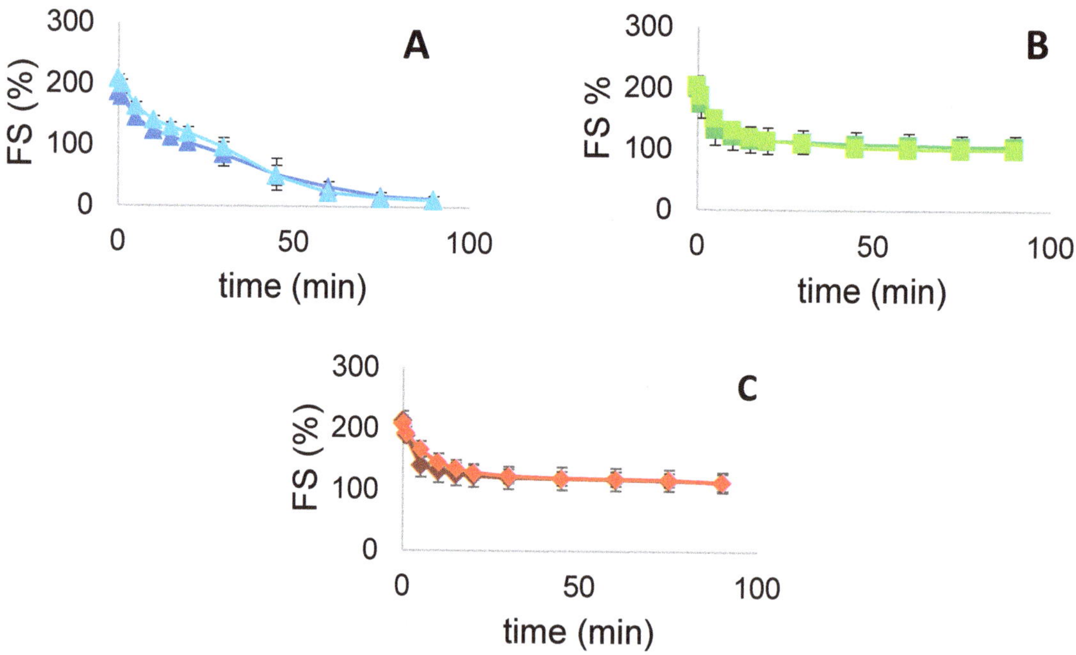


**Figure 4.** *FS* (%) over time (min) for (**A**) FG (■ untreated, ■ treated), (**B**) PPI (■ untreated, ■ treated) and (**C**) MIX (■ untreated, ■ treated). Values are here reported as the average of three replicates and their standard deviation is shown in the error bars.

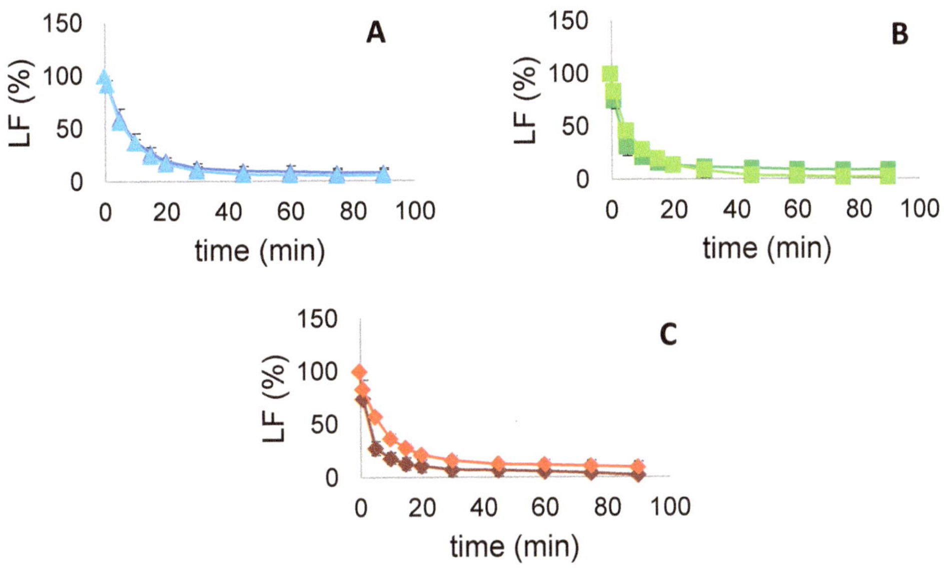


**Figure 5.** *LF* (%) over time (min) for (**A**) FG (untreated, treated), (**B**) PPI (untreated, treated) and (**C**) MIX (untreated, treated). Values are here reported as the average of three replicates and their standard deviation is shown in the error bars.

3.5.4. Turbiscan Tower

Immediately after the whipping process, FG, PPI and MIX foams were transferred into a 55 mm high tube and loaded into the Turbiscan Tower instrument. In particular, the two-phase separation occurs, where the drained liquid drops to the bottom whereas the foam stays at the top. Samples were run during 90 min, the same period of time considered by the visual analysis of their foaming properties. Over the time, 60 scans were performed (1 every 1.5 min) measuring the backscattering value (BS) all over the height of the tube. BS values over time, as obtained from the instrument, are presented in Figure S3.

BS% is caused by air bubbles of the foam. BS% value decrease over time due to foam aging and its gradual collapse. Foam collapse is mainly described by two effects: liquid drainage and bubbles coalescence. Liquid drainage effect is represented by the continuous shift of BS peak over the height of the tube sample, while coalescence is described by the steady decrease of BS% value. Both of these effects are influenced by the size of the air bubbles. Over time, bubbles tend to coalesce together and their size increase until it reaches the size limit. Over this limit the bubble bursts and the liquid that was trapped is now released, generating a drainage phase at the bottom and decreasing BS%. Turbiscan software, based on the Mie equation, allows the automatic computation of mean diameter of air bubbles from the BS level on the foam phase. The results are presented in Figure 6.

All of the samples start with a similar mean diameter, generally over 115 μm, and during the period of time considered they evolve in a different way. In particular, FG both untreated and sonicated increase rapidly in bubble size until they burst around 30 min with a diameter above 500 μm. On the other hand, PPI and MIX samples present a similar trend: they steadily increase their mean diameter, overcoming FG limit size and reaching a size around 600 μm, without bursting over the 90 min considered. Furthermore, we can observe that the US treatment changed the evolution of the bubbles size of FG and PPI samples, by increasing their diameter. In particular, it caused the collapse of FG sonicated sample earlier than the untreated one. Nevertheless, this change did not cause the earlier burst for sonicated PPI, which maintained a good stability over time. Regarding MIX samples, the US treatment did not change significantly their behaviour and showed the same evolution over time. The results are overall in agreement with the visual analysis on foaming properties. Especially with *FS* from which it results that FG samples stability constantly decreases with foam aging, here shown with the bubbles burst, and the US treatment even decreased it. While PPI and MIX samples reached a good stability of the foam and the US treatment did not improve it significantly.

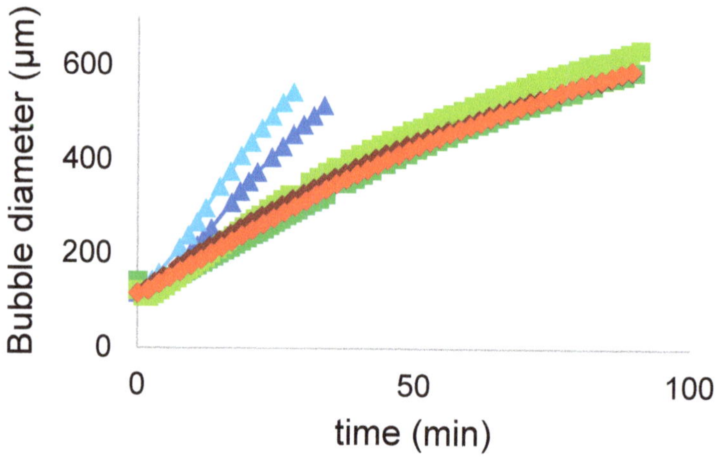


**Figure 6.** Bubble mean diameter (μm) evolution over time (min) of FG (untreated, treated), PPI (untreated, treated) and MIX (untreated, treated).

### *3.6. Confocal Laser Scanning Microscopy (CLSM)*

CLSM was employed to evaluate the distribution and behaviour of proteins at the A-W interface and to better understand foam microstructure. However, due to the relative instability of FG foams, only PPI and MIX samples were analyzed by this technique. Indeed, FG foams drained too much liquid into the sample well, covering all of the air bubbles and hiding them from the microscope vision.

From the movies recorded by the instrument, frames of the sample bubbles were collected and Figure 7, was designed. From Figure 7, it can be clearly seen the fluorescent film layer made from protein aggregates, as suggested by DLS measurements. Ultimately, the protein layer was confirmed with the image comparison with a bathroom soap foam, free of proteins in its composition. PPI and MIX foam frames were then compared and no significant difference was seen between untreated and sonicated proteins. Moreover, MIX protein layers did not differ from PPI, suggesting that pea proteins only contributed to the film formation when mixed with fish proteins, giving a similar result to PPI foam. Therefore, it was not possible to detect an interaction between FG and PPI aggregates at the A-W interface.

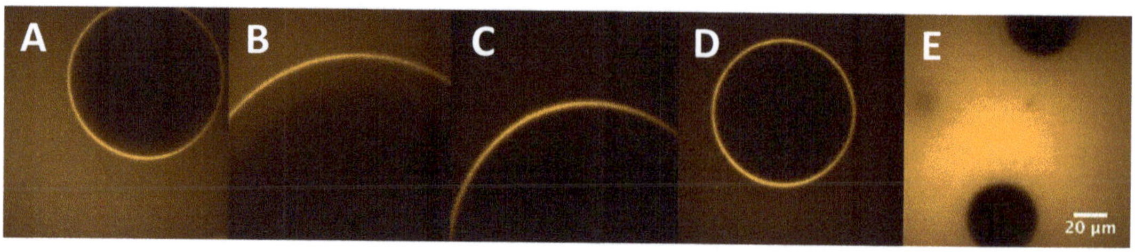


**Figure 7.** CLSM images of PPI (**A**,**B**) and MIX (**C**,**D**) samples. To discriminate the fluorescent protein layer around each air bubble, a comparison with soap bubbles with no protein content (**E**) has been carried out.

On the contrary, Jarpa-Parra et al. [25] employed CLSM on foams stabilized by legumin-like protein and polysaccharides and the technique succeed to visualize an interaction between the two biopolymers due to different fluorescent excitation wavelengths. The authors concluded that the interaction was allowed by the opposite charges of the polymers and the formation of weak electrostatic attractions between them.

### 3.7. Interfacial Tension (IFT)

IFT at the A-W interface was measured with the pendant drop method. The instrument was calibrated monitoring the IFT of water before measure. IFT values as a function of time of FG, PPI and MIX solutions is presented in Figure 8. After 1000 s we can observe a decrease only for FG samples after US treatment. Probably this could be caused by the different protein size of FG and PPI. In addition, FG stabilized at 52.2 ± 0.3 mN/m whereas PPI and MIX present stable values at 48.9 ± 0.1 mN/m and 47.2 ± 0.2 mN/m. Therefore, combining the two proteins allowed to reach the lowest IFT values. A similar study was realized by Xiong et al. [10]. The authors applied 30 min of US treatment at different amplitudes (30, 60, 90%) and observed a general decrease of IFT values for sonicated PPI when compared to control. These results could be explained by the improved affinity of PPI to the dispersed phase, as more hydrophobic regions were exposed after the sonication. Indeed, an increase in hydrophobicity can result in a reduction of the energy barrier at the A-W interface, allowing a facilitated adsorption [26]. In our study, after 5 min of treatment, no significant differences were observed for both PPI and MIX samples when compared to their controls. Perhaps the treatment did not last enough to increase the exposed hydrophobic regions of PPI aggregates and thus to change the rate adsorption at the interface. Moreover, as explained by ζ-potential values, the US treatment increased surface charges of both PPI and MIX samples resulting in higher electrostatic barriers at the A-W interface and thus causing the decrease of their adsorption rate [27].

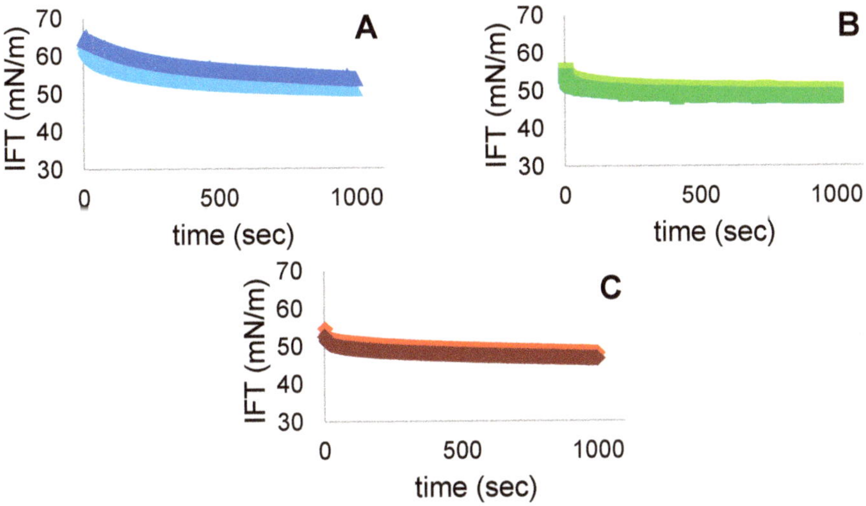


**Figure 8.** IFT values (mN/m) over time (sec) for (**A**) FG (untreated, treated), (**B**) PPI (untreated, treated) and (**C**) MIX (untreated, treated).

## 4. Conclusions and Perspectives

The interaction between FG and PPI in solution and at A-W interface, before and after US treatment was investigated. We focused on bulk rheology and the interfacial tension contribution of the protein solutions and then we evaluated the respective foaming properties. The treatment was responsible for a viscosity decrease of both FG and PPI samples, whereas MIX showed the same flow properties, in between FG and PPI. IFT value at the A-W interface of the solutions revealed that US treatment only affected FG proteins while PPI and MIX did not show a significant change. However, lower IFT values were obtained when the proteins were combined. Regarding the foaming properties, four main parameters were evaluated: foam capacity (*FC*), foam stability (*FS*), liquid fraction (*LF*) and bubble size evolution over time. Results showed that this specific ultrasound treatment (275 W, 5 min) only improved *FC* of FG while PPI and MIX were not significantly affected. *FS* was not improved by the sonication for FG and PPI, while MIX samples had the highest *FS* values during all the tests conducted, suggesting a possible interaction and improvement in stabilizing the foam system compared to the single protein foams. In addition, US allowed MIX sample to have higher *LF* values than its control meaning it could slow the liquid drainage and the foam to dry over the time, while FG and PPI on the other hand slightly decreased in value with the treatment. In the end, bubble size evolution over time confirmed the better stability achieved by both PPI and MIX while FG bubbles rapidly grew in size until they busted, causing the foams to collapse. Moreover, US treatment not only did not improve stability of FG samples but caused an earlier collapse of their foams. Based on these considerations we can conclude that the specific US treatment employed cannot significantly improve these protein sources foaming properties. Overall, MIX samples presented the most encouraging interfacial properties, improving slightly the characteristics of PPI only, as if PPI and FG had developed a weak interaction between them. To better inspect a possible interaction, a visual analysis of the protein layer around each air bubble in foam was employed using CLSM, but unfortunately, no differences were noticed between PPI and MIX layers. Further investigations are therefore needed to demonstrate this possible interaction between FG and PPI. In addition, different US treatments could be applied to the samples as presented by Xiong et al., 2018 [10], changing the combination of power and time. In fact, the treatment submitted in this study seemed to be too weak to significantly affect PPI and MIX behavior at the A-W interface. Finally, CLSM could also be employed to calculate the thickness of the protein layer and the aggregates adsorption ratio at the A-W interface.

**Supplementary Materials:** The following are available online at https://www.mdpi.com/article/10.3390/foods11050659/s1, Figure S1: Calibration line of PPI supernatant soluble protein content (%). 5 different starting concentrations were analysed by Dumas method after the removal of the insoluble part, Figure S2: Viscoelastic properties for all the protein solutions at a frequency range between 1–10 Hz. For all the graphs, elastic modulus (G′) is represented in orange squares while viscous modulus (G″) is represented in blue dots, Figure S3: BS values (%) of sonicated PPI over the height of the tube and over time as reported by the instrument Turbiscan Tower. On the right side, scan time is reported with a gradient of colors, Table S1: Soluble protein content of different starting solution concentrations.

**Author Contributions:** Conceptualization, F.C.; Data curation, D.O., K.S., R.M., M.V.-l. and F.C.; Formal analysis, D.O., K.S., A.S., R.M. and M.V.-l.; Funding acquisition, F.J. and F.C.; Investigation, A.S.; Methodology, R.M.; Project administration, M.A.M. and F.J.; Supervision, F.J.; Validation, M.A.M., A.F.d.C. and M.V.; Visualization, G.S. and A.F.d.C.; Writing—original draft, D.O., K.S. and A.S.; Writing—review & editing, D.O., M.A.M., F.J., G.S., M.V.-l., A.F.d.C., M.V. and F.C. All authors have read and agreed to the published version of the manuscript.

**Funding:** The authors acknowledge the support provided by the Green Development and Demonstration Program (GUDP) and Ministry of Environment and Food of Denmark (J. Nr. 34009-17-1299).

**Data Availability Statement:** Data is contained within the article or supplementary material.

**Acknowledgments:** Data was generated though accessing research infrastructure at DTU National Food Institute, including FOODHAY (Food and Health Open Innovation Laboratory, Danish Roadmap for Research Infrastructure). The authors acknowledge Roquette® to kindly provided the pea protein sample. The authors acknowledge the funding program "Tesi all'Estero" furnished by Università Cattolica del Sacro Cuore, Milano, Italia.

**Conflicts of Interest:** The authors declare no conflict of interest.

## References


1. Zayas, J.F. Functionality of Proteins in Food. In *Functionality of Proteins in Food*; Springer: Berlin/Heidelberg, Germany, 1997.
2. Kralova, I.; Sjöblom, J. Surfactants used in food industry: A review. *J. Dispers. Sci. Technol.* **2009**, *30*, 1363–1383. [CrossRef]
3. Tomadoni, B.; Capello, C.; Valencia, G.A.; Gutiérrez, T.J. Self-assembled proteins for food applications: A review. *Trends Food Sci. Technol.* **2020**, *101*, 1–16. [CrossRef]
4. Benjakul, S.; Kittiphattanabawon, P.; Regenstein, J.M. Fish Gelatin. In *Food Biochemistry and Food Processing*, 2nd ed.; Wiley-Blackwell: Oxford, UK, 2012. [CrossRef]
5. Benjakul, S.; Kittiphattanabawon, P. Gelatin. In *Encyclopedia of Food Chemistry*; Elsevier: Amsterdam, The Netherlands, 2018; pp. 121–127. [CrossRef]
6. Tabata, Y.; Ikada, Y. Protein release from gelatin matrices. *Adv. Drug Deliv. Rev.* **1998**, *31*, 287–301. [CrossRef]
7. Casanova, F.; Mohammadifar, M.A.; Jahromi, M.; Petersen, H.O.; Eybye, K.L.; Kobbelgaard, S.; Jakobsen, G.; Jessen, F. Physico-chemical, structural and techno-functional properties of gelatin from saithe (*Pollachius virens*) skin. *Int. J. Biol. Macromol.* **2020**, *156*, 918–927. [CrossRef]
8. Hinderink, E.B.A.; Münch, K.; Sagis, L.; Schroën, K.; Berton-Carabin, C.C. Synergistic stabilisation of emulsions by blends of dairy and soluble pea proteins: Contribution of the interfacial composition. *Food Hydrocoll.* **2019**, *97*, 105206. [CrossRef]
9. O'Sullivan, J.; Murray, B.; Flynn, C.; Norton, I. The effect of ultrasound treatment on the structural, physical and emulsifying properties of animal and vegetable proteins. *Food Hydrocoll.* **2016**, *53*, 141–154. [CrossRef]
10. Xiong, T.; Xiong, W.; Ge, M.; Xia, J.; Li, B.; Chen, Y. Effect of high intensity ultrasound on structure and foaming properties of pea protein isolate. *Food Res. Int.* **2018**, *109*, 260–267. [CrossRef]
11. Hinderink, E.B.A.; Sagis, L.; Schroën, K.; Berton-Carabin, C.C. Behavior of plant-dairy protein blends at air-water and oil-water interfaces. *Colloids Surf. B Biointerfaces* **2020**, *192*, 111015. [CrossRef]
12. Vall-llosera, M.; Jessen, F.; Henriet, P.; Marie, R.; Jahromi, M.; Sloth, J.J.; Mohammadifar, M.A.; Petersen, H.O.; Jørgensen, B.M.; Casanova, F. Physical Stability and Interfacial Properties of Oil in Water Emulsion Stabilized with Pea Protein and Fish Skin Gelatin. *Food Biophys.* **2020**, *16*, 139–151. [CrossRef]
13. Lionberger, R.A.; Russel, W.B. A Smoluchowski theory with simple approximations for hydrodynamic interactions in concentrated dispersions. *J. Rheol.* **1997**, *41*, 399–425. [CrossRef]
14. Lock, J.A.; Gouesbet, G. Generalized Lorenz-Mie theory and applications. *J. Quant. Spectrosc. Radiat. Transf.* **2009**, *110*, 800–807. [CrossRef]
15. Sakurai, K.; Konuma, T.; Yagi, M.; Goto, Y. Structural dynamics and folding of β-lactoglobulin probed by heteronuclear NMR. *Biochim. Biophys. Acta Gen. Subj.* **2009**, *1970*, 527–537. [CrossRef] [PubMed]
16. Shevkani, K.; Singh, N.; Kaur, A.; Rana, J.C. Structural and functional characterization of kidney bean and field pea protein isolates: A comparative study. *Food Hydrocoll.* **2015**, *43*, 679–689. [CrossRef]
17. Klost, M.; Giménez-Ribes, G.; Drusch, S. Enzymatic hydrolysis of pea protein: Interactions and protein fractions involved in fermentation induced gels and their influence on rheological properties. *Food Hydrocoll.* **2020**, *105*, 105793. [CrossRef]
18. Silva, M.; Zisu, B.; Chandrapala, J. Influence of low-frequency ultrasound on the physico-chemical and structural characteristics of milk systems with varying casein to whey protein ratios. *Ultrason. Sonochem.* **2018**, *49*, 268–276. [CrossRef]
19. Zhang, Z.; Liu, Y. Recent progresses of understanding the viscosity of concentrated protein solutions. *Curr. Opin. Chem. Eng.* **2017**, *16*, 48–55. [CrossRef]
20. Ahmed, J.; Ptaszek, P.; Basu, S. Food Rheology: Scientific Development and Importance to Food Industry. In *Advances in Food Rheology and Its Applications*; Woodhead Publishing: Sawston, UK, 2017. [CrossRef]
21. Sharma, V.; Jaishankar, A.; Wang, Y.C.; McKinley, G.H. Rheology of globular proteins: Apparent yield stress, high shear rate viscosity and interfacial viscoelasticity of bovine serum albumin solutions. *Soft Matter* **2011**, *7*, 5150–5160. [CrossRef]
22. Ge, J.; Sun, C.X.; Corke, H.; Gul, K.; Gan, R.Y.; Fang, Y. The health benefits, functional properties, modifications, and applications of pea (Pisum sativum L.) protein: Current status, challenges, and perspectives. *Compr. Rev. Food Sci. Food Saf.* **2020**, *19*, 1835–1876. [CrossRef]
23. Morales, R.; Martínez, K.D.; Pizones Ruiz-Henestrosa, V.M.; Pilosof, A.M.R. Modification of foaming properties of soy protein isolate by high ultrasound intensity: Particle size effect. *Ultrason. Sonochem.* **2015**, *26*, 2416–2420. [CrossRef]
24. Li, R.; Chen, X.E.; Chang, Y.; Zhang, L.; Zhang, Y.; Zhu, Y.; Wang, T. Increase of bubble size playing a critical role in foam-induced protein aggregation: Aggregation of BSA in foam fractionation. *Chem. Eng. Sci.* **2017**, *174*, 387–395. [CrossRef]
25. Jarpa-parra, M.; Tian, Z.; Temelli, F.; Zeng, H.; Chen, L. Understanding the stability mechanisms of lentil legumin-like protein and polysaccharide foams. *Food Hydrocoll.* **2016**, *61*, 903–913. [CrossRef]
26. Delahaije, R.J.B.M.; Wierenga, P.A.; Giuseppin, M.L.F.; Gruppen, H. Improved emulsion stability by succinylation of patatin is caused by partial unfolding rather than charge effects. *J. Colloid Interface Sci.* **2014**, *430*, 69–77. [CrossRef] [PubMed]
27. Wierenga, P.A.; Meinders, M.B.J.; Egmond, M.R.; Voragen, A.G.J.; De Jongh, H.H.J. Quantitative description of the relation between protein net charge and protein adsorption to air-water interfaces. *J. Phys. Chem. B* **2005**, *109*, 16946–16952. [CrossRef] [PubMed]

*foods* MDPI

*Article*

# Impact of Enzymatic Hydrolysis and Heat Inactivation on the Physicochemical Properties of Milk Protein Hydrolysates

**Alice Gruppi [1,2], Maria Dermiki [2,3], Giorgia Spigno [1,*] and Richard J. FitzGerald [2]**

1 Department for Sustainable Food Process (DiSTAS), Università Cattolica del Sacro Cuore, Via Emilia Parmense 84, 29122 Piacenza, Italy; alice.gruppi@unicatt.it
2 Department of Biological Sciences, School of Natural Sciences, University of Limerick, V94 T9PX Limerick, Ireland; Dermiki.Maria@itsligo.ie (M.D.); dick.fitzgerald@ul.ie (R.J.F.)
3 Faculty of Science, Institute of Technology Sligo, F91 YW50 Sligo, Ireland
* Correspondence: giorgia.spigno@unicatt.it

**Abstract:** This study determined the physicochemical properties (apparent viscosity ($\eta_{app}$), turbidity ($A_{550nm}$), particle size and molecular mass distribution) of hydrolysates generated from whey protein concentrate (WPC), milk protein concentrate (MPC) and sodium caseinate (NaCN), following incubation with Debitrase HYW20™ and Prolyve™ at 50 °C, pH 7.0 for 1 and 4 h, before and after heat inactivation (80 °C for 10 min). The degree of hydrolysis (DH) increased with incubation time, giving values of 6.56%, 8.17% and 9.48%, following 1 h hydrolysis of WPC, MPC and NaCN with Debitrase HYW20™, and 12.04%, 15.74% and 17.78%, respectively, following 4 h incubation. These DHs were significantly higher compared to those obtained following 4 h incubation with Prolyve™. Hydrolysis with Debitrase HYW20™ gave >40% of peptides with molecular masses < 1 kDa for all substrates, which was higher than the value obtained following hydrolysis with Prolyve™. The effect of hydrolysis on the physicochemical properties was substrate dependent, since $\eta_{app}$ decreased in WPC and NaCN hydrolysates, particle size decreased for all the substrates, with aggregate formation for MPC, and turbidity decreased in WPC and MPC hydrolysates, while it increased in NaCN hydrolysates. The physical properties of the hydrolysates were influenced by the enzyme thermal inactivation step in a DH-dependent manner, with no significant effect on turbidity and viscosity for hydrolysates at higher DHs.

**Keywords:** degree of hydrolysis; milk protein concentrate; molecular mass distribution; sodium caseinate; turbidity; viscosity; whey protein

**Citation:** Gruppi, A.; Dermiki, M.; Spigno, G.; FitzGerald, R.J. Impact of Enzymatic Hydrolysis and Heat Inactivation on the Physicochemical Properties of Milk Protein Hydrolysates. *Foods* **2022**, *11*, 516. https://doi.org/10.3390/foods11040516

Academic Editors: Federico Casanova and Michela Verni

Received: 10 January 2022
Accepted: 8 February 2022
Published: 11 February 2022



## 1. Introduction

Dairy proteins are commonly used as ingredients in complex food systems, either for structuring or for nutritional purposes [1]. The enzymatic hydrolysis of dairy proteins can enhance their techno- and bio-functional properties [2–5]. Whey protein concentrate (WPC), milk protein concentrate (MPC) and sodium caseinate (NaCN) represent some of the most used, and therefore, the most extensively studied dairy protein ingredients. Specifically, NaCN is utilised due to its foaming and emulsifying properties [6], while whey protein ingredients are utilised for their gelation and emulsification properties, and their good solubility at acid pH [7,8]. The application of MPCs [9], with high protein contents (>70%), may be impacted by the variability in their solubility properties [10–12].

Milk protein hydrolysates exhibit improved techno-functional properties compared to the intact proteins, thereby enhancing their applications as ingredients in different food preparations. Several studies have demonstrated an increase in protein solubility following enzymatic hydrolysis [13–17], as reported by Ryan et al. [18] who found that the hydrolysis of milk protein isolate, with Flavourzyme, Neutrase and Protamex, led to increased solubility at pH 4.0–7.0. Moreover, hydrolysis can lead to significant improvements in the foaming, gelling and emulsifying properties of whey protein hydrolysates [19–22] of casein hydrolysates [23] and milk protein hydrolysates [24,25] in comparison with the intact proteins. This improvement in functional properties may be attributed to changes in the secondary structure and to a decrease in molecular mass following hydrolysis [26]. In many cases, enzymatic hydrolysis of milk proteins can lead to improved bio-functional properties, due to the formation of peptides with, e.g., angiotensin converting enzyme-inhibitory activity, antidiabetic or antimicrobial activities [2], which are not biologically active when they are in the parent protein [2]. Whey proteins can also give rise to bioactive peptides once the primary structure is hydrolysed; these peptides can display different bioactivities, such as antioxidative activity, be more effective in treating tumours in some cancers and can inhibit ACE activity in vitro [27].

The properties of milk protein hydrolysates depend on the conditions under which they have been generated [3,4], such as the enzymes used, since different enzymes, depending on their specificity, will result in the formation of peptides with varying molecular masses and hydrophobicity [3]. Furthermore, hydrolysate properties are dependent on the pH employed during hydrolysis [28], hydrolysis time, incubation temperature [29], enzyme to substrate ratio [30] and total solids [31]. During their manufacture, hydrolysates are usually subjected to an enzyme heat inactivation step, prior to concentration and spray-drying. Therefore, when generating hydrolysates, it is important to consider the impact of both the hydrolysis and heat inactivation conditions on the properties of the hydrolysates generated, prior to and following drying. This is due to the fact that changes in the viscosity, turbidity and the formation of aggregates, during hydrolysis and subsequent enzyme thermal inactivation, can affect the heat and mass transfer properties during various processing steps. Therefore, it is expected that hydrolysis parameters, such as the enzyme employed, the duration of hydrolysis along with heat inactivation, will impact the properties of the hydrolysates generated using different substrates.

The aim of this study was to investigate the impact of the enzyme preparation and incubation time on the physicochemical properties (DH, $\eta_{app}$, turbidity ($A_{550nm}$), particle size and molecular mass distribution) of hydrolysates, generated from WPC, MPC and NaCN using Debitrase™ and Prolyve™. The impact of enzyme heat inactivation on hydrolysate particle size, $\eta_{app}$ and turbidity was also investigated.

## 2. Materials and Methods

### 2.1. Materials

The milk protein substrates NaCN (Arrabawn Co-Operative Creamery, Nenagh, Ireland), MPC (Kerry Ingredients, Listowel, Ireland) and WPC (Carbery, Ballineen, Ireland) had protein contents of 85%, 84% and 82% (*w*/*w*), respectively, and were kindly provided by the above manufacturers. The food-grade proteolytic preparation Debitrase HYW20™ was kindly provided by Rhodia Ltd. (Cheshire, UK), and Prolyve 1000™ was provided by Lyven Enzymes Industrielles (Caen, France). 2,4,6-Trinitrobenzenesulfonic acid (TNBS) was obtained from Pierce Biotechnology (Medical Supply, Dublin, Ireland) and all other chemicals and reagents were purchased from Sigma Chemical Company Ltd. (Dublin, Ireland).

### 2.2. Methods

#### 2.2.1. Generation of Enzymatic Hydrolysates

The hydrolysis procedure was based on the protocol employed by Dermiki and FitzGerald [29] with some adaptations. Substrate suspensions with different concentrations were prepared based on their solubility. MPC and NaCN were dissolved at 8% (*w*/*w*) while WPC was reconstituted at 10% (*w*/*w*) in distilled water at room temperature and then gently stirred at 5 °C for 16 h to aid hydration. Before hydrolysis, the temperature of the substrate samples was adjusted to 50 °C and maintained at this temperature for 1 h prior to initiating the hydrolysis reaction. The pH of the substrate solutions was adjusted to pH 7.0 using 0.5 M NaOH. The enzyme (Debitrase HYW20™ or Prolyve 1000™) was added at an enzyme to substrate ratio (E:S) of 0.5% (*v*/*w*) while hydrolysis was carried out at 50 °C, using a magnetic stirrer (Fisherbrand, Fisher Scientific, Dublin, Ireland) set at

500 rpm and the pH of the hydrolysis reaction was controlled at pH 7.0 using a pH-stat (Titrando 843, Tiamo 1.4 Metrohm, Dublin, Ireland) by the addition of 0.5 M NaOH. Control samples of MPC, WPC and NaCN, without enzymes, were included in the experimental plan. The volume of all samples (hydrolysates and unhydrolyzed protein samples without the addition of enzyme) was adjusted with distilled water to achieve the same final total solids concentration (7.99% (*w*/*w*) for NaCN and MPC; and 9.99% for WPC). Enzyme inactivation was conducted by heating the solutions at 80 °C for 10 min in a water bath as reported by Dermiki and FitzGerald [29]. Hydrolysates were then frozen and stored at −20 °C until further analysis, unless stated otherwise. Figure 1 provides a schematic overview of the experimental approach employed.

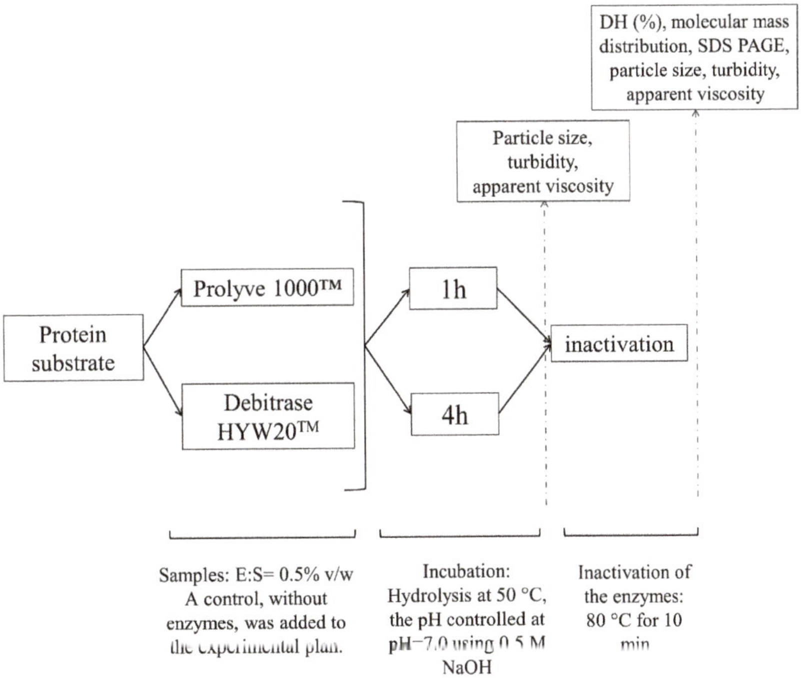


**Figure 1.** Schematic representation of the experimental approach. The initial substrate concentration for milk protein concentrate (MPC) and sodium caseinate (NaCN) was 8% (*w*/*w*) total solids and for whey protein concentrate (WPC) it was 10% (*w*/*w*) total solids in distilled water. Note: DH: degree of hydrolysis; E:S: enzyme to substrate ratio (*v*/*w*), SDS PAGE: Sodium dodecyl sulphate polyacrylamide gel electrophoresis.

Particle size distribution, turbidity and $\eta_{app}$ analyses were determined before and after heat inactivation on the day of hydrolysate production while DH and molecular mass distribution analyses were performed on the heat-treated hydrolysates (Figure 1).

2.2.2. Determination of DH

The spectrophotometric TNBS assay was used to determine the DH (%) as reported by Adler-Nissen [32] with the modifications of Le Maux et al. [28]. Samples (hydrolysed and unhydrolyzed control samples after heat inactivation) were diluted in 1% (*w*/*v*) SDS to a final protein concentration/protein equivalent of 5% (*w*/*v*) and prior to incubation at 50 °C for 30 min. Then,10 µL of the test hydrolysate samples and leucine standards (0, 2, 5, 7, 14, 21, 28 and 56 mg of nitrogen/L) were loaded onto a 96-well plate with 160 µL TNBS working solution (0.05% TNBS (*w*/*v*) and a 1:1 mixture of preheated (50 °C) water and

0.2125 M phosphate buffer pH 8.2). The plate was incubated at 50 °C for 1 h in a Synergy™ HT plate reader (BioTek Instruments Limited, Bedfordshire, UK) and the absorbance at 420 nm was recorded. The DH was calculated using Equation (1), as follows:

$$\text{DH }(\%) = \frac{\text{AN}_{\text{sample}} - \text{AN}_{\text{unhydrolysed sample}}}{\text{Npb}} \quad (1)$$

where $\text{AN}_{\text{sample}}$ is the amino nitrogen content of the protein hydrolysate (mg/g protein), $\text{AN}_{\text{unhydrolyzed sample}}$ is the amino nitrogen content of the protein substrate before hydrolysis (mg/g protein) and Npb is the nitrogen content of the peptide bonds in the protein substrate (mg/g protein); values of 100, 123.3 and 112.1, respectively, were used for MPC [24], WPC [7] and NaCN [33]. Analysis was conducted in triplicate (n = 3).

#### 2.2.3. Particle Size Distribution

Particle size distribution of the hydrolysates and controls was determined by laser light scattering as described by Le Maux et al. [34].

A Mastersizer 2000 with a Hydro 2000S dispersion system (Malvern Instruments, Worcestershire, UK) was used to analyse the particle sizes of the samples. Laser obscuration between 5–10% was obtained using the dispersion unit before each measurement. The particle and the dispersant refractive index used were 1.52 and 1.33, respectively. Each sample was measured in triplicate. Analysis of the results was performed using the general-purpose model available from the Malvern software. The particle size distributions were expressed as the cumulative weight (%) per volume moment mean diameter of the particles ($D_{3,2}$ μm).

#### 2.2.4. Molecular mass Distribution

Gel permeation high-performance liquid chromatography (GP-HPLC) was used to determine the molecular mass distribution of the samples at a concentration of 0.25% (*w*/*v*) protein/protein equivalent as described by Nongonierma and FitzGerald [35]. Samples were filtered through 0.2 μm PTFE filters prior to injection. Aliquots (20 μL) of diluted sample were injected onto a TSK G2000 SW separating column (600 × 7.5 mm ID) (Tosoh 157 Bioscience, Stuttgart, Germany). Separation took place by isocratic elution using 0.1% (*v*/*v*) trifluoroacetic acid (TFA), 30% (*v*/*v*) acetonitrile in $H_2O$, at a flow rate of 1 mL/min. Detector response was monitored at 214 nm [35]. A calibration curve was generated using the average retention times of the standards (BSA (67,500 Da), β-lactoglobulin (36,000 Da), α-lactalbumin (14,200 Da), aprotinin (6500 Da), bacitracin (1400 Da), Leu-Trp-Met-Arg-OH (605 Da), Asp-Glu (262 Da), Tyr (181 Da)). Molecular mass distributions were obtained by integrating the area under the curve corresponding to the average retention time of the different molecular masses at 10, 5 and 1 kDa.

#### 2.2.5. Turbidity ($A_{550nm}$) Measurements

Turbidity was evaluated as described by O'Loughlin et al. [36]. Samples were diluted to 0.1% (*w*/*v*) protein/protein equivalent with distilled water and vortexed to prevent immediate separation, the absorbance (200 μL sample volume) was then read at 550 nm at room temperature using a Synergy™ HT plate reader. All samples were analysed in triplicate (n = 3).

#### 2.2.6. Determination of Apparent Viscosity ($\eta_{app}$)

The samples were equilibrated at 50 °C and the $\eta_{app}$ was measured using a Brookfield DVII+ LV (Brookfield Engineering Laboratories, Middleboro, MA, USA) viscometer, fitted with an ultra-low adaptor (ULA). Measurements were conducted at a defined shear rate, i.e., 112 $s^{-1}$ (or rotational speed of the spindle at 100 rpm). The ULA adaptor was connected to a Brookfield refrigerated circulating water bath (model TC-500) by an ULA-40Y water jacket in order to control the temperature at 50 °C during measurements as described by Dermiki and FitzGerald [29].

2.2.7. Statistical Analysis

All statistical analyses were conducted using XLStat statistical software (XLStat, 2020.1.3.65324, Addinsoft, New York, NY, USA) [37]. Values presented are the mean of three replicates ± standard deviation, unless otherwise stated. The standardised residuals were calculated and were normally distributed. Normality was tested using the Shapiro–Wilk test which is best suited for small sample sizes. Data were also tested for homogeneity by plotting scatterplots of the residuals against predictors. These tests were all conducted as part of the ANOVA analysis using XLStat [37]. One-way analysis of variance (ANOVA), two-way ANOVA or full factorial design was used to test the effect of one, two or more factors (enzyme, incubation time, heated or unheated) on the responses studied, respectively. When significance was noted, comparison of means was conducted by employing a Tukey post-hoc test. Significance was determined at $p < 0.05$.

## 3. Results and Discussion

### *3.1. Hydrolysis of Milk Protein Substrates*

The following two commercially available proteolytic preparations were used in the current study: Debitrase HYW20™ and Prolyve 1000™. Debitrase HYW20™ is an enzyme preparation derived from *Aspergillus oryzae* and *Bacillus spp.* (rich in exopeptidase and with proteases), which has been shown to produce hydrolysates with reduced bitterness [38,39], while Prolyve 1000™ is a *Bacillus licheniformis* proteinase, which does not cause gelation [31] and has also been reported to show decreased bitterness compared to Alcalase™, another commonly used *Bacillus licheniformis* proteinase [35].

According to Cui et al. [17], the treatment with Protamex, which has both endo- and exo-protease activity, led to the hydrolysates with the lowest levels of bitterness, compared with Alcalase (which preferentially hydrolyses peptide bonds containing aromatic amino acid residues), and Flavouryme (which is produced from *Aspergillus oryzae,* as for Debitrase HYW20™). However, MPC treated with Alcalase presented higher DH than those obtained by Protamex and Flavourzyme.

Table 1 shows the DHs obtained for the three substrates (WPC, MPC and NaCN) after incubation for 1 and 4 h with Debitrase and Prolyve. A two-way ANOVA was conducted to investigate the effect of enzyme and incubation time on the DH for each substrate. As expected from previous studies [40], there was an effect of incubation time on DH, with higher DHs being observed after 4 h incubation for all substrates with both enzymes. When considering the effect of the enzyme, no effect was observed after 1 h incubation with a DH of approximately 7% being reached with both enzymes for WPC and MPC, whereas in the case of NaCN, a higher DH % (9.48 ± 0.88) was observed on incubation with Debitrase. When comparing the three substrates, a higher DH was observed for NaCN and MPC after 4 h incubation (17.78 ± 1.00 and 15.74 ± 1.36, respectively, with Debitrase), possibly due to the fact that caseins are more susceptible to hydrolysis compared to whey proteins [22] and these two substrates predominantly contain casein. The higher susceptibility of casein to hydrolysis could be attributed to its open or disordered structure, while whey proteins have a globular structure in their native state [41].

Prolyve is a *Bacillus licheniformis* proteolytic preparation with a broad substrate specificity, which may explain the relatively high DH (around 7%) obtained with all three substrates, even after 1 h of incubation. This enzyme preparation preferentially cleaves at the carboxyl side of hydrophobic amino acid residues [42]. Debitrase HYW20 contains *Aspergillus oryzae* exopeptidase activity, allowing it to generate hydrolysates rich in free amino acids. Moreover, due to the presence of proteases from *Bacillus* spp., Debitrase has broad specificity [43], resulting in relatively high DH values. When comparing the DH of the hydrolysates generated after 4 h of incubation, Debitrase resulted in higher DH values compared to Prolyve with all substrates. These findings are in agreement with those of Spellman et al. [7], who reported higher DH values for whey protein hydrolysates generated using Debitrase, compared to those generated using Alcalase, which is also a *Bacillus licheniformis* proteinase like Prolyve. In other studies [44], the hydrolysis of NaCN with Debitrase resulted in a relatively low DH compared to a range of other enzymes; however, the DH was calculated from the volume of NaOH consumed during hydrolysis, whereas in the current study, it was determined using the TNBS method. Moreover, different reaction times and pH values were employed in the two studies. There appears to be no information on the hydrolysis of MPC using Debitrase or Prolyve. As previously mentioned, Cui et al. [17] hydrolysed MPC using Alcalase, Protamex and Flavourzyme, and obtained DH around 15.3% with Alcalase after 2 h of hydrolysis. This value is significantly higher than the DH obtained in the current study using Prolyve, an enzyme with similar activity to Alcalase. These differences could be due to the different ways of measuring the DH (TNBS vs. o-phthaldialdehyde (OPA)) or possibly different hydrolysis conditions of pH and temperature.

**Table 1.** Degree of hydrolysis (DH, %) of milk protein concentrate (MPC), whey protein concentrate (WPC) and sodium caseinate (NaCN) hydrolysates as a function of enzyme preparation, Prolyve 1000™ (Pro) or Debitrase HYW20™ (Deb), and incubation time. Values presented are mean ± standard deviation (n = 3).

| Substrate | Enzyme | DH (%) at Different Incubation Times | |
|---|---|---|---|
| | | 1 h | 4 h |
| MPC | Control | $0.92 \pm 0.21^{e}$ | $1.04 \pm 0.37^{e}$ |
| | Pro | $7.05 \pm 0.40^{c}$ | $11.65 \pm 0.98^{b}$ |
| | Deb | $8.17 \pm 0.58^{c}$ | $15.74 \pm 1.36^{a}$ |
| WPC | Control | $2.11 \pm 0.15^{d}$ | $2.26 \pm 0.07^{d}$ |
| | Pro | $7.12 \pm 0.50^{c}$ | $9.54 \pm 1.09^{b}$ |
| | Deb | $6.56 \pm 0.38^{c}$ | $12.04 \pm 0.73^{a}$ |
| NaCN | Control | $0.30 \pm 0.26^{e}$ | $0.66 \pm 0.26^{e}$ |
| | Pro | $7.18 \pm 0.67^{d}$ | $11.67 \pm 0.67^{b}$ |
| | deb | $9.48 \pm 0.88^{c}$ | $17.78 \pm 1.00^{a}$ |

Note: $^{a, b, c, d, e}$ superscripts with a different letter across each substrate represent means which are significantly different at a significance level $p < 0.05$ as determined using two-way ANOVA and post-hoc Tukey test.

Protein hydrolysis was further confirmed using GP-HPLC, as seen in Figure 2, showing the breakdown of high molecular mass (>10 kDa) components representing the intact protein during the hydrolysis process. The molecular mass distribution of the peptides generated during hydrolysis may impact the nutritional properties of the hydrolysates, e.g., their bioavailability and digestibility [45], along with techno-functional properties, such as solubility, emulsification, foaming and gelation. This has been related to lower molecular weight components exhibiting better interfacial diffusivity compared to large biopolymers [46], even though higher DH values may result in the loss of emulsifying properties [47,48].

As seen in Figure 2, enzymatic treatment clearly results in a decrease in molecular mass, with higher percentages of lower molecular mass components in the hydrolysates compared to the corresponding controls. As expected from the results of the DH analyses presented in Table 1, all the substrates show a higher proportion of lower molecular mass components after 4 h incubation. As already mentioned, Debitrase contains exopeptidase activity, and its hydrolysates are expected to contain higher concentrations of short peptides/free amino acids. On the other hand, hydrolysis by *B. licheniformis* proteases, such as Prolyve, is expected to release peptides without free amino acids [49]. Debitrase hydrolysates had a higher proportion of components with lower molecular mass (<1 kDa) compared to Prolyve hydrolysates, e.g., in the case of NaCN (56% versus 49%) and WPC (43.5% versus 40%). However, in the case of the MPC hydrolysates, both enzymes produced a similar amount (~50%) of components, with a molecular mass < 1 kDa. The observation that the percentage of molecules with low molecular mass is lower in WPC hydrolysates is in agreement with the differences observed in DH (see Table 1) and the fact that casein is more susceptible to hydrolysis compared to whey proteins. Moreover, the relatively high percentage of low molecular weight compounds in the NaCN and MPC hydrolysates is in agreement with the findings of McDonagh and FitzGerald [43], who reported high percentages of low molecular weight components (<3 kDa) when Debitrase was used.

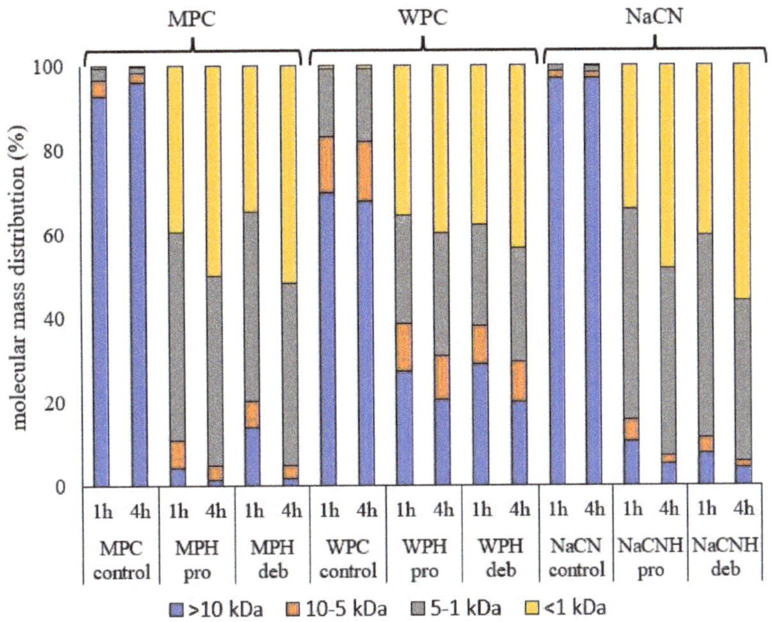


**Figure 2.** Molecular mass distribution profiles of the unhydrolyzed whey protein concentrate (WPC), sodium caseinate (NaCN) and milk protein concentrate (MPC) control samples and their corresponding hydrolysates (respectively, WPH, NaCNH and MPH) generated using Prolyve 1000™ (WPH pro, NaCNH pro, MPH pro) and Debitrase HYW20™ (WPH deb, NaCNH deb, MPH deb) following 1 and 4 h incubation at 50 °C. kDa: kilo Dalton.

### 3.2. Physicochemical Characteristics of the Hydrolysates

#### 3.2.1. Particle Size

Figure 3 represents the volume moment mean diameter D[3,2] of particles in suspension for WPC, MPC and NaCN, before and after hydrolysis, and after heat inactivation showing an influence of both substrate and enzyme type. For MPC, the particle size decreased with incubation time for both the hydrolysates and the intact protein. The latter may be related to the increased solubility of MPC with increasing incubation time. Moreover, at 1 h incubation, the particle size of the MPC hydrolysates was higher compared to the intact protein, possibly due to aggregation between the peptides in solution, while subsequent heating resulted in a decrease in D[3,2]. This was more evident at low DH and short incubation times for both enzymes. Cui et al. [17] reported increased particle size of hydrolysates of MPC, generated using Alcalase, Protamex and Flavourzyme, compared to the intact protein. In the study by Cui et al. [17] changes in particle size during incubation were enzyme-dependent. They had not reported, however, the effect of heat inactivation on the particle size of MPC hydrolysates. In the case of NaCN, the particle size decreased on hydrolysis, regardless of the enzyme used, even after 1 h incubation. For the WPC hydrolysate, particle size decreased with incubation time for both enzymes but with no clear trend, as seen in Figure 3. This may be attributed to the observation that in most cases, the particle size distribution was neither normal nor bimodal, especially for the MPC and WPC hydrolysates, for which Figure A1 shows how the particle size distributions changed during hydrolysis. For WPC, it was evident that the particle size decreased without aggregation during hydrolysis, with no differences due to treatment time and enzyme type. However, the heat inactivation step modified the particle size distributions, as seen in Figure 3. In the case of NaCN, no aggregate formation was observed for the hydrolysates generated using Prolyve, especially after 4 h of incubation, while subsequent heating resulted in the formation of aggregates. However, in the case of NaCN treated with Debitrase HYW20™, the particle size distributions were not affected by heating, as seen in Figure 3. The formation of aggregates during hydrolysis can impact the turbidity and the behaviour of hydrolysates as ingredients in complex food matrices [50].

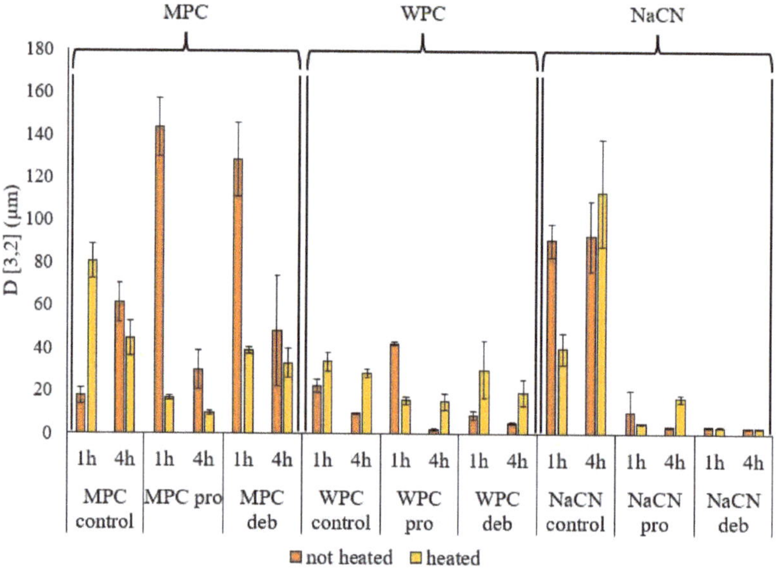


**Figure 3.** Particle size distribution expressed as volume moment mean diameter D[3,2] for the unhydrolyzed (control) protein substrates, milk protein concentrate (MPC), whey protein concentrate (WPC) and sodium caseinate (NaCN) and their corresponding hydrolysates generated using Prolyve 1000™ (pro) and Debitrase HYW20™ (deb) following 1 and 4 h of incubation at 50 °C before (not heated) and after heat inactivation at 80 °C for 10 min (heated).

### 3.2.2. Turbidity

The turbidity of protein solutions depends on protein concentration, the presence of non-dissolved particles, the particle size and particle number per volume unit [51]. Turbidity analysis of the samples, expressed as absorbance at 550 nm (Figure 4), indicated that the presence of aggregates (as confirmed by the particle size as measured using light scattering, seen in Figures 3 and A1) increased the turbidity of NaCN after hydrolysis, as reported by Ewert et al. [52], while for WPC and MPC, the turbidity decreased after hydrolysis. The heat inactivation treatment significantly increased the turbidity of unhydrolyzed whey protein concentrate and the whey protein hydrolysate (WPH) generated using Debitrase after 1 h of hydrolysis. This could be attributed to the fact that Debitrase is an exopeptidase containing preparation, which at low DH, resulted in the formation of low molecular mass compounds, though there was still a significant amount of intact protein present (~30%), as shown in Figure 2. This result was further confirmed on SDS PAGE analysis (Figure A2), where a band of intact β-lactoglobulin was evident. Previous research has shown that the presence of β-lactoglobulin (β-lg) could lead to the formation of heat-induced aggregates, which in turn contribute to increased turbidity [31].

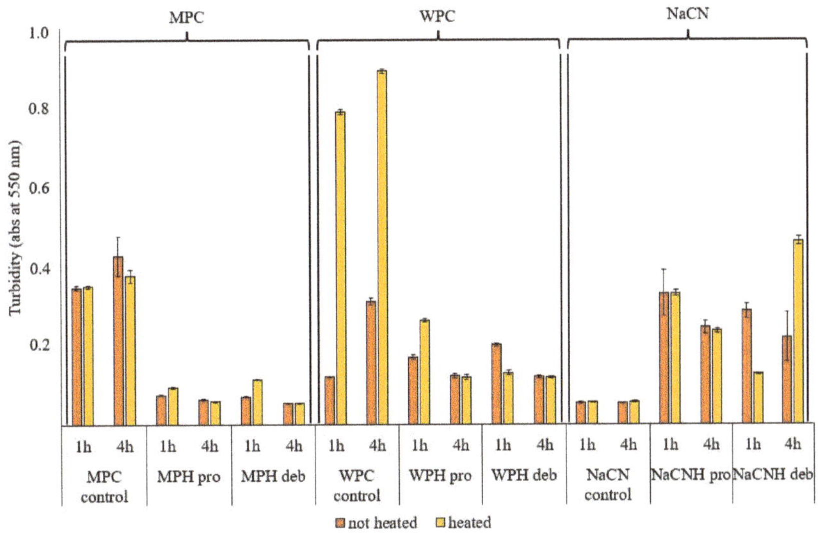


**Figure 4.** Turbidity expressed as absorbance at 550 nm for unhydrolyzed milk protein concentrate (MPC control), whey protein concentrate (WPC control) and sodium caseinate (NaCN control) control samples and their corresponding hydrolysates (H) incubated with Prolyve 1000™ (Pro) (MPH pro, WPH pro, NACNH pro) and Debitrase HYW20™ (Deb) (MPH deb, WPH deb, NaCNH deb) after 1 and 4 h of incubation at 50 °C before (not heated) and after heat inactivation at 80 °C for 10 min (heated).

In the case of the WPH generated with Prolyve, heating had no impact on the turbidity in the 4 h hydrolysates, while at low DH (after 1 h of incubation), there was a decrease in turbidity with heating. This is an indication that there were no aggregates in the hydrolysate after heating, which corroborates further the findings from the particle size distribution analysis (Figure 3). The absence of aggregates, herein, is in agreement with Spellman et al. [38], who showed no aggregation in the WPHs generated using Prolyve in contrast to those generated with Alcalase, which is a *Bacillus licheniformis* enzyme preparation that has been reported to lead to WPC hydrolysate aggregation. Heating only affected the turbidity of unhydrolyzed WPC, as heating of whey at 80 °C can lead to aggregation, due to the presence of β-lg, which is subject to thermal denaturation and, consequently, the formation of aggregates [53]. SDS PAGE analysis (Figure A2) showed that β-lg was hydrolysed extensively in the case of the WPH generated using Prolyve, and this could be the reason why heat inactivation did not increase the turbidity of these hydrolysates.

In the case of MPC, turbidity was lower for the hydrolysates compared to intact MPC. After 4 h of hydrolysis, where highest DH was observed, the turbidity was lower, regardless of the enzyme used, possibly due to the low percentages of high molecular weight components, an indication that the intact protein had been hydrolysed (Figure 2). At low DH (after 1 h of incubation), the turbidity was affected by the inactivation treatment for both enzymes. Interestingly, there was a small increase in turbidity for the Prolyve MPC hydrolysates and a larger increase for the Debitrase MPC hydrolysates, which may be explained by the higher percentage of high MW components in Debitrase vs. Prolyve hydrolysates after 1 h incubation, as seen in Figure 2 (13.8% vs. 4.2%, respectively).

In the case of NaCN hydrolysates, turbidity increased for the hydrolysates compared to the unhydrolyzed control samples. This was also evident from the particle size distribution profiles (Figure A1), showing that the particle sizes increased for the hydrolysates generated using Prolyve, which could be due to the formation of aggregates. However, the significant increase in turbidity of the heated 4 h hydrolysates generated using Debitrase cannot be explained, taking into consideration the particle size distribution and the particle size of these samples, as seen in Figures 3 and A1, respectively. This increase in turbidity could, however, be due to changes in solubility.

Changes in turbidity and particle size are indicative of the presence of aggregates or a decrease in solubility, which could be the case for the sodium caseinate hydrolysates (NaCNH) herein. Previous research has shown a decreased nitrogen solubility index at pH 7.0 for NaCN hydolysates generated using Protamex, a *Bacillus* proteinase, at low DH [14]. Low solubility at pH 6.0 and 7.0 (in the current study, all analyses were conducted at pH 7) was observed for NaCNHs generated with different enzymes [26,54]. As described by Flanagan and FitzGerald [14], this low solubility of the hydrolysates, compared to the intact protein, could be attributed to the formation of peptides with different pIs. High turbidity, due to low solubility, can affect the further processing of hydrolysates, which need, e.g., to be pumped and spray dried. Moreover, this can affect the food products that contain them, such as in the case of juices, or beverages in general, where turbidity can impact consumer acceptability [55]. Moreover, in the case of yoghurts, the presence of large protein aggregates could result in the formation of products with low storage moduli, yield stress, firmness and thickness [56], or could lead to the gelation of acid milk gels, as described by Gélebart et al. [57].

#### 3.2.3. Apparent Viscosity

Figure 5 represents the findings on the $\eta_{app}$ for the different samples. A decrease in viscosity after enzymatic hydrolysis has been widely demonstrated for most substrates in previous research [58,59]. The hypothesis that hydrolysis leads to a decrease in $\eta_{app}$ has been confirmed for WPC and NaCN (Figure 5). A higher decrease in $\eta_{app}$, compared to in MPC and WPC hydrolysates, was observed for the NaCN hydrolysates, for both enzymes used, while there were no significant differences in $\eta_{app}$ of the MPC hydrolysate compared to the original MPC. In the case of NaCN, previous research reported a decrease in the apparent viscosity of hydrolysates having DH values ~10%, compared to intact NaCN [14]. The viscosity of NaCN and its hydrolysates generated using a *Bacillus* proteinase differed significantly at pH values close to the isoelectric point (pH = 4.0), while in the current study, all samples were tested at pH 7.0. In the case of whey proteins, previous studies observed gel formation of whey after limited hydrolysis with a *Bacillus licheniformis* protease [60], on extensive hydrolysis [61] or on heat treatment of whey protein hydrolysates [31]. However, this was not the case in the current study, where Prolyve, a *Bacillus licheniformis* enzyme, which does not induce gelation, was used [35].

Heating also had no effect on the viscosity of the WPC and NaCN hydrolysates, regardless of enzyme and incubation time. This is an important result, considering that an enzyme inactivation step, typically by heating, is required before the hydrolysates can be further processed. While careful control of the viscosity is important for processing, it is also of relevance for the application of hydrolysate ingredients in a range of products. Maintaining low viscosity can enhance heat and mass transfer during subsequent hydrolysate processing steps, such as pumping, concentration and drying. In terms of final product characteristics, careful control of the viscosity may lead to the development of desirable mouthfeel and textural properties.

In relation to MPC at low DH, the $\eta_{app}$ increased after heat inactivation, while there were no significant differences in $\eta_{app}$ after heating of the 4 h-hydrolysates regardless of the enzyme used. This could partly be explained by the molecular mass distribution profile shown in Figure 2 where in the case of 1 h incubation there was a significant amount of intact protein remaining compared to that in the 4 h hydrolysates. These results are in line with the MPH turbidity findings, as seen in Figure 4.

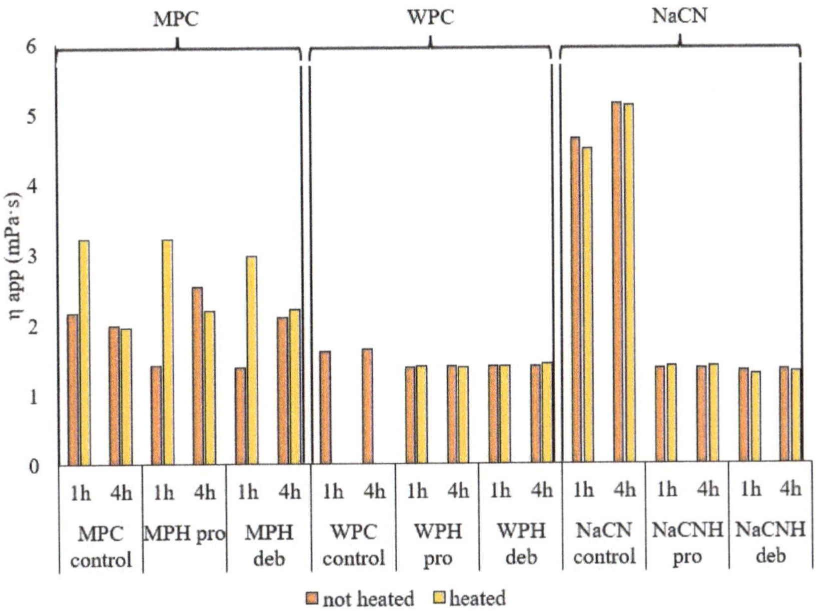


**Figure 5.** Apparent viscosity (η app) in mPa·s, measured at a shear rate of 122 $s^{-1}$, of unhydrolyzed milk protein concentrate (MPC control), whey protein concentrate (WPC control) and sodium caseinate (NaCN control) control samples and their corresponding hydrolysates (H) before (no heated) and after heating at 80 °C for 10 min (heated). Hydrolysates were generated using Prolyve 1000™ (pro) and Debitrase HYW20™ (deb) after 1 and 4 h of incubation at 50 °C. Note: WPC unhydrolyzed control gelled on heating; therefore, the $\eta_{app}$ could not be measured.

Since no previous studies have been conducted on the hydrolysis of MPC with Debitrase or Prolyve, the findings of the current study may only be discussed in relation to a study testing the hydrolysis of MPI using different enzymes [24]. Ryan et al. [24], who studied the changes in $\eta_{app}$ for hydrolysates of milk protein isolate, at three different temperatures (25, 45 and 90 °C) and three different pH values (6.2, 6.8 and 7.2). These authors reported an effect of temperature and pH on $\eta_{app}$. Moreover, the changes observed depended on the DH of the hydrolysates, which ranged from 15 to 37%, while the highest DH achieved in the current study with MPC was 15% (Table 1). It is difficult to compare the apparent viscosities of the current study with the findings of Ryan et al. [24], because of the different enzymes used and the different pHs at which the viscosity was measured. However, at lower DH, in the present study, we observed a decrease in $\eta_{app}$ of MPH compared to MPC. Heating at 80 °C for 10 min resulted in an increase in $\eta_{app}$, possibly due to a decreased heat stability (as reported by Ryan et al. [24], between pH 6.2 and 7.4), possibly due to changes in casein micelle structure. These changes may be linked to the observed changes in turbidity. Further hydrolysis (4 h) increased the viscosity of the unheated MPC hydrolysates. However, heating did not alter the viscosity of these samples. The $\eta_{app}$ in this case for the heat-treated samples was measured after the samples were heated at 80 °C and were then cooled to 50 °C, to conduct all measurements at the same temperature. This cooling step may impact the $\eta_{app}$ values, as shown by Ryan et al. [24], who reported a viscosity increase upon the cooling of milk protein isolate (MPI) control samples and low DH MPI hydrolysates, generated using a variety of enzymes (i.e., by ~15% with Neutrase and Flavourzyme and by ~17% with Protamex).

## 4. Conclusions

The current study explored the impact of hydrolysis conditions, such as enzyme preparation, incubation time and heat inactivation, on the physicochemical properties of hydrolysates, generated from different milk protein substrates (WPC, MPC and NaCN). The two enzymes used, Debitrase and Prolyve, were chosen on the basis that they may lead to the generation of hydrolysates with reduced bitterness. Following 1 h incubation, the degree of hydrolysis was higher for NaCN compared to the other substrates and, while there was no enzyme effect on the DHs of WPC and MPC hydrolysates, hydrolysis with Debitrase resulted in a higher DH for NaCN. The effect of hydrolysis on the physicochemical properties depended on the substrate. For example, $\eta_{app}$ decreased in WPC and NaCN hydrolysates, particle size decreased for all the substrates, with aggregate formation for MPC, and turbidity decreased in WPC and MPC hydrolysates but increased in NaCN hydrolysates. Viscosity, turbidity and particle size changes were substrate- and incubation time-dependent and, to a lesser extent, also enzyme-dependent. Heat inactivation, an essential step during the processing of hydrolysates, impacted the hydrolysate physicochemical properties in different ways, depending on both the DH and the starting substrate. At longer incubation times, which also resulted in higher DHs, heat inactivation of the hydrolysates did not significantly impact the turbidity and viscosity of the hydrolysates, which is of significant industrial importance, when considering the further processing of hydrolysates. The findings reported herein, may help in the design of enzymatic processing approaches for the generation of hydrolysate ingredients from different milk protein starting substrates.

**Author Contributions:** Conceptualization, R.J.F., M.D. and G.S.; methodology, R.J.F., M.D., A.G. and G.S.; formal analysis, A.G. and M.D.; resources, R.J.F. and G.S.; data curation, A.G., M.D. and G.S.; writing—original draft preparation, A.G. and M.D.; writing—review and editing, R.F and G.S.; supervision, R.J.F. and G.S.; funding acquisition, R.J.F. and G.S. All authors have read and agreed to the published version of the manuscript.

**Funding:** This research received no external funding.

**Institutional Review Board Statement:** Not applicable.

**Informed Consent Statement:** Not applicable.

**Data Availability Statement:** The data that support the findings of this study are available from the corresponding author upon reasonable request.

**Acknowledgments:** This research was supported by Doctoral School on the Agro-Food System of the Università Cattolica del Sacro Cuore (PhD grantship from Fondazione di Piacenza e Vigevano to Alice Gruppi, XXXII cycle).

**Conflicts of Interest:** The authors declare no conflict of interest.

## Appendix A

SDS polyacrylamide gel electrophoresis (SDS PAGE): For determination of the electrophoretic profile, the samples (hydrolysates and unhydrolyzed control after heat inactivation) were resuspended in a solution of 1% (*w*/*v*) SDS, in order to dissociate the proteins. The sample buffer contained β-mercaptoethanol to carry out the electrophoresis under reducing conditions. The running buffer used was Tris-glycine-SDS (25 mM Tris, 192 mM glycine and 0.1% SDS, pH 8.6). Gels were run at a constant voltage of 150 mV for 60 min. Mini-PROTEAN TGX precast gels (4–20% resolving gel, Bio-Rad Laboratories Inc., Hercules, CA, USA) were used on a Mini Protean II system (Bio-Rad), according to the manufacturer's instructions. The quantity of protein added to each well was 15 µg. The molecular masses of the proteins were estimated by reference to the relative migration of the molecular mass standards, which had a wide range of molecular weights (6500–200,000 Da, Sigma-Aldrich, St. Louis, MO, USA). The gels were stained with Coomassie Blue R-250.

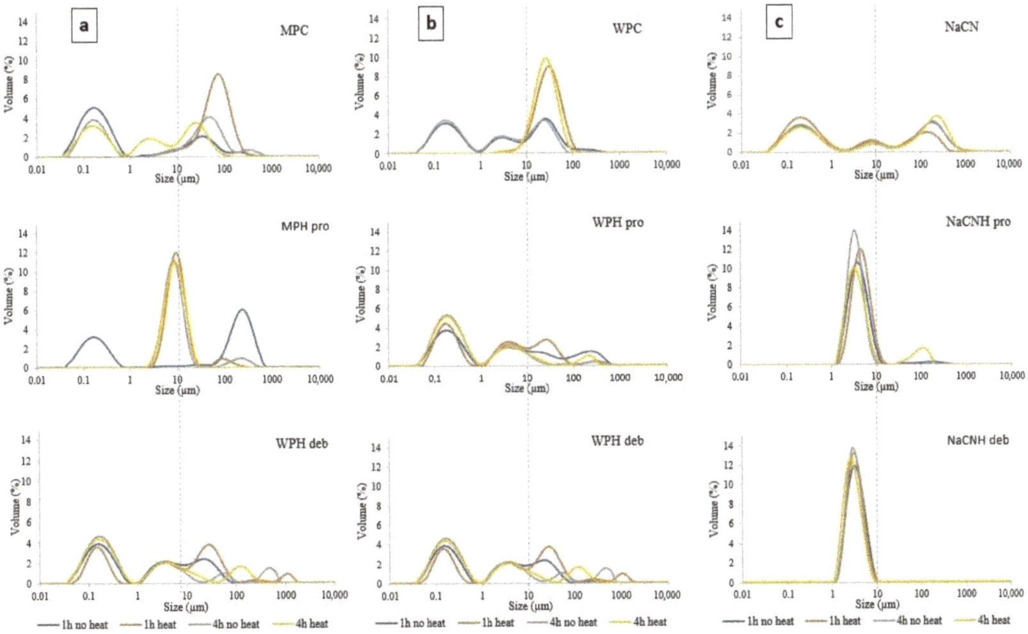


**Figure A1.** Particle size distribution profiles of intact protein substrates and their corresponding hydrolysates (H). ((**a**): milk protein concentrate (MPC); (**b**): whey protein concentrate (WPC); (**c**): sodium caseinate (NaCN)) generated using Prolyve 1000™ (pro) (MPH pro, WPH pro, NaCNH pro) and Debitrase HYW20™ (deb) (MPH Deb, WPH deb, NaCNH deb) after 1 and 4 h of incubation at 50 °C before (no heat) and after heat inactivation at 80 °C for 10 min (heat).

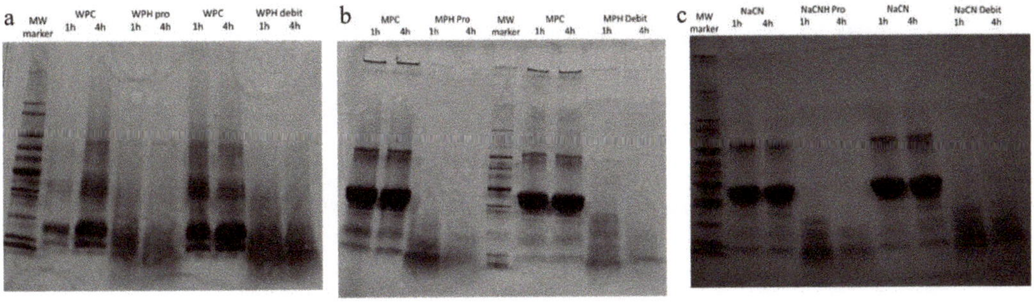


**Figure A2.** Sodium Dodecyl Sulphate Polyacrylamide Gel Electrophoresis (SDS page) of the intact protein and the corresponding hydrolysates. ((**a**): WPC; (**b**): MPC; (**c**): NaCN) generated using Prolyve (Pro) and Debitrase (Debit) after 1 h and 4 h of incubation.

## References

1. Carr, A.; Golding, M. Functional milk proteins production and utilization: Casein-based ingredients. In *Advanced Dairy Chemistry*, 4th ed.; McSweeney, P.L.H., O'Mahony, J.A., Eds.; Springer: Cork, Ireland, 2016; Volume 1B, pp. 35–66.
2. Nongonierma, A.B.; FitzGerald, R.J. The scientific evidence for the role of milk protein-derived bioactive peptides in humans: A Review. *J. Funct. Foods* **2015**, *17*, 640–656. [CrossRef]
3. Cheison, S.C.; Kulozik, U. Impact of the environmental conditions and substrate pre-treatment on whey protein hydrolysis: A review. *Crit. Rev. Food Sci. Nutr.* **2017**, *57*, 418–453. [CrossRef]
4. Abd El-Salam, M.H.; El-Shibiny, S. Preparation, properties, and uses of enzymatic milk protein hydrolysates. *Crit. Rev. Food Sci. Nutr.* **2017**, *57*, 1119–1132. [CrossRef]

5. Izadi, A.; Khedmat, L.; Mojtahedi, S.Y. Nutritional and therapeutic perspectives of camel milk and its protein hydrolysates: A review on versatile biofunctional properties. *J. Funct. Foods* **2019**, *60*, 103441. [CrossRef]
6. Mulvihill, D.M. Production, functional properties and utilization of milk protein products. In *Advanced Dairy Chemistry*; Fox, P.F., Ed.; Elsevier Science Publishers: London, UK, 1992; Volume 1, pp. 369–404. [CrossRef]
7. Spellman, D.; McEvoy, E.; O'Cuinn, G.; FitzGerald, R.J. Proteinase and exopeptidase hydrolysis of whey protein: Comparison of the TNBS, OPA and pH stat methods for quantification of degree of hydrolysis. *Int. Dairy J.* **2003**, *13*, 447–453. [CrossRef]
8. Pelegrine, D.H.G.; Gasparetto, C.A. Whey proteins solubility as function of temperature and pH. *LWT-Food Sci. Technol.* **2005**, *38*, 77–80. [CrossRef]
9. Agarwal, S.; Beausire, R.L.; Patel, S.; Patel, H. Innovative uses of milk protein concentrates in product development. *J. Food Sci.* **2015**, *80*, A23–A29. [CrossRef]
10. De Castro-Morel, M.; Harper, W.J. Basic functionality of commercial milk protein concentrates. *Milchwissenschaft* **2002**, *57*, 367–370.
11. Meena, G.S.; Singh, A.K.; Panjagari, N.R.; Arora, S. Milk protein concentrates: Opportunities and challenges. *J. Food Sci. Technol.* **2017**, *54*, 3010–3024. [CrossRef]
12. Khalesi, M.; FitzGerald, R.J. Insolubility in milk protein concentrates: Potential causes and strategies to minimize its occurrence. *Crit. Rev. Food Sci. Nutr.* **2021**, *15*, 1–17. [CrossRef]
13. Panyam, D.; Kilara, A. Enhancing the functionality of food proteins by enzymatic modification. *Trends Food Sci. Technol.* **1996**, *71*, 120–125. [CrossRef]
14. Flanagan, J.; FitzGerald, R.J. Functionality of *Bacillus* proteinase hydrolysates of sodium caseinate. *Int. Dairy J.* **2002**, *12*, 737–748. [CrossRef]
15. Tuncturk, Y.; Zorba, O. The effects of enzymatic hydrolysis of casein on apparent yield stress and some emulsion properties. *Food Hydrocoll.* **2006**, *20*, 475–482. [CrossRef]
16. De Castro, R.J.S.; Bagagli, M.P.; Sato, H.H. Improving the functional properties of milk proteins: Focus on the specificities of proteolytic enzymes. *Curr. Opin. Food Sci.* **2015**, *1*, 64–69. [CrossRef]
17. Cui, Q.; Sun, Y.; Zhou, Z.; Cheng, J.; Guo, M. Effects of enzymatic hydrolysis on physicochemical properties and solubility and bitterness of milk protein hydrolysates. *Foods* **2021**, *10*, 2462. [CrossRef]
18. Ryan, G.; Nongonierma, A.B.; O'Regan, J.; FitzGerald, R.J. Functional properties of bovine milk protein isolate and associated enzymatic hydrolysates. *Int. Dairy J.* **2018**, *81*, 113–121. [CrossRef]
19. Turgeon, S.L.; Gauthier, S.F.; Paquin, P. Interfacial and Emulsifying Properties of Whey Peptide Fractions Obtained with a Two-step Ultrafiltration Process. *J. Agric. Food Chem.* **1991**, *39*, 70–73. [CrossRef]
20. Mutilangi, W.A.M.; Panyam, D.; Kilara, A. Functional Properties of Hydrolysates from Proteolysis of Heat-denatured Whey Protein Isolate. *J. Food Sci.* **1996**, *61*, 270–274. [CrossRef]
21. Severin, S.; Xia, W.S. Enzymatic hydrolysis of whey proteins by two different proteases and their effect on the functional properties of resulting protein hydrolysates. *J. Food Biochem.* **2006**, *30*, 77–97. [CrossRef]
22. Cermeño, M.; Felix, M.; Connolly, A.; Brennan, E.; Coffey, B.; Ryan, E.; FitzGerald, R.J. Role of carbohydrate conjugation on the emulsification and antioxidant properties of intact and hydrolysed whey protein concentrate. *Food Hydrocoll.* **2019**, *88*, 170–179. [CrossRef]
23. Walsh, D.J.; Russell, K.; FitzGerald, R.J. Stabilisation of sodium caseinate hydrolysate foams. *Food Res. Int.* **2008**, *41*, 43–52. [CrossRef]
24. Banach, J.C.; Lin, Z.; Lamsal, B.P. Enzymatic modification of milk protein concentrate and characterization of resulting functional properties. *LWT-Food Sci. Technol.* **2013**, *54*, 397–403. [CrossRef]
25. McIntyre, I.; Carolan, A.; O'Sullivan, M.; Jacquier, J.; Hutchings, S.; Murray, B.; O'Riordan, D. Incorporation of bioactive dairy hydrolysate influences the stability and digestion behaviour of milk protein stabilised emulsions. *Food Funct.* **2018**, *9*, 5813–5823. [CrossRef]
26. Chobert, J.; Bertrand-Harb, C.J.; Nicolas, M. Solubility and Emulsifying Properties of Caseins and Whey Proteins Modified Enzymatically by Trypsin. *J. Agric. Food Chem.* **1988**, *36*, 883–892. [CrossRef]
27. Zhao, C.; Chen, N.; Ashaolu, T.J. Whey proteins and peptides in health-promoting functions—A review. *Int. Dairy J.* **2021**, *126*, 105269. [CrossRef]
28. Le Maux, S.; Nongonierma, A.B.; Barre, C.; FitzGerald, R.J. Enzymatic generation of whey protein hydrolysates under pH-controlled and non pH-controlled conditions: Impact on physicochemical and bioactive properties. *Food Chem.* **2016**, *199*, 246–251. [CrossRef] [PubMed]
29. Dermiki, M.; FitzGerald, R.J. Physicochemical and gelling properties of whey protein hydrolysates generated at 5 and 50 °C using Alcalase® and Neutrase®, effect of total solids and incubation time. *Int. Dairy J.* **2020**, *110*, 104792. [CrossRef]
30. Le Maux, S.; Nongonierma, A.B.; FitzGerald, R.J. Peptide composition and dipeptidyl peptidase IV inhibitory properties of β-lactoglobulin hydrolysates having similar extents of hydrolysis while generated using different enzyme-to-substrate ratios. *Food Res. Int.* **2017**, *99*, 84–90. [CrossRef]
31. Spellman, D.; Kenny, P.; O'Cuinn, G.; FitzGerald, R.J. Aggregation properties of whey protein hydrolysates generated with *Bacillus licheniformis* proteinase activities. *J. Agric. Food Chem.* **2005**, *53*, 1258–1265. [CrossRef]
32. Adler-Nissen, J. Control of the Proteolytic Reaction and of the Level of Bitterness in Protein Hydrolysis Processes. *J. Chem. Technol. Biotechnol.* **1984**, *34*, 215–222. [CrossRef]

33. Rajarathnam, E.; Nongonierma, A.B.; O'Sullivan, D.; Flynn, C.; FitzGerald, R.J. Impact of enzyme preparation and degree of hydrolysis on peptide profile and nitrogen solubility of sodium caseinate hydrolysates. *Int. J. Food Sci. Technol.* **2016**, *51*, 2123–2131. [CrossRef]
34. Le Maux, S.; Nongonierma, A.B.; Lardeux, C.; FitzGerald, R.J. Impact of enzyme inactivation conditions during the generation of whey protein hydrolysates on their physicochemical and bioactive properties. *Int. J. Food Sci. Technol.* **2018**, *53*, 219–227. [CrossRef]
35. Nongonierma, A.B.; FitzGerald, R.J. Tryptophan-containing milk protein-derived dipeptides inhibit xanthine oxidase. *Peptides* **2012**, *37*, 263–272. [CrossRef] [PubMed]
36. O'Loughlin, I.B.; Murray, B.A.; Kelly, P.M.; FitzGerald, R.J.; Brodkorb, A. Enzymatic Hydrolysis of Heat-Induced Aggregates of Whey Protein Isolate. *J. Agric. Food Chem.* **2012**, *60*, 4895–4904. [CrossRef] [PubMed]
37. Addinsoft. XLSTAT Statistical and Data Analysis Solution 2020. New York, NY, USA. Available online: https://www.xlstat.com (accessed on 16 May 2020).
38. Spellman, D.; O'Cuinn, G.; FitzGerald, R.J. Physicochemical and sensory characteristics of whey protein hydrolysates generated at different total solids levels. *J. Dairy Res.* **2005**, *72*, 138–143. [CrossRef] [PubMed]
39. FitzGerald, R.J.; O'Cuinn, G. Enzymatic debittering of food protein hydrolysates. *Biotechnol. Adv.* **2006**, *24*, 234–237. [CrossRef]
40. Whitehurst, R.J.; Van Oort, M. *Enzymes in Food Technology*, 2nd ed.; Wiley Blackwell: Singapore, 2009.
41. Power, O.; Jakeman, P.; FitzGerald, R.J. Antioxidative peptides: Enzymatic production, in vitro and in vivo antioxidant activity and potential applications of milk-derived antioxidative peptides. *Amino Acids* **2013**, *44*, 797–820. [CrossRef]
42. Svendsen, I.; Breddam, K. Isolation and amino acid sequence of a glutamic acid specific endopeptidase from *Bacillus licheniformis*. *Eur. J. Biochem.* **1992**, *204*, 165–171. [CrossRef]
43. Nongonierma, A.B.; FitzGerald, R.J. Enzymes exogenous to milk in dairy technology. Proteinases. In *Encyclopedia of Dairy Science*, 2nd ed.; Fuquay, J.W., Fox, P.F., Mc Sweeney, P.J.H., Eds.; Academic Press: San Diego, CA, USA, 2011; pp. 289–296.
44. McDonagh, D.; FitzGerald, R.J. Production of caseinophosphopeptides (CPPs) from sodium caseinate using a range of commercial protease preparations. *Int. Dairy J.* **1998**, *8*, 39–45. [CrossRef]
45. Bhandari, D.; Rafiq, S.; Gat, Y.; Gat, P.; Waghmare, R.; Kumar, V. A review on bioactive peptides: Physiological functions, bioavailability and safety. *Int. J. Pept. Res. Ther.* **2020**, *26*, 139–150. [CrossRef]
46. Adjonu, R.; Doran, G.; Torley, P.; Agboola, S. Whey protein peptides as components of nanoemulsions: A review of emulsifying and biological functionalities. *J. Food Eng.* **2014**, *122*, 15–27. [CrossRef]
47. Ding, Y.; Chen, L.; Shi, Y.; Akhtar, M.; Chen, J.; Ettelaie, R. Emulsifying and emulsion stabilizing properties of soy protein hydrolysates, covalently bonded to polysaccharides: The impact of enzyme choice and the degree of hydrolysis. *Food Hydrocoll.* **2021**, *113*, 106519. [CrossRef]
48. Karami, Z.; Akbari-Adergani, B. Bioactive food derived peptides: A review on correlation between structure of bioactive peptides and their functional properties. *J. Food Sci. Technol.* **2019**, *56*, 535–547. [CrossRef] [PubMed]
49. Morais, H.A.; Silvestre, M.P.C.; Viviane, D.M.; Mauro, R.S.; Silva, A.S.; Silvera, J.N. Correlation between the degree of hydrolysis and the peptide profile of whey protein of hydrolysis and peptide profile of whey protein concentrate hydrolysate: Effect of the enzyme type and reaction time. *Am. J. Food Technol.* **2013**, *8*, 1–16. [CrossRef]
50. Gani, A.; Broadway, A.A.; Ahmad, M.; Ashwar, B.A.; Wani, A.A. Enzymatic hydrolysis of whey and casein protein- effect on functional, rheological, textural and sensory properties of breads. *J. Food Sci. Technol.* **2015**, *52*, 7697–7709. [CrossRef] [PubMed]
51. Mahler, H.; Friess, W.; Grauschopf, U.; Kiese, S. Protein Aggregation: Pathways, Induction Factors and Analysis. *J. Pharm. Sci.* **2008**, *98*, 2909–2934. [CrossRef]
52. Ewert, J.; Luz, A.; Volk, V.; Stressler, T.; Fischer, L. Enzymatic production of emulsifying whey protein hydrolysates without the need of heat inactivation. *J. Sci. Food Agric.* **2019**, *99*, 3443–3450. [CrossRef]
53. Iametti, S.; De Gregori, B.; Vecchio, G.; Bonomi, F. Modifications occur at different structural levels during the heat denaturation of β-lactoglobulin. *Eur. J. Biochem.* **1996**, *237*, 106–112. [CrossRef]
54. Slattery, H.; Fitzgerald, R.J. Functional Properties and Bitterness of Sodium Caseinate Hydrolysates Prepared with a *Bacillus* Proteinase. *J. Food Sci.* **1998**, *63*, 418–422. [CrossRef]
55. Nam, S.-H.; Wagh, A.; Martini, S.; Walsh, M.K. Sensory characterisation of a high-protein beverage. *Int. J. Dairy Technol.* **2017**, *70*, 432–438. [CrossRef]
56. Lesme, H.; Rannou, C.; Loisel, C.; Famelart, M.H.; Bouhallab, S.; Prost, C. Controlled whey protein aggregates to modulate the texture of fat-free set-type yoghurts. *Int. Dairy J.* **2019**, *92*, 28–36. [CrossRef]
57. Gélébart, P.; Riaublanc, A.; Famelart, M.H.; Jonchère, C.; Beaumal, V.; Anton, M.; Garnier, C. Protein aggregates modulate the texture of emulsified and acidified acid milk gels. *Food Hydrocoll.* **2019**, *93*, 176–188. [CrossRef]
58. Jung, S.; Murphy, P.A.; Johnson, L.A. Physicochemical and Functional Properties of Soy Protein Substrates Modified by Low Levels of Protease Hydrolysis. *J. Food Sci.* **2006**, *70*, C180–C185. [CrossRef]
59. Diniz, F.M.; Martin, A.M. Effects of the Extent of Enzymatic Hydrolysis on Functional Properties of Shark Protein Hydrolysate. *LWT-Food Sci. Technol.* **1997**, *30*, 266–272. [CrossRef]

60. Ju, Z.Y.; Otte, J.; Madsen, J.S.; Qvist, K.B. Effects of limited proteolysis on gelation and gel properties of whey protein isolate. *J. Dairy Sci.* **1995**, *78*, 2119–2128. [CrossRef]
61. Otte, J.; Ju, Z.Y.; Skriver, A.; Qvist, K.B. Effects of limited proteolysis on the microstructure of heat-induced whey protein gels at varying pH. *J. Dairy Sci.* **1996**, *79*, 782–790. [CrossRef]

foods

MDPI

Article

# Antioxidant Potential of Cookies Formulated with Date Seed Powder

Zein Najjar [1], Jaleel Kizhakkayil [2], Hira Shakoor [2], Carine Platat [2], Constantinos Stathopoulos [3,*] and Meththa Ranasinghe [1]

[1] Department of Food Science, College of Food and Agriculture, United Arab Emirates University, Al-Ain P.O. Box 15551, United Arab Emirates; zeinrnajjar@gmail.com (Z.N.); 201990013@uaeu.ac.ae (M.R.)
[2] Department of Nutrition and Health, College of Medicine and Health Sciences, United Arab Emirates University, Al-Ain P.O. Box 15551, United Arab Emirates; JaleelK@uaeu.ac.ae (J.K.); 201890012@uaeu.ac.ae (H.S.); PlatatCarine@uaeu.ac.ae (C.P.)
[3] Faculty of Agrobiology, Food and Natural Resources, Czech University of Life Sciences Prague, 165 00 Prague, Czech Republic
* Correspondence: stathopoulos@af.czu.cz

**Abstract:** Utilising major waste products from the food industry can have both a great environmental impact and be a means to improve consumer health. Date seed is a food industry byproduct that has been proven to have high nutritional value. The aim of this work was to measure the total polyphenolic content (TPC), flavonoids, and antioxidant activity of the seeds of six date fruit varieties, *Fard*, *Khalas*, *Khinaizi*, *Sukkary*, *Shaham*, and *Zahidi*, and to use those seeds to enhance the antioxidant value of cookies by partially substituting flour with ground date seed. Date seed powder (DSP) was extracted at three levels of sample to solvent ratio (5:1, 10:1 and 15:1 mg/mL). Cookies were prepared using three substitution levels of wheat flour (2.5, 5.0, and 7.5%, *w*/*w*) by DSP and two types of flour (white and whole wheat), and were baked at two different temperatures, 180 and 200 °C. The composite cookies were found to contain a significant amount of TPC and flavonoids, and showed increased antioxidant activity compared with the control samples.

**Keywords:** antioxidant; cookies; date seed; flavonoids; polyphenols

**Citation:** Najjar, Z.; Kizhakkayil, J.; Shakoor, H.; Platat, C.; Stathopoulos, C.; Ranasinghe, M. Antioxidant Potential of Cookies Formulated with Date Seed Powder. *Foods* **2022**, *11*, 448. https://doi.org/10.3390/foods11030448

Academic Editors: Michela Verni and Federico Casanova

Received: 21 December 2021
Accepted: 28 January 2022
Published: 3 February 2022



## 1. Introduction

The date palm tree is widely grown in regions of southwest Asia and North Africa [1–3], with more than 600 varieties currently cultivated [3,4]. The edible part of date fruit contains high amounts of sugars and dietary fibres as well as small amounts of protein, fat, ash, and polyphenols [5]. The inedible part, the date seed, represents a major waste in the date processing industry accounting for more than 6% of the fruit [6]; it has significant amounts of fibres, minerals, vitamins, lipids and proteins [6,7], and is rich in antioxidants [1,6,7]. Date seed has been proven to have the potential to ameliorate liver damage and to protect against hepatotoxicity in rats [8].

Baked foods are popular among consumers because of their taste and widespread availability in the form of biscuits, cookies, muffins, cakes and more. Cookies and biscuits are among the most consumed bakery products, as they are ready to eat, cheap, and available in a wide range of flavours [9–13]; on the other hand, as with most bakery items, they are high in sugar and have low amounts of antioxidants, fibre, and minerals [14]. Consumers today tend to choose and eat healthier food [15,16], and there is an increasing trend to make cookies and biscuits utilising functional ingredients [12,13,17]. The use of refined flour results in products lacking the nutritive value of grain in terms of dietary fibres and phytochemicals [18,19]. At the same time, achieving the sensory parameters of cookies (taste, texture and colour) that meet consumers expectations [20], especially while avoiding the use of synthetic additives, can be challenging [21]. Antimicrobials, antioxidants, and anti-browning agents are currently used in the food industry as preservatives [22]. It has been reported that the use of synthetic molecules may be linked to carcinogenesis, and this has led to some restraints on their use [23–25]; synthetic preservatives can be replaced by natural extracts from plant origin, which provide bioactive properties and thus increase the nutritive value of the final product [26–28]. Natural extracts from spices, fruit powder, and aromatic plants for antioxidant purposes have been incorporated in bakery, dairy, and meat products [20,25,29–31]. As the food industry generates high amounts of by-products rich in valuable constituents which can be utilised as functional ingredients, such as antioxidants, incorporating these substances in food would be beneficial environmentally and economically while providing healthier options to consumers [32]. Consumer demand towards functional foods enriched with by-products [33] is increasing due to enhanced awareness of the health and environmental implications of current food choices. Hence, development of cookies fortified with date seeds, the major by-product of the date industry, can generate significant impact environmentally through the reduction of waste financially through the development of a new product and the reduction of waste disposal costs, and societally through the enhanced nutrition made available to consumers.

The objective of this study, therefore, is to quantify the total phenolic and flavonoid content and the antioxidant activity of six varieties of date seed powder (DSP), and to evaluate these characteristics in composite cookies formulated with partial substitution of wheat flour by DSP.

## 2. Materials and Methods

### 2.1. Chemicals

Follin–Ciocalteu's reagent, gallic acid, 2,2-diphenyl-1-picrylhydrazyl (DPPH), 6-hydroxy-2,5,7,8-tetramethylchroman-2-carboxylic acid (Trolox), 2,2′-azinobis-3-ethylbenzothiazoline-6-sulfonic acid (ABTS), glacial acetic acid, hydrochloric acid, ferric chloride, 2,4,6-tripirydyl-S-triazine, aluminium chloride, sodium hydroxide, quercetin (Sigma-Aldrich, Steinheim, Germany), sodium carbonate, sodium phosphate, methanol, ethanol, potassium persulfate, sodium acetate trihydrate, and sodium nitrite (Merck, Darmstadt, Germany) were used; all chemicals were of analytical grade.

### 2.2. Date Seed Powder Production

The seeds of six varieties of Emirati dates, *Fard*, *Khalas*, *Khinaizi*, *Sukkari*, *Shaham*, and *Zahidi* (2 kg of each), were obtained from local farms in Al Ain, United Arab Emirates. Date seeds were soaked in water for six hours, air-dried for three days, ground using a Teeba Date Seed Grinding Machine (Teeba Engineering Industries LLC, Dubai, United Arab Emirates) and sieved to obtain date seed powder with particles of less than 250 μm diameter.

### 2.3. Production of Formulated Cookies with 2.5%, 5%, and 7.5% Date Seed Powder

Cookie dough was made according to the AACC method [34] with some modifications. Ingredients were acquired from the local market (80 g white/whole wheat flour, 30 g palm oil (raised to 40 g when using whole wheat flour), 35 g sucrose, 0.8 g $NaHCO_3$, 1.0 g salt, 0.4 g $NH_4HCO_3$, and 17.6 g water). Cookies were produced by replacing 2.5%, 5%, and 7.5% (*w*/*w*) of flour with date seed powder. Two types of flour (white, WF, and whole wheat, WW) were used to make cookie dough, which was kneaded to a uniform 3 mm thickness with the aid of a pasta machine and cut with a stainless-steel circular cutter of 4.5 cm diameter. The cookies were baked at two different temperatures (A: 180 °C for 10 min; B: 200 °C for 8 min), making twelve combinations for each date seed variety. Baked cookies were packed in sealed plastic bags and stored at −20 °C until extraction.

### 2.4. Sample Preparation and Extraction

#### 2.4.1. Powder

Powder was extracted at three sample to solvent ratios, 5:1, 10:1, and 25:1, following the procedure of Khanavi et al. [35]. Specifically, 25, 50, and 75 mg of date seed powder were mixed with 5 mL (40–50 °C) of deionised water. The mixture was placed on a shaker for 5 min and left at 4 °C for 30 min prior to 10 min centrifugation at 10,000× g, the supernatant was collected, and one additional extraction was made with 4 mL (40–50 °C) deionised water. The supernatants were pooled and stored at −20 °C until analysis.

#### 2.4.2. Cookie

Composite cookies (1 g) were mixed with 5 mL of solvent (deionised water) [35]. Composite cookies were prepared at three levels of substitution of flour by DSP (2.5%, 5%, 7.5%). The mixture was placed on a shaker for 5 min and left at 4 °C for 30 min prior to 10 min centrifugation at 10,000× g, the supernatant was collected, and one additional extraction was made with 4 mL (40–50 °C) deionised water. The supernatants were pooled and stored at −20 °C until analysis.

### 2.5. Determination of TPC and Flavonoids

TPC was determined using Folin–Ciocalteu reagent and external calibration with gallic acid as mentioned in [36] with some modifications; 20 µL of the sample was diluted in 180 µL deionised water and mixed with 200 µL Folin–Ciocalteu reagent, vortexed, and 800 µL sodium carbonate was added and incubated for 30 min at 40 °C. Absorbance was measured at 765 nm. The concentrations of TPC were calculated from the standard curve of gallic acid in the range of 0 to 25 µg/mL; the results are expressed as Gallic acid equivalent in (µg/g) of the sample.

Flavonoids were measured by the aluminium chloride colorimetric method as described in [37]; in a vial, 150 µL of the sample was mixed with 150 µL methanol and 75 µL 5% $NaNO_2$ and allowed to react for 5 min at room temperature. Then, 1.25 mL 7% $AlCl_3$ and 0.5 mL 5% NaOH were added and left for an extra 5 min, and absorbance was measured at 510 nm. The concentrations of flavonoids were calculated from the standard curve of quercetin in the range of 25 to 200 µg/mL, with the results expressed as a percentage (*w*/*w*).

### 2.6. Determination of Antioxidant Activity

The DPPH scavenging ability of the extracts was determined according to the modified method of [38]; 100 µL of sample was mixed with 400 µL 0.1 mM DPPH, and free radical scavenging activity was determined by measuring the absorbance at 520 nm after 30 min of incubation at room temperature.

ABTS scavenging ability was determined according to the procedure developed in a previous study [39]. Specifically, 5 mL of 7 mM ABTS was added to 88 µL of 140 mM potassium persulphate solution and left to react for 24 h at room temperature in the dark. Then, ABTS solution was diluted (1:88) with 10 mM sodium phosphate buffer of pH 7.4. A 600 µL volume of the ABTS was added to the 50 µL of the sample. Absorbance was measured at 734 nm after 5 s of incubation at room temperature. The results were expressed as mmol of Trolox equivalents per mg of sample.

A modified assay method of the Ferric-reducing antioxidant power (FRAP) reported in [40] was performed; 100 µL of the sample was mixed with a prewarmed mixture of 500 µL 300 mM acetate buffer (pH 3.6), 50 µL 10 mM 2,4,6-tripirydyl-S-triazine, and 50 µL 20 mM $FeCl_3$, incubated at 37 °C for 5 min, and absorbance was measured at 593 nm.

### 2.7. Statistical Analysis

All experiments were performed with more than three replications; the results obtained are expressed as mean ± standard deviation (SD). Data were assessed by one way ANOVA followed by Tukey's test using Minitab 19, State College, Pennsylvania, USA statistical software package; $p < 0.05$ determined the level of significance.

## 3. Results and Discussion

### 3.1. Analysis of Date Seed Powder

The results regarding TPC are presented in Figure 1a. TPC measurements ranged from 389.03 to 2365.75 µg GAE/g at 5:1 and 15:1 sample to solvent ratio levels, respectively. The data clearly show that higher sample to solvent ratios had a higher content of polyphenols. The results of our study did not correlate with previous investigations of the TPC of certain date seed varieties. The TPC of *Khalas* seed was 2194 µg/g [41]; for *Khinaizi* and *Zahidi* it was 9540 and 11,610 µg/g [42], respectively. For *Sukkary* it was 37.1 µg/g [43], while in another study it was 2058 µg/g for *Zahidi* [44]. This variation appears related to different extraction methods [42], and the use of water as opposed to other solvents in extraction, which has been proven to have low solubility for phenolics and flavonoids [1]. For example, in Ardekani et al., higher TPCs were detected when the polar aprotic solvent dimethyl sulfoxide (DMSO) was used in comparison to common polar solvents (water, aqueous methanol, and methanol) [42]. This underlines the importance of the extraction solvent when determining TPC.

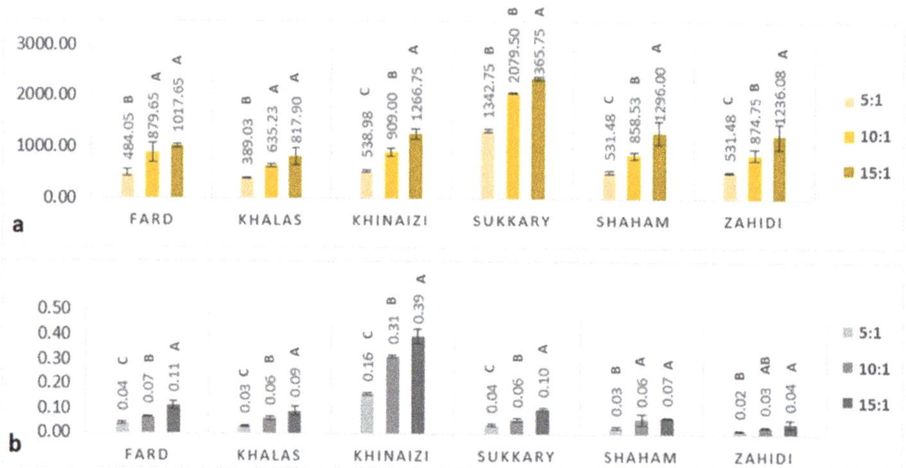


**Figure 1.** (**a**) TPC (Gallic acid equivalent per gram (µg/g) of the sample); (**b**) Flavonoid content (*w*/*w*) % of six varieties of date seed powder extracted at three sample to solvent ratios (5:1, 10:1, and 15:1). Different uppercase superscript letters in a variety denote significant differences, $p < 0.05$.

Figure 1b presents the data on flavonoid content, which ranged from 0.02 to 0.39 (*w*/*w*) %, and the *Khinaizi* variety had not less than three-fold higher the amount of flavonoid content among the other varieties, with values of 0.16, 0.31, and 0.39 (*w*/*w*) % at 5:1, 10:1, and 15:1 sample to solvent ratio levels, respectively. At the same time, *Zahidi* had the lowest flavonoid content with 0.04 (*w*/*w*) %, at the highest sample to solvent ratio level. Though the TPC was high in *Sukkary*, flavonoid content was comparatively lower than *Khinaizi*.

The variations in TPC and flavonoid content in date seed powder might be due to differences in date seed variety, maturity, growing conditions, season, fertiliser, soil type and storage conditions [42,45]. For instance, higher contents of phenolics were reported in the seeds of Omani date fruits [45], whereas much lower amounts were shown in Saudi Arabian varieties [43]. Ahfaiter et al. showed variances in flavonoid content in *Sukkary* date seed powder ranging from 1.2–1.4%, depending on the geographical origin [46]. Apart from this, the reason for the observed differences in results from previous studies is the difference in the extraction procedures used [42,47]. Several previous studies used several solvents for extraction, especially organic solvents or organic and aqueous solution mixtures. For example, Parry et al. used acetone for pumpkin seed flour [48], whereas Mistrello et al. [44]

used an acetone and water mixture for date seeds. Several studies have used aqueous methanol or ethanol for extraction of seed flour from date seeds, mango seeds, and sesame seeds [46,49–52]. In our study, even with water as the solvent the extraction efficiency was significant, showing considerable amounts of phenolic compounds in alignment with Al Juhaimi et al. [43] and Ifesan et al. [11], who used aqueous extracts of seed flour as well.

The antioxidant properties in terms of radical scavenging activity of date seed powder using DPPH and ABTS assay and antioxidant power using FRAP assay are shown in Figure 2. In DPPH assay (Figure 2a), the data range from 193.1 to 212.03 µg/g of date seed powder, and in ABT from 0.43 to 2.03 % per mg of date seed powder. At the same time, in FRAP the range was from 385.53 and 1368.25 µg/g. The phenol and flavonoid assays showed similar trends, the with higher sample to solvent ratio levels being correlated with higher antioxidant activity. It was noted that the *Sukkary* variety, shown in Figures 1a and 2c, had the highest phenolic content and the highest FRAP values among all varieties by a significant margin. However, although the *Sukkary* variety had the highest FRAP activity, a plateau was observed in that above a 5:1 sample to solvent ratio level there was no increase.

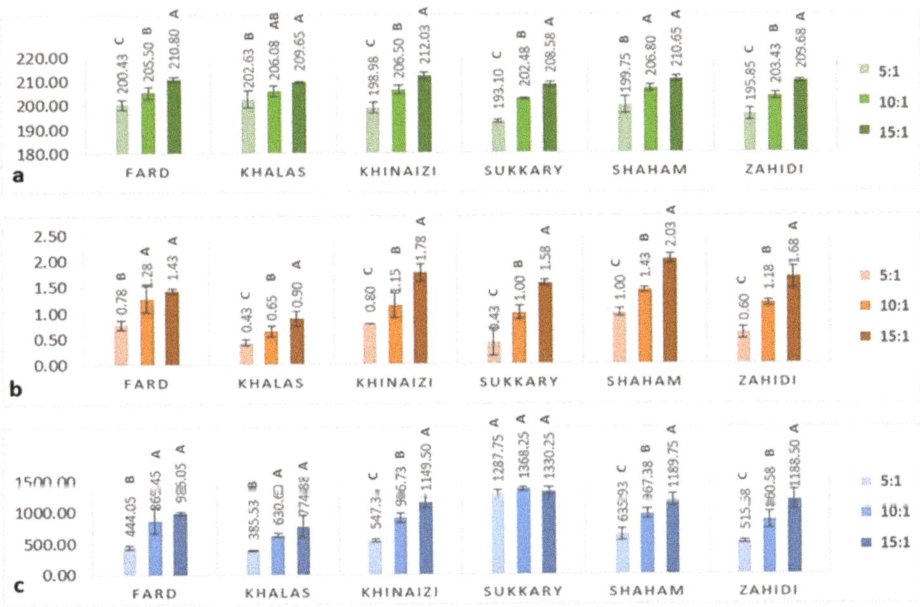


**Figure 2.** (**a**) DPPH radical scavenging activity (µg Trolox equivalent/g sample); (**b**) ABTS radical scavenging (% Scavenging effect with ABTS/mg) %; (**c**) Ferric reducing assay, (µg equivalent Trolox/g) of six varieties of date seed powder extracted at three sample to solvent ratios (5:1, 10:1, and 15:1). Different uppercase superscript letters in a variety denote significant differences, $p < 0.05$.

The majority of the total phenolic content was assumed to be composed of flavonoids, with 0.39% at the highest sample to solvent ratio level. These dietary phenolics may be the ones most responsible for the high antioxidant capacity reported in date seeds with increasing sample to solvent ratio levels. Hence, this suggests that date seeds can be considered a potential raw material as a natural, active ingredient for food applications such as bakery products. As explained in Al-Farsi et al. [45], when compared to the other by-products of dates, the seeds have the highest antioxidant activity thanks to their high phenolic content.

### 3.2. Total Phenolic Content (TPC) and Flavonoid Content of Composite Cookies

The TPC of cookies enriched with date seed powder was measured; the contents are listed in Figure 3. The highest TPC of the control cookies (zero date seed powder substitution) was 265.77 µg/g. Almost all the composite cookie formulations showed higher TPC levels above 265.77 µg/g, except *Zahidi* with 2.5% substitution level of white flour at both 180 and 200 °C temperature conditions. Cookies formulated with 7.5% *Fard* variety made with white flour at either temperature had significantly higher TPC content, more than 962.17 µg/g, and at substitution level of 7.5% the TPC level were more than three-fold higher compared to the control cookies. The TPC of *Khalas* cookies at 7.5% substitution level was the highest when made with white flour and baked at 180 °C and with whole wheat flour baked at 200 °C. The values of *Khinaizi* and *Shaham* cookies were mostly similar. The highest value of TPCs, observed in *Sukkary* date seed powder (Figure 1a) at 7.5% substitution level, explains the highest values of *Sukkary*-made cookies, which range from 940.37 to 1213.00 µg/g, almost four times higher than control cookies, and which differs from the TPC of *Zahidi*-made cookies, which presented the lowest effect among all varieties with value of 621.87 µg/g when made with 7.5% substitution level, whole wheat flour and baked at 200 °C. Some previous studies on the total phenolic content of date seed powder exhibited high polyphenol content, for example, in *Sukkary* [43,44,46], strengthening the results of this study.

On the other hand, it was expected that the high flavonoid content found in *Khinaizi* date seed powder (Figure 1b) would lead to *Khinaizi* cookies having the highest values (Figure 4); instead, composite cookies of *Shaham* and *Zahidi* had the highest content of flavonoids, up to 0.0393 ($w/w$) %. Cookies formulated with *Fard, Khalas, Khinaizi,* and *Sukkary* had mostly similar measurements, ranging from 0.0077 ($w/w$) % at 2.5% substitution level to 0.0317 ($w/w$) % at 7.5%. A similar pattern was observed between *Sukkary* composite cookies and *Sukkary* flour observations; i.e., despite having the highest TPC level, the flavonoid content was comparatively lower than the other date seed varieties, although higher than the control cookies. Though the TPC and flavonoid content of cookies increased with higher levels of date seed flour, a substantial decrease in those levels in cookies in comparison to flour was observed upon baking. The reason behind this could be attributed to the loss of phenolic components during baking [53].

However, it is clearly shown that the increase in TPCs and flavonoids was noticeable regardless of the flour type and baking temperature. Both polyphenols and flavonoids have a strong contribution to human health. Polyphenols are known to have antioxidant activity, anti-inflammatory actions and anticarcinogenic activities [54,55] while flavonoids have the ability to interfere with the formation and propagation of free radicals and protect low density lipoproteins from oxidation [56]. The variability of their therapeutic potential depends on the structure of the particular flavonoid [57–59]. The results indicate that date seeds can be effectively used to enhance the antioxidant potential of cookies.

### 3.3. Antioxidant Activity

Cookies produced at substitution levels of 2.5%, 5% and 7.5% were tested for their radical scavenging activity using the DPPH and ABTS assays. The DPPH results are shown in Figure 5; the data show that substitution of either type of flour resulted in increased antioxidant activity in the cookies. Similar results can be seen in the data from the ABTS assay, shown in Figure 6.

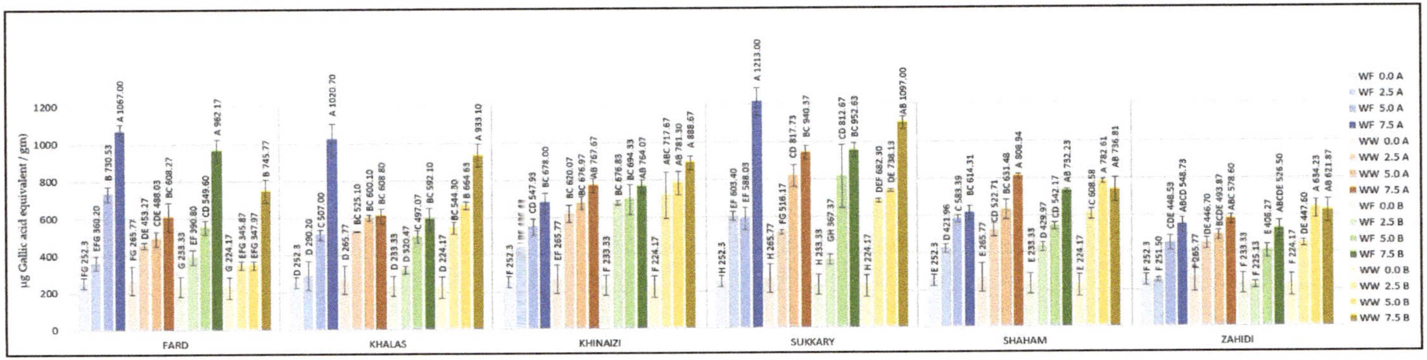


**Figure 3.** TPC of composite cookies formulated with 2.5%, 5%, and 7.5% substitution of flour by date seed powder. (Data expressed as Gallic acid equivalent per gram (µg/g) of the cookie sample. Different uppercase superscript letters in a variety denote significant differences, $p < 0.05$. (A—180 °C, B—200 °C).

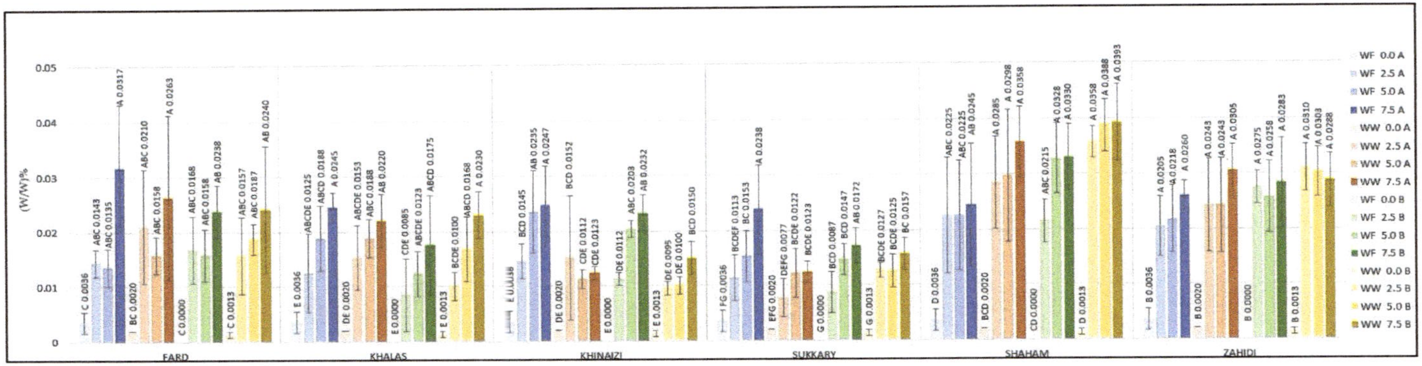


**Figure 4.** Total contents of Flavonoids of composite cookies formulated with 2.5%, 5%, and 7.5% substitution of flour by date seed powder. (Data expressed as Gallic acid equivalent per gram (µg/g) of the cookie sample. Different uppercase superscript letters in a variety denote significant differences, $p < 0.05$. (A—180 °C, B—200 °C).

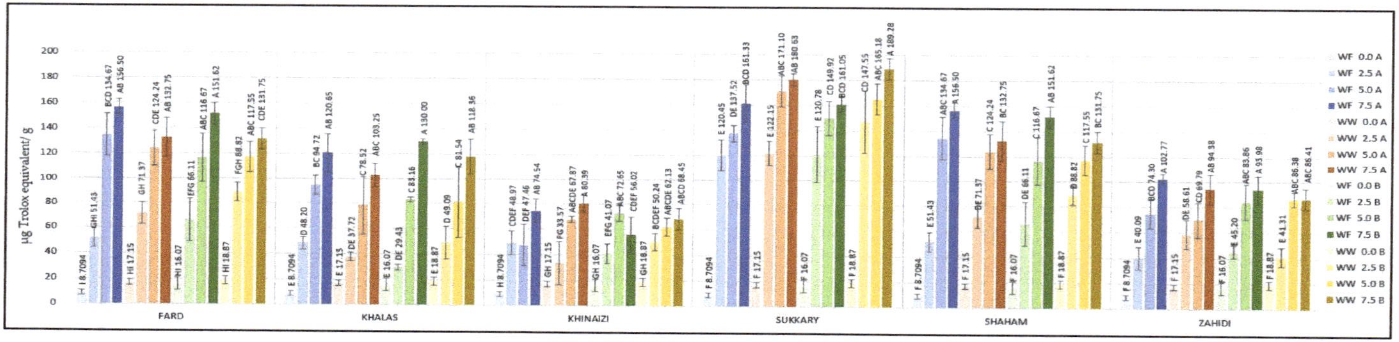


**Figure 5.** DPPH radical scavenging activity (µg Trolox equivalent/g) of composite cookies formulated with 2.5%, 5%, and 7.5 % substitution of flour by date seed powder Different uppercase superscript letters in a variety denote significant differences, $p < 0.05$. (A—180 °C, B—200 °C).

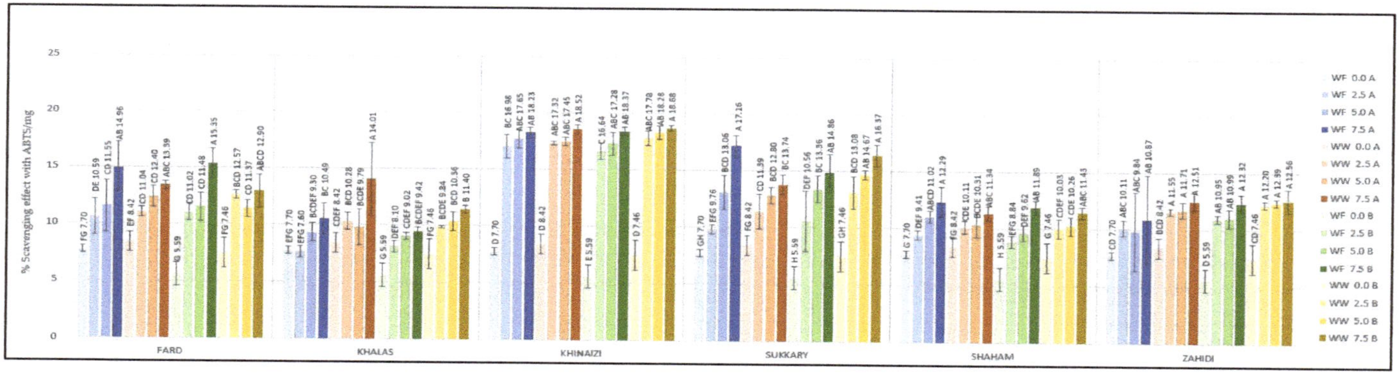


**Figure 6.** ABTS radical scavenging (% Scavenging effect with ABTS/mg) of composite cookies formulated with 2.5%, 5%, and 7.5 % substitution of flour by date seed powder. Different superscript letters in a variety denote significant differences, $p < 0.05$. (A—180 °C, B—200 °C).

Regarding DPPH, the control cookies had values of less than 18.87 μg/g, and even the cookies with the lowest substitution level had values more than four-fold higher than the control, except for *Khalas-* and *Khinaizi-* made cookies (regardless of flour type and temperature), where the values only double. Even though *Sukkary* date seed powder was insignificantly different from other varieties the *Sukkary* cookies had the highest antioxidant activity, with more than 161.05 μg/g at 7.5% substitution level, while *Khinaizi* cookies had the lowest at all substitution levels, with no more than 80.39 μg/g, which is opposite from the ABTS results. *Khinaizi* cookies had the highest measurements, with more than 18.23% ABTS/g at the highest substitution level, slightly higher than *Sukkary* cookies (17.16% ABTS/g). Mistrello et al., Al Juhaimi et al., and Ardekani et al., showed high antioxidant activity in *Sukkary* date seed powder, where it seems that it contributes to the high antioxidant activity in composite cookies as well [43,44,46]. Very slight differences were observed between the *Fard*, *Khalas*, *Shaham* and *Zahidi* varieties; ABTS ranged from 7.6% to 12.7%, 9.02% to 12.40%, and from 9.42% to 15.35% ABTS/g at 2.5%, 5.0%, and 7.5% respectively.

The antioxidant power of the cookies was determined using the FRAP procedure (Figure 7). The data show that the maximum antioxidant power was achieved at 7.5% substitution level of date seed powder, with the highest value of 262.94 μg/g for *Sukkary* composite cookies. *Fard* substitution was affected by baking temperature; cookies baked at 180 °C have 240.43 and 128.29 μg/g using white flour and whole wheat flour, respectively, compared with less than 73.75 μg/g for cookies baked at 200 °C, while the temperature did not influence FRAP results of the other varieties. On the other hand, the *Khinaizi* and *Sukkary* varieties exhibited higher FRAP values regardless of the flour type and baking temperature, and similar values to *Fard* cookies at 7.5% substitution and 180 °C baking temperature. Almost all of the varieties showed the highest reducing antioxidant power with white flour at 7.5% substitution level and baking temperature at 180 °C, except for the *Shaham* composite cookies. Although FRAP value increased with increasing levels of date seed powder in the cookies, the values were comparatively lower than the FRAP values of flour. The antioxidant capacity is based on preventing free radical formation by breaking free radical chain reactions through donation of hydrogen atoms [60]. As the polyphenols are responsible for this reaction, the loss of polyphenols might be the reason for this observation of reduced FRAP values upon baking, as there was a greater reduction in antioxidant power at the higher temperature of 200 °C. Overall, while an increase in antioxidant capacity was observed upon substitution of date seed flour, the increment was lower at the higher temperature.

The differences in antioxidant levels obtained from the assays might be a reflection of a relative difference in the activity of antioxidant compounds in the extract [61]. Date seeds contain very high levels of phenolic antioxidants [42–45]. Because date seeds are rich in dietary fibre [45,62–65], they are considered a good source of natural antioxidants due to their richness in phenolic compounds. Mrabet et al.'s explanation of date dietary fibre concentrate indicates the presence of significant amount of bound phenolics in dietary fibre [66], which adds additional health benefits to the antioxidant properties of date seed. In this study, water was used as the extraction solvent for extraction of TPCs and for assays of antioxidant action, ABTS, and FRAP, yet the results were significantly higher. Most of the studies on bakery items incorporated seeds including grape seed, flaxseed, sunflower seed and fig seed, and used organic solvents such as ethanol and methanol [32,67–71], ending up with high levels of TPCs, antioxidants, ABTS and FRAP levels.

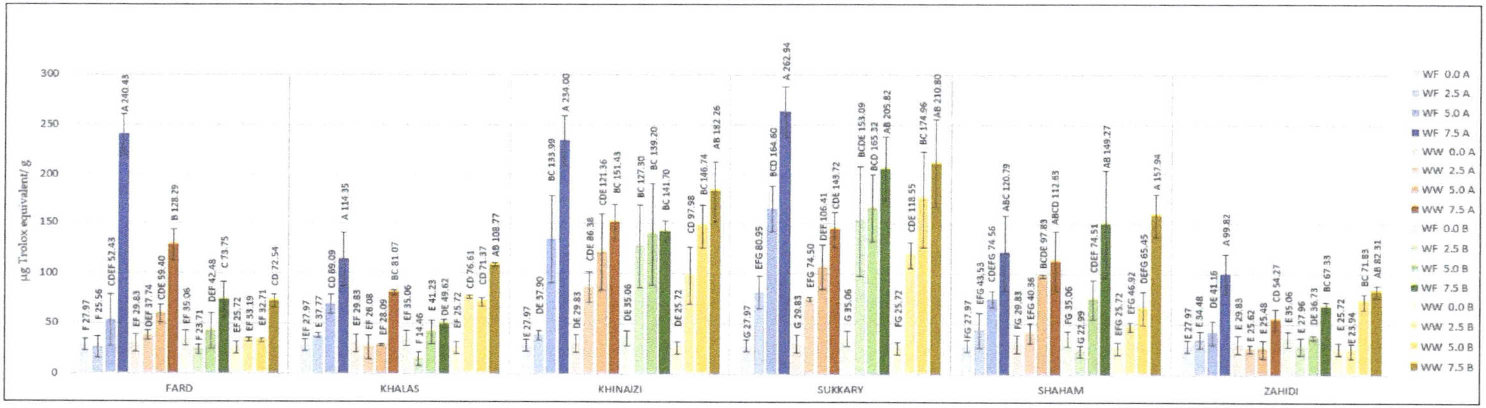


**Figure 7.** Ferric reducing antioxidant power, (µg equivalent Trolox/g) of composite cookies formulated with 2.5%, 5%, and 7.5 % substitution of flour by date seed powder. Different superscript letters in a variety denote significant differences, $p < 0.05$ (A—180 °C, B—200 °C).

High levels of antioxidants in date seed powder as well as the improved antioxidant properties of composite cookies with increasing substitution levels show that date seed powder can be used as functional ingredient in the food industry to improve the quality of baked goods. Our results show that the substitution of flour with date seed powder significantly upgraded the content of polyphenols and flavonoids as well as the antioxidant activity. A number of studies have reported similar results regarding the increased nutritive value of composite bakery products, including cookies and biscuits fortified with several seed varieties. For example, Acun and Gul [13] observed high antioxidant and TPC in cookies formulated with grape seed flour, while Kaur et al. found rich antioxidant capacity which was evident from higher TPCs [67]. Another study showed enhanced antioxidant potential and flavonoid levels in cookies formulated with chia and quinoa seed flour [12]. Aksoylu et al. [68] found higher TPC in biscuits fortified with blueberry and grape seeds, while Gbenga-Fabusiwa et al. [72] and Grasso et al. [14] reported higher phenolic and antioxidant levels in biscuits incorporated with pigeon-pea and sunflower seed flour, respectively, than the biscuits made only with wheat flour.

As wheat flour has a very low polyphenol content [73], incorporation of date seed powder would be expected to increase the nutritional value of bakery products such as cookies. Moreover, increasing the natural antioxidant content in bakery products leads to the extension of shelf life by lowering the fat oxidation, which is a factor determining food quality [74]. The obtained results reveal the importance of incorporating date seed flour to achieve better functionality of produced cookies.

## 4. Conclusions

This study investigated date seed powder to determine its natural polyphenolic compound and flavonoid contents and showed it to have high antioxidant potential. The highest amount, 2365.75 µg GAE/g TPC at a 15:1 sample to solvent ratio level, was in *Sukkary* date seed powder, while *Khinaizi* exhibited the highest flavonoid content. Mostly similar DPPH values were observed in all the varieties, with the highest value at 15:1 sample to solvent ration level. *Shaham* seed flour showed a higher ABTS activity, whereas *Sukkary* was significantly higher in FRAP, having quite similar values in all three sample to solvent ration levels. Cookies formulated with 7.5% *Fard* variety made with white flour at either temperature had a significantly higher TPC content, more than 962.17 µg/g. Although the highest flavonoid content was found in *Khinaizi* date seed powder, composite cookies of *Shaham* and *Zahidi* had high flavonoid contents of up to 0.0393 µg/g as well. The increase in TPCs and flavonoids was observable regardless of the flour type and baking temperature. Both DPPH and FRAP results showed the highest antioxidant activity in *Sukkary* cookies at 7.5% substitution level, with amounts of 189.28 µg/g at 200 °C and 262.94 µg/g at 180 °C, respectively. *Khinaizi* cookies had the highest ABTS measurements, with more than 18.23% ABTS/g at 7.5% substitution level, slightly higher than *Sukkary* cookies with 17.16% ABTS/g. Moreover, higher TPCs and flavonoids resulted even under less efficient conditions of extraction where water was used as the solvent.

Based on the results for the composite cookies in this study, it can be concluded that the TPC and antioxidant activity of date seeds successfully enhanced the quality of final product; hence it can be recommended as a flour fortifier and as an ingredient in functional foods. These findings improve our knowledge on the value of utilising date seeds, which are considered a waste product in the food industry. Future work could focus on clinical trials where conclusive observations can be made as to how these fortified cookies might positively impact consumer health.

**Author Contributions:** Conceptualisation, C.S.; Formal analysis, Z.N. and M.R.; Funding acquisition, C.S.; Investigation, Z.N., J.K. and H.S.; Methodology, J.K., C.P. and C.S.; Supervision, C.S.; Validation, C.S.; Writing—original draft, Z.N.; Writing—review & editing, C.P., C.S. and M.R. All authors have read and agreed to the published version of the manuscript.

**Funding:** This research was funded by United Arab Emirates University start up grant G00002958.

**Institutional Review Board Statement:** Not applicable.

**Informed Consent Statement:** Informed consent was obtained from all participants in the study.

**Data Availability Statement:** The datasets generated for this study are available on request to the corresponding author.

**Acknowledgments:** The authors acknowledge the support from the University farm in the production of date seed powder.

**Conflicts of Interest:** The authors declare no conflict of interest.

## References

1. Al-Farsi, M.A.; Lee, C.Y. Optimization of phenolics and dietary fibre extraction from date seeds. *Food Chem.* **2008**, *108*, 977–985. [CrossRef] [PubMed]
2. Dowson, V.H.W. *Date Production and Protection with Special Reference to North Africa and the Near East*; No. 35, FAO Technical Bulletin; FAO: Rome, Italy, 1982.
3. Zaid, A. *Date Palm Cultivation*; FAO No. 156; FAO: Rome, Italy, 1999.
4. Ahmed, A.W.K.; Ahmed, A.I.; Robinson, R.K. Chemical composition of date varieties as influenced by the stage of ripening. *Food Chem.* **1995**, *54*, 305–309. [CrossRef]
5. Al-Farsi, M.; Alasalvar, C.; Morris, A.; Baron, M.; Shahidi, F. Comparison of antioxidant activity, anthocyanins, carotenoids, and phenolics of three native fresh and sun-dried date (*Phoenix dactylifera* L.) varieties grown in Oman. *J. Agric. Food Chem.* **2005**, *53*, 7592–7599. [CrossRef] [PubMed]
6. Habib, H.M.; Ibrahim, W.H. Nutritional quality evaluation of eighteen date pit varieties. *Int. J. Food Sci. Nutr.* **2009**, *60*, 99–111. [CrossRef]
7. Habib, H.M.; Kamal, H.; Ibrahim, W.H.; Al Dhaheri, A.S. Carotenoids, fat soluble vitamins and fatty acid profiles of 18 varieties of date seed oil. *Ind. Crop. Prod.* **2013**, *42*, 567–572. [CrossRef]
8. Abdel-Rahman, H.; El-Mougy, S.A. Protective effect of extracts from dates (*Phoenix dactylifera* L.) on carbon tetrachloride-induced hepatotoxicity in rats. *J. Appl. Res. Vet. Med.* **2004**, *2*, 176–180.
9. Caleja, C.; Barros, L.; Antonio, A.L.; Oliveira, M.B.P.; Ferreira, I.C. A comparative study between natural and synthetic antioxidants: Evaluation of their performance after incorporation into biscuits. *Food Chem.* **2017**, *216*, 342–346. [CrossRef]
10. Elhassaneen, Y.; Ragab, S.; Mashal, R. Improvement of bioactive compounds content AND Antioxidant properties in crackers with the incorporation of prickly pear and potato peels powder. *Int. J. Nutr. Food Sci.* **2016**, *5*, 53. [CrossRef]
11. Ifesan, B.O.; Femi-Ajayi, O.; Adeloye, J.B.; Ifesan, B.T. Quality assessment and consumer acceptability of cookies from blends of wheat flour and pumpkin (*Cucurbita* spp.) seed flour. *Himal. J. Appl. Med. Sci. Res.* **2020**, *1*, 1–7.
12. Goyat, J.; Passi, S.J.; Suri, S.; Dutta, H. Development of chia (*Salvia hispanica* L.) and quinoa (*Chenopodium quinoa* L.) seed flour substituted cookies-physicochemical, nutritional and storage studies. *Curr. Res. Nutr. Food Sci.* **2018**, *6*, 757–769. [CrossRef]
13. Acun, S.; Gül, H. Effects of grape pomace and grape seed flours on cookie quality. *Qual. Assur. Saf. Crop. Foods* **2014**, *6*, 81–88. [CrossRef]
14. Heo, Y.; Kim, M.J.; Lee, J.W.; Moon, B. Muffins enriched with dietary fiber from kimchi by-product: Baking properties, physical–chemical properties, and consumer acceptance. *Food Sci. Nutr.* **2019**, *7*, 1778–1785. [CrossRef] [PubMed]
15. Ndife, J.; Abbo, E. Functional foods: Prospects and challenges in Nigeria. *J. Sci. Technol.* **2009**, *1*, 1–6.
16. Baba, M.D.; Manga, T.A.; Daniel, C.; Danrangi, J. Sensory evaluation of toasted bread fortified with banana flour: A preliminary study. *Am. J. Food Sci. Nutr.* **2015**, *2*, 9–12.
17. Dewettinck, K.; Van Bockstaele, F.; Kühne, B.; Van de Walle, D.; Courtens, T.M.; Gellynck, X. Nutritional value of bread: Influence of processing, food interaction and consumer perception. *J. Cereal Sci.* **2008**, *48*, 243–257. [CrossRef]
18. Fardet, A. New hypotheses for the health-protective mechanisms of wholegrain cereals: What is beyond fiber? *Nutr. Res. Rev.* **2010**, *23*, 65–134. [CrossRef]
19. Sozer, N.; Cicerelli, L.; Heiniö, R.L.; Poutanen, K. Effect of wheat bran addition on in vitro starch digestibility, physico-mechanical and sensory properties of biscuits. *J. Cereal Sci.* **2014**, *60*, 105–113. [CrossRef]
20. Bajaj, S.; Urooj, A.; Prabhasankar, P. Effect of incorporation of mint on texture, colour and sensory parameters of biscuits. *Int. J. Food Prop.* **2006**, *9*, 691–700. [CrossRef]
21. Carocho, M.; Morales, P.; Ferreira, I.C.F.R. Natural food additives: Quo vadis? *Trends Food Sci. Technol.* **2015**, *45*, 284–295. [CrossRef]
22. Carocho, M.; Barreiro, M.F.; Morales, P.; Ferreira, I.C.F.R. Adding molecules to food, pros and cons: A review of synthetic and natural food additives. *Comp. Rev. Food Sci. Food Saf.* **2014**, *13*, 377–399. [CrossRef]
23. Branen, A.L. Toxicology and biochemistry of butylated hydroxyl anisole and butylated hydroxytoluene. *J. Am. Oil Chem. Soc.* **1975**, *52*, 59–63. [CrossRef] [PubMed]
24. Ito, N.; Fukushima, S.; Haqlwara, A.; Shibata, M.; Ogiso, T. Carcinogenicity of butylated hydroxyanisole in F344 rats. *J. Natl. Cancer Inst.* **1983**, *70*, 343–347. [CrossRef] [PubMed]
25. Reddy, V.; Urooj, A.; Kumar, A. Evaluation of antioxidant activity of some plant extracts and their application in biscuits. *Food Chem.* **2005**, *90*, 317–321. [CrossRef]

26. Pasqualone, A.; Bianco, A.M.; Paradiso, V.M.; Summo, C.; Gambacorta, G.; Caponio, F.; Blanco, A. Production and characterization of functional biscuits obtained from purple wheat. *Food Chem.* **2015**, *180*, 64–70. [CrossRef] [PubMed]
27. Rasooli, I. Food preservation—A biopreservative approach. *Food* **2007**, *1*, 111–136.
28. Ye, H.; Shen, S.; Xu, J.; Lin, S.; Yuan, Y.; Jones, G.S. Synergistic interaction of cinnemaldehyde in combination with carvacrol against food-borne bacterial. *Food Contr.* **2013**, *34*, 619–623. [CrossRef]
29. Caleja, C.; Barros, L.; Antonio, A.L.; Ciric, A.; Barreira, J.C.; Sokovic, M.; Oliveira, M.B.P.; Santos-Buelga, C.; Ferreira, I.C. Development of a functional dairy food: Exploring bioactive and preservation effects of chamomile (*Matricaria recutita* L.). *J. Funct. Foods* **2015**, *16*, 114–124. [CrossRef]
30. Caleja, C.; Barros, L.; Antonio, A.L.; Ciric, A.; Soković, M.; Oliveira, M.B.P.P.; Santos-Buelga, C.; Ferreira, I.C.F.R. *Foeniculum vulgare* Mill. as natural conservation enhancer and health promoter by incorporation in cottage cheese. *J. Funct. Foods* **2015**, *12*, 428–438. [CrossRef]
31. Shah, M.A.; Bosco, S.J.D.; Mir, S.A. Plant extracts as natural antioxidants in meat and meat products. *Meat Sci.* **2014**, *98*, 21–33. [CrossRef]
32. Antonic, B.; Dordevic, D.; Jancikova, S.; Holeckova, D.; Tremlova, B.; Kulawik, P. Effect of grape seed flour on the antioxidant profile, textural and sensory properties of waffles. *Processes* **2021**, *9*, 131. [CrossRef]
33. Helkar, P.B.; Sahoo, A. Review: Food industry by-products used as a functional food ingredients. *Int. J. Waste Resour.* **2016**, *6*, 1000248. [CrossRef]
34. Approved Methods of Analysis. AACC Method 10-54.01. Baking Quality of Cookie Flour-Micro Wire-Cut Formulation; 11th ed.; Cereals & Grains Association: St. Paul, MN, USA. Available online: https://methods.aaccnet.org/methods/10-54.pdf (accessed on 21 November 2020).
35. Khanavi, M.; Saghari, Z.; Mohammadirad, A.; Khademi, R.; Hadjiakhoondi, A.; Abdollahi, M. Comparison of antioxidant activity and total phenols of some date varieties. *Daru J. Pharm. Sci.* **2009**, *17*, 104–107.
36. Singleton, V.L.; Rossi, J.A. Colorimetry of total phenolics with phosphomolybdic phosphotungstic acid reagents. *Am. J. Enol. Viticul.* **1965**, *16*, 144–158.
37. Zhishen, J.; Mengcheng, T.; Jianming, W. The determination of flavonoid contents in mulberry and their scavenging effect on superoxide radicals. *Food Chem.* **1999**, *64*, 555–559. [CrossRef]
38. Blois, M.S. Antioxidant determinations by the use of a stable free radical. *Nature* **1958**, *181*, 1199–1200. [CrossRef]
39. Re, R.; Pellegrini, N.; Proteggente, A.; Pannala, A.; Yang, M.; Rice-Evans, C. Antioxidant activity applying an improved ABTS radical cation decolonization assay. *Free Radic. Biol. Med.* **1999**, *26*, 1231–1237. [CrossRef]
40. Benzie, I.F.F.; Strain, J.J. The ferric reducing ability of plasma (FRAP) as a measure of "antioxidant power": The FRAP assay. *Anal. Biochem.* **1996**, *239*, 70–76. [CrossRef]
41. Habib, H.M.; Platat, C.; Meudec, E.; Cheynier, V.; Ibrahim, W.H. Polyphenolic compounds in date fruit seed (*Phoenix dactylifera*): Characterisation and quantification by using UPLC-DAD-ESI-MS. *J. Sci. Food Agric.* **2014**, *94*, 1084–1089. [CrossRef]
42. Ardekani, M.R.S.; Khanavi, M.; Hajimahmoodi, M.; Jahangiri, M.; Hadjiakhoondi, A. Comparison of antioxidant activity and total phenol contents of some date seed varieties from Iran. *Iran. J. Pharm. Res. IJPR* **2010**, *9*, 141–146.
43. Al Juhaimi, F.; Ghafoor, K.; Ozcan, M.M. Physical and chemical properties, antioxidant activity, total phenol and mineral profile of seeds of seven different date fruit (*Phoenix dactylifera* L.) varieties. *Int. J. Food Sci. Nutr.* **2012**, *63*, 84–89. [CrossRef]
44. Mistrello, J.; Sirisena, S.D.; Ghavami, A.; Marshall, R.J.; Krishnamoorthy, S. Determination of the antioxidant capacity, total phenolic and flavonoid contents of seeds from three commercial varieties of culinary dates. *Int. J. Food Stud.* **2014**, *3*, 34–44. [CrossRef]
45. Al-Farsi, M.; Alasalvar, C.; Al-Abid, M.; Al-Shoaily, K.; Al-Amry, M.; Al-Rawahy, F. Compositional and functional characteristics of dates, syrups, and their by-products. *Food Chem.* **2007**, *104*, 943–947. [CrossRef]
46. Ahfaiter, H.; Zeitoun, A.; Abdallah, A.E. Physicochemical properties and nutritional value of Egyptian date seeds and its applications in some bakery products. *J. Adv. Agric. Res.* **2018**, *23*, 260–279.
47. Suresh, S.; Guizani, N.; Al-Ruzeiki, M.; Al-Hadhrami, A.; Al-Dohani, H.; Al-Kindi, I.; Rahman, M.S. Thermal characteristics, chemical composition and polyphenol contents of date-pits powder. *J. Food Eng.* **2013**, *119*, 668–679. [CrossRef]
48. Parry, J.W.; Cheng, Z.; Moore, J.; Yu, L.L. Fatty acid composition, antioxidant properties, and antiproliferative capacity of selected cold-pressed seed flours. *J. Am. Oil Chem. Soc.* **2008**, *85*, 457–464. [CrossRef]
49. Ashoush, I.S.; Gadallah, M.G. Utilization of mango peels and seed kernels powders as sources of phytochemicals in biscuit. *World J. Dairy Food Sci.* **2011**, *6*, 35–42.
50. Song, Z.; Li, Y.; Gao, B.; Lee, J.; Wu, Y.; Sun, J.; Whent, M.; Chen, P.; Lee, S.-H.; Yu, L. The chemical composition, antioxidant activity, and antiproliferative activity of selected seed flours. *J. Food Bioact.* **2020**, *10*, 77–85. [CrossRef]
51. Ben Othman, S.; Katsuno, N.; Kanamaru, Y.; Yabe, T. Water-soluble extracts from defatted sesame seed flour show antioxidant activity in vitro. *Food Chem.* **2014**, *175*, 306–314. [CrossRef]
52. Biswas, R.; Bains, A.; Chawla, P. Antioxidant potential of *Cucumis melo* and *Citrullus vulgaris* seed flours. *Int. J. Res. Analyt. Rev.* **2018**, *5*, 1171–1179.
53. Jonsson, L. Thermal degradation of carotenes and influence on their physiological functions. *Nutr. Toxicol. Conseq. Food Process.* **1991**, *289*, 75–82. [CrossRef]
54. Surh, Y.-J. Anti-tumor promoting potential of selected spice ingredients with antioxidative and anti-inflammatory activities: A short review. *Food Chem. Toxicol.* **2002**, *40*, 1091–1097. [CrossRef]

55. Willcox, J.K.; Ash, S.L.; Catignani, G.L. Antioxidants and prevention of chronic disease. *Crit. Rev. Food Sci. Nutr.* **2004**, *44*, 275–295. [CrossRef] [PubMed]
56. Frankel, E. Chemistry of free radical and singlet oxidation of lipids. *Prog. Lipid Res.* **1984**, *23*, 197–221. [CrossRef]
57. Agrawal, A.D. Pharmacological activities of flavonoids: A review. *Int. J. Pharm. Sci. Nanotechnol.* **2011**, *4*, 1394–1398. [CrossRef]
58. Kumar, S.; Pandey, A.K. Chemistry and biological activities of flavonoids: An overview. *Sci. World J.* **2013**, *2013*, 162750. [CrossRef]
59. Thaipong, K.; Boonprakob, U.; Crosby, K.; Cisneros-Zevallos, L.; Byrne, D.H. Comparison of ABTS, DPPH, FRAP, and ORAC assays for estimating antioxidant activity from guava fruit extracts. *J. Food Compos. Anal.* **2006**, *19*, 669–675. [CrossRef]
60. Sharma, P.; Gujral, H.S. Effect of sand roasting and microwave cooking on antioxidant activity of barley. *Food Res. Int.* **2011**, *44*, 235–240. [CrossRef]
61. Platat, C.; Habib, H.; Hashim, I.B.; Kamal, H.; AlMaqbali, F.; Souka, U.; Ibrahim, W.H. Production of functional pita bread using date seed powder. *J. Food Sci. Technol.* **2015**, *52*, 6375–6384. [CrossRef]
62. Ambigaipalan, P.; Shahidi, F. Date seed flour and hydrolysates affect physicochemical properties of muffin. *Food Biosci.* **2015**, *12*, 54–60. [CrossRef]
63. Almana, H.; Mahmoud, R. Palm date seeds as an alternative source of dietary fiber in Saudi bread. *Ecol. Food Nutr.* **1994**, *32*, 261–270. [CrossRef]
64. Bouaziz, F.; Ben Abdeddayem, A.; Koubaa, M.; Ghorbel, R.E.; Chaabouni, S.E. Date Seeds as a Natural Source of Dietary Fibers to Improve Texture and Sensory Properties of Wheat Bread. *Foods* **2020**, *9*, 737. [CrossRef] [PubMed]
65. Mrabet, A.; Hammadi, H.; Rodríguez-Gutiérrez, G.; Jimenez-Araujo, A.; Sindic, M. Date Palm Fruits as a Potential Source of Functional Dietary Fiber: A Review. *Food Sci. Technol. Res.* **2019**, *25*, 1–10. [CrossRef]
66. Vaher, M.; Matso, K.; Levandi, T.; Helmja, K.; Kaljurand, M. Phenolic compounds and the antioxidant activity of the bran, flour and whole grain of different wheat varieties. *Procedia Chem.* **2010**, *2*, 76–82. [CrossRef]
67. Kaur, M.; Singh, V.; Kaur, R. Effect of partial replacement of wheat flour with varying levels of flaxseed flour on physicochemical, antioxidant and sensory characteristics of cookies. *Bioact. Carbohydr. Diet. Fibre* **2017**, *9*, 14–20. [CrossRef]
68. Grasso, S.; Omoarukhe, E.; Wen, X.; Papoutsis, K.; Methven, L. The use of upcycled defatted sunflower seed flour as a functional ingredient in biscuits. *Foods* **2019**, *8*, 305. [CrossRef] [PubMed]
69. Bölek, S. Effects of waste fig seed powder on quality as an innovative ingredient in biscuit formulation. *J. Food Sci.* **2020**, *86*, 55–60. [CrossRef] [PubMed]
70. Grasso, S.; Pintado, T.; Pérez-Jiménez, J.; Ruiz-Capillas, C.; Herrero, A.M. Characterisation of muffins with upcycled sunflower flour. *Foods* **2021**, *10*, 426. [CrossRef] [PubMed]
71. Soto, M.U.R.; Brown, K.; Ross, C.F. Antioxidant activity and consumer acceptance of grape seed flour-containing food products. *Int. J. Food Sci. Technol.* **2011**, *47*, 592–602. [CrossRef]
72. Aksoylu, Z.; Çağındı, Ö.; Köse, E. Effects of blueberry, grape seed powder and poppy seed incorporation on physicochemical and sensory properties of biscuit. *J. Food Qual.* **2015**, *38*, 164–174. [CrossRef]
73. Nascimento, K.D.O.D.; Paes, S.D.N.D.; Augusta, I.M. A review 'clean labeling': Applications of natural ingredients in bakery products. *J. Food Nutr. Res.* **2018**, *6*, 285–294. [CrossRef]
74. Gbenga-Fabusiwa, F.J.; Oladele, E.P.; Oboh, G.; Adefegha, S.A.; Oshodi, A.A. Polyphenol contents and antioxidants activities of biscuits produced from ginger-enriched pigeon pea-wheat composite flour blends. *J. Food Biochem.* **2018**, *42*, e12526. [CrossRef]

*foods*

MDPI

*Article*

# New Advances in the Phenolic Composition of Tiger Nut (*Cyperus esculentus* L.) By-Products

**María del Carmen Razola-Díaz [1,2], Ana María Gómez-Caravaca [2,3], Eduardo J. Guerra-Hernández [1], Belén Garcia-Villanova [1] and Vito Verardo [1,2,*]**

1 Department of Nutrition and Food Science, Campus of Cartuja, University of Granada, 18071 Granada, Spain; carmenrazola@ugr.es (M.d.C.R.-D.); ejguerra@ugr.es (E.J.G.-H.); belenv@ugr.es (B.G.-V.)
2 Institute of Nutrition and Food Technology 'José Mataix', Biomedical Research Centre, University of Granada, Avd. Conocimiento s/n, 18100 Granada, Spain; anagomez@ugr.es
3 Department of Analytical Chemistry, Faculty of Sciences, University of Granada, Avd. Fuentenueva s/n, 18071 Granada, Spain
* Correspondence: vitoverardo@ugr.es

**Abstract:** "Horchata" is a well-known Spanish beverage obtained from pressing tiger nuts. Its by-product is a potential source of sugar and fiber but also contains polyphenols; thus, it could be used as a new ingredient in the food industry. The aim of this work is to determine the phenolic compounds and compare the phenolic profile of two tiger nut by-products. A Box–Behnken design has been carried out to optimize the extraction of phenolic compounds from tiger nut by-products by ultrasound technology. The independent factors were time (min), ethanol/water (% *v/v*), and solvent/sample ratio (*v/w*). The model was validated and confirmed by ANOVA. A Protected Designation of Origin (PDO) of Valencia and a non-Protected Designation of Origin (n-PDO) tiger nut by-products were extracted under the optimal conditions and were characterized by HPLC-DAD-ESI-TOF-MS (High Performance Liquid Chropatography coupled to a photodiode array time-of-flight mass detector). Moreover, their antioxidant activities measured by three different methods (DPPH (2,2-diphenyl-1-picrylhydrazyl), ABTS (2,2′-Azinobis [3-ethylbenzothiazoline-6-sulfonic acid]-diammonium salt) and FRAP (ferric reducing antioxidant power)) were compared. A total of 45 polar compounds were identified, and the phenolic ones were quantified, some of them for the first time. PDO tiger nut by-product has been demonstrated to be richer in phenolic acids and other polyphenols and has higher antioxidant activity; meanwhile, n-PDO tiger nut by-product is richer in phenol precursors.

**Keywords:** ultrasound-assisted extraction; phenolic compounds; antioxidant activity; chufa; HPLC-ESI-TOF-MS

**Citation:** Razola-Díaz, M.d.C.; Gómez-Caravaca, A.M.; Guerra-Hernández, E.J.; Garcia-Villanova, B.; Verardo, V. New Advances in the Phenolic Composition of Tiger Nut (*Cyperus esculentus* L.) By-Products. *Foods* **2022**, *11*, 343. https://doi.org/10.3390/foods11030343

Academic Editor: Victor Rodov

Received: 4 December 2021
Accepted: 20 January 2022
Published: 25 January 2022



## 1. Introduction

Tiger nut (*Cyperus esculentus* L.) is a tuber mainly used to obtain a tiger nut beverage with a milky appearance, called "horchata". It is typically from Spain, where the horchata production industry supposes €60 million per year. The by-products generated from this industry are up to 60% of the tiger nut used, which equals 1.8 million kg per year. Tiger nut by-products are mainly used for animal feed or as organic matter for combustion [1], as a carbon source for the growth of probiotic bacteria [2], or as a sugar source for microalgae [3]. Besides, there are several studies that have evaluated the potential of this by-product as a source of fiber to enrich meat products such as pork [4], beef [5] burgers, and pork sausages [6], to enhance cooking performance, moisture and fat retention. However, there are only a few studies concerning the potential use of this by-product for food and nutraceutical industries. Some studies have evaluated enzyme pretreatments combined with high pressure [7] or different mixtures of solvents at different temperatures [8] to extract phenolic compounds from tiger nut by-product. Nevertheless, they only used spectrophotometric methods to quantify total phenolic compounds, but individual components were not quantified nor chemically characterized. More recent studies have characterized the phenolic profile of the tiger nut by-product extracted by conventional extraction [9] or of the tiger nut by-product oil by supercritical fluid extraction with $CO_2$ [10]. Clemente-Villalba et al. (2021) [11] have recently published a comparison of the sensory profile, volatile composition, and consumer's acceptance of PDO and n-PDO tiger nut milks. Razola Díaz et al. (2020) [3] compared the chemical composition and sugar content in PDO and n-PDO tiger nut by-products. Thus, the objective of the present study is to compare, for the first time, the phenolic profiles of two tiger nut by-products from different origins (PDO and n-PDO) that were obtained during production at an industrial scale. The extraction of phenolic compounds was optimized using ultrasound technology with a Box–Behnken design. All the polar compounds were tentatively identified by HPLC-ESI-TOF-MS, and the phenolics were quantified. The antioxidant activity of both by-products was measured by three different methods (DPPH, ABTS, and FRAP).

## 2. Materials and Methods

### 2.1. Chemicals and Samples

HPLC (High Performance Liquid Chromatography)-grade water and other reagents and solvents were purchased from Merck KGaA (Darmstadt, Germany).

Tiger nut by-product samples were provided by Puleva company located in Granada, Spain, in May 2019. Two types of samples were collected, one from Protected Designation of Origin (PDO) of Valencia and another one without appellation of origin (non-Protected Designation of Origin (n-PDO)) from Ivory Coast. The PDO by-product was obtained after three successive pressings of the tiger nut, while the n-PDO by-product was obtained after only two presses due to its original composition and the final target product of the company as the non-PDO tiger nut was richer in fat. The samples were dried in a ventilated oven at 40 °C until obtaining 5.57 ± 0.29 and 5.46 ± 0.67 % of humidity in the PDO and n-PDO tiger nut by-product, respectively, and milled and sieved to 100 μm; after that, they were frozen at −18 °C until the analyses.

### 2.2. Experimental Design

A Box–Behnken design composed of 15 experiments in three levels (−1,0,1) was established to optimize the extraction of phenolic compounds from tiger nut by-products. All the parameters of the model were established according to the previous experience of the group and previous trials. The independent variables were time (5, 45 and 85 min), ratio ethanol/water (0, 50 and 100% *v/v*), and ratio solvent/sample (20, 40 and 100 *v/w*), and the response was the total phenolic content measured using the Folin–Ciocalteu method. The model was fitted to a second-order polynomial equation, and its adjustment was evaluated and confirmed by ANOVA. To select the optimal conditions, response surface methodology (RSM) was used. All the data was processed using STATISTICA 7.0 (2002, StatSoft, Tulsa, OK, USA).

### 2.3. Ultrasound Bath Extraction

Briefly, 0.125 g of tiger nut by-product powder was dissolved in a 10 mL solution of ethanol/water 40/60, *v/v*. The mixture was placed in an ultrasonic bath for 50 min, and then it was centrifuged for 10 min at 9000 rpm. The supernatant was collected, evaporated, and reconstituted in 1 mL of methanol/water (50:50, *v/v*). The final extracts were filtered with regenerated cellulose filters 0.2 μm (Millipore, Bedford, MA, USA) and stored at −18 °C until the analyses.

### 2.4. Determination of Total Phenolic Content (TPC)

Folin–Ciocalteu spectrophotometric method was used to determine the TPC in tiger nut by-product [12]. Thus, 100 μL of the extract was added to 500 μL of the Folin–Ciocalteu reagents and 6 mL of bi-distilled water. The flask was agitated for a minute. After that, 2 mL of 15% (*w/v*) $Na_2CO_3$ and made up to 10 mL with bi-distilled water. The flasks were kept in darkness. After 2 h, the measures were carried out at 750 nm and 25 °C with a UV-visible spectrophotometer (Spectrophotometer 300 Array, UV-Vis, single beam, Shimadzu, Duisburg, Germany). The calibration curve was developed with Gallic acid from 1 to 100 ppm, and the obtained equation was y = 0.0012x + 0.0227 ($R^2$ = 0.9995). Results are expressed as µg gallic acid equivalents (GAE)/g dry weight (d.w.).

### 2.5. Determination of Polar Compounds by HPLC-ESI-QTOF-MS

The phenolic profile characterization and quantification of the two tiger nut by-products extracted by the optimal conditions were performed according to a previously described method [13]. The analyses were carried out in duplicate on an ACQUITY Ultra Performance LC system (Waters Corporation, Milford, MA, USA) coupled to an electrospray ionization (ESI) source operating in the negative mode and a time-of-flight (TOF) mass detector (Waters Corporation, Milford, MA, USA). The compounds of interest were separated on an ACQUITY UPLC BEH Shield RP18 column (1.7 µm, 2.1 mm × 100 mm; Waters Corporation, Milford, MA, USA) at 40 °C using a gradient previously stated by Verni et al. [13] using water containing 1% acetic acid as mobile phase A and acetonitrile as mobile phase B. The gradient was: from 0 to 2.3 min, 1% B; 4.4 min, 7% B; 8.1 min, 14% B; 12.2 min, 24% B; 16 min, 40% B; 18.3 min, 100% B, 21 min, 100% B; 22.4 min, 1% B; 25 min, 1% B. Flow rate was established to 0.6 mL/min. The volume injection was 2 µL.

Finally, external calibration curves were prepared for the quantification of phenolic compounds: vanillic acid (y = 99.369x + 40.864, $R^2$ = 0.9995), ferulic acid (y = 173.03x + 64.326, $R^2$ = 0.9958), rutin (y = 1137.8x + 74.815, $R^2$ = 0.9943), quercetin (y = 1272.5x + 84.49, $R^2$ = 0.9938), and catechin (y = 317.69x + 105.65, $R^2$ = 0.9955). The data were elaborated using MassLynx 4.1 software (Waters Corporation, Milford, MA, USA).

### 2.6. Antioxidant Assays in Tiger Nut By-Products

The antioxidant capacity in the two tiger nut by-products has been evaluated in the extract obtained by the optimal ultrasound bath conditions using three different methods.

The ABTS method was carried out according to Re et al. (1999) [14]. The monocation $ABTS^{\bullet+}$ is generated by oxidation of the ABTS with potassium persulfate in the dark at room temperature for 12–24 h. For each extract, 1 mL of this ABTS solution was added to 0.01 mL of the extract and the detriment of absorbance during 30 min at 734 nm was measured.

The DPPH radical scavenging activity was assayed with a method proposed by several authors [15,16]. In total, 100 µL of each extract was added to 2.9 mL of DPPH, and after rapid stirring, the bleaching power of the extract was observed in a time interval from 0 to 30 min at 517 nm.

The FRAP assay was carried out following the procedure developed by Pulido et al. (2000) [17]. It is based on the reduction of $Fe^{3+}$ to $Fe^{2+}$ by the antioxidant substances. A total of 30 µL of each extract was added to 90 µL of distilled water and 900 µL of the FRAP reagent. It was kept for 30 min at 37 °C and measured in the spectrophotometer at 595 nm.

Standard curves of Trolox equivalents (TE) (1, 5, 10, 20, 50, 80, 100, 150, 200 ppm) were elaborated for each assay and the equations obtained were y = 0.0009x + 0.0258 ($R^2$ = 0.9971), y = 0.0026x + 0.0374 ($R^2$ = 0.9963), and y = 0.0039x + 0.0915 ($R^2$ = 0.9972), for the ABTS, DPPH, and FRAP assays, respectively. Results are expressed as mg TE/g d.w.

## 3. Results and Discussion

### 3.1. Fitting the Model

The extraction step is the most important to reach the highest amount of the target compounds, in this case, phenolic compounds. A Box–Behnken design coupled to RSM was used to find the optimal conditions of time ($X_1$), ethanol/water ratio ($X_2$), and solvent/sample ratio ($X_3$) to extract phenolic compounds from the tiger nut by-product using ultrasound bath technology. The experimental values of TPC obtained for each run are shown in Table 1.

The lowest recovery (42.80 ± 1.08 μg GAE/g d.w.) was at 45 min, with the ethanol/water 100% and ratio solvent/sample 20 *v/w*, and the highest (383.11 ± 1.76 μg GAE/g d.w.) at 85 min, was ethanol/water 50% and ratio solvent/sample 100 *v/w*. This by-product has a low relation weight/volume (<1); therefore, those experiments carried out with the lower value of ratio solvent/sample led to the lowest results due to the saturation of the solvent and consequently the reduced contact surface between sample and solvent. Similar results were found when the tendency using ethanol was 100%; therefore, it demonstrates that some water is needed as surfactant for the extraction of phenolic compounds to satisfactorily take place.

**Table 1.** Box–Behnken design with natural and coded values (parenthesis) of the conditions of extraction and the experimental results obtained for total phenolic content (TPC) expressed with the average and the standard deviation.

| Run | Independent Factors | | | Response |
|---|---|---|---|---|
| | $X_1$ | $X_2$ | $X_3$ | TPC (μg GAE/g d.w.) |
| 1 | 5 (−1) | 0 (−1) | 40 (0) | 124.02 ± 1.60 |
| 2 | 85 (1) | 0 (−1) | 40 (0) | 142.00 ± 1.46 |
| 3 | 5 (−1) | 100 (1) | 40 (0) | 57.95 ± 0.91 |
| 4 | 85 (1) | 100 (1) | 40 (0) | 93.64 ± 0.40 |
| 5 | 5 (−1) | 50 (0) | 20 (−1) | 135.34 ± 1.44 |
| 6 | 85 (1) | 50 (0) | 20 (−1) | 100.73 ± 1.74 |
| 7 | 5 (−1) | 50 (0) | 100 (1) | 278.72 ± 1.67 |
| 8 | 85 (1) | 50 (0) | 100 (1) | 383.11 ± 1.76 |
| 9 | 45 (0) | 0 (−1) | 20 (−1) | 77.80 ± 1.11 |
| 10 | 45 (0) | 100 (1) | 20 (−1) | 42.80 ± 1.08 |
| 11 | 45 (0) | 0 (−1) | 100 (1) | 281.54 ± 1.48 |
| 12 | 45 (0) | 100 (1) | 100 (1) | 221.53 ± 1.12 |
| 13 | 45 (0) | 50 (0) | 40 (0) | 305.47 ± 1.06 |
| 14 | 45 (0) | 50 (0) | 40 (0) | 291.97 ± 1.93 |
| 15 | 45 (0) | 50 (0) | 40 (0) | 298.69 ± 1.75 |

$X_{1\text{-}3}$: Time (min), ethanol/water (% *v/v*), and solvent/sample ratio (*v/w*).

The experimental data were adjusted to a second-order polynomial equation, and all the estimated regression effects are shown in Table 2. The model was analyzed with a significance level of $p < 0.05$, and all the lineal terms ($\beta_1$, $\beta_2$, and $\beta_3$), all the quadratic terms ($\beta_{11}$, $\beta_{22}$ and $\beta_{33}$), and the crossed values between time ($X_1$) and ratio solvent/sample ($X_3$), had significance, but the rest of the terms were discarded. Therefore, the model was recalculated, and the ANOVA test was performed. As shown in Table 2 the model revealed a high correlation coefficient ($R^2 = 0.9891$), a significant regression model ($p < 0.05$), and a non-significant lack of fit ($p > 0.05$). According to Bezerra et al. [18], the adequacy of the model was confirmed.

Optimal conditions were selected using RSM concerning the three-dimensional graphs presented in Figure 1. A compromise has been made between the independent factors to establish them at the minimal possible values. The optimal conditions chosen were: 50 min, 40% ethanol/water, and ratio solvent/sample 80 *v/w* that gave a predicted value of 403.61 ± 49.54 μg GAE/g d.w. The coefficient of variation between the obtained (400.43 ± 4.63) and predicted values was lower than 1%, so the model was validated.

The optimal result obtained was better than the one reported by Roselló-Soto et al. [8] that used conventional extraction with a similar solvent (ethanol 25%) but with differences in terms of time (3 h) and temperature (60 °C). Therefore, ultrasound technology has been demonstrated to be a non-thermal and lower time-consuming technique that allows the satisfactorily recovery of phenolic compounds from tiger nut by-products. However, the authors reported slightly higher results using 50% ethanol, 35 °C, and pH 2.5 for 3 h [9]. In our case, the pH has not been changed in order to avoid the hydrolysis of bound compounds.

**Table 2.** Estimated regression effects of the fitted second-order polynomial equation and ANOVA of the fitted model.

| Regression Coefficients | Response | | | |
|---|---|---|---|---|
| | Effect | Standard Error | $t$-Value | $p$-Value |
| $\beta_0$ * | 178.4055 | 1.9924 | 89.5448 | 0.0001 |
| Lineal | | | | |
| $\beta_1$ * | 46.8713 | 5.0403 | 9.2993 | 0.0114 |
| $\beta_2$ * | −52.3614 | 4.7817 | −10.9505 | 0.0082 |
| $\beta_3$ * | 201.9806 | 4.7817 | 42.2407 | 0.0006 |
| Quadratic | | | | |
| $\beta_{11}$ * | 62.9065 | 3.5192 | 17.8752 | 0.0031 |
| $\beta_{22}$ * | 131.3867 | 3.5192 | 37.3342 | 0.0007 |
| $\beta_{33}$ * | 61.8885 | 3.7167 | 16.6515 | 0.0036 |
| Crossed | | | | |
| $\beta_{12}$ | 8.8587 | 6.7623 | 1.3100 | 0.3204 |
| $\beta_{13}$ * | 63.7294 | 6.3755 | 9.9959 | 0.0099 |
| $\beta_{23}$ | −12.5084 | 6.7623 | −1.8497 | 0.2056 |
| $R^2$ | | 0.9891 | | |
| $p$ model | | 0.0000 | | |
| $p$ lack of fit | | 0.2304 | | |

*: Significant at 0.05.

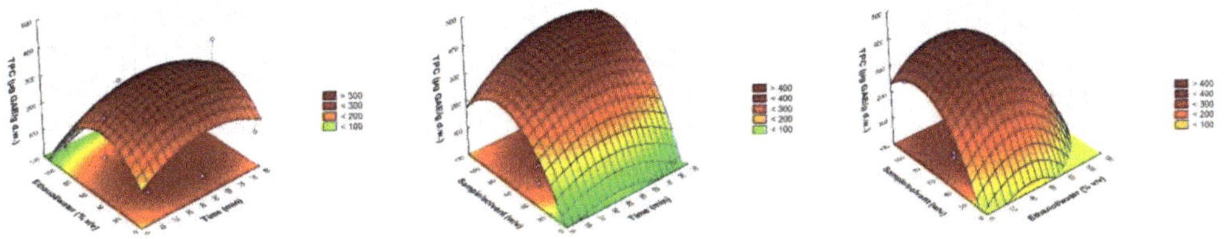

**Figure 1.** Response surface plots showing combined effects of process variables.

### 3.2. Identification of Polar Compounds by HPLC-ESI-TOF-MS

The samples of PDO and n-PDO tiger nut by-products were characterized by HPLC-MS, and 44 compounds were identified, 18 were phenolic compounds and 26 other polar compounds were identified as hydroxyl fatty acids. To our knowledge, a total of 27 polar compounds were identified for the first time in tiger nut by-products.

All the identified compounds are described in Table 3 with their retention time, molecular formula, experimental and calculated $m/z$ fragments, and score and error (ppm). To ensure mass accuracy, the tolerances chosen had a score higher than 90% and an error lower than 5 ppm (part per million) between the experimental and calculated $m/z$. To identify the compounds, the generated molecular formula and some in source fragments were checked and studied also comparing with different databases such as PubChem, Phenol-Explorer, and the literature.

Five phenolic acids were identified at times 1.34, 2.57, 3.88, 5.51, and 5.73 min (peaks 4, 6, 7, 9, and 19). They were named $p$-hydroxybenzoic acid, vanillic acid, ethyl vanillin, ferulic acid, and $p$-coumaric acid, respectively. In addition, peaks 2, 8, 15, and 21 were assigned to scopoletin (a coumarin), 4-vinylphenol (a hydroxy styrene), sinensetin (a methylated flavone), and cyanidin (an anthocyanidin), with $m/z$ 191, 119, 371, and 286, respectively. These compounds are in agreement with Roselló-Soto et al. (2019) [9,10] who previously described them in tiger nut by-product. As they described some vanillin and $p$-coumaric acid derivates, the compounds found at 9.47 and 12.67 min (peaks 13 and 17) were named dehydrovanillin and $p$-coumaric acid ethyl ester, respectively. Overall vanillic and $p$-coumaric acid derivates described here were attributed to have good bioavailability and bio-accessibility and to reduce the risk of developing several disorders such as cancer and cardiovascular diseases [19].

**Table 3.** Polar compounds tentatively identified by HPLC-ESI-TOF-MS.

| No. | Compound | Retention Time (min) | Molecular Formula | *m*/*z* Experimental | *m*/*z* Calculated | Fragments | Score (%) | Error (ppm) |
|---|---|---|---|---|---|---|---|---|
| 1 | 2-*O*-Galloyl-1,4-galactarolactone | 0.274 | $C_{13}H_{12}O_{11}$ | 343.0308 | 343.0301 | 201.0249 | 90.7 | 2.0 |
| 2 | Scopoletin | 0.520 | $C_{10}H_8O_4$ | 191.1686 | 191.1680 | 111.0071; 174.0401; 160.8401 | 100 | −3.1 |
| 3 | Imbricantonol | 1.155 | $C_{26}H_{20}O_7$ | 443.1111 | 443.1131 | 214.9512; 229.0186; 570.0947 | 93.9 | −4.5 |
| 4 | *p*-hydroxybenzoic acid | 1.341 | $C_7H_6O_3$ | 137.0234 | 137.0239 | - | 100 | −3.6 |
| 5 | L-leucic acid | 1.916 | $C_6H_{12}O_3$ | 131.0702 | 131.0708 | 85.0646 | 100 | −4.6 |
| 6 | Vanillic acid | 2.570 | $C_8H_8O_4$ | 167.0338 | 167.0344 | - | 100 | −3.6 |
| 7 | Ethyl vanillin | 3.877 | $C_9H_{10}O_3$ | 165.0556 | 165.0552 | 151.8809; 136.9292 | 100 | 2.4 |
| 8 | 4-vinylphenol | 4.643 | $C_8H_8O$ | 119.0492 | 119.0497 | - | 100 | −4.2 |
| 9 | Ferulic acid | 5.515 | $C_{10}H_{10}O_4$ | 193.0504 | 193.0501 | 134.0359; 166.9101 | 100 | 1.6 |
| 10 | *p*-coumaric acid | 5.734 | $C_9H_8O_3$ | 163.0403 | 163.0395 | 119.0499; 117.0318; 149.0263 | 100 | 4.9 |
| 11 | 3-Hydroxyphloretin 2′-*O*-glucoside (aspalathin) | 7.505 | $C_{21}H_{24}O_{11}$ | 451.3276 | 451.3271 | 225.1289; 337.1716; 433.2887; 291.9893; 189.9602; | 100 | 1.1 |
| 12 | Kaempferol 3,7-diglucoside, sophoraflavonoloside or luteolin-7,3′-di-*O*-glucoside | 8.246 | $C_{27}H_{30}O_{16}$ | 609.1453 | 609.1456 | 297.0602; 153.0205; 507.1086; 285.0299; 447.0884 | 93.4 | −0.5 |
| 13 | Dehydrodivanillin | 9.467 | $C_{16}H_{14}O_6$ | 301.0698 | 301.0712 | 286.0476; 166.9093; 215.0308; 239.0385 | 99.9 | −4.7 |
| 14 | Veronicafolin 3-glucosyl-(1->3)-galactoside | 9.652 | $C_{30}H_{36}O_{18}$ | 683.1799 | 683.1823 | 721.1418; 593.1249; 563.1607 | 94.7 | −3.5 |
| 15 | Sinensetin | 10.368 | $C_{20}H_{20}O_7$ | 371.1117 | 371.1131 | 175.0426; 193.0490; 145.0287; 161.0240; 161.9289 | 97.6 | −3.8 |
| 16 | Sinapyl alcohol | 11.642 | $C_{11}H_{14}O_4$ | 209.0807 | 209.0814 | 193.0477; 175.0416 | 100 | −3.3 |
| 17 | *p*-Coumaric acid ethyl ester | 12.672 | $C_{11}H_{12}O_3$ | 191.0699 | 191.0708 | 165.0252; 195.0582; 119.0503; 116.9892; 179.0338; | 100 | −4.7 |
| 18 | Trihydroxy octadecenoic acid isomer a | 14.400 | $C_{18}H_{34}O_5$ | 329.2334 | 329.2328 | 229.1451; 211.1347; 171.1032; 183.1397 | 99.8 | 1.8 |
| 19 | Trihydroxy octadecenoic acid isomer b | 14.550 | $C_{18}H_{34}O_5$ | 329.2332 | 329.2328 | 229.1451; 211.1347; 171.1032; 183.1397 | 95.3 | 1.2 |
| 20 | Trihydroxy octadecenoic acid isomer c | 15.295 | $C_{18}H_{34}O_5$ | 329.2325 | 329.2328 | 229.1451; 211.1347; 171.1032; 183.1397 | 99.6 | −0.9 |
| 21 | Cyanidin | 15.853 | $C_{15}H_{11}O_6$ | 286.2408 | 286.2400 | 265.0123; 116.1101 | 92.3 | 0.7 |
| 22 | Benzoic acid | 16.131 | $C_{19}H_{20}O_7$ | 359.1118 | 359.1131 | 311.1693; 183.0196; 163.0394; 149.0360 | 98.4 | −3.6 |
| 23 | Dihydroxyoleic acid acid isomer a | 16.487 | $C_{18}H_{34}O_4$ | 313.237 | 313.2379 | 183.1375;295.2259; 269.0670 | 100 | −2.9 |
| 24 | Dihydroxyoleic acid isomer b | 16.570 | $C_{18}H_{34}O_4$ | 313.2374 | 313.2379 | 201.1127; 223.0940; 171.1013 | 100 | −1.6 |
| 25 | Dihydroxyoleic acid isomer c | 16.830 | $C_{18}H_{34}O_4$ | 313.2367 | 313.2379 | 171.1015; 277.2166; 295.2282 | 100 | −3.8 |
| 26 | Dihydroxyoleic acid isomer d | 16.917 | $C_{18}H_{34}O_4$ | 313.2376 | 313.2379 | 157.0854; 171.1016; 187.0971 | 100 | −1.0 |
| 27 | Dihydroxystearic acid | 16.975 | $C_{18}H_{36}O_4$ | 315.2531 | 315.2535 | 297.2422; 279.2317 | 100 | −1.3 |
| 28 | Hydroxylinoleic acid isomer a | 17.091 | $C_{18}H_{32}O_3$ | 295.2263 | 295.2273 | 277.2172; 187.0946; 171.1006 | 100 | −3.4 |
| 29 | Hydroxypalmitic acid | 17.120 | $C_{16}H_{32}O_3$ | 271.2262 | 271.2273 | 187.0974; 152.9931; 125.0954 | 100 | −4.1 |
| 30 | Hydroxylinoleic acid isomer b | 17.202 | $C_{18}H_{32}O_3$ | 295.2266 | 295.2273 | 277.2162; 171.1014; 195.1375 | 100 | −2.4 |
| 31 | Hydroxyoleic acid isomer a | 17.302 | $C_{18}H_{34}O_3$ | 297.2418 | 297.2430 | 279.2306 | 100 | −4.0 |
| 32 | Hydroxyoleic acid isomer b | 17.339 | $C_{18}H_{34}O_3$ | 297.2418 | 297.2430 | 279.2319 | 100 | −4.0 |

**Table 3.** *Cont.*

| No. | Compound | Retention Time (min) | Molecular Formula | $m/z$ Experimental | $m/z$ Calculated | Fragments | Score (%) | Error (ppm) |
|---|---|---|---|---|---|---|---|---|
| 33 | Hydroxyoleic acid isomer c | 17.384 | $C_{18}H_{34}O_3$ | 297.2421 | 297.2430 | 279.2316 | 100 | −3.0 |
| 34 | Hydroxylinoleic acid isomer c | 17.442 | $C_{18}H_{32}O_3$ | 295.2263 | 295.2273 | 277.2166; 171.1017 | 100 | −3.4 |
| 35 | Hydroxystearic acid | 17.538 | $C_{18}H_{36}O_3$ | 299.2575 | 299.2586 | 281.2489; 253.2527 | 96.7 | −3.7 |
| 36 | Hydroxyoleic acid isomer d | 17.621 | $C_{18}H_{34}O_3$ | 297.2422 | 297.2430 | 279.2321 | 100 | −2.7 |
| 37 | Linolenic acid | 17.753 | $C_{18}H_{30}O_2$ | 277.2156 | 277.2168 | 279.2310; 255.2310 | 100 | −4.3 |
| 38 | Myristic acid | 17.782 | $C_{14}H_{28}O_2$ | 227.2001 | 227.2011 | 152.9942; 209.0703 | 100 | −4.4 |
| 39 | Palmitoleic acid | 17.865 | $C_{16}H_{30}O_2$ | 253.2161 | 253.2168 | 152.9945 | 100 | −2.8 |
| 40 | Linoleic acid | 17.923 | $C_{18}H_{32}O_2$ | 279.2319 | 279.2324 | 152.9952; 255.2314; 241.0059 | 100 | −1.8 |
| 41 | Methylpalmitic acid | 18.009 | $C_{17}H_{32}O_2$ | 267.2319 | 267.2324 | 255.2336; 253.2151; 152.9937 | 96.2 | −1.9 |
| 42 | Palmitic acid | 18.088 | $C_{16}H_{32}O_2$ | 255.2316 | 255.2324 | 152.9935 | 99.6 | −3.1 |
| 43 | Oleic acid | 18.121 | $C_{18}H_{34}O_2$ | 281.2472 | 281.2481 | 255.2308; 152.9945 | 99.2 | −3.2 |
| 44 | Heptadecanoic acid | 18.216 | $C_{17}H_{34}O_2$ | 268.2473 | 268.2481 | 269.2477; 255.2326; 152.9957 | 98.7 | −3.0 |
| 45 | Stearic acid | 18.344 | $C_{18}H_{36}O_2$ | 283.2629 | 283.2637 | 255.2336; 279.2322; 152.9945 | 99.3 | −2.8 |

At time 8.25 min (peak 8), the compound kaempferol 3,7-diglucoside has been identified according to its mass fragments 285 and 447. This compound could also be named sophoraflavonoloside or luteolin-7,3′-di-*O*-glucoside. Peak 1 had been previously detected but not quantified in fruits. It was named 2-*O*-galloyl-1,4-galactarolactone [20]. With the fragment *m/z* 85 corresponding with peak 5, L-leucic acid was tentatively identified. At time 1.16 (peak 3), with molecular formula $C_{26}H_{20}O_7$ and the fragment with *m/z* 229, imbricantonol was found according to Byrne et al. (1987) [21], who found this compound for the first time in *Stypandra glauca* (blind grass). According to the authors, the fragments found for this compound indicate the possible couple of dianellidin (*m/z* 215) with stypandrone (*m/z* 229) and the presence of three hydroxyl substituents (*m/z* 570). Peak 11 was identified as aspalathin (3-hydroxyphloretin 2′-*O*-glucoside), previously described as the major compound in *Aspalathus linearis* (rooibos) by Kreuz et al. (2008) [22]. At *m/z* 683 (peak 14) and fragments at *m/z* 721, 593 and 563, the veronicafolin 3-glucosyl-(1->3)-galactoside has been tentatively identified according to Tan et al. (2020) [23] and previously described in *Clinacanthus nutans*, a popular herbal plant in the Southeast Asian region. Previously some studies have found flavonoids as kaempferol glycosides to have several roles in human health, including antioxidant, anti-inflammatory, antimicrobial, antidiabetic, anticancer, anti-osteoporotic, cardioprotective, anxiolytic, analgesic, and neuroprotective, among others [24].

Sinapyl alcohol and benzoic acid were identified at 11.64 and 16.13 min, corresponding with peaks 16 and 22. Sinapyl alcogol has been described as a precursor to various stilbenes and coumarins, and benzoic acid as a precursor of phenolic acids [25].

The compounds with molecular formula $C_{18}H_{34}O_5$ and fragments at *m/z* 229, 211 and 171 (peaks 18, 19 and 20) were identified as trihydroxy octadecenoic acid isomers. They also could be named pinellic acid isomers, according to Ruan et al. (2019), who described them in *Pluchea indica* aerial parts [26]. Peaks 37–40 and 42–45 were identified as fatty acids, linolenic, myristic, palmitoleic, linoleic, palmitic, oleic, heptadecanoic, and stearic, according to several authors that found them in tiger nut oil [27–29]. Hydroxyoleic acid isomers (peaks 31–33, 35) and dihydroxyoleic acids isomers (peaks 23–26), derived from oleic acid, have also been found. Other derivates were found at different times: Hydroxystearic and dihydroxystearic acids at 17.54 and 16.98 min, hydroxylinoleic acid isomers at 17.09, 17.20, and 17.44 min, and hydroxypalmitic and methylpalmitic acids at 17.12 and 18.00 min. Another fatty acid tentatively identified was trihydroxy octadecenoic acid isomers at peaks 18–20. None of which were previously described in tiger nut by-products.

### 3.3. Quantification of Phenolic Compounds by HPLC-ESI-TOF-MS and Antioxidant Activity

A total of 18 phenolic compounds were quantified, and they are summarized in Table 4.

Comparing the phenolic compounds (Table 4) in both tiger nut by-products, the PDO sample clearly has a higher amount of 2-*O*-galloyl-1,4-galactarolactone, scopoletin, imbricantonol, p-hydroxybenzoic acid, vanillic acid, ethyl vanillin, 4-vinylphenol, ferulic acid, *p*-coumaric acid, and *p*-coumaric acid ethyl ester. On the other hand, n-PDO by-product is sightly richer in cyanidin, veronicafolin 3-glucosyl-(1->3)-galactoside, and the two phenolic compounds precursors, benzoic acid and sinapyl alcohol. Similar amounts of 3-hydroxyphloretin 2′-*O*-glucoside (aspalathin), kaempferol 3,7-diglucoside, and sinensetin were found in both samples.

The phenolic composition, mainly based on phenolic acids, justified the optimal conditions of extraction obtained in this work. In fact, according to Waszkowiak and Gliszczynska-Swigło, the phenolic acid content decreases when ethanol concentration is higher than 60 to 70 % [30]. Similar data were reported by Roselló-Soto et al. [9], obtaining the highest recovery of phenolic acids in tiger nuts using 41.4 % ethanol.

Compared with other authors, we have achieved a higher amount of cyanidin, 4-vinylphenol, ethyl vanillin, and p-coumaric acid than Roselló-Soto et al. [9]. This could be because they used a conventional extraction while we used ultrasound bath assisted extraction; however, the difference between the samples must also be considered. No other references were found in the bibliography regarding phenolic compounds in tiger nut by-products.

**Table 4.** Phenolic compounds and their precursors quantified in PDO and n-PDO tiger nut by-product expressed as mean ± standard deviation (μg/g dry weight (d.w.)).

| No. | Compound | PDO by-Product (μg/g d.w.) | n-PDO by-Product (μg/g d.w.) |
|---|---|---|---|
| 1 | 2-*O*-Galloyl-1,4-galactarolactone | 29.17 ± 2.55 [b] | 5.00 ± 0.28 [a] |
| 2 | Scopoletin | 3.58 ± 0.05 [b] | 1.44 ± 0.03 [a] |
| 3 | Imbricantonol | 0.63 ± 0.04 [b] | <LOQ [a] |
| 4 | *p*-hydroxybenzoic acid | 0.62 ± 0.03 [b] | <LOQ [a] |
| 5 | Vanillic acid | 1.26 ± 0.01 [b] | 0.97 ± 0.02 [a] |
| 6 | Ethyl vanillin | 5.60 ± 0.13 [b] | 0.55 ± 0.01 [a] |
| 7 | 4-vinylphenol | 3.73 ± 0.00 [b] | 0.88 ± 0.02 [a] |
| 8 | Ferulic acid | 1.84 ± 0.02 [b] | 1.19 ± 0.06 [a] |
| 9 | *p*-coumaric acid | 0.77 ± 0.15 [b] | 0.16 ± 0.08 [a] |
| 10 | 3-Hydroxyphloretin 2′-*O*-glucoside (aspalathin) | 0.19 ± 0.03 [a] | 0.18 ± 0.00 [a] |
| 11 | Kaempferol 3,7-diglucoside, sophoraflavonoloside or luteolin-7,3′-di-*O*-glucoside | 0.14 ± 0.01 [a] | 0.10 ± 0.02 [a] |
| 12 | Dehydrodivanillin | 2.27 ± 0.04 [b] | 0.38 ± 0.07 [a] |
| 13 | Veronicafolin 3-glucosyl-(1->3)-galactoside | 0.23 ± 0.00 [a] | 0.31 ± 0.01 [b] |
| 14 | Sinensetin | 0.03 ± 0.00 [a] | 0.03 ± 0.00 [a] |
| 15 | Sinapyl alcohol | 1.37 ± 0.41 [a] | 2.49 ± 0.58 [a] |
| 16 | *p*-Coumaric acid ethyl ester | 0.60 ± 0.19 [a] | <LOQ [a] |
| 17 | Cyanidin | 0.67 ± 0.09 [a] | 1.01 ± 0.03 a |
| 18 | Benzoic acid | 1.37 ± 0.58 [a] | 2.47 ± 0.02 [a] |
| | Sum of phenolic compounds | 54.07 ± 4.33 [b] | 17.16 ± 1.23 [a] |

Different letters (a, b) in the same line indicate significant differences ($p < 0.05$).

In addition, PDO and n-PDO tiger nut by-products were revealed to have antioxidant activity (Table 5), being around 15% higher for the PDO by-products with the three performed methods. Antioxidant activity has been reported previously in tiger nut oils [31] and tiger nut beverages [32]. However, no references were found in tiger nut by-products apart from those reported by Roselló-Soto et al. (2019) [9], but the method used was different (total antioxidant capacity).

**Table 5.** Antioxidant activity of PDO and n-PDO tiger nut by-product expressed as mean ± standard deviation (μg Trolox equivalent (TE)/g dry weight (d.w.)).

| Antioxidant Assay | PDO by-Product (μg/g d.w.) | n-PDO by-Product (μg/g d.w.) |
|---|---|---|
| DPPH | 434.41 ± 4.73 [b] | 356.90 ± 5.56 [a] |
| ABTS | 834.61 ± 9.62 [b] | 726.22 ± 7.27 [a] |
| FRAP | 757.48 ± 6.55 [b] | 618.31 ± 7.89 [a] |

Different letters (a, b) in the same line indicate significant differences ($p < 0.05$).

Roselló-Soto et al. (2018) [33] reported a total of 30 phenolic compounds that were previously found and quantified in tiger nut tubers. Parts of these phenols naturally remian in the horchata after the pressing steps; therefore, in the by-product, fewer compounds remain and in lower amounts. This corroborates the results obtained in this work as 18 phenolic compounds were identified and quantified in tiger nut industrial by-products. Besides, tiger nut by-products have a lower amount of phenolic compounds with lower antioxidant activity than tiger nut oils [31,34] and tiger nut by-product oils [10]. This is mainly because of the oil extraction techniques that hydrolysate the compounds. However, the use of ultrasound technology seems to be a promising technique to apply to tiger nut by-products that allow obtaining phenolic compounds with antioxidant activity. This makes it clear that the PDO tiger nut by-product contains higher amounts of phenolic compounds than the n-PDO tiger nut by-product. This result was realized as tiger nuts cultivated in Valencia (Spain) contain higher amounts of these antioxidants compared to the samples from Ivory Coast that contain more precursors. This difference could be justified with the growth differentiation balance framework. According to this hypothesis, when there are high levels of nitrogen and good environmental conditions, the growth of the plant is favored. On the contrary, if the environmental conditions are not favorable and the availability of essential components is low, the secondary metabolism is favored increasing the phenolic content. However, to confirm this trend, further analyses on the most robust sampling are needed.

## 4. Conclusions

The ultrasound-assisted extraction of phenolic compounds from tiger nut by-products has been established using a Box–Behnken design combined with RSM. The optimal selected conditions were 50 min, 40% ethanol, and ratio solvent/sample 80 *v/w*. The results highlighted that ultrasound technology permits a high recovery of polyphenols from tiger nut by-product. Moreover, PDO and n-PDO tiger nut by-products were compared for the first time according to their phenolic composition and antioxidant activity. Both were characterized by HPLC-MS, and a total of 45 free polar compounds were identified, from which 27 polar compounds were identified for the first time in tiger nut by-products. The predominant compounds were phenolic acids, the major ones being 2-*O*-galloyl-1,4-galactarolactone and vanillic acid derivatives. The quality attributed to the Protected Designation of Origin of Valencia has been confirmed in the PDO tiger nut by-product as it showed higher amounts of phenolic acids and other polyphenols also exhibiting higher antioxidant activity than the n-PDO tiger nut by-product. Further research, with a more robust sampling, could confirm the added value of Protected Designation of Origin of Valencia tiger nut by-products compared to n-PDO options.

To conclude, taking into account that the use of tiger nut flour for gluten-free products was studied, tiger nut by-products could be used as cheaper ingredients/flours for the formulation of bakery products.

**Author Contributions:** Conceptualization, E.J.G.-H. and V.V.; formal analysis, M.d.C.R.-D.; investigation, M.d.C.R.-D.; data curation, V.V., B.G.-V. and A.M.G.-C.; writing—original draft preparation, M.d.C.R.-D.; writing—review and editing, V.V. and A.M.G.-C.; supervision, E.J.G.-H. and V.V.; funding acquisition, V.V. and A.M.G.-C. All authors have read and agreed to the published version of the manuscript.

**Funding:** This research received no external funding.

**Informed Consent Statement:** Not applicable.

**Data Availability Statement:** Not applicable.

**Acknowledgments:** Vito Verardo thanks the Spanish Ministry of Economy and Competitiveness (MINECO) for "Ramon y Cajal" contract (RYC-2015-18795). The authors would like to thank the Puleva company for providing "horchata" by-products.

**Conflicts of Interest:** The authors declare no conflict of interest.

## References

1. Sánchez-Zapata, E.; Fernández-López, J.; Pérez-Alvarez, J.Á. Tiger Nut (*Cyperus esculentus*) commercialization: Health aspects, composition, properties, and food applications. *Compr. Rev. Food Sci. Food Saf.* **2012**, *11*, 366–377. [CrossRef]
2. Sánchez-Zapata, E.; Fernández-López, J.; Pérez-Alvarez, J.A.; Soares, J.; Sousa, S.; Gomes, A.M.P.; Pintado, M.M.E. In vitro evaluation of "horchata" co-products as carbon source for probiotic bacteria growth. *Food Bioprod. Process.* **2013**, *91*, 279–286. [CrossRef]
3. Razola-Díaz, M.d.C.; Verardo, V.; Martín-García, B.; Díaz-De-Cerio, E.; García-Villanova, B.; Guerra-Hernández, E.J. Establishment of acid hydrolysis by box-behnken methodology as pretreatment to obtain reducing sugars from tiger nut byproducts. *Agronomy* **2020**, *10*, 477. [CrossRef]

4. Sánchez-Zapata, E.; Muñoz, C.M.; Fuentes, E.; Fernández-López, J.; Sendra, E.; Sayas, E.; Navarro, C.; Pérez-Alvarez, J.A. Effect of tiger nut fibre on quality characteristics of pork burger. *Meat Sci.* **2010**, *85*, 70–76. [CrossRef] [PubMed]
5. Bobreneva, I.V.; Baioumy, A.A. Effect of using tiger nuts (*Cyperus esculentus*) on nutritional and organoleptic characteristics of beef burger. *Biosci. Res.* **2018**, *15*, 1424–1432.
6. Sánchez-Zapata, E.; Zunino, V.; Pérez-Alvarez, J.A.; Fernández-López, J. Effect of tiger nut fibre addition on the quality and safety of a dry-cured pork sausage ("Chorizo") during the dry-curing process. *Meat Sci.* **2013**, *95*, 562–568. [CrossRef] [PubMed]
7. Ezeh, O.; Niranjan, K.; Gordon, M.H. Effect of enzyme pre-treatments on bioactive compounds in extracted tiger nut oil and sugars in residual meals. *J. Am. Oil Chem. Soc.* **2016**, *93*, 1541–1549. [CrossRef]
8. Roselló-Soto, E.; Barba, F.J.; Putnik, P.; Kovačević, D.B.; Lorenzo, J.M.; Cantavella-Ferrero, Y. Enhancing bioactive antioxidants' extraction from "horchata de chufa" by-products. *Foods* **2018**, *7*, 161. [CrossRef]
9. Roselló-Soto, E.; Martí-Quijal, F.J.; Cilla, A.; Munekata, P.E.S.; Lorenzo, J.M.; Remize, F.; Barba, F.J. Influence of temperature, solvent and pH on the selective extraction of phenolic compounds from tiger nuts by-products: Triple-TOF-LC-MS-MS characterization. *Molecules* **2019**, *24*, 797. [CrossRef]
10. Roselló-Soto, E.; Barba, F.J.; Lorenzo, J.M.; Munekata, P.E.S.; Gómez, B.; Moltó, J.C. Phenolic profile of oils obtained from "horchata" by-products assisted by supercritical-$CO_2$ and its relationship with antioxidant and lipid oxidation parameters: Triple TOF-LC-MS-MS characterization. *Food Chem.* **2019**, *274*, 865–871. [CrossRef]
11. Clemente-Villalba, J.; Cano-Lamadrid, M.; Issa-Issa, H.; Hurtado, P.; Hernández, F.; Carbonell-Barrachina, Á.A.; López-Lluch, D. Comparison on sensory profile, volatile composition and consumer's acceptance for PDO or non-PDO tigernut (*Cyperus esculentus* L.) milk. *LWT* **2021**, *140*, 110606. [CrossRef]
12. Singleton, V.L.; Orthofer, R.; Lamuela-Raventós, R.M. Analysis of Total Phenols and other oxidation substrates and antioxidants by means of Folin-Ciocalteu Reagent. In *Oxidants and Antioxidants Part A, Methods in Enzymology*; Accademic Press: Boston, MA, USA, 1999; Volume 299, pp. 152–178.
13. Verni, M.; Pontonio, E.; Krona, A.; Jacob, S.; Pinto, D.; Rinaldi, F.; Verardo, V.; Díaz-de-Cerio, E.; Coda, R.; Rizzello, C.G. Bioprocessing of brewers' spent grain enhances its antioxidant activity: Characterization of phenolic compounds and bioactive peptides. *Front. Microbiol.* **2020**, *11*, 1–15. [CrossRef] [PubMed]
14. Re, R.; Pellegrini, N.; Proteggente, A.; Pannala, A.; Yang, M.; Rice-Evans, C. Antioxidant activity applying an improved ABTS radical cation decolorization assay. *Free Radic. Biol. Med.* **1999**, *26*, 1231–1237. [CrossRef]
15. Brand-Williams, W.; Cuvelier, M.E.; Berset, C. Use of a free redical method to evaluate antioxidant activity. *LWT-Food Sci. Technol.* **1995**, *28*, 25–30. [CrossRef]
16. Parejo, I.; Codina, C.; Petrakis, C.; Kefalas, P. Evaluation of scavenging activity assessed by Co(II)/EDTA-induced luminol chemiluminescence and DPPH· (2,2-diphenyl-1-picrylhydrazyl) free radical assay. *J. Pharmacol. Toxicol. Methods* **2000**, *44*, 507–512. [CrossRef]
17. Pulido, R.; Bravo, L.; Saura-Calixto, F. Antioxidant activity of dietary polyphenols as determined by a modified ferric reducing/antioxidant power assay. *J. Agric. Food Chem.* **2000**, *48*, 3396–3402. [CrossRef]
18. Bezerra, M.A.; Santelli, R.E.; Oliveira, E.P.; Villar, L.S.; Escaleira, L.A. Response surface methodology (RSM) as a tool for optimization in analytical chemistry. *Talanta* **2008**, *76*, 965–977. [CrossRef]
19. Pei, K.; Ou, J.; Huang, J.; Ou, S. p-Coumaric acid and its conjugates: Dietary sources, pharmacokinetic properties and biological activities. *J. Sci. Food Agric.* **2016**, *96*, 2952–2962. [CrossRef]
20. Yannai, S. *Dictionary of Food Compounds with CD-ROM: Additives, Flavors, and Ingredients*; Chapman & Hall/CRC: Boca Raton, FL, USA, 2004.
21. Byrne, L.T.; Colegate, M.S.; Darling, P.R.; Huxtable, C.R. Imbricatonol, a naphthol-naphthoquinone dimer isolated from stypandra imbricata and dianella revoluta. *Aust. J. Chem.* **1987**, *40*, 1315–1320. [CrossRef]
22. Kreuz, S.; Joubert, E.; Waldmann, K.H.; Ternes, W. Aspalathin, a flavonoid in Aspalathus linearis (rooibos), is absorbed by pig intestine as a C-glycoside. *Nutr. Res.* **2008**, *28*, 690–701. [CrossRef]
23. Tan, H.M.; Leong, K.H.; Song, J.; Mohd Sufian, N.S.F.; Mohd Hazli, U.H.A.; Chew, L.Y.; Kong, K.W. Antioxidant and LC-QToF-MS/MS analysis of polyphenols in polar and non-polar extracts from Strobilanthes crispus and Clinacanthus nutans. *Int. Food Res. J.* **2020**, *27*, 903–914.
24. Calderón-Montaño, J.M.; Burgos-Morón, E.; Pérez-Guerrero, C.; López-Lázaro, M. A review on the dietary flavonoid kaempferol | BenthamScience. *Mini Rev. Med. Chem.* **2011**, *11*, 298–344. [CrossRef] [PubMed]
25. Boerjan, W.; Ralph, J.; Baucher, M. Lignin Biosynthesis. *Annu. Rev. Plant Biol.* **2003**, *54*, 519–546. [CrossRef] [PubMed]
26. Ruan, J.; Yan, J.; Zheng, D.; Sun, F.; Wang, J.; Han, L.; Zhang, Y.; Wang, T. Comprehensive chemical profiling in the ethanol extract of *Pluchea indica* aerial parts by liquid chromatography/mass spectrometry analysis of its silica gel column chromatography fractions. *Molecules* **2019**, *24*, 2784. [CrossRef] [PubMed]
27. Duman, E. Some physico-chemical properties, fatty acid compositions, macro-micro minerals and sterol contents of two variety tigernut tubers and oils harvested from east mediterranean region. *Food Sci. Technol.* **2019**, *39*, 610–615. [CrossRef]
28. Aljuhaimi, F.; Ghafoor, K.; Özcan, M.M.; Miseckaite, O.; Babiker, E.E.; Hussain, S. The effect of solvent type and roasting processes on physico-chemical properties of tigernut (*Cyperus esculentus* L.) tuber oil. *J. Oleo Sci.* **2018**, *67*, 823–828. [CrossRef]

29. Roselló-Soto, E.; Barba, F.J.; Lorenzo, J.M.; Dominguez, R.; Pateiro, M.; Mañes, J.; Moltó, J.C. Evaluating the impact of supercritical-$CO_2$ pressure on the recovery and quality of oil from "horchata" by-products: Fatty acid profile, α-tocopherol, phenolic compounds, and lipid oxidation parameters. *Food Res. Int.* **2019**, *120*, 888–894. [CrossRef]
30. Waszkowiak, K.; Gliszczynska-Swigło, A. Binary ethanol–water solvents affect phenolic profile and antioxidant capacity of flaxseed extracts. *Eur. Food Res. Technol.* **2016**, *242*, 777–786. [CrossRef]
31. Guo, T.; Wan, C.; Huang, F.; Wei, C. Evaluation of quality properties and antioxidant activities of tiger nut (*Cyperus esculentus* L.) oil produced by mechanical expression or/with critical fluid extraction. *LWT* **2021**, *141*, 110915. [CrossRef]
32. Badejo, A.A.; Olawoyin, B.; Salawu, S.O.; Fasuhanmi, O.S.; Boligon, A.A.; Enujiugha, V.N. Antioxidative potentials and chromatographic analysis of beverages from blends of gluten-free acha (*Digitaria exilis*) and tigernut (*Cyperus esculentus*) extracts. *J. Food Meas. Charact.* **2017**, *11*, 2094–2101. [CrossRef]
33. Roselló-Soto, E.; Poojary, M.M.; Barba, F.J.; Lorenzo, J.M.; Mañes, J.; Moltó, J.C. Tiger nut and its by-products valorization: From extraction of oil and valuable compounds to development of new healthy products. *Innov. Food Sci. Emerg. Technol.* **2018**, *45*, 306–312. [CrossRef]
34. Koubaa, M.; Barba, F.J.; Mhemdi, H.; Grimi, N.; Koubaa, W.; Vorobiev, E. Gas assisted mechanical expression (GAME) as a promising technology for oil and phenolic compound recovery from tiger nuts. *Innov. Food Sci. Emerg. Technol.* **2015**, *32*, 172–180. [CrossRef]

foods

MDPI

*Communication*

# The Establishment of Ultrasonic-Assisted Extraction for the Recovery of Phenolic Compounds and Evaluation of Their Antioxidant Activity from *Morus alba* Leaves

**Beatriz Martín-García [1], María José Aznar-Ramos [1], Vito Verardo [1,2,*] and Ana María Gómez-Caravaca [2,3]**

[1] Department of Nutrition and Food Science, Campus of Cartuja s/n, University of Granada, 18071 Granada, Spain; bea91mg@ugr.es (B.M.-G.); mariajoseaznar@ugr.es (M.J.A.-R.)
[2] Institute of Nutrition and Food Technology 'José Mataix', Biomedical Research Center, University of Granada, Avda del Conocimiento sn., 18100 Granada, Spain; anagomez@ugr.es
[3] Department of Analytical Chemistry, Faculty of Sciences, University of Granada, Avd. Fuentenueva s/n, 18071 Granada, Spain
* Correspondence: vitoverardo@ugr.es; Tel.: +34-958243864

**Abstract:** Phenolic compounds of *Morus alba* leaves are bioactive compounds with beneficial properties for human health. Therefore, in this study, an optimization of ultrasonic assisted extraction by Box–Behnken design was used for the first time to optimize factors such as the percentage of ethanol, ratio solvent/sample ($v/w$) and extraction time to reach the highest phenolic compound amounts (evaluated by HPLC-MS) while also evaluating in vitro antioxidant activity using DPPH, ABTS and FRAP assays. The optimal extraction conditions were 40% ethanol, 1/400 ($w/v$) and 35 min. Applying these optimal conditions, which were identified and quantified by HPLC-MS, resulted in the extraction of 21 phenolic compounds. According to these results, the main phenolic compounds in *Morus alba* leaves are the phenolic glycoside and phenolic acid named protocatechuic acid-glucoside and caffeoylquinic. In addition, *Morus alba* leaf extract contains flavonols such quercetin-3-O-6-acetylglucoside and rutin, which represent more than 7% of its total phenolic content.

**Keywords:** *Morus alba* leaves; phenolic compounds; Box-Behnken; HPLC-MS; antioxidant activity

**Citation:** Martín-García, B.; Aznar-Ramos, M.J.; Verardo, V.; Gómez-Caravaca, A.M. The Establishment of Ultrasonic-Assisted Extraction for the Recovery of Phenolic Compounds and Evaluation of Their Antioxidant Activity from *Morus alba* Leaves. *Foods* **2022**, *11*, 314. https://doi.org/10.3390/foods11030314

Academic Editor: Filipa V. M. Silva

Received: 4 January 2022
Accepted: 19 January 2022
Published: 24 January 2022



## 1. Introduction

Mulberry (*Morus alba* L.) is a plant from the family of *Moraceae* that is native to Asia and is usually employed as food for silkworm breeding. It is cultivated in subtropical and tropical regions and the plant adapts to different pedo-climatic conditions [1]. Mulberry leaves are rich in several bioactive compounds such as phenolic compounds, alkaloids, polysaccharides and vitamins [2]. Mulberry leaves contain calcium, carbohydrates, iron, proteins, vitamin B1, vitamin D and β-carotene, which are also considered to be a nutritious and could be used as foods or for their medicinal activity [3,4]. Mulberry leaves also contain phenolic compounds including flavonols such as rutin and other quercetin derivatives which all possess anti-diabetic, hypolipidemic, antihypertensive, anti-atherosclerotic and anticonvulsant properties [2,5,6]. These beneficial effects are related in part to the antioxidant activity of these phenolic compounds [3,7].

The extraction technique is highly important when it comes to the quantity of phenolic compounds obtained before analysis. It depends on several factors such as the solvent composition, the structure of the matrix and the technique used for the extraction [8]. Conventional extraction techniques in plant materials such as heating, refluxing or using Soxhlet apparatus could activate oxidation or hydrolysis of phenolic compounds, while maceration and percolation require longer extracting time. The recent development of new automated high throughput extractors such as microwave-assisted extraction (MAE), ultrasonic-assisted extraction (UAE) and pressurized-assisted extraction (PLE), have resulted in increased attention as these techniques have shown to be efficient in the recovery of bioactive phytochemicals [9]. However, MAE and PLE require investment in high cost instruments [10]. The advantages of ultrasonic-assisted extraction are simple, easy to handle and inexpensive compared with the others. In ultrasonic-assisted extraction, a shear force is produced by ultrasonic cavitation which breaks the plant cell wall, which in turn accelerates the transfer of bioactive compounds so as to extract solvent in shorter time than maceration or percolation [9,11]. There are previous studies about the extraction of phenolic compounds in *Morus alba* leaves by using ultrasonic-assisted extraction [12,13]. Nevertheless, there is very little information about the optimization of extraction conditions in *Morus alba* leaves in the previous analyses. In addition, a wide variation in phenolic recovery in plants has been reported by using different extraction conditions. With ultrasonic-assisted extraction, the most influential factors on the yield of phenolic compounds are the solid/liquid ratio, extraction time and solvent concentration [14]. Therefore, in this paper an optimization of extraction conditions by ultrasonic-assisted extraction bath was proposed in order to obtain the highest phenolic content and the highest antioxidant activity by DPPH, ABTS and FRAP assays in *Morus alba* leaf extracts. This extract obtained under optimal conditions, was characterized by using HPLC-MS.

## 2. Materials and Methods

### 2.1. Samples

Samples were collected from Granada (Spain). Two cultivars of *Morus alba* leaves were collected from two different fields in Granada (Spain). Forty leaves from 3 different trees were picked-up in two different fields. The leaves were air dried in dark conditions at room temperature and they were milled using a 10 basic miller (IKA, Staufen, Germany) and they were sieved to obtain a particle size of 0.2 mm.

### 2.2. Chemicals

All solvents were purchased from Merck KGaA (Darm-stadt, Germany), whereas water was obtained in situ using a Milli-Q system (Millipore, Bedford, MA, USA). Chemical standards of the phenolic compounds were acquired from Sigma-Aldrich (St. Louis, MO, USA).

### 2.3. Experimental Design

A Box–Behnken design (BBD) is more efficient than other experimental designs such as central composite design and the three-level full factorial designs where the efficiency of one experimental design is the number of coefficients in the estimated model divided by the number of experiments. In addition, BBD does not contain combinations for which all parameters are simultaneously at their highest or lowest levels. Therefore, these designs avoid unsatisfactory results, which occur when the experiments are performed under extreme conditions [15]. For all these reasons, in this study the optimization of the ultrasonic-assisted extraction to obtain the maximum phenolic recovery in *Morus alba* leaves was obtained with a BBD with 3 independent factors ($X_1$: ethanol/water ratio ($v/v$), $X_2$: solvent/sample ratio ($v/w$) and $X_3$: extraction time (min) with 3 levels for each variable. The dependent variables (Y) were the sum of phenolic compounds (SPC) determined by HPLC-MS, and the antioxidant capacity obtained by DPPH, ABTS and FRAP assays (Table 1). Food and Drug Administration (FDA) has labeled ethanol as a generally recognized safe solvent to use in food products, for this reason this solvent was chosen for the extraction of phenolic compounds [16]. The percentage of ethanol/water was 0–100% ($v/v$), the solid-to-solvent ratio was from 1/20 to 1/500 ($w/v$) and the extraction time was from 10 min to 90 min and, these parameters were chosen based on the extraction conditions employed by previous studies for the recovery of phenolic compounds in *Morus alba* leaves [12,17,18]. The design comprised 15 experiments with 3 center points (Table 1).

**Table 1.** Values for the dependent factors, and response variables obtained in the Box-Behnken design (BBD).

| Run | Dependent Factors | | | Response Variables | | | |
|---|---|---|---|---|---|---|---|
| | $X_1$ | $X_2$ | $X_3$ | SPC | DPPH | ABTS | FRAP |
| 1 | 100 | 500 | 50 | 12.17 | 3.38 | 9.05 | 30.41 |
| 2 | 0 | 20 | 50 | 14.08 | 2.84 | 5.92 | 6.87 |
| 3 | 50 | 500 | 90 | 31.40 | 22.35 | 27.32 | 27.39 |
| 4 | 0 | 500 | 50 | 30.07 | 12.29 | 6.39 | 26.63 |
| 5 | 50 | 500 | 10 | 32.65 | 22.21 | 31.29 | 32.54 |
| 6 | 50 | 260 | 50 | 32.73 | 24.60 | 29.08 | 32.58 |
| 7 | 50 | 260 | 50 | 33.04 | 27.55 | 27.76 | 35.34 |
| 8 | 100 | 260 | 90 | 12.69 | 5.89 | 8.12 | 13.30 |
| 9 | 0 | 260 | 90 | 23.23 | 7.93 | 3.40 | 19.37 |
| 10 | 0 | 260 | 10 | 22.85 | 7.89 | 10.30 | 23.67 |
| 11 | 100 | 20 | 50 | 5.37 | 1.84 | 10.06 | 5.38 |
| 12 | 50 | 20 | 90 | 26.44 | 17.00 | 15.20 | 20.69 |
| 13 | 50 | 20 | 10 | 24.18 | 15.67 | 14.39 | 23.41 |
| 14 | 100 | 260 | 10 | 6.65 | 3.89 | 6.69 | 9.79 |
| 15 | 50 | 260 | 50 | 33.76 | 25.95 | 27.51 | 32.24 |

$X_1$: Ethanol/water ratio ($v/v$), $X_2$: solvent/sample ratio ($v/w$) and $X_3$: extraction time (min). The sum of phenolic compounds (SPC) was given in mg/g d.w. DPPH, ABTS and FRAP were expressed as mg Trolox eq./g d.w.

Response surface methodology (RSM) is the most relevant multivariate technique employed in analytical optimization. The relationships between the response and independent variables is described as a second-order polynomial equation. The data were processed with the statistical software STATISTICA 7.0 (2002, StatSoft, Tulsa, OK, USA).

### 2.4. *Ultrasound-Assisted Extraction of Phenolic Compounds in Morus alba Leaves*

To extract the phenolic compounds from *Morus alba* leaves an ultrasonic bath (Bandelin, Sonorex, RK52, Berlin, Germany) was used, which operates at a frequency of 35 kHz. Powdered *Morus alba* leaves was placed with 10 mL of solvent extraction using the experimental conditions of the model. After centrifugations for 10 min at 1000 g, the solvent was evaporated by Buchi R-205 rotavapor and reconstituted in 2 mL of methanol/water (1:1, $v/v$). Finally, the extracts were filtered before the analysis using a 0.2 μm nylon syringe filter.

### 2.5. *Antioxidant Capacity*

The determination of antioxidant activities of *Morus alba* leaf extracts was carried out by three different assays. The results were expressed as mg Trolox equivalent/g of dry weight leaves. Three replicates of each sample were processed.

#### 2.5.1. DPPH Radical Scavenging

The protocol of Brand-Williams et al., 1995 [19] was used to develop the DPPH assay. Briefly, 0.1 mL of the extract was added to 2.9 mL of 100 μM DPPH solution in MeOH/$H_2O$ 1/1 ($v/v$) and the absorbance was determined after 30 min at 517 nm (25 °C).

#### 2.5.2. ABTS Cation Radical Scavenging

The ABTS assay was undertaken according to Re et al., 1999 [20]. ABTS radical cation ($ABTS^+$) was added to EtOH to reach an absorbance of 0.7 ± 0.02 at 734 nm and 30 °C. After that, 10 μL of extract was added to 1 mL of ABTS reagent and its absorbance was measurement after 10 min.

#### 2.5.3. Ferric Reducing Antioxidant Power (FRAP)

This assay was done following the process described by Pulido et al., 2000 [21]. Under this procedure, 30 μL of the extracts was added of 0.9 mL of water and 0.9 mL of FRAP reagent. The absorbance measurement at 595 nm was undertaken after 30 min.

### 2.6. Analysis of Phenolic Compounds in *Morus alba* Leaf Extracts by HPLC-ESI-TOF-MS

The analysis of *Morus alba* leaf extracts was performed by HPLC-ESI-TOF-MS as previously reported by Verni et al., 2020 [22]. Three replicates of each sample were processed. The equipment consists of a UPLC system ACQUITY (Waters Corporation, Milford, MA, USA) coupled to a time-of-flight analyzer (TOF) (Waters Corporation, Milford, MA, USA). The phenolic compounds were separated using a BEH Shield RP18 column (1.7 µm, 2.1 mm × 100 mm; Waters Corporation, Milford, MA, USA). The analysis was carried out at 40 °C and the data were processed using MassLynx 4.1 software (Waters Corporation, Milford, MA, USA).

## 3. Results and Discussion

### 3.1. Fitting the Model

Table 2 shows the values obtained for each variable response in experimental extraction conditions, according to the Box–Behnken design.

**Table 2.** Coefficients of regression, effects and analysis of variance (ANOVA) of the model for the response variables.

| | SPC | | DPPH | | ABTS | | FRAP | |
|---|---|---|---|---|---|---|---|---|
| | Coefficients | Effects | Coefficients | Effects | Coefficients | Effects | Coefficients | Effects |
| $\beta_0$ | 10.85230 * | 20.1500 | −3.42399 | 10.26411 | −1.56191 | 12.34545 | 8.764200 * | 19.95608 |
| Linear | | | | | | | | |
| $\beta_1$ | 0.48233 * | −13.3355 | 0.66784 * | −3.98835 | 0.69638 * | 1.97903 | 0.389353 * | −4.41331 |
| $\beta_2$ | 0.05756 * | 9.0560 | 0.05799 * | 5.72199 | 0.05052 * | 7.12160 | 0.059015 * | 15.15499 |
| $\beta_3$ | 0.12328 * | 1.8605 | 0.17588 ** | 0.87715 | 0.19197 * | −2.15594 | 0.195844 ** | −2.16710 |
| Cross product | | | | | | | | |
| $\beta_{12}$ | −0.00019 * | −4.5953 | −0.00016 | −3.95706 | −0.00003 | −0.73700 | 0.000110 | 2.63046 |
| $\beta_{13}$ | 0.00071 * | 2.8316 | 0.00025 | 0.98140 | 0.00104 * | 4.16383 | 0.000977 | 3.90610 |
| $\beta_{23}$ | −0.00009 ** | −1.7544 | −0.00003 | −0.58807 | −0.00012 | −2.38837 | −0.000063 | −1.21595 |
| Quadratic | | | | | | | | |
| $\beta_{11}$ | −0.00601 * | 15.0324 | −0.00677 * | 16.92810 | −0.00721 * | 18.01628 | −0.005108 * | 12.77022 |
| $\beta_{22}$ | −0.00005 * | 2.7197 | −0.00007 * | 4.02061 | −0.00005 * | 3.09327 | −0.000057 ** | 3.29606 |
| $\beta_{33}$ | −0.00112 * | 1.7866 | −0.00169 ** | 2.70756 | −0.00239 * | 3.81799 | −0.002553 * | 4.08468 |
| $R^2$ | 0.98932 | | 0.99476 | | 0.92168 | | 0.86893 | |
| *p* (Lack of fit) | 0.054291 | | 0.461657 | | 0.052922 | | 0.083124 | |

* Significant at $p < 0.05$ level, ** Significant at $p < 0.1$ level.

The model was assessed in accordance with the regression significance coefficients, quadratic correlation coefficients ($R^2$), quadratic correlation coefficients adjusted ($R^2$ adjusted), coefficient of variation (CV) and lack of fit (Table 2). The level of significance established was $\alpha < 0.1$ in accordance with previous studies [23,24]. The significant variables on the response of SPC were the linear effect of ethanol/water % (*v*/*v*) ($X_1$) ($p = 0.000899$) and its quadratic effect ($X_{11}$) ($p = 0.000332$), linear effect of solvent-to solid ratio ($X_2$) ($p = 0.002941$) and its quadratic effect ($X_{22}$) ($p = 0.009983$), the linear effect of time ($X_3$) ($p = 0.025838$) and its quadratic effect ($X_{33}$) ($p = 0.022689$) and the cross effect between ethanol/water % (*v*/*v*) with solvent-to-solid ratio (*v*/*w*) ($X_{12}$) ($p = 0.012855$), the cross effect between ethanol/water % (*v*/*v*) with time ($X_{13}$) ($p = 0.032829$) and the cross effect between ratio and time ($X_{23}$) ($p = 0.079367$). The significant variables on the variable response of DPPH were the linear effect of ethanol/water % (*v*/*v*) ($X_1$) ($p = 0.003672$) and its quadratic effect ($X_{11}$) ($p = 0.002050$), linear effect of solvent-to solid ratio (*v*/*w*) ($X_2$) ($p = 0.022121$) and its quadratic effect ($X_{22}$) ($p = 0.034571$), the linear effect of time ($X_3$) ($p = 0.089993$) and its quadratic effect ($X_{33}$) ($p = 0.071824$). In addition, the significant effects on the response of ABTS were the following: ethanol/water % (*v*/*v*) ($X_1$) ($p = 0.002055$) and its quadratic effect ($X_{11}$) ($p = 0.001100$), linear effect of solvent-to solid ratio (*v*/*w*) ($p = 0.017817$) and its quadratic effect ($X_{22}$) ($p = 0.035413$), the linear effect of time ($X_3$) ($p = 0.049034$) and its quadratic effect ($X_{33}$) ($p = 0.023673$), the cross effect between ethanol/water % (*v*/*v*) and time ($X_{13}$) ($p$ = 0.068455). Finally, the significant effects on the response FRAP were ethanol/water % ($v/v$) ($X_1$) ($p$ = 0.014110) and its quadratic effect ($X_{11}$) ($p$ = 0.004760), linear effect of solvent-to solid ratio ($v/w$) ($X_2$) ($p$ = 0.028083) and its quadratic effect ($X_{22}$) ($p$ = 0.065024), the linear effect of time ($X_3$) ($p$ = 0.095433) and its quadratic effect ($X_{33}$ = 0.043803).

Statistical significance was set at the 95% of confidence level to establish all the effects. A high correlation between independent and dependent factors was obtained with quadratic correlation coefficient ($R^2$) from 92.17–99.48%, which, with the exception of the FRAP, provided a good correlation but lower than the other ones ($R^2$ = 86.89%). According to a previous study, $R^2$ should be at least 0.80 for a good fit [25]. In addition, the verification of the suitability of the model was carried out according to the $p$-value obtained, it being non-significant ($p$ > 0.05) means that the model fits well (Table 3). Moreover, as the $p$-value was lower than 0.05 for all cases, all models were considered statistically acceptable.

**Table 3.** Optimal conditions for ultrasonic-assisted extraction.

| Optimal Conditions | SPC | DPPH | ABTS | FRAP |
|---|---|---|---|---|
| Ethanol/water % ($v/v$) | 40 | 40 | 40 | 40 |
| Solvent-to-solid ratio ($v/w$) | 400 | 400 | 400 | 400 |
| Time (min) | 35 | 35 | 35 | 35 |
| Predicted | 36 ± 2 | 25 ± 3 | 29 ± 3 | 36 ± 4 |
| Observed | 37.3 ± 0.7 | 27.6 ± 0.9 | 30.5 ± 0.3 | 36.8 ± 0.2 |
| Significant differences | N.S. | N.S. | N.S. | N.S. |

N.S.: no significant differences. SPC was expressed as mg/g d.w. DPPH, ABTS and FRAP were expressed as mg trolox/g sample d.w.

### 3.2. Response Surfaces Methodology Analysis

Figures 1 and 2 plot the three-dimensional response surfaces, which show the effects of % EtOH ($X_1$) with solvent-to-solid ratio ($v/w$) ($X_2$) (a, d, g), %EtOH ($X_1$) with time (min) ($X_3$) (b, e, h) and time ($X_3$) with solvent-to-solid ratio ($v/w$) ($X_2$) (c, f, i) on the SPC, DPPH, ABTS and FRAP.

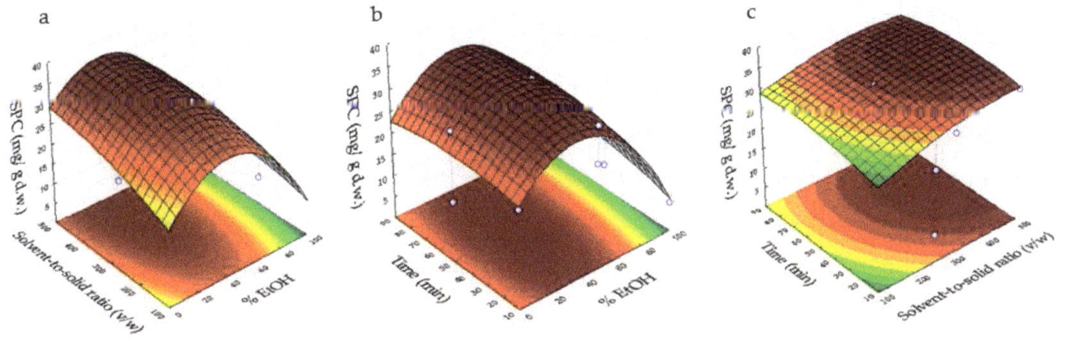


**Figure 1.** Response surfaces of combined effects for sum of phenolic compounds (SPC). (**a**) Solvent-to-solid ratio ($v/w$) vs. % EtOH; (**b**) Time vs. %EtOH; and (**c**) time vs. solvent-to-solid ratio ($v/w$).

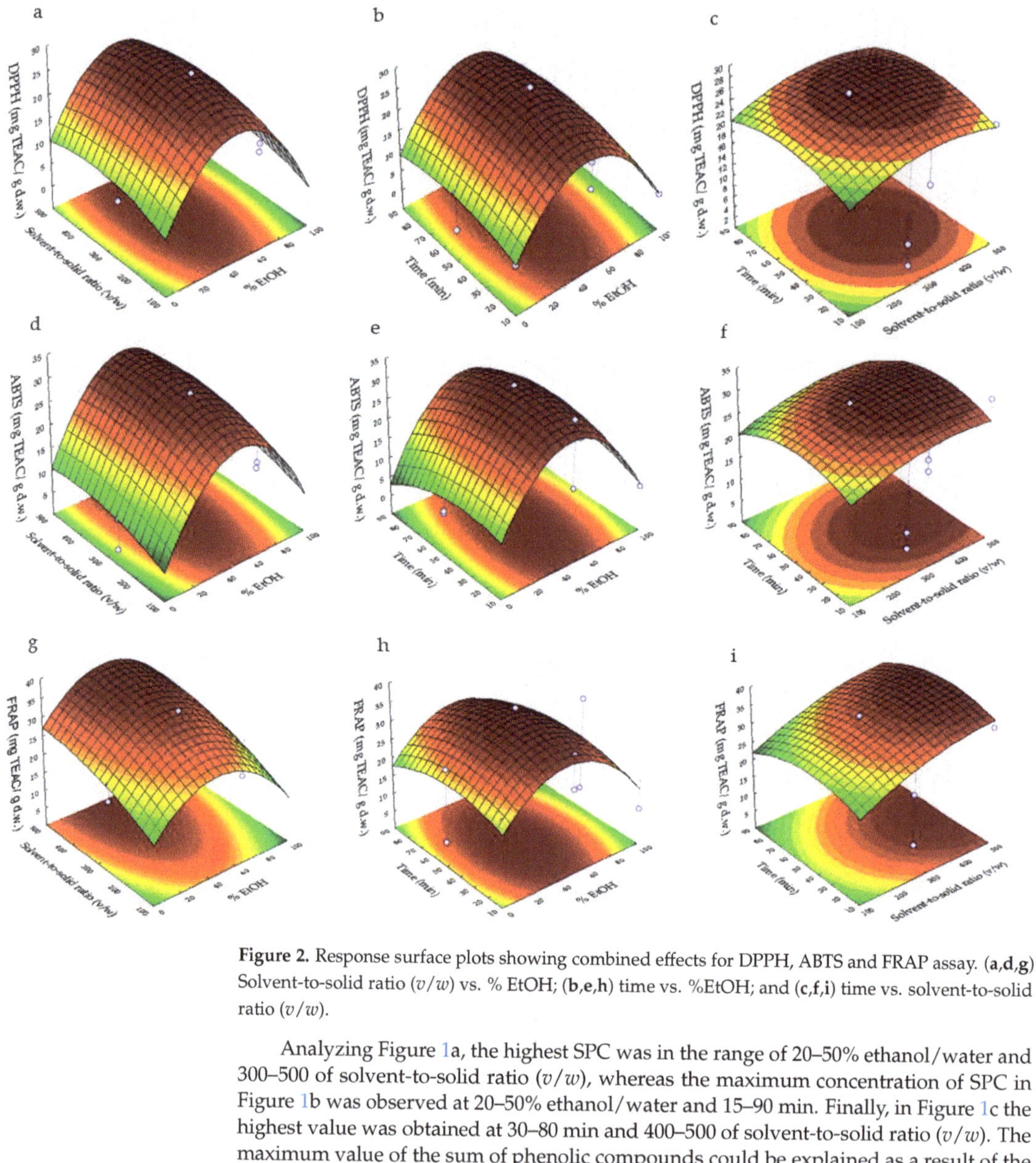


**Figure 2.** Response surface plots showing combined effects for DPPH, ABTS and FRAP assay. (**a**,**d**,**g**) Solvent-to-solid ratio ($v/w$) vs. % EtOH; (**b**,**e**,**h**) time vs. %EtOH; and (**c**,**f**,**i**) time vs. solvent-to-solid ratio ($v/w$).

Analyzing Figure 1a, the highest SPC was in the range of 20–50% ethanol/water and 300–500 of solvent-to-solid ratio ($v/w$), whereas the maximum concentration of SPC in Figure 1b was observed at 20–50% ethanol/water and 15–90 min. Finally, in Figure 1c the highest value was obtained at 30–80 min and 400–500 of solvent-to-solid ratio ($v/w$). The maximum value of the sum of phenolic compounds could be explained as a result of the positive influence of the quadratic effect of EtOH and linear effect of the solvent-to-solid ratio. In addition, the decrease of this response could be mainly due to the linear negative effect of EtOH.

In respect of DPPH, its maximum content was obtained in a range of 30–60% ethanol/water at 150–500 of solvent-to-solid ratio (Figure 2a), whereas the highest DDPH value shows in the range of 30–70 min and 40–60% ethanol/water in Figure 2b and 250–400 of solvent-to-solid ratio ($v/w$) and 35–70 min in Figure 2c. The increase in the DPPH response could be due to the positive quadratic effect of EtOH as this variable exerts the highest effect on this response in comparison with the rest of the variables. In addition, with regard to ABTS response, its maximum value was shown between 40–60 % ethanol/water at 200–500 of solvent-to-solid ratio ($v/w$) (Figure 2d), 25–65 min and 40–60% ethanol/water in Figure 2e and 250–500 of solvent-to-solid ratio ($v/w$) at 25–65 min in Figure 2f. The quadratic of EtOH and linear solvent-to-solid ratio were the variables which provide a higher effect on the ABTS response, which could explain the increase in this response. Finally, the highest value of FRAP can be observed at 40–60% of ethanol/water and 450–500 of solvent-to-solid ratio ($v/w$) (Figure 2g), 30–55 min and 40–55% of ethanol/water (Figure 2h), whereas the maximum content of FRAP in Figure 2i was obtained 400–500 of solvent-to-solid ratio ($v/w$) and 25–55 min. Variables which exert the highest effect on this response were the quadratic of EtOH and solvent-to-solid ratio, which explain the maximum value obtained in FRAP response.

### 3.3. *Optimization of Ultrasonic-Assisted Extraction*

#### 3.3.1. Optimal Ultrasonic-Assisted Extraction Conditions

After the analysis of the 3-D plots and the choice of the optimal conditions, accuracy of the mathematical model was established comparing the predicted and experimental data.

Table 3 shows the results of the sum phenolic compounds and in vitro antioxidant activity by the three different assays from *Morus alba* leaf extract obtained at optimum conditions. The same following optimal conditions were established for all responses: 40% ethanol/water, 35 min and 400 of solvent-to-solid ratio ($v/w$), obtaining predictable values of 36 ± 2 mg/g d.w. for the sum of phenolic compounds, 25 ± 3, 29 ± 3, 36 ± 4 mg TEAC/ g d.w. for DPPH, ABTS and FRAP. According to the results, the extraction time was lower than that reported by a previous study, whose extraction conditions in *Morus alba* leaves were methanol (each 2 L) for 4 h at 60 °C obtaining 23.2 and 55.4 mg gallic acid equivalent/g d.w. [26]. In addition, another study reported an extraction of phenolic compounds from *Morus alba* leaves with 80% aqueous methanol acidified with formic acid (1%) sonicated at 25 ± 5 °C for 60 min [27]. However, Kim et al., 2020 [13] reported a similar extraction time of 30 min in *Morus* samples using 70% methanol with the ultrasonic extractor. Another study reported an ultrasonic-assisted extraction by using methanol/water mixture at a shorter extraction time than in the present study (10 min) and $\frac{1}{4}$ of solid-to-solvent ratio ($w/v$) to obtain 0.31 mg GAE/g d.w. and 0.19 mg TEAC/g d.w. for total phenolic compounds and DPPH [12]. Nevertheless, this DPPH value was 99% lower than that obtained by the present study [12]. Another study reported a similar concentration range for the sum of phenolic compounds 19.17–58.47 mg/g obtained by pressurized liquid extraction in *Morus alba* leaves using a similar mixture $EtOH/H_2O$ 50:50 ($v$:$v$) as solvent, at 200 °C for 20 min in static cycle [18]. Therefore, it has been proven that ultrasonic-assisted extraction at optimum conditions is a process as efficient in the phenolic recovery from *Morus alba* leaves as pressurized liquid extraction. Therefore, the application of ultrasonic-assisted extraction could be an efficient alternative to other green techniques reducing the cost of extraction operations. Nevertheless, there is a wide variability among the phenolic contents and antioxidant activities in mulberry leaves due to the different cultivars of *Morus alba* leaves, different extraction techniques used and different analytical methods used in each studies [12].

#### 3.3.2. Determination of Phenolic Compounds in *Morus alba* Leaf Extracts by HPLC-MS

Phenolic compounds of *Morus alba* leaf extract were identified by HPLC-ESI-TOF-MS according to their mass data and by comparing them with literature, the co-elution with commercial standards (when possible) and with several databases. Mass data, experimental and calculated $m/z$, error and Fit Conf %, mainly in source fragments and molecular formulae $(M-H)^-$, were considered.

As reported in Table 4, 21 phenolic compounds were detected in the *Morus alba* leaf extracts, including seven phenolic acid derivatives and fourteen flavonols. Supplementary Figure S1 shows the base peak chromatogram obtained by HPLC-ESI-TOF-MS for each compound in the *Morus alba* leaf extract obtained at optimal ultrasonic-assisted extraction conditions. Peak 1 at 2.02 min and $m/z$ 315.0714 showed a molecular formula of $C_{13}H_{15}O_9$ and fragment ions at *m/z* 153.0162 and 109.0279; according to a previous study [18] it was assigned to protocatechuic acid-glucoside. Peak 2 at 3.78 min with *m/z* 353.0870 with a molecular formula of $C_{16}H_{17}O_9$ and fragments at *m/z* 179.0336, 191.0551, 135.0436 and 173.0419 was identified as 3-caffeoylquinic acid, which has been identified previously in *Morus alba* leaf extracts [12,18,28,29]. Peak 3 at 5.32 min with *m/z* 515.1406 and fragments at *m/z* 341.0868, 191.0472, 179.0314 with a molecular formula of $C_{22}H_{27}O_{14}$, was proposed to be chlorogenic acid hexose [18]. Peak 4 at 5.52 min with *m/z* 353.0866 with a molecular formula of $C_{16}H_{17}O_9$ and fragment ions at *m/z* 191.0551, 173.0455 and 135.0436, this compound was proposed to be 5-caffeoylquinic acid (chlorogenic acid) [28]. Peak 5 at 5.75 min with *m/z* 353.0873 and with a molecular formula of $C_{16}H_{17}O_9$ presented a fragment ion at *m/z* 191.0553, 173.0432 and 179.0428, which correspond with 4-caffeoylquinic acid (cryptochlorogenic acid) [28]. Peak 6 at 6.05 min with *m/z* 771.1996 and fragment ions with *m/z* 609.1451, 463.0819, 299.0175 and 300.0279 and a molecular formula of $C_{33}H_{39}O_{21}$ was proposed as quercetin rhammosyl hexoside, which has been identified in mulberry samples [30]. Peak 7 with *m/z* 625.1411 and fragment ions at *m/z* 300.0234 and 301.0336 with a molecular formula of $C_{27}H_{29}O_{17}$ correspond with quercetin di-hexoside, which has been previously identified in mulberry fruit and leaves [18,30]. At 7.35 min (peak 8) with *m/z* 609.1456 and fragment ions *m/z* 285.0388 and 447.0918 with $C_{27}H_{29}O_{16}$ was identified as kaempferol-hexoside-hexoside, which has been identified previously in mulberry leaves [27]. Peak 9 at 7.47 min with a *m/z* 711.1434 and a fragment ion *m/z* 667.1544 with $C_{30}H_{31}O_{20}$ was proposed to be quercetin malonyl di-hexoside, which has been identified previously in white and black mulberry leaf extracts [27]. Peak 10 at 8.37 min with *m/z* 695.1463 and ion fragments 651.1573, 489.1035, 531.1118 was detected as kaempferol-malonyl-dihexoside [27]. Peak 11 at 9.64 min with *m/z* 755.2035 with fragment ions *m/z* at 300.0264 and 271.0244 and a molecular formula of $C_{33}H_{39}O_{20}$ was detected as kaempferol rutinoside hexoside, which has been identified in *Morus alba* leaf extract [18,31]. Peaks 12 and 13 (10.07 and 10.32 min) with a molecular formula of $C_{27}H_{29}O_{16}$ and *m/z* 609.1456 and as fragment ion *m/z* 301.03 correspond with isomers of rutin [18,28]. Peak 14 at 10.53 min with *m/z* 463.0894 with a molecular formula of $C_{21}H_{19}O_{12}$ and fragment ions *m/z* 255.0298 and 300.0277 was detected as isoquercitrin (quercetin-3-glucoside) [28]. Peak 15 (10.79 min) with *m/z* 593.1511 and fragment ions *m/z* 285.0381 with a molecular formula of $C_{27}H_{29}O_{15}$ was detected as kaempferol-3-rutinoside [28]. Peak 16 (11.2 min) with *m/z* 593.1519 and fragment ions *m/z* 353.0872 and 473.2368 with a molecular formula of $C_{27}H_{29}O_{15}$ was detected as vicenin-2 [28]. Peaks 17, 19 and 20 (11.41, 11.89 and 12.1 min) with *m/z* 505.0984 with a molecular formula $C_{23}H_{21}O_{13}$ and fragments *m/z* 255, 271, 300 and 301 were proposed to be isomers of quercetin-3-O-(6-acetylglucoside) [18]. Peak 18 at 11.51 min with a molecular formula of $C_{21}H_{19}O_{11}$ and *m/z* 447.0916 with a fragment ion *m/z* 284.0318 was proposed to be kaempferol 3-o-glucoside [28]. Peak 21 at 12.48 min with a *m/z* 489.1051 ($C_{23}H_{21}O_{12}$) and a ion fragments *m/z* 285.0398 and 191.0552 was detected as kaempferol-3-O-6″-O-acetyl-β-D-glucopyranoside [27,28].

**Table 4.** Table of identification of phenolic compounds from optimum *Morus alba* leaf extract by HPLC-MS.

| Peak | RT | *m/z* Experimental | *m/z* Calculated | Tolerance (ppm) | Error (ppm) | Fit Conf % | In Source Fragments | Molecular Formula | Compound |
|---|---|---|---|---|---|---|---|---|---|
| 1 | 2.02 | 315.0714 | 315.0716 | 10 | 1.3 | 99.96 | 153.0162, 109.0279 | $C_{13}H_{15}O_9$ | Protocatechuic acid-glucoside |
| 2 | 3.78 | 353.087 | 353.0873 | 10 | −0.8 | 99.98 | 179.0336, 191.0551, 135.0436, 173.0419 | $C_{16}H_{17}O_9$ | 3-Caffeoylquinic acid (neochlorogenic acid) |
| 3 | 5.32 | 515.1405 | 515.1401 | 10 | 0.8 | 99.85 | 341.0868, 191.0472, 179.0314 | $C_{22}H_{27}O_{14}$ | Chlorogenic acid hexoside |

**Table 4.** *Cont.*

| Peak | RT | *m/z* Experimental | *m/z* Calculated | Tolerance (ppm) | Error (ppm) | Fit Conf % | In Source Fragments | Molecular Formula | Compound |
|---|---|---|---|---|---|---|---|---|---|
| **4** | 5.52 | 353.0866 | 353.0873 | 10 | −2 | 99.96 | 191.0551, 179.0337, 173.0455 | $C_{16}H_{17}O_9$ | 5-caffeoylquinic acid (chlorogenic acid) |
| **5** | 5.75 | 353.0873 | 353.0873 | 10 | 0.0 | 96.51 | 191.0553 173.0432 179.0428 | $C_{16}H_{17}O_9$ | 4-Caffeoylquinic acid (cryptochlorogenic acid) |
| **6** | 6.05 | 771.1996 | 771.1984 | 10 | 1.6 | 98.28 | 609.1451, 463.0819, 300.0279 299.0175 | $C_{33}H_{39}O_{21}$ | Quercetin rhammosyl hexoside |
| **7** | 6.64 | 625.1411 | 625.1405 | 10 | 1 | 99.49 | 300.0234, 301.0336 | $C_{27}H_{29}O_{17}$ | Quercetin dihexoside |
| **8** | 7.35 | 609.1446 | 609.1456 | 10 | −1.6 | 94.92 | 285.0388, 447.0918 | $C_{27}H_{29}O_{16}$ | Kaempferol-hexoside-hexoside |
| **9** | 7.47 | 711.1434 | 711.1409 | 10 | 3.5 | 98.63 | 667.1544 | $C_{30}H_{31}O_{20}$ | Quercetin malonyl di-hexoside |
| **10** | 8.37 | 695.1463 | 695.146 | 10 | 0.4 | 98.62 | 651.1573, 489.1035, 531.1118 | $C_{30}H_{31}O_{19}$ | Kaempferol-malonyl-dihexoside |
| **11** | 9.64 | 755.2037 | 755.2035 | 10 | 0.3 | 99.55 | 300.0264, 271.0244 | $C_{33}H_{39}O_{20}$ | Kaempferol rutinoside hexoside |
| **12** | 10.07 | 609.1467 | 609.1456 | 10 | 1.8 | 99.32 | 301.0321 | $C_{27}H_{29}O_{16}$ | Rutin isomer a |
| **13** | 10.32 | 609.1483 | 609.1456 | 10 | 4.4 | 84.77 | 301.0343 | $C_{27}H_{29}O_{16}$ | Rutin isomer b |
| **14** | 10.53 | 463.0894 | 463.0877 | 10 | 3.7 | 94.13 | 255.0298, 300.0277 | $C_{21}H_{19}O_{12}$ | Isoquercitrin (Quercetin-3-glucoside) |
| **15** | 10.79 | 593.1511 | 593.1506 | 10 | 0.8 | 99.78 | 285.0381 | $C_{27}H_{29}O_{15}$ | Kaempferol-3-rutinoside |
| **16** | 11.20 | 593.1519 | 593.1506 | 10 | 2.2 | 99.99 | 353.0872, 473.2368 | $C_{27}H_{29}O_{15}$ | Vicenin-2 |
| **17** | 11.41 | 505.0984 | 505.0982 | 10 | 0.4 | 99.71 | 255.0289, 271.0237, 300.0265 301.0332 | $C_{23}H_{21}O_{13}$ | Quercetin-3-O-(6-acetylglucoside) isomer a |
| **18** | 11.51 | 447.0916 | 447.0927 | 10 | −2.5 | 88.67 | 284.0318 | $C_{21}H_{19}O_{11}$ | Kaempferol 3-o-glucoside |
| **19** | 11.89 | 505.0967 | 505.0982 | 10 | −3 | 99.96 | 255.0277, 271.0230, 301.0303, 300.0256 | $C_{23}H_{21}O_{13}$ | Quercetin-3-O-(6-acetylglucoside) isomer b |
| **20** | 12.11 | 505.0983 | 505.0982 | 10 | 0.2 | 98.9 | 255.0367, 271.0314, 300.0284, 301.0421 | $C_{23}H_{21}O_{13}$ | Quercetin-3-O-(6-acetylglucoside) isomer c |
| **21** | 12.48 | 489.1051 | 489.1033 | 10 | 3.7 | 87.34 | 285.0398, 191.0552 | $C_{23}H_{21}O_{12}$ | Kaempferol-3-O-6″-O-acetyl-β-D-glucopyranoside |

Twenty-one phenolic compounds were also quantified. The quantification of phenolic compounds in *Morus Alba* leaf optimum extracts was done by using the calibration curves of standards. A good linearity was obtained in all calibration curves ($r^2 > 0.9954$). The limit of detection (LOD) and limit of quantification (LOQ) were 0.04–0.47 mg/L, and 0.14–1.57 mg/L, respectively.

The quantitative results of phenolic compounds from the two *M. alba* leaf extracts by HPLC-MS are shown in Table 5. According to the results, phenolic acid derivatives were the most abundant phenolic compounds in *Morus alba* leaf optimum extract. These results are in agreement with previous studies [18,32]. Protocatechuic acid-glucoside was the most concentrated phenolic acid derivative, followed by 4-caffeoylquinic acid (cryptochlorogenic acid), which represent more than 33% of total phenolic acid derivatives. These results are similar to those obtained by a previous study, which reported cryptochorogenic acid as the most concentrated phenolic compound in a range of 4.6–16.5 mg/g d.w. in different *Morus alba* leaves genotypes [18]. Another study reported a similar concentration of chlorogenic acid in *Morus alba* leaf extract (1.7–2.3 mg/g d.w.) [33]. The total content of caffeoylquinic acids was in the same order of magnitude as that obtained by a previous study in white and black mulberry leaves (6.43–10.05 mg/g d.w.) [32]. In addition, the most concentrated flavonol was quercetin-3-O-(6-acetylglucoside) followed by rutin isomer b. These results are in concordance with previous studies that reported rutin as the most abundant flavonol in a different genotype of *Morus alba* leaves, whose content was in a similar order of magnitude (0.58–2.98 mg/g d.w.) than that obtained in the present study [18,32]. Another study reported a similar content of rutin obtained in *Morus alba* leaf collected from different regions (3.10 mg/g d.w.) [26]. The total flavonols content was 5.2 ± 0.8 mg/g d.w., which was in a similar order of magnitude as that reported by Sanchez-Salcedo et al., 2015 in different white and black *Morus alba* leaves (3.66–9.75 mg/g d.w.) [32]. In addition, the total phenolic content in *Morus alba* leaf extract was 16.4 ± 0.6 mg/g d.w., which is in a similar order of magnitude as that provided by a previous study in different *Morus alba* leaves genotypes (19.171–58.474 mg/g d.w.) [18]. Therefore, the difference in the phenolic content obtained in the present study in comparison with previous research may be caused by the different climatic conditions and environmental conditions (temperature, altitude, soil, cultivar, humidity) [12,34].

**Table 5.** Quantification of phenolic compounds in the two cultivars of *Morus alba* leaves (MAL1 and MAL2) by HPLC-MS expressed as mg/g d.w and antioxidant activity expressed as mg Trolox/g d.w.

| Compound | MAL1 | MAL2 |
|---|---|---|
| Protocatechuic acid-glucoside | 9.3 ± 0.3 | 4.1 ± 0.4 |
| 3-Caffeoylquinic acid (neochlorogenic acid) | 3.4 ± 0.1 | 1.5 ± 0.07 |
| Chlorogenic acid hexoside | 0.95 ± 0.05 | 0.4 ± 0.02 |
| 5-caffeoylquinic acid (chlorogenic acid) | 3.4 ± 0.2 | 1.5 ± 0.2 |
| 4-Caffeoylquinic acid (cryptochlorogenic acid) | 8.4 ± 0.7 | 3.7 ± 0.2 |
| Quercetin rhammosyl hexoside | 0.08 ± 0.01 | 0.037 ± 0.008 |
| Quercetin dihexoside | 0.61 ± 0.05 | 0.27 ± 0.03 |
| Kaempferol-hexoside-hexoside | 0.57 ± 0.03 | 0.25 ± 0.02 |
| Quercetin malonyl di-hexoside | 0.11 ± 0.01 | 0.05 ± 0.02 |
| Kaempferol-malonyl-dihexoside | 0.022 ± 0.002 | 0.010 ± 0.001 |
| Kaempferol rutinoside hexoside | 0.29 ± 0.03 | 0.13 ± 0.01 |
| Rutin isomer a | 0.81 ± 0.01 | 0.36 ± 0.04 |
| Rutin isomer b | 1.8 ± 0.1 | 0.8 ± 0.1 |
| Isoquercitrin (Quercetin-3-glucoside) | 1.8 ± 0.2 | 0.79 ± 0.06 |
| Kaempferol-3-rutinoside | 0.25 ± 0.03 | 0.11 ± 0.03 |
| Vicenin-2 | 0.41 ± 0.05 | 0.18 ± 0.02 |
| **Quercetin-3-O-(6-acetylglucoside) isomer a** | 2.44 ± 0.06 | 1.08 ± 0.05 |
| Kaempferol-3-O-glucoside | 0.20 ± 0.02 | 0.09 ± 0.02 |
| **Quercetin-3-O-(6-acetylglucoside) isomer b** | 0.16 ± 0.01 | 0.07 ± 0.04 |
| **Quercetin-3-O-(6-acetylglucoside) isomer c** | 0.072 ± 0.004 | 0.03 ± 0.01 |
| **Kaempferol-3-O-6″-O-acetyl-β-D-glucopyranoside** | 2.3 ± 0.2 | 1.0 ± 0.3 |
| Sum flavonols | 11.9 ± 0.3 | 5.2 ± 0.8 |
| Sum phenolic acid derivatives | 25.4 ± 0.5 | 11.2 ± 0.8 |
| Sum of phenolic compounds (SPC) | 37.3 ± 0.7 | 16.4 ± 0.6 |
| DPPH | 27.6 ± 0.9 | 16.4 ± 0.4 |
| ABTS | 30.5 ± 0.3 | 21.6 ± 0.1 |
| FRAP | 36.8 ± 0.2 | 20.1 ± 0.3 |

## 4. Conclusions

In conclusion, an optimization of an ultrasonic-assisted extraction parameters (percentage of ethanol, solvent/sample ratio ($v/w$) and extraction time) for the phenolic recovery of *Morus alba* leaves was established for the first time by using a mathematical model. This is an important step in assessing the quality control of mulberry leaves. The highest amounts of phenolic compounds and the correspondent antioxidant activity evaluated by DPPH, ABTS and FRAP were obtained at 40% ethanol (water solution), 35 min and a solvent/sample ratio ($v/w$) of 400. Among phenolic compounds, 21 compounds were identified by HPLC-ESI-TOF-MS and the most concentrated were protocatechuic acid-glucoside, cryptochlorogenic acid, quercetin-3-O-(6-acetylglucoside) and rutin, which represent more than 55% of the total phenolic content. As demonstrated, *Morus alba* leaf extracts obtained in these optimum conditions reported high variability on phenolic content, thus, the proposed method helps the quality control of the *Morus alba* leaves for nutra-pharmaceutical purposes.

**Supplementary Materials:** The following supporting information can be downloaded at: https://www.mdpi.com/article/10.3390/foods11030314/s1, Figure S1: Base peak chromatogram (BPC) obtained from HPLC-ESI-TOF-MS analysis of *Morus alba* leaf extract obtained by optimal ultrasonic-assisted extraction conditions. Peaks have been numbered according to the elution order.

**Author Contributions:** Investigation, B.M.-G. and M.J.A.-R.; conceptualization, V.V. and A.M.G.-C.; resources, V.V.; writing—original draft preparation, B.M.-G.; writing—review and editing, V.V. and A.M.G.-C.; supervision, V.V. and A.M.G.-C. All authors have read and agreed to the published version of the manuscript.

**Funding:** This research received no external funding.

**Data Availability Statement:** Data is contained within the article and Supplementary Material.

**Acknowledgments:** Vito Verardo would like to thank the Spanish Ministry of Economy and Competitiveness (MINECO) for "Ramon y Cajal" contract (RYC-2015-18795).

**Conflicts of Interest:** The authors declare no conflict of interest.

## References

1. Donno, D.; Mellano, M.G.; Cerutti, A.K.; Beccaro, G.L. Nutraceuticals in alternative and underutilized fruits as functional food ingredients: Ancient species for new health needs. *Altern. Replace. Foods* **2018**, 261–282. [CrossRef]
2. Wang, Z.; Tang, C.; Dai, F.; Xiao, G.; Luo, G. HPLC determination of phenolic compounds in different solvent extracts of mulberry leaves and antioxidant capacity of extracts. *Int. J. Food Prop.* **2021**, *24*, 544–552. [CrossRef]
3. Baranwal, V.K.; Negi, N.; Khurana, P. Auxin response factor genes repertoire in mulberry: Identification, and structural, functional and evolutionary analyses. *Genes* **2017**, *8*, 202. [CrossRef] [PubMed]
4. Wang, Z.; Tang, C.; Xiao, G.; Dai, F.; Lin, S.; Li, Z.; Luo, G. Comparison of free and bound phenolic compositions and antioxidant activities of leaves from different mulberry varieties. *BMC Chem.* **2021**, *15*, 21.
5. Doi, K.; Kojima, T.; Makino, M.; Kimura, Y.; Fujimoto, Y. Studies on the constituents of the leaves of *Morus alba* L. *Chem. Pharm. Bull.* **2001**, *49*, 151–153. [CrossRef] [PubMed]
6. Zhong, Y.; Song, B.; Zheng, C.; Zhang, S.; Yan, Z.; Tang, Z.; Kong, X.; Duan, Y.; Li, F. Flavonoids from mulberry leaves alleviate lipid dysmetabolism in high fat diet-fed mice: Involvement of gut Microbiota. *Microorganisms* **2020**, *8*, 860. [CrossRef]
7. Becerril-sánchez, A.L.; Quintero-salazar, B.; Dublán-garcía, O.; Escalona-buendía, H.B. Phenolic compounds in honey and their relationship with antioxidant activity, botanical origin, and color. *Antioxidants* **2021**, *10*, 1700. [CrossRef]
8. Jahromi, S.G. Extraction techniques of phenolic compounds from plants. In *Plant Physiological Aspects of Phenolic Compounds Basic*; IntechOpen: London, UK, 2019; pp. 1–18.
9. Zhang, Q.W.; Lin, L.G.; Ye, W.C. Techniques for extraction and isolation of natural products: A comprehensive review. *Chin. Med.* **2018**, *13*, 20. [CrossRef]
10. Sungthong, B.; Butiman, C.; Jitsaeng, K. Optimized ultrasonic-assisted extraction of antioxidant from mulberry (*Morus alba* L.) leaves using multiple linear regression analysis. *Int. J. Pharm. Pharm. Sci.* **2014**, *6*, 914–917.
11. Mohamed Ahmed, I.A.; Al-Juhaimi, F.; Adisa, A.R.; Adiamo, O.Q.; Babiker, E.E.; Osman, M.A.; Gassem, M.A.; Ghafoor, K.; Alqah, H.A.S.; Elkareem, M.A. Optimization of ultrasound-assisted extraction of phenolic compounds and antioxidant activity from Argel (*Solenostemma argel Hayne*) leaves using response surface methodology (RSM). *J. Food Sci. Technol.* **2020**, *57*, 3071–3080. [CrossRef]
12. Polumackanycz, M.; Wesolowski, M.; Viapiana, A. *Morus alba* L. and *Morus nigra* L. Leaves as a promising food source of phenolic compounds with antioxidant activity. *Plant Foods Hum. Nutr.* **2021**, *76*, 458–465. [CrossRef] [PubMed]
13. Kim, J.H.; Doh, E.J.; Lee, G. Quantitative comparison of the marker compounds in different medicinal parts of *Morus alba* L. Using high-performance liquid chromatography-diode array detector with chemometric analysis. *Molecules* **2020**, *25*, 5592. [CrossRef] [PubMed]
14. Elshreef, H.S.; Mirghani, M.E.S.; Sulaiman, S.; Jami, M.S. A review of the effect of UAE optimization parameters on antioxidant activity. *IOP Conf. Ser. Earth Environ. Sci.* **2021**, *765*, 012109. [CrossRef]
15. Ferreira, S.L.C.; Bruns, R.E.; Ferreira, H.S.; Matos, G.D.; David, J.M.; Brandão, G.C.; da Silva, E.G.P.; Portugal, L.A.; dos Reis, P.S.; Souza, A.S.; et al. Box-Behnken design: An alternative for the optimization of analytical methods. *Anal. Chim. Acta* **2007**, *597*, 179–186. [CrossRef] [PubMed]
16. Menezes Rodrigues, G.; Cardozo-Filho, L.; da Silva, C. Pressurized liquid extraction of oil from soybean seeds. *Can. J. Chem. Eng.* **2017**, *95*, 2383–2389. [CrossRef]
17. Aiyarakanchanakun, P.; Palanuvej, C.; Ruangrungsi, N.; Phanumartwiwath, A. Pharmacognostic specifications, RP-HPLC analysis of chlorogenic acid content and antioxidant activity of morus alba linn. leaves in Thailand. *Pharmacogn. J.* **2021**, *13*, 1186–1194. [CrossRef]

18. Leyva-Jiménez, F.J.; Ruiz-Malagón, A.J.; Molina-Tijeras, J.A.; Diez-Echave, P.; Vezza, T.; Hidalgo-García, L.; Lozano-Sánchez, J.; Arráez-Román, D.; Cenis, J.L.; Lozano-Pérez, A.A.; et al. Comparative study of the antioxidant and anti-inflammatory effects of leaf extracts from four different *Morus alba* genotypes in high fat diet-induced obesity in mice. *Antioxidants* **2020**, *9*, 733. [CrossRef]
19. Brand-Williams, W.; Cuvelier, M.E.; Berset, C. Use of a free radical method to evaluate antioxidant activity. *LWT—Food Sci. Technol.* **1995**, *28*, 25–30. [CrossRef]
20. Re, R.; Pellegrini, N.; Proteggente, A.; Pannala, A.; Yang, M.; Rice-Evans, C. Antioxidant activity applying an improved ABTS radical cation decolorization assay. *Free Radic. Biol. Med.* **1999**, *26*, 1231–1237. [CrossRef]
21. Pulido, R.; Bravo, L.; Saura-Calixto, F. Antioxidant activity of dietary polyphenols as determined by a modified ferric reducing/antioxidant power assay. *J. Agric. Food Chem.* **2000**, *48*, 3396–3402. [CrossRef]
22. Verni, M.; Pontonio, E.; Krona, A.; Jacob, S.; Pinto, D.; Rinaldi, F.; Verardo, V.; Díaz-de-Cerio, E.; Coda, R.; Rizzello, C.G. Bioprocessing of brewers' spent grain enhances its antioxidant activity: Characterization of phenolic compounds and bioactive peptides. *Front. Microbiol.* **2020**, *11*, 1831. [CrossRef] [PubMed]
23. Díaz-de-cerio, E.; Tylewicz, U.; Verardo, V.; Fernández-Gutiérrez, A.; Segura-Carretero, A.; Romani, S. Design of sonotrode ultrasound-assisted extraction of phenolic compounds from *Psidium guajava* L. leaves. *Food Anal. Methods* **2017**, *10*, 2781–2791. [CrossRef]
24. Martín-García, B.; Pasini, F.; Verardo, V.; Díaz-De-cerio, E.; Tylewicz, U.; Gómez-Caravaca, A.M.; Caboni, M.F. Optimization of sonotrode ultrasonic-assisted extraction of proanthocyanidins from brewers' spent grains. *Antioxidants* **2019**, *8*, 282. [CrossRef] [PubMed]
25. Okçu, G.D.; Pakdil, N.B.; Ökten, H.E.; Yalçuk, A. A Box–Behnken design (BBD) optimization of the photocatalytic degradation of 2,4-dichlorophenoxyacetic acid (2,4-D) using $TiO_2/H_2O_2$. *Desalin. Water Treat.* **2018**, *123*, 188–195. [CrossRef]
26. Kim, D.-S.; Kang, Y.M.; Jin, W.Y.; Sung, Y.-Y.; Choi, G.; Kim, H.K. Antioxidant activities and polyphenol content of *Morus alba* leaf extracts collected from varying regions. *Biomed. Rep.* **2014**, *2*, 675–680. [CrossRef]
27. Sánchez-Salcedo, E.M.; Tassotti, M.; Del Rio, D.; Hernández, F.; Martínez, J.J.; Mena, P. (Poly)phenolic fingerprint and chemometric analysis of white (*Morus alba* L.) and black (*Morus nigra* L.) mulberry leaves by using a non-targeted UHPLC-MS approach. *Food Chem.* **2016**, *212*, 250–255. [CrossRef]
28. Cao, X.; Yang, L.; Xue, Q.; Yao, F.; Sun, J.; Yang, F.; Liu, Y. Antioxidant evaluation-guided chemical profiling and structure-activity analysis of leaf extracts from five trees in Broussonetia and Morus (*Moraceae*). *Sci. Rep.* **2020**, *10*, 4808. [CrossRef]
29. Zhou, P.; Wang, X.; Liu, P.; Huang, J.; Wang, C.; Pan, M.; Kuang, Z. Enhanced phenolic compounds extraction from Morus alba L. leaves by deep eutectic solvents combined with ultrasonic-assisted extraction. *Ind. Crops Prod.* **2018**, *120*, 147–154. [CrossRef]
30. Jin, Q.; Yang, J.; Ma, L.; Wen, D.; Chen, F.; Li, J. Identification of polyphenols in mulberry (genus Morus) cultivars by liquid chromatography with time-of-flight mass spectrometer. *J. Food Compos. Anal.* **2017**, *63*, 55–64. [CrossRef]
31. Memete, A.R.; Timar, A.V.; Vuscan, A.N.; Groza, F.M.; Venter, A.C.; Vicas, S.I. Phytochemical composition of different botanical parts of Morus species, health benefits and application in Food Industry. *Plants* **2022**, *11*, 152. [CrossRef]
32. Sánchez-Salcedo, E.M.; Mena, P.; García-Viguera, C.; Hernández, F.; Martínez, J.J. (Poly)phenolic compounds and antioxidant activity of white (*Morus alba*) and black (*Morus nigra*) mulberry leaves: Their potential for new products rich in phytochemicals. *J. Funct. Foods* **2015**, *18*, 1039–1046. [CrossRef]
33. Zou, Y.; Liao, S.; Shen, W.; Liu, F.; Tang, C.; Chen, C.Y.O.; Sun, Y. Phenolics and antioxidant activity of mulberry leaves depend on cultivar and harvest month in southern China. *Int. J. Mol. Sci.* **2012**, *13*, 16544–16553. [CrossRef] [PubMed]
34. Rodrigues, E.L.; Marcelino, G.; Silva, G.T.; Figueiredo, P.S.; Garcez, W.S.; Corsino, J.; Guimarães, R.D.C.A.; Freitas, K.D.C. Nutraceutical and medicinal potential of the *Morus* species in metabolic dysfunctions. *Int. J. Mol. Sci.* **2019**, *20*, 301. [CrossRef] [PubMed]

foods

MDPI

*Article*

# Comparison the Structural, Physicochemical, and Prebiotic Properties of Litchi Pomace Dietary Fibers before and after Modification

Yina Li [1,2], Yuanshan Yu [1,*], Jijun Wu [1], Yujuan Xu [1], Gengsheng Xiao [1], Lu Li [1] and Haoran Liu [1]

1 Sericultural & Argi-Food Research Institute, Guangdong Academy of Agricultural Sciences/Key Laboratory of Functional Foods, Ministry of Agriculture and Rural Affairs/Guangdong Key Laboratory of Agricultural Products Processing, Guangzhou 510610, China; yara1ne@163.com (Y.L.); guoshuwujijun@163.com (J.W.); guoshuxuyujuan@163.com (Y.X.); guoshuxgs@163.com (G.X.); lilu045@163.com (L.L.); winstonliu111@163.com (H.L.)

2 College of Food Sciences, South China Agricultural University, Guangzhou 510642, China

* Correspondence: yuyuanshan@gdaas.cn; Tel.: +86-159-7559-6649

**Abstract:** Litchi pomace, a by-product of litchi processing, is rich in dietary fiber. Soluble and insoluble dietary fibers were extracted from litchi pomace, and insoluble dietary fiber was modified by ultrasonic enzymatic treatment to obtain modified soluble and insoluble dietary fibers. The structural, physicochemical, and functional properties of the dietary fiber samples were evaluated and compared. It was found that all dietary fiber samples displayed typical polysaccharide absorption spectra, with arabinose being the most abundant monosaccharide component. Soluble dietary fibers from litchi pomace were morphologically fragmented and relatively smooth, with relatively high swelling capacity, whereas the insoluble dietary fibers possessed wrinkles and porous structures on the surface, as well as higher water holding capacity. Additionally, soluble dietary fiber content of litchi pomace was successfully increased by 6.32 ± 0.14% after ultrasonic enzymatic modification, and its arabinose content and apparent viscosity were also significantly increased. Further, the soluble dietary fibers exhibited superior radical scavenging ability and significantly stimulated the growth of probiotic bacterial species. Taken together, this study suggested that dietary fiber from litchi pomace could be a promising ingredient for functional foods industry.

**Keywords:** litchi pomace; dietary fiber; monosaccharide composition; structure; prebiotic activity

**Citation:** Li, Y.; Yu, Y.; Wu, J.; Xu, Y.; Xiao, G.; Li, L.; Liu, H. Comparison the Structural, Physicochemical, and Prebiotic Properties of Litchi Pomace Dietary Fibers before and after Modification. *Foods* **2022**, *11*, 248. https://doi.org/10.3390/foods11030248

Academic Editors: Michela Verni and Federico Casanova

Received: 8 December 2021
Accepted: 14 January 2022
Published: 18 January 2022



## 1. Introduction

Litchi (*Litchi chinensis* Sonn.) is an evergreen tree of the *Sapindaceae* family, with the main growing countries being China, Vietnam, Thailand, and India [1]. In China, Huaizhi and Heiye are the main litchi cultivars used for further processing to produce value-added products such as dried litchi, litchi juice, and litchi wine [2]. However, the resulting pomace, pericarp, and seeds are frequently discarded, which have been reported to contain diverse nutrients and bioactive compounds, including polysaccharides, crude fiber, and polyphenols [3]. Therefore, increasing research has focused on studying the utilization of nutrients and bioactive compounds from litchi fruit waste for the reduction of environmental burden and the potential application in food and/or functional food industry [4–6].

Litchi pomace produced by the process is more often used for feed production. At present, research on the structural and functional properties of dietary fiber extracted from litchi pomace is limited compared to polysaccharides and polyphenols [7–9]. Dietary fiber is a broad category of indigestible food ingredients and roughly classified as water-soluble dietary fiber and water-insoluble dietary fiber according to their solubility in water. Specifically, water-soluble dietary fiber mainly includes pentosan, soluble hemicellulose, gum, and pectin, while cellulose, insoluble hemicellulose and lignin are the major compositions of IDF [10]. Multiple animal studies and human trials have demonstrated that dietary fiber possess several biological activities, such as improving intestinal absorption, increasing satiety, enhancing immune function, and promoting colon health, which are determined by the source of dietary fiber with different chemical structures and compositions [11]. Moreover, many studies have been conducted to modify the structure and composition of dietary fiber through different treatments to enhance and modify its function. The single modification method has its limitations, whereas the combined modification possesses a stronger modifying effect [12,13]. Nowadays, ultrasonic enzymatic treatment is more applied to modify dietary fiber, owing to its safety, low damage to the molecular structure of dietary fiber, and its effectiveness in increasing the water-soluble dietary fiber content [14].

To better utilize the litchi fruit waste and explore the potential application of litchi dietary fiber in food and/or functional food industry, the current study aimed to compare the differences in structural and functional properties of dietary fibers obtained from litchi pomace, so as to provide ideas and theoretical basis for the further processing of litchi pomace.

## 2. Materials and Methods

### 2.1. Materials

Litchi pomace (cv. Huaizhi) was provided by Guangzhou Shunchangyuan Green Food Co., Ltd. (Guangzhou, Guangdong Province, China). Prior to dietary fiber extraction, litchi pomace was desugared by *Saccharomyces cerevisiae* fermentation. Briefly, litchi pomace was heated in a water bath at 80 °C for 20 min, 0.02% (*w*/*w*) activated *Saccharomyces cerevisiae* was added after returning to room temperature, and the mixture was placed in a constant temperature incubator for fermentation at 30 °C for 24 h, then dried with the moisture content of 8 to 12% by weight using the hot air vacuum drying. The dried sample was crushed and passed through a 60-mesh sieve to obtain the desugared litchi pomace powder, which was kept in a desiccator for later use. The monosaccharide standards were HPLC grade, and were purchased from Shanghai Yuanye Biological Technology Co., LTD (Shanghai, China). Other chemicals used in this study were of analytical grade.

### 2.2. Preparation of Dietary Fiber from Litchi Pomace

Dietary fiber from litchi pomace was prepared as previously described by Wu et al. [15] with minor modifications. Initially, the desugared litchi pomace powder was mixed with distilled water (1:20, *w*:*v*) and hydrolyzed by 0.2% (*w*/*w*, 2000 U/g) α-amylase in a shaking bath (150 r/min) at pH 6.0, 60 °C for 1 h. Thereafter, 1.0% (*w*/*w*, 100 U/mg) neutral protease was added and incubated at pH 7.0, 50 °C for 1.5 h. The treated mixture was centrifuged at 6000× *g* for 10 min after the inactivation of enzyme in boiling water bath for 10 min. The resulting residue was washed with distilled water, 95% (*v*/*v*) ethanol, and acetone and then freeze-dried in a vacuum freezer to obtain the water-insoluble dietary fiber (IDF), whereas the supernatant was vacuum-concentrated (50 °C, 40 rpm) and precipitated with four times volumes of 95% ethanol at room temperature for 12 h. After centrifugation at 6000× *g* for 10 min, the precipitate was collected and dissolved in distilled water to remove ethanol by vacuum rotary evaporation before lyophilization to obtain the water-soluble dietary fiber (SDF). The yields of SDF and IDF obtained from litchi pomace were 7.28 ± 0.13% and 69.30 ± 0.78%, respectively.

### 2.3. Ultrasonic Enzymatic Modification of IDF from Litchi Pomace

Ultrasonic enzymatic modification was conducted as the following steps: Initially, IDF was mixed with distilled water (1:20, *w*:*v*) and sonicated at 450 W, 50 °C for 30 min using the ultrasonic bath (DL-800B, Zhixin, China). Subsequently, 50 μL/g of cellulase (700 EGU/g) was added, and the mixture was incubated for 3 h at 50 °C with constant shaking (150 rpm). After the incubation, the cellulase was inactivated, followed by the centrifugation of mixture, and the resulting supernatant and precipitation were treated by aforementioned process. The water-soluble dietary fiber sample obtained was marked as M-SDF, and the water-insoluble dietary fiber sample obtained was marked as M-IDF. The yields of M-SDF and M-IDF obtained from litchi pomace were 6.32 ± 0.14% and 55.67 ± 0.53%, respectively.

### 2.4. Monosaccharide Composition

The monosaccharide composition of dietary fibers from litchi pomace was measured by 1-phenyl-3-methyl-5-pyrazolone (PMP) derivatization and high-performance liquid chromatography (HPLC, Agilent 1200 series, Mundelein, IL, USA) with a Wondasil C18 column (4.6 mm × 250 mm i.d., 5 μm particle size) according to the method of Liu et al. [16]. The standard monosaccharides, including mannose, ribose, rhamnose, galactose acid, glucose, galactose, xylose, and arabinose, were used to analyze the monosaccharides in hydrolyzed samples. Chromatographic separation was carried out using 0.05 M sodium phosphate buffer (pH 6.85) and acetonitrile at a ratio of 82:18 (*v:v*) as mobile phase at the flow rate of 1 mL/min. The detection wavelength was set as 250 nm.

### 2.5. Scanning Electron Microscopy (SEM)

The dietary fiber samples were mounted on an aluminum stub with double-sided stick tape and coated with a 10 nm gold layer. An electron microscope (S-3000N, Hitachi Ltd., Tokyo, Japan) captured the scanning images at an accelerating voltage of 20.0 kV. The micrographs were taken at ×1000 magnification (scale bar 50 μm).

### 2.6. Fourier Transform Infrared Spectroscopy (FTIR)

Changes in the molecular structure of samples were obtained at room temperature using FTIR (VERTEX 70, Bruker, Karlsruhe, Germany). The dry sample was mixed with KBr powder under infrared irradiation and pressed into a tablet. Spectra were collected in the range of 4000–400 $cm^{-1}$ with a resolution of 4 $cm^{-1}$ [17].

### 2.7. Hydration Properties

Hydration properties, including the water holding capacity (WHC), swelling capacity (SC), and water solubility (WS), were determined by the methods described by Shen et al. [18]. The WHC, SC, and WS were reported as mean value of three determinations per sample.

### 2.8. Rheological Property

The viscosity of water-soluble dietary fiber samples was measured by an AR1500EX rheometer (TA Instruments Ltd., New Castle, DE, USA) equipped with a cone and plane geometry system (40 mm diameter, 1 mm gap). In steady shear tests, the steady flow behaviors of water-soluble dietary fiber samples with different concentrations (4, 8, and 12%, *w*/*v*) were measured at a 0.1–100 $s^{-1}$ shear rate range and 25 °C.

### 2.9. The Radical Scavenging Activity

The antioxidant active ingredient was extracted on the basis of the previous method [19], and the radical scavenging activity was measured via ABTS assay as per the method of Ruiz-Torralba et al. [20] with slight modifications. Briefly, 20 μL of each sample extract/sample aqueous solution with different mass concentrations were individually placed in a 96-well polystyrene microplate, after which 200 μL of the working ABTS solution was added. The mixture was stored at room temperature in darkness for 10 min, and the absorbance was measured by the microplate reader at 734 nm. The percent scavenging activity was calculated as follows:

$$\text{ABTS radical scavenging activity (\%)} = 100 \times [1 - (A_1 - A_2)/A_0]$$

where $A_0$ is the absorbance of the ABTS solution with ethanol instead of sample, $A_1$ is the absorbance of the ABTS solution with sample, and $A_2$ is the absorbance of the sample with anhydrous methanol instead of ABTS solution

### 2.10. In Vitro Probiotic Activity

In this study, both *Lactobacillus acidophilus* and *Lactobacillus plantarum* were provided by the microbiology laboratory of Sericultural and Agri-Food Research Institute (Guangdong, China) and were used to evaluate the probiotic activity of SDF and M-SDF. The preparation of inoculants was determined in accordance with steps described by Okolie et al. [21]. First, the sugar-free MRS base medium was prepared as described by Safoura et al. [22]. Then, glucose (negative control), inulin (positive control), SDF, and M-SDF were added to sugar-free MRS base medium separately at concentrations of 2% ($w/v$), while sugar-free MRS basal medium served as a blank control. All MRS media were autoclaved at 121 °C for 20 min and cooled to 50 °C, followed by the inoculation of 2% ($v/v$) above bacterial suspensions and incubation at 37 °C for 24 h. Growth of each strain was monitored by measuring pH and the optical density (OD) of 10-fold diluted at 600 nm.

### 2.11. Statistical Analysis

All measurements were performed at least in triplicate, and the results were expressed as the mean values with their standard deviation. Statistical analyses were carried using the SPSS 22.0 software (SPSS Inc., Chicago, IL, USA). Data were analyzed using one-way ANOVA and Duncan's test. $p < 0.05$ was considered to be statistically significant.

## 3. Results and Discussion

### 3.1. Structural Characterization

#### 3.1.1. Monosaccharide Composition Analysis

A total of eight monosaccharides were detected in all four dietary fiber samples, and the monosaccharide content varied among four different dietary fiber samples (Table 1). The highest content was arabinose (83.14 ± 0.39–233.11 ± 1.84 mg/g), which was higher than that reported in defatted coconut flour (11.84 ± 0.82 mg/g) and pear pomace dietary fiber (20.40 ± 6.30 mg/g) [23,24]. Wang et al. [25] found that glucose was the main monosaccharide in orange SDF (64.00 ± 1.73 mg/g), while the main monosaccharide of the SDFs from grapefruit and lemon was arabinose (100.72 ± 2.43 mg/g and 58.82 ± 1.84 mg/g). The abundance of arabinose is characteristic of dietary fiber from litchi pomace, which has been reported to have physiological activities such as regulating lipid metabolism, affecting intestinal microbiota and metabolism [26,27]. Further, SDF and M-SDF contained significantly higher levels of galacturonic acid, arabinose, rhamnose, and galactose, which were typical constituents of pectin [28], and thus it may be concluded that arabinose-rich pectin is the key component of soluble dietary fiber from litchi pomace. The high contents of arabinose, glucose, and xylose in IDF and M-IDF indicated that arabinoxylan and cellulose may be the main components of insoluble dietary fiber from litchi pomace [29]. In addition, M-SDF contained more galacturonic acid, glucose, and arabinose glucose than SDF, whereas the concentrations of galacturonic acid, xylose, and arabinose in M-IDF were lower than that in IDF, indicating that ultrasonic enzymatic modification may lead to the hydrolysis of cellulose in the cell wall and the release of some xylan hemicellulose, thus facilitating the conversion of insoluble dietary fiber to soluble dietary fiber [30].

#### 3.1.2. SEM Analysis

As depicted in Figure 1, the morphological characterization of the dietary fiber samples at magnification times of 1000 was observed using SEM. These micro images clearly indicated the significant variations of surface topography among the dietary fiber samples. Compared to SDF, IDF had rougher topography with obvious wrinkles, which might have been the result of different dietary fibers types. In the study of Lyu et al. [31], the tissue state of high-purity insoluble fiber from soybean dregs showed hierarchical and low aggregation degree. The finding of Chen et al. [32] revealed that wheat-soluble dietary fiber had a relatively flat and loose structure with a certain gap between fibers, whereas the surface of wheat-insoluble dietary fiber was irregular with a large number of cracks and small clumps. In addition, the microstructure of the dietary fibers obtained by ultrasonic enzymatic modification showed significant changes in morphology, with M-SDF appearing more fragmented than SDF and M-IDF, presenting cellular network structure with stronger porosity than IDF, which indicated that ultrasonic enzymatic modification can destroy the matrix of cellulose, hemicellulose, and lignin, thus promoting the transformation of IDF to M-SDF [33]. Similar results were obtained by Ma et al. [17]. The physiochemical properties of dietary fiber are determined by its microstructure. Porous and folded structure can increase the specific surface area and expose more polar groups, consequently promoting the adsorption and binding of water, which further affects its application in food [34].

**Table 1.** Monosaccharide compositions of SDF, M-SDF, IDF, and M-IDF.

| mg/g | SDF | M-SDF | IDF | M-IDF |
|---|---|---|---|---|
| Mannose | $37.49 \pm 0.16^{a}$ | $27.22 \pm 0.69^{b}$ | $26.63 \pm 1.27^{b}$ | $25.44 \pm 1.29^{b}$ |
| Ribose | $0.74 \pm 0.19^{a}$ | $0.16 \pm 0.01^{b}$ | $0.41 \pm 0.04^{b}$ | $0.25 \pm 0.02^{b}$ |
| Rhamnose | $17.81 \pm 0.13^{a}$ | $18.63 \pm 0.65^{a}$ | $8.43 \pm 0.49^{b}$ | $5.86 \pm 0.18^{c}$ |
| Galacturonic acid | $111.87 \pm 4.54^{b}$ | $133.27 \pm 6.46^{a}$ | $25.38 \pm 1.27^{c}$ | $11.27 \pm 0.48^{d}$ |
| Glucose | $44.40 \pm 1.87^{c}$ | $79.17 \pm 0.55^{a}$ | $50.14 \pm 0.98^{b}$ | $49.09 \pm 1.88^{b}$ |
| Galactose | $41.26 \pm 0.14^{a}$ | $30.73 \pm 1.09^{b}$ | $20.55 \pm 0.27^{c}$ | $15.75 \pm 0.15^{d}$ |
| Xylose | $3.21 \pm 0.06^{d}$ | $7.03 \pm 0.10^{c}$ | $21.90 \pm 0.38^{a}$ | $12.32 \pm 0.86^{b}$ |
| Arabinose | $187.02 \pm 2.17^{b}$ | $233.11 \pm 1.84^{a}$ | $102.00 \pm 0.47^{c}$ | $83.14 \pm 0.39^{d}$ |

Different letters (a, b, c, d) in the same row indicate significantly different means at $p < 0.05$ (Duncan's test).

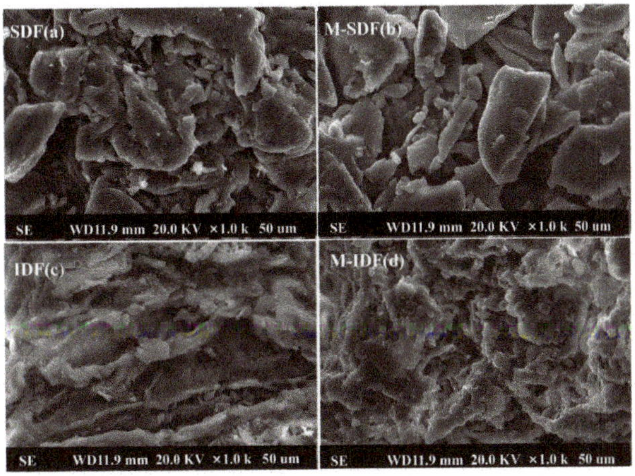


**Figure 1.** SEM images of SDF (**a**), M-SDF (**b**), IDF (**c**), and M-IDF (**d**). Magnification: 1000× (scale bar 50 μm).

### 3.1.3. FTIR Analysis

The infrared spectra of all dietary fiber samples displayed the characteristic absorption peaks of cellulose polysaccharides (Figure 2). The overall peak intensity of SDF and M-SDF was significantly higher than that of IDF and M-IDF, indicating that SDF and M-SDF had more typical polysaccharide complex structures. The wide absorption band near 3400 $cm^{-1}$ was caused by the stretching of O–H groups in polysaccharides. The peak intensity of M-SDF was relatively high at this wavelength, indicating that there were more hydrogen bonds in the associative state, which may have been related to M-SDF containing more gels (galactose acid) and hemicelluloses (mannose, glucose, galactose, and arabinose) (Table 1) [30]. The absorption peak near 1620 $cm^{-1}$ was the stretching vibration absorption of C=O, suggesting the presence of amide groups. The absorption peak in the range of 1200–1400 cm$^{-1}$ was caused by the angular vibrations of C–H, representing the typical structure of a carbohydrate skeleton. The absorption peak in the range of 1200–1000 cm$^{-1}$ corresponded to the stretching vibrations of C–C, C–O, and C–O–C, which were frequently reported as the presence of sugar aldehyde groups [35]. The modified dietary fibers presented stronger absorption peaks at the wavelength of these characteristic romaegions, which indicated that the supramolecular structure and reactive groups of litchi pomace dietary fiber were changed by the ultrasonic enzymatic modification. In addition, IDF and M-IDF were observed to have characteristic absorption peaks of aromatic lignin hydrocarbons near 1530 cm$^{-1}$ [36], while SDF and M-SDF showed clearly characteristic absorption peaks of β-glucoside bond, furanose, and α- and β- pyranose in the range of 1000–700 cm$^{-1}$ [37].

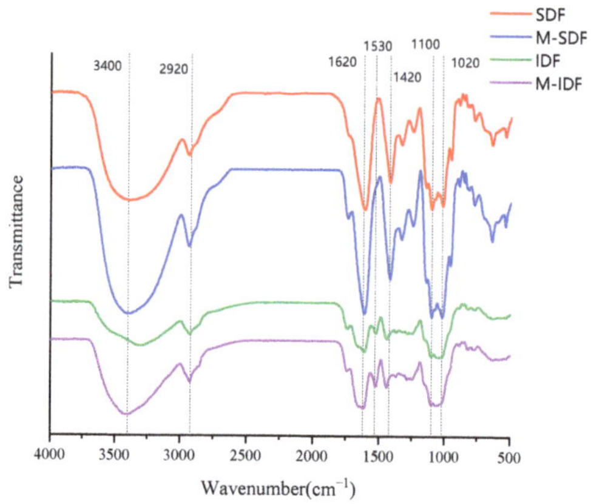


**Figure 2.** FTIR of SDF, M-SDF, IDF, and M-IDF.

### 3.2. Physicochemical Properties

#### 3.2.1. Hydration Properties

Hydration properties of the dietary fiber samples are presented in Table 2. WHC refers to the ability of wet materials to retain water under external force (centrifugal force or compression, etc.). SC measures the change of the total volume of materials and water, while WS reflects the degree to which materials can be dissolved by water at a certain temperature [38]. As observed from the results, SDF had the highest SC and WS, while the WHC of IDF was higher than that of SDF. These results were consistent with the findings of Zhu et al. [39], which might have been due to greater fiber porosity and hydrogen bond content in IDF. After ultrasonic enzymatic modification, the SC of M-IDF was 7.18 ± 0.25 mL/g, whereas the WHC decreased to 9.07 ± 0.09 g/g, which were higher relative to that of apple fiber (6.89 ± 0.11 mL/g and 6.12 ± 0.11 g/g) [40]. A previous work found that the SC and WHC were decreased in rice bran dietary fiber after cellulase modification [33], which was attributed to cellulase treatment changing the contents of hydrophilic components such as hemicellulose and cellulose. Hydration properties of dietary fiber are closely related to its source, structural morphology, porosity, and processing parameters [41]. Typically, the high WHC and SC indicate that litchi pomace dietary fiber may be considered for use in food processing to modify texture and avoid dehydration in formula foods.

**Table 2.** Hydration properties of SDF, M-SDF, IDF, and M-IDF.

| mg/g | SDF | M-SDF | IDF | M-IDF |
|---|---|---|---|---|
| WHC (g/g) | 6.75 ± 0.09 $^{c}$ | 6.9 ± 0.13 $^{c}$ | 9.69 ± 0.2 $^{a}$ | 9.07 ± 0.09 $^{b}$ |
| SC (mL/g) | 8.63 ± 0.18 $^{a}$ | 7.75 ± 0.35 $^{b}$ | 6.65 ± 0.28 $^{c}$ | 7.18 ± 0.25 $^{b,c}$ |
| WS (%) | 83.1 ± 1.76 $^{a}$ | 78.34 ± 0.2 $^{b}$ | 10.22 ± 0.17 $^{c}$ | 9.29 ± 0.32 $^{c}$ |

Different letters (a, b, c) in the same row indicate significantly different means at $p < 0.05$ (Duncan's test).

3.2.2. Rheological Properties

The steady flow test curves of SDF and M-SDF are plotted in Figure 3. The apparent viscosity of SDF and M-SDF decreased gradually with the increase of shear rate at room temperature, presenting shear dilution phenomenon, which showed that SDF and M-SDF belonged to pseudoplastic non-Newtonian fluids. Moreover, with the increase of concentration, the apparent viscosities of SDF and M-SDF increased accordingly. It can be seen that the concentration of the sample had a direct and nonlinear effect on the viscosities of the constant temperature solution, which was consistent with the results of Feng et al. [42]. This might have been due to the fact that the interaction force among the SDFs molecules elevated at high concentration, which increased the degree of cross-linking and polymerization of substances. Additionally, at the same concentration and shear rate, the apparent viscosity of M-SDF was greater than that of SDF, which may have been associated with the greater pectin structures and longer the molecular chain length of M-SDF [24,43]. Previous studies have shown that viscosity is one of the important properties of SDF, which is closely related to the ability of SDF to delay and reduce the absorption of food components in the digestive tract and reduce the postprandial blood glucose response [44]. Thus, we speculated that M-SDF may have a better in vivo physiological activity than SDF.

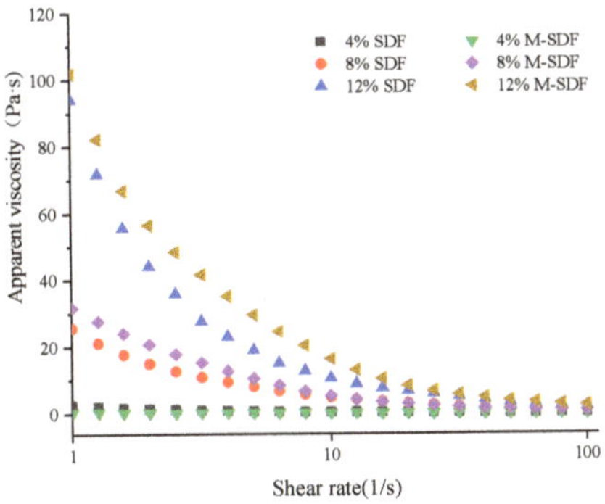


**Figure 3.** Viscosity curves of SDF and M-SDF at 4, 8, and 12% ($w/v$).

## 3.3. *Functional Properties*

### 3.3.1. The Radical Scavenging Activity

As shown in Figure 4, both types of the dietary fiber samples exhibited radical scavenging activity with a clear dose-dependent manner. At the dose of 5.0 mg/mL, the ABTS radical scavenging activity of SDF, M-SDF, IDF, and M-IDF extracts were, respectively, 50.24%, 19.08%, 58.92%, and 59.95% (Figure 4a), which was similar to that reported in black mulberry [45]. Consistent with the findings of a previous work [46], IDF extract was observed to possess higher radical scavenging as compared to SDF extract, which can be attributed to a higher amount of phenols that were co-extracted with the molecule [47]. The hydroxyl groups of phenolic compounds could donate electrons or hydrogen atoms to enhance the antioxidant activity [48]. In addition, at the dose of 3.0 mg/mL, the ABTS radical scavenging activities of dietary fiber aqueous solution were 82.24%, 65.20%, 27.90%, and 18.20%, respectively (Figure 4b). The radical scavenging activity of SDF and M-SDF aqueous solution was significantly higher than that of IDF and M-IDF aqueous solution, which was probably due to its higher content of uronic acid and the presence of hydroxyl groups in the main backbone [32]. SDF possessed relatively higher radical scavenging capacity than M-SDF, which may be related to its glycosidic bonding and the molecular chain binding substances [49].

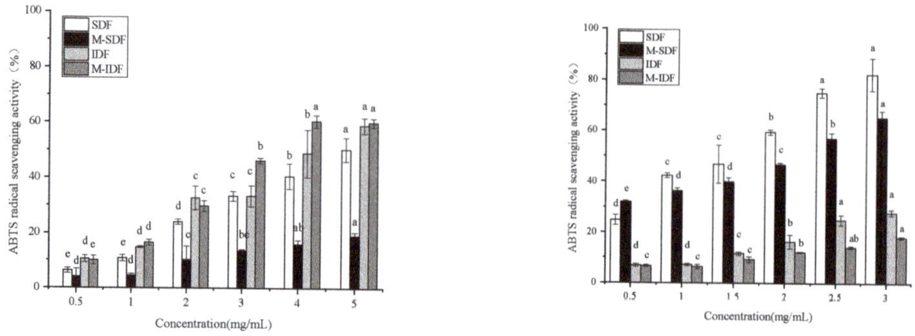


(**a**) (**b**)

**Figure 4.** (**a**) ABTS radical scavenging activities of sample extracts; (**b**) ABTS radical scavenging activities of sample aqueous solutions. Different letters (a, b, c, d, e) in the same row indicate significantly different means at $p < 0.05$ (Duncan's test).

### 3.3.2. In Vitro Probiotic Activity

In this study, we evaluated the ability of *Lactobacillus acidophilus* and *Lactobacillus plantarum* to utilize SDF and M-SDF as their carbon sources in glucose-free MRS base medium in comparison with inulin or glucose as sole carbon source. As showed in Figure 5, the growth of *Lactobacillus acidophilus* and *Lactobacillus plantarum* in the MRS medium supplemented with glucose was significantly higher than other media after 24 h of fermentation. The depletion of carbon sources promoted the growth of probiotics and correspondingly lowered the pH of the medium. Our results were similar to that of Chen et al. [50]. SDF and M-SDF significantly improved the growth of two bacterial species as compared to the sugar-free MRS medium, but its probiotic effect was weaker than that of inulin. In addition, M-SDF promoted the growth of *Lactobacillus plantarum* better than SDF at the concentration of 2.0%, which may have been due to the complex microstructure and the different monosaccharide ratios that affect accessibility and fermentability of dietary fiber. In living organisms, most of glucose is expected to be absorbed into the blood, rather than stimulating bacterial growth in the large intestine [51]. Related studies have shown that dietary fiber can resist the hydrolysis by digestive enzymes in the upper gut and undergo fermentation in the lower gut by the gut microbiota, stimulating the growth and activity of beneficial bacterial populations [52,53]. Therefore, it could be speculated that SDF and M-SDF are good substrates for promoting probiotic growth.

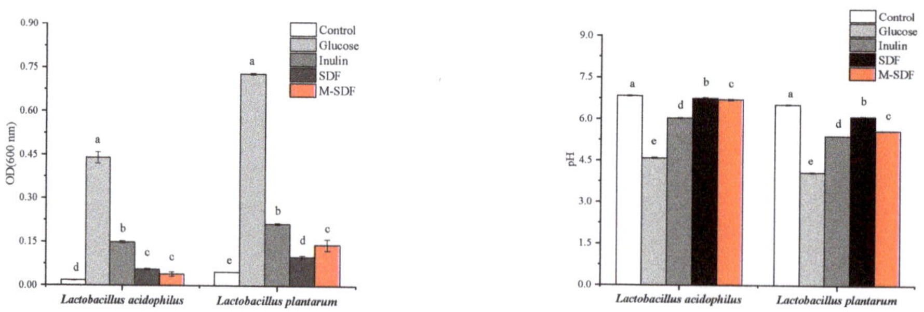


(**a**) (**b**)

**Figure 5.** (**a**) OD value of MRS medium; (**b**) pH value of MRS medium. Different letters (a, b, c, d, e) in the same row indicate significantly different means at $p < 0.05$ (Duncan's test).

## 4. Conclusions

This study reports the first data in the literature for the dietary fiber samples obtained from litchi pomace and modified by ultrasonic enzymatic treatment, contributing to better understand the structural and functional properties of litchi pomace dietary fibers. The results indicated that a total of eight monosaccharides were detected in the four dietary fiber samples, with arabinose being the most abundant. All dietary fiber samples had the typical polysaccharide functional groups and different morphology. Additionally, SDF exhibited superior SC and WS, while IDF exhibited greater WHC. Compared to other fibers, the polysaccharide constituents of litchi pomace dietary fibers were strongly hydrophilic, and those materials might be well suited for formulated foods to reduce calories, avoid syneresis, prolong stability, and modify the texture. Moreover, soluble dietary fiber content of litchi pomace was successfully increased with enhanced arabinose content and apparent viscosity after ultrasonic enzymatic modification. Both SDF and M-SDF exhibited superior radical scavenging ability and potential as a source of prebiotics. Considering their ideal functional properties in vitro, we suggest soluble dietary fiber from litchi pomace could be exploited as an ingredient in functional foods. The functional activity in vivo of litchi pomace dietary fiber and its physicochemical and nutritional interactions with other ingredients in the food system are worthy of further study.

**Author Contributions:** Conceptualization, Y.Y. and Y.L.; methodology, Y.Y.; software, Y.L. and L.L.; validation, G.X., J.W. and Y.X.; formal analysis, Y.Y.; investigation, Y.L.; resources, Y.X.; data curation, Y.Y., L.L. and Y.L.; writing—original draft preparation, Y.L.; writing—review and editing, Y.L. and H.L.; visualization, H.L.; supervision, G.X. and J.W.; project administration, Y.Y.; funding acquisition, Y.X. All authors have read and agreed to the published version of the manuscript.

**Funding:** This research was funded by agricultural advantage industry discipline team projects of Guangdong Academy of Agricultural Sciences in the 14th Five-Year Period (no. 202109TD), Guangdong Provincial Special Fund for Litchi Industry Innovation Team of Modern Agriculture Industry Technology Innovation Teams (no. 2021KJ107-3), and personnel training program of Guangdong Academy of Agricultural Sciences (no. R2020PY-JG010).

**Institutional Review Board Statement:** Not applicable.

**Informed Consent Statement:** Not applicable.

**Data Availability Statement:** All data from this study have been reported in the manuscript.

**Acknowledgments:** The authors would like to thank all those who contributed directly or indirectly to the project.

**Conflicts of Interest:** The authors declare no conflict of interest and that the funder had no role in the study design and implementation.

## References

1. Pareek, S. Nutritional and Biochemical Composition of Lychee (*Litchi chinensis* Sonn.) Cultivars. In *Nutritional Composition of Fruit Cultivars*; Simmonds, M.S.J., Preedy, V.R., Eds.; Elsevier: Amsterdam, The Netherlands, 2016. [CrossRef]
2. Jiang, Y.M.; Wang, Y.; Song, L.; Liu, H.; Lichter, A.; Kerdchoechuen, O.; Joyce, D.C.; Shi, J. Postharvest characteristics and handling of litchi fruit—An overview. *Aust. J. Exp. Agric.* **2006**, *46*, 1541–1556. [CrossRef]
3. Zhu, X.-R.; Wang, H.; Sun, J.; Yang, B.; Duan, X.-W.; Jiang, Y.-M. Pericarp and seed of litchi and longan fruits: Constituent, extraction, bioactive activity, and potential utilization. *J. Zhejiang Univ. Sci. B* **2019**, *20*, 503–512. [CrossRef]
4. Pop, C.; Suharoschi, R.; Pop, O. Dietary Fiber and Prebiotic Compounds in Fruits and Vegetables Food Waste. *Sustainability* **2021**, *13*, 7219. [CrossRef]
5. Emanuele, S.; Lauricella, M.; Calvaruso, G.; D'Anneo, A.; Giuliano, M. Litchi chinensis as a Functional Food and a Source of Antitumor Compounds: An Overview and a Description of Biochemical Pathways. *Nutrients* **2017**, *9*, 992. [CrossRef]
6. Su, D.; Zhang, R.; Zhang, C.; Huang, F.; Xiao, J.; Deng, Y.; Wei, Z.; Zhang, Y.; Chi, J.; Zhang, M. Phenolic-rich lychee (*Litchi chinensis Sonn.*) pulp extracts offer hepatoprotection against restraint stress-induced liver injury in mice by modulating mitochondrial dysfunction. *Food Funct.* **2016**, *7*, 508–515. [CrossRef]
7. Huang, F.; Zhang, R.; Liu, Y.; Xiao, J.; Liu, L.; Wei, Z.; Yi, Y.; Zhang, M.; Liu, D. Dietary litchi pulp polysaccharides could enhance immunomodulatory and antioxidant effects in mice. *Int. J. Biol. Macromol.* **2016**, *92*, 1067–1073. [CrossRef] [PubMed]

8. Su, D.; Zhang, R.; Hou, F.; Zhang, M.; Guo, J.; Huang, F.; Deng, Y.; Wei, Z. Comparison of the free and bound phenolic profiles and cellular antioxidant activities of litchi pulp extracts from different solvents. *BMC Complement. Altern. Med.* **2014**, *14*, 9. [CrossRef]
9. Zhao, L.; Wang, K.; Wang, K.; Zhu, J.; Hu, Z. Nutrient components, health benefits, and safety of litchi (*Litchi chinensis Sonn.*): A review. *Compr. Rev. Food Sci. Food Saf.* **2020**, *19*, 2139–2163. [CrossRef]
10. Perry, J.R.; Ying, W. A Review of Physiological Effects of Soluble and Insoluble Dietary Fibers. A Review of Physiological Effects of Soluble and Insoluble Dietary Fibers. *J. Nutr. Food Sci.* **2016**, *6*, 476. [CrossRef]
11. Fuller, S.; Beck, E.; Salman, H.; Tapsell, L. New Horizons for the Study of Dietary Fiber and Health: A Review. *Plant Foods Hum. Nutr.* **2016**, *71*, 1–12. [CrossRef]
12. Ain, H.B.U.; Saeed, F.; Ahmed, A.; Khan, M.A.; Niaz, B.; Tufail, T. Improving the physicochemical properties of partially enhanced soluble dietary fiber through innovative techniques: A coherent review. *J. Food Process. Preserv.* **2019**, *43*, e13917. [CrossRef]
13. Ozyurt, V.H.; Ötles, S. Effect of food processing on the physicochemical properties of dietary fibre. *Acta Sci. Pol. Technol. Aliment.* **2016**, *15*, 233–245. [CrossRef] [PubMed]
14. Gan, J.; Xie, L.; Peng, G.; Xie, J.; Chen, Y.; Yu, Q. Systematic review on modification methods of dietary fiber. *Food Hydrocoll.* **2021**, *119*, 106872. [CrossRef]
15. Wu, W.; Hu, J.; Gao, H.; Chen, H.; Fang, X.; Mu, H.; Han, Y.; Liu, R. The potential cholesterol-lowering and prebiotic effects of bamboo shoot dietary fibers and their structural characteristics. *Food Chem.* **2020**, *332*, 127372. [CrossRef] [PubMed]
16. Liu, J.; Wang, Z.; Wang, Z.; Hao, Y.; Wang, Y.; Yang, Z.; Li, W.; Wang, J. Physicochemical and functional properties of soluble dietary fiber from different colored quinoa varieties (Chenopodium quinoa Willd). *J. Cereal Sci.* **2020**, *95*, 103045. [CrossRef]
17. Ma, R.; Chen, J.-N.; Zhou, X.-J.; Lin, H.; Gao, Q.; Peng, X.; Tanokura, M.; Xue, Y.-L. Effect of chemical and enzymatic modifications on the structural and physicochemical properties of dietary fiber from purple turnip (*Brassica rapa* L.). *LWT* **2021**, *145*, 111313. [CrossRef]
18. Shen, M.; Weihao, W.; Cao, L. Soluble dietary fibers from black soybean hulls: Physical and enzymatic modification, structure, physical properties, and cholesterol binding capacity. *J. Food Sci.* **2020**, *85*, 1668–1674. [CrossRef]
19. Chandrasekara, A.; Shahidi, F. Content of Insoluble Bound Phenolics in Millets and Their Contribution to Antioxidant Capacity. *J. Agric. Food Chem.* **2010**, *58*, 6706–6714. [CrossRef]
20. Ruiz-Torralba, A.; Guerra-Hernández, E.J.; García-Villanova, B. Antioxidant capacity, polyphenol content and contribution to dietary intake of 52 fruits sold in Spain. *CyTA -J. Food* **2018**, *16*, 1131–1138. [CrossRef]
21. Okolie, C.L.; Mason, B.; Mohan, A.; Pitts, N.; Udenigwe, C.C. The comparative influence of novel extraction technologies on in vitro prebiotic-inducing chemical properties of fucoidan extracts from Ascophyllum nodosum. *Food Hydrocoll.* **2018**, *90*, 462–471. [CrossRef]
22. Akbari-Alavijeh, S.; Soleimanian-Zad, S.; Sheikh-Zeinoddin, M.; Hashmi, S. Pistachio hull water-soluble polysaccharides as a novel prebiotic agent. *Int. J. Biol. Macromol.* **2018**, *107*, 808–816. [CrossRef] [PubMed]
23. Du, X.; Wang, L.; Huang, X.; Jing, H.; Ye, X.; Gao, W.; Bai, X.; Wang, H. Effects of different extraction methods on structure and properties of soluble dietary fiber from defatted coconut flour. *LWT* **2021**, *143*, 111031. [CrossRef]
24. Yan, L.; Li, T.; Liu, C.; Zheng, L. Effects of high hydrostatic pressure and superfine grinding treatment on physicochemical/functional properties of pear pomace and chemical composition of its soluble dietary fibre. *LWT* **2019**, *107*, 171–177. [CrossRef]
25. Wang, L.; Xu, H.; Yuan, F.; Pan, Q.; Fan, R.; Gao, Y. Physicochemical characterization of five types of citrus dietary fibers. *Biocatal. Agric. Biotechnol.* **2015**, *4*, 250–258. [CrossRef]
26. Hao, L.; Lu, X.; Sun, M.; Li, K.; Shen, L.; Wu, T. Protective effects of L-arabinose in high-carbohydrate, high-fat diet-induced metabolic syndrome in rats. *Food Nutr. Res.* **2015**, *59*, 28886. [CrossRef]
27. Tamura, M.; Kurusu, Y.; Hori, S. Effect of Dietary L-arabinose on the Intestinal Microbiota and Metabolism of Dietary Daidzein in Adult Mice. *Biosci. Microbiota Food Heal.* **2012**, *31*, 59–65. [CrossRef] [PubMed]
28. Wandee, Y.; Uttapap, D.; Mischnick, P. Yield and structural composition of pomelo peel pectins extracted under acidic and alkaline conditions. *Food Hydrocoll.* **2019**, *87*, 237–244. [CrossRef]
29. Qi, J.; Yokoyama, W.; Masamba, K.G.; Majeed, H.; Zhong, F.; Li, Y. Structural and physico-chemical properties of insoluble rice bran fiber: Effect of acid–base induced modifications. *RSC Adv.* **2015**, *5*, 79915–79923. [CrossRef]
30. Wang, K.; Li, M.; Wang, Y.; Liu, Z.; Ni, Y. Effects of extraction methods on the structural characteristics and functional properties of dietary fiber extracted from kiwifruit (Actinidia deliciosa). *Food Hydrocoll.* **2020**, *110*, 106162. [CrossRef]
31. Lyu, B.; Wang, H.; Swallah, M.S.; Fu, H.; Shen, Y.; Guo, Z.; Tong, X.; Li, Y.; Yu, H.; Jiang, L. Structure, Properties and Potential Bioactivities of High-purity Insoluble Fibre from Soybean Dregs (Okara). *Food Chem.* **2021**, *364*, 130402. [CrossRef] [PubMed]
32. Chen, H.; Xiong, M.; Bai, T.; Chen, D.; Zhang, Q.; Lin, D.; Liu, Y.; Liu, A.; Huang, Z.; Qin, W. Comparative study on the structure, physicochemical, and functional properties of dietary fiber extracts from quinoa and wheat. *LWT* **2021**, *149*, 111816. [CrossRef]
33. Wen, Y.; Niu, M.; Zhang, B.; Zhao, S.; Xiong, S. Structural characteristics and functional properties of rice bran dietary fiber modified by enzymatic and enzyme-micronization treatments. *LWT–Food Sci. Technol.* **2017**, *75*, 344–351. [CrossRef]
34. Chen, Y.; Ye, R.; Yin, L.; Zhang, N. Novel blasting extrusion processing improved the physicochemical properties of soluble dietary fiber from soybean residue and in vivo evaluation. *J. Food Eng.* **2014**, *120*, 1–8. [CrossRef]
35. Lettow, M.; Grabarics, M.; Mucha, E.; Thomas, D.A.; Polewski, L.; Freyse, J.; Rademann, J.; Meijer, G.; Von Helden, G.; Pagel, K. IR action spectroscopy of glycosaminoglycan oligosaccharides. *Anal. Bioanal. Chem.* **2020**, *412*, 533–537. [CrossRef]

36. Zhao, X.; Chen, J.; Chen, F.; Wang, X.; Zhu, Q.; Ao, Q. Surface characterization of corn stalk superfine powder studied by FTIR and XRD. *Colloids Surf. B Biointerfaces* **2013**, *104*, 207–211. [CrossRef]
37. Li, N.; Feng, Z.; Niu, Y.; Yu, L. Structural, rheological and functional properties of modified soluble dietary fiber from tomato peels. *Food Hydrocoll.* **2018**, *77*, 557–565. [CrossRef]
38. Alfredo, V.-O.; Gabriel, R.-R.; Luis, C.-G.; David, B.-A. Physicochemical properties of a fibrous fraction from chia (*Salvia hispanica* L.). *LWT-Food Sci. Technol.* **2008**, *42*, 168–173. [CrossRef]
39. Zhu, K.; Huang, S.; Peng, W.; Qian, H.; Zhou, H. Effect of ultrafine grinding on hydration and antioxidant properties of wheat bran dietary fiber. *Food Res. Int.* **2010**, *43*, 943–948. [CrossRef]
40. Rosell, C.M.; Santos, E.; Collar, C. Physico-chemical properties of commercial fibres from different sources: A comparative approach. *Food Res. Int.* **2008**, *42*, 176–184. [CrossRef]
41. Twarogowska, A.; Van Poucke, C.; Van Droogenbroeck, B. Upcycling of Belgian endive (Cichorium intybus var. foliosum) by-products. Chemical composition and functional properties of dietary fibre root powders. *Food Chem.* **2020**, *332*, 127444. [CrossRef] [PubMed]
42. Feng, Z.; Dou, W.; Alaxi, S.; Niu, Y.; Yu, L.L. Modified soluble dietary fiber from black bean coats with its rheological and bile acid binding properties. *Food Hydrocoll.* **2017**, *62*, 94–101. [CrossRef]
43. Cheng, L.; Zhang, X.; Hong, Y.; Li, Z.; Li, C.; Gu, Z. Characterisation of physicochemical and functional properties of soluble dietary fibre from potato pulp obtained by enzyme-assisted extraction. *Int. J. Biol. Macromol.* **2017**, *101*, 1004–1011. [CrossRef] [PubMed]
44. Dhital, S.; Dolan, G.; Stokes, J.R.; Gidley, M.J. Enzymatic hydrolysis of starch in the presence of cereal soluble fibre polysaccharides. *Food Funct.* **2014**, *5*, 579–586. [CrossRef] [PubMed]
45. Chen, H.; Pu, J.; Liu, D.; Yu, W.; Shao, Y.; Yang, G.; Xiang, Z.; He, N. Anti-Inflammatory and Antinociceptive Properties of Flavonoids from the Fruits of Black Mulberry (*Morus nigra* L.). *PLoS ONE* **2016**, *11*, e0153080. [CrossRef]
46. Deng, J.-S.; Chi, C.-S.; Huang, S.-S.; Shie, P.-H.; Lin, T.-H.; Huang, G.-J. Antioxidant, analgesic, and anti-inflammatory activities of the ethanolic extracts of Taxillus liquidambaricola. *J. Ethnopharmacol.* **2011**, *137*, 1161–1171. [CrossRef] [PubMed]
47. Zhang, R.; Zeng, Q.; Deng, Y.; Zhang, M.; Wei, Z.; Zhang, Y.; Tang, X. Phenolic profiles and antioxidant activity of litchi pulp of different cultivars cultivated in Southern China. *Food Chem.* **2013**, *136*, 1169–1176. [CrossRef]
48. Guo, W.; Beta, T. Phenolic acid composition and antioxidant potential of insoluble and soluble dietary fibre extracts derived from select whole-grain cereals. *Food Res. Int.* **2013**, *51*, 518–525. [CrossRef]
49. Hu, S.; Yin, J.; Nie, S.; Wang, J.; Phillips, G.O.; Xie, M.; Cui, S.W. In vitro evaluation of the antioxidant activities of carbohydrates. *Bioact. Carbohydr. Diet. Fibre* **2016**, *7*, 19–27. [CrossRef]
50. Chen, G.; Chen, X.; Yang, B.; Yu, Q.; Wei, X.; Ding, Y.; Kan, J. New insight into bamboo shoot (Chimonobambusa quadrangularis) polysaccharides: Impact of extraction processes on its prebiotic activity. *Food Hydrocoll.* **2019**, *95*, 367–377. [CrossRef]
51. Tadayoni, M.; Sheikh-Zeinoddin, M.; Soleimanian-Zad, S. Isolation of bioactive polysaccharide from acorn and evaluation of its functional properties. *Int. J. Biol. Macromol.* **2015**, *72*, 179–184. [CrossRef]
52. Bai, Y.; Zhao, J.; Tao, S.; Zhou, X.; Pi, Y.; Gerrits, W.J.; Johnston, L.J.; Zhang, S.; Yang, H.; Liu, L.; et al. Effect of dietary fiber fermentation on short-chain fatty acid production and microbial composition in vitro. *J. Sci. Food Agric.* **2020**, *100*, 4282–4291. [CrossRef] [PubMed]
53. Linares, D.M.; Ross, P.; Stanton, C. Beneficial Microbes: The pharmacy in the gut. *Bioengineered* **2016**, *7*, 11–20. [CrossRef] [PubMed]

foods

*Article*

# Extraction, Purification and In Vitro Antioxidant Activity Evaluation of Phenolic Compounds in California Olive Pomace

**Hefei Zhao [1], Roberto J. Avena-Bustillos [2] and Selina C. Wang [1,*]**

1 Department of Food Science and Technology, University of California, Davis, CA 95616, USA; hefzhao@ucdavis.edu
2 Western Regional Research Center, Healthy Processed Foods Research, Albany, CA 94710, USA; roberto.avena@usda.gov
* Correspondence: scwang@ucdavis.edu

**Abstract:** Olive pomace (OP) is a valuable food byproduct that contains natural phenolic compounds with health benefits related to their antioxidant activities. Few investigations have been conducted on OP from the United States while many studies on European OP have been reported. OP of Arbequina, the most common cultivar from California, was collected and extracted by water, 70% methanol and 70% ethanol, followed by purification using macroporous absorbing resin. Results showed that the extractable total phenolic content (TPC) was 36–43 mg gallic acid equivalents (GAE)/g in pitted, drum-dried defatted olive pomace (DOP), with major contributions from hydroxytyrosol, oleuropein, rutin, verbascoside, 4-hydroxyphenyl acetic acid, hydroxytyrosol-glucoside and tyrosol-glucoside. Macroporous resin purification increased TPC by 4.6 times the ethanol crude extracts of DOP, while removing 37.33% total sugar. The antioxidant activities increased 3.7 times Trolox equivalents (TrE) by DPPH and 4.7 times TrE by ferric reducing antioxidant power (FRAP) in the resin purified extracts compared to the ethanol crude extracts. This study provided a new understanding of the extraction of the bioactive compounds from OP which could lead to practical applications as natural antioxidants, preservatives and antimicrobials in clean-label foods in the US.

**Keywords:** olive pomace; natural phenolic compounds; antioxidant activity; macroporous absorbing resin; hydroxytyrosol

**Citation:** Zhao, H.; Avena-Bustillos, R.J.; Wang, S.C. Extraction, Purification and In Vitro Antioxidant Activity Evaluation of Phenolic Compounds in California Olive Pomace. *Foods* **2022**, *11*, 174. https://doi.org/10.3390/foods11020174

Academic Editors: Michela Verni and Federico Casanova

Received: 4 December 2021
Accepted: 5 January 2022
Published: 10 January 2022



## 1. Introduction

One of the most iconic and representative ingredients in the Mediterranean diet are the oils and fats of *Olea europaea* L.; the bioactive and phenolic compounds of which have been identified as the major contributions to their antioxidant and health-promoting effects by analytical and technical approaches of modern medicine as well as food science and chemistry.

Olive pomace (OP), the main residue derived from olive oil extraction, is a valuable food byproduct containing natural phenolic compounds (phenols and polyphenols) with health benefits related to their antioxidant activities. According to the Food and Agriculture Organization (FAO), the United States (mainly in California) produced 151,950 tons of olives in 2019. Based on the literature (~67.5% water and ~17.5% lipids) [1] and our preliminary experimental results of California OP that has an oil content of ~11% and total phenolic content (TPC) of 43 mg gallic acid equivalents (GAE)/g hexane defatted olive pomace (DOP), there could be as much as ~872 tons of bioactive compounds in the US OP, an equivalence of ~$79 million per year as raw material, assuming benchmark price to ~$90 per kg food-grade gallic acid. As a commercialized health food, the price of hydroxytyrosol in capsules form is currently up to $4100 per kg; therefore, the total value of the US olive pomace (fruit) extracts could be as high as about $3.6 billion per year.

Many research interests have been focused on the extraction and purification of natural phenolic compounds (NPCs) from various waste-streams of the European olive oil industry, such as olive mill wastewater (OMWW) and OP and on the bioactive abilities of the crude or purified extracts. Water among other solvents was found to be the best media for the extraction from residual oil extracted and exhausted olive pomace (EOP) in Spain, and the water-extractable TPC was 44.5 mg GAE/g EOP [2]. Aqueous extraction of OP in Portugal was successfully concentrated by reverse osmosis (RO) membranes with the highest 100% membrane retention of TPC and the lowest fouling index among two other nanofiltration (NF) membranes [3]. The TPC was increased from 0.1098 mg/mL in the feed aqueous to 1.2343 mg/mL in the concentrates, but free sugars were not removed and separated from phenols. OMWW in Italy was clarified by microfiltration (MF) and then concentrated by NF and RO, respectively [4]. While the MF slightly decreased hydroxytyrosol from 0.3733 mg/mL in feed to 0.3201 mg/mL in permeate, the hydroxytyrosol of MF permeate was increased to 1.0175 and 1.5222 mg/mL by NF and RO, respectively. In Turkey, olive leaf phenols were extracted by macroporous resins, and the Amberlite® XAD7HP resin performed the highest adsorption (91%) and desorption (97%) rate of phenols based on oleuropein contents among other resins (Amberlites® XAD2, XAD4 and XAD16); however, 1 g of resin was applied for the purification which was an analytical scale [5]. The purification of OP phenols coupled with the removal of sugar or other impurities has yet to be comprehensively determined and monitored. In general, the potential utilization of the bioactive compounds in OP has been far less investigated in the US than in European countries.

The objective of this study was to identify phenolic compounds, compare TPC and antioxidant activities of crude extracts from water, 70% methanol, 70% ethanol and purified extracts from pilot-scale [6] (hundreds mL of resin) and preparative chromatography columns of macroporous adsorption resin, using the most common olive cultivar in the United States. The efficacy of the removal of sugar was also monitored and reported. It is necessary to decrease sugar content in the potential OP natural ingredients because the association between sugars and the risk of obesity has been ubiquitously acknowledged by the public [7].

## 2. Materials and Methods

### 2.1. Materials and Chemicals

#### 2.1.1. Olive Material

Fresh Arbequina olive pomace (OP) from first olive oil extraction was collected at California Olive Ranch at their facility in Artois, CA, during the 2019 harvest season and was stored at room temperature in sealed buckets for 4 h for transportation to our lab in Albany, CA, USA. Then, it was processed by the following steps reported by our research group [8]:

(1). Steam blanching. Fresh first oil extraction OP was steam-blanched for enzymatic inactivation to reduce phenolics losses. Blanching was conducted using a steam blancher at atmospheric pressure over 0.25″ thick olive pomace to a final temperature of 80 °C after 3 min.

(2). Pit and skin separation. The separation of skins and pits was conducted using a 150 Langsenkamp Laboratory Separator (Warner Bodies, Elwood, IN, USA). The pomace was passed through the separator in two stages. First using a 0.060 inch hole diameter S.S. screen and then using a 0.027 inch hole diameter S.S. screen.

(3). Drum-drying. The pitted olive pomace was drum-dried on a Buflovaks Atmospheric Double Drum Dryer (Hebeler Process Solutions, Tonawanda, NY, USA), with a space of 9-10/1000″ at 135 °C. Drum-drying treatments were differentiated by rotational drum speeds of 92 s/rev.

(4). Milling. To obtain smaller particle sizes, drum-dried OP samples were milled for 6 s, with a KRUPS F203 (KRUPS, Parsippany, NJ, USA) coffee mill.

#### 2.1.2. Chemicals

Amberlite® XAD7HP (DuPont, Wilmington, DE, USA), chromatographic grade chemicals of hexane, water, acetic acid, sodium acetate trihydrate, methanol, ethanol, acetonitrile and analytical grade of chemicals of 2,2′-diphenyl-1-picrylhydrazyl radical (DPPH), Trolox®, 2,4,6-tripirydyl-Striazine (TPTZ) and ferric chloride were purchased from Fisher Scientific (Waltham, MA, USA). In addition, 96–98% (g/g) concentrated sulfuric acid, Folin–Ciocalteu reagents, sodium carbonate, phenolic compound standards of gallic acid, hydroxytyrosol, tyrosol, 4-hydroxyphenylacetic acid (4-HPA), vanillic acid, vanillin, o-coumaric acid, oleuropein, pinoresinol, cinnamic acid, caffeic acid, p-coumaric acid, ferulic acid, apigenin-7-glucoside, apigenin, luteolin-7-glucoside and luteolin were purchased from Sigma-Aldrich (St. Louis, MI, USA). Verbascoside was bought from the HWI group (Ruelzhelm, Germany). Rutin was bought from PhytoLab GmbH & Co. KG (Vestenbergsgreuth, Germany).

## 2.2. *Basic Chemical Composition Analysis*

### 2.2.1. Moisture Content

Five gram OP samples or extracts samples in triplicate were added into 5 cm diameter aluminum pans and placed in an isotherm oven (Thermo Fisher Scientific, Waltham, MA, USA) at 105 °C for 48 h till constant weight [9]. A desiccator was used for 30 min to allow dry samples to reach an ambient temperature before weighing. The final weights of each sample were recorded and used to determine the total amount of water in dry samples, considering that the amount of water is equal to the initial weight of OP samples or extracts minus the final weight of oven-dry OP samples or extracts, and then dividing by the initial weight of OP samples or extracts and multiplying by 100, its moisture content was estimated as % of the original OP samples or extracts.

### 2.2.2. Crude Protein Content of Olive Pomace

Protein content olive pomace was measured using an FP-628 TrueSpec N analyzer (LECO Corporation, St. Joseph, MI, USA) which referred to the method of Zhao et al. [9]. One gram of OP sample was placed into a ceramic boat for the combustion process, and each sample was analyzed by triplicate, and the nitrogen content of the OP samples was multiplied by a factor of 6.25 to convert it into protein content.

### 2.2.3. Oil Content

In total, 10 g of OP was determined by hexane Soxhlet extraction in a Universal Extractor (BÜCHI, New Castle, DE, USA) [9]. Each sample was performed by 30 cycles of solvent extraction and in triplicate.

### 2.2.4. Ash Content

In total, 3 g of each sample was placed by triplicate in crucibles and introduced into a Lindberg/Blue M box furnace (Thermo Fisher Scientific, Waltham, MA, USA) at 550 °C for 16 h, which referred to Zhao et al. [9] with some modifications. Before the final weighting, the crucibles containing the ash were carefully relocated in a desiccator to reach room temperature.

### 2.2.5. Total Carbohydrates

Total carbohydrates were estimated by difference.

### 2.2.6. Total Sugar Content

The determination of total sugar content (TSC) of OP extracts referred to Zhang et al. [10] with some modifications. In brief, to 0.5 mL of 0.2 mg/L sample solution, 0.5 mL of 5% (g/g) phenol solutions was added, followed by 2.5 mL of 96% (g/g) concentrated sulfuric acid and then vortexed. The sample mixture was maintained for 10 min and then cooled in a water bath. The absorbance was measured at 490 nm by a UV-vis spectrophotometer. The sugar concentration of an unknown sample can be obtained from a calibration curve of glucose solutions prepared by a series of diluted solutions from 0 to 0.2 mg/mL, the results of total sugar content were expressed as mg glucose equivalents (GE)/g of sample. 0.2 mg/mL gallic acid standard was used as a control.

### 2.3. Extraction and Purification of Phenolic Compounds from OP

The flow diagram of the extraction and purification of the pitted drum-dried olive pomace powder is shown below in Figure 1.

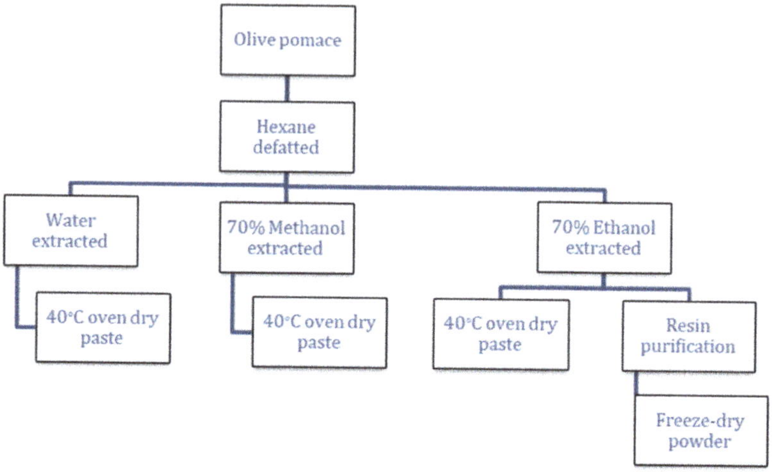


**Figure 1.** Flow diagram of the extraction and purification of pitted drum-dried olive pomace.

#### 2.3.1. Extractable Phenols

The extraction of water, 70% methanol and 70% ethanol extractable phenols of OP referred to the methods by De Bruno et al. [11] and Xu et al. [12] with some modifications. To 200 g of olive pomace (OP) in a 1000 mL beaker, 500 mL of hexane was added with gentle stirring every 30 min and stayed in a fume hood in the dark with a cover for two hours. Then, the supernatant hexane was decanted for removing oil and fat, and the hexane extractions were repeated two more times. Some of the small particles of OP inevitably outflowed from the beaker with the hexane. Then, the beaker was covered with a paper towel for overnight drying to obtain 163.8 g defatted olive pomace (DOP). Then, three of 0.25 g DOPs were extracted at room temperature (RT) by 5 mL extraction solvent of water, 70% methanol and 70% ethanol under sonication for 5 min, then stored in the dark for 3 h with shaking every 30 min. After the supernatants were poured out and collected, the precipitate was extracted again with the same procedure, which rendered the solvent to solid (from beginning) ratio as 10 mL/0.25 g, which was equal to 40:1. The extraction was centrifuged at 4 kG for 10 min at 4 °C, then the supernatant was separated and collected. All extracts were stored at RT in the dark until further analysis (normally within 5 days). Extractions were applied in triplicate for each individual OP powder-solvent combination.

#### 2.3.2. Preparation of Dry Paste of Crude Extracts of OP

To three of 60 g DOP in a 1000 mL beaker, 300 mL of DI water, 70% methanol and 70% ethanol were added, respectively, to each of the beakers and sonicated with cover for 10 min. The three beakers were placed in a fume hood in the dark for 1 h with gentle stirring every 30 min. Then, the three upper supernatants were decanted and collected into three individual bottles and the extractions were repeated two times more to acquire a total of ~900 mL extracts for each of the three extractions media. Then, only the water extract was vacuum-filtered by the filter layer of a stomacher bag (otherwise, it was difficult to be filtrated by Whatman paper directly, due to the higher viscosity of water and more pomace swelling than those extracts in 70% ethanol and 70% methanol). The water, 70% methanol and 70% ethanol extracts were vacuum-filtered by double-layer Whatman filter papers, respectively. The three clear filtrates were collected in three 1000 mL bottles, separately and respectively, and placed in a refrigerator.

Then, the three ~900 mL extracts were concentrated by rotary evaporation at 40 °C to remove alcohol and water and decrease the volume to ~100 mL, by a Rotavapor® R-300 (BÜCHI, New Castle, DE, USA). For the two extracts by 70% methanol and 70% ethanol, 100 mL water was added in each and concentrated by the rotary evaporation to ~100 mL again to thoroughly remove the alcohol out of the azeotrope; otherwise, residual alcohol was an antifreezing solution which would result in difficulty during the freeze-drying process. Low water-soluble phenols attached to the inner wall of the evaporating flask were carefully detached by a sonication bath. Then, each of the extracts was adjusted to 120 mL by DI water, and the extracts were collected into three bottles, respectively, and placed in a refrigerator. Each extract of 30 mL was poured into a 5 cm diameter aluminum pan and placed in an isotherm oven (Thermo Fisher Scientific, Waltham, MA, USA) at 40 °C for 48 h till constant weight to obtain dry pastes of OP crude extracts: 6.86 g of water extract, 6.84 g of 70% methanol extract and 7.14 g of 70% ethanol extract.

Preparation of freeze-dried powders of OP phenols extracted by 70% ethanol and purified by preparative chromatography of macroporous absorbing resin. Amberlite® XAD7HP macroporous resin purification referred to the method of Jiang et al. [13]. In total, 90 mL of 70% ethanol extract after removal of ethanol was reconstituted by DI water to 600 mL and vacuum-filtered by double-layer Whatman filter papers. The clear filtrates were submitted to the XAD7HP resin column with 262.5 mL (4.8 cm diameter and 14.5 cm height) bed volumes (BV), at 1.9 BV/hour speed. The low water-soluble fractions were directedly submitted onto the top of the resin bed at the end of filtrated extracts loading absorption stage. Then, the resin bed was rinsed with 3.8 BV DI water, and the first 1/2 BV water rinsing elutes were emerged into loading absorption elutes, and those elutes were rotatory-evaporated at 40 °C and concentrated to ~100 mL and freeze-dried to obtain ~40 mL heavy syrup. The moisture content of the heavy syrup of resin elutes during absorption was tested as the previous description. To the DI water-rinsed resin bed, 3.8 BV of 70% ethanol [14] was loaded to desorb the OP phenols. Then, the desorbing elutes were rotatory-evaporated at 40 °C and concentrated to remove alcohol and decreased to ~100 mL. Then, 100 mL water was added, and the elutes were concentrated to ~100 mL again to thoroughly remove the alcohol. The concentrates were then reconstituted to 120 mL and submitted to a freeze-dryer to obtain 2.78 g resin purified extracts of OP.

### 2.4. *High-Performance-Liquid-Chromatography (HPLC)-Diode-Array Detector (DAD) and HPLC-Electrospray Ionization (ESI)-Quadrupole-Time of Flight-Mass Spectrometry (Q-ToF-MS$^n$) Analysis*

The identification of individual phenolic compounds was implemented by an Agilent 1290 high-performance-liquid-chromatography (HPLC) system (Santa Clara, CA, USA) with an Agilent 1290 diode-array detector (DAD) which referenced Sinrod et al. [8]. An analytical C18 column (Eclipse Plus, 4.6 mm ×250 mm, 5 μm, Agilent Technologies) was used for separation. Elution was applied using mobile phase A (3% acetic acid aqueous solution) and mobile phase B (50% methanol and 50% acetonitrile). The following linear gradient was used: 0 min starting from 5% B (while 95% A, similarly hereinafter); linear increase to 30% B at 25 min; to 35% B at 35 min; to 40% B at 40 min; to 70% B at 50 min; to 100% B at 55 min, then decreasing to 5% B at 60 min and holding at 5% B for 5 min for the column equilibrium for the next injection. The flow rate was 1.0 mL/min. The injection volume was 20 μL. The DAD was set to absorbance wavelengths at 280 nm for hydroxytyrosol, tyrosol, 4-hydroxyphenylacetic acid (4-HPA), vanillic acid, vanillin, o-coumaric acid, oleuropein, pinoresinol, cinnamic acid, at 320 nm for caffeic acid, p-coumaric acid, ferulic acid, apigenin-7-glucoside, apigenin, verbascoside and at 365 nm for rutin, luteolin-7-glucoside and luteolin, respectively. Standard curves were made by each of the standard chemicals at concentrations of 10, 20, 40, 60, 80 and 100 mg/L, respectively.

Because not all the olive phenolic standards were commercially available, an Agilent 1290 HPLC coupled to an Agilent 6530 quadrupole time-of-flight mass spectrometer (Q-ToF-$MS^n$) were applied to identify unknown peaks by negative mode, with instrument control and data acquisition in Agilent MassHunter Acquisition (ver. 6). The chromatographic conditions were identical to the HPLC-DAD. The collision energy (CE) was 15 eV. The unknown phenolic compounds were identified based on their MS and MS/MS fragments compared to the report of Peralbo-Molina et al. [15] and the MassBank at https://massbank.eu/MassBank.asp (accessed on 12 September 2021). Only the XAD7HP resin purified freeze-dried powders were diluted 500 times in 70% methanol and injected to HPLC- Q-ToF-$MS^n$, because most of sugar and polar impurities from 0 to 14 min were eliminated by the resin from the purified extracts which facilitated the mass identification.

### 2.5. Total Phenol Content and Antioxidant Activities Analysis

#### 2.5.1. Total Phenolic Content

The total phenol content (TPC) was determined by Folin–Ciocalteu assay which referred to the method of Zhao et al. [16]. In total, 50 µL extracted sample was added to 3 mL DI Water, 250 µL Folin–Ciocalteu reagent, 750 µL 20% sodium carbonate and 950 µL DI water. This total of 5 mL solution was incubated for 30 min in 40 °C water base, and then 200 µL solution out of 5 mL was pipetted into a microplate plate with 96 wells (Corning Incorporated, Corning, NY, USA), then determined by absorption at 760 nm in a Synergy H1 microplate reader (BioTek, Winooski, VT, USA). Here, 0.1–1 mg/mL gallic acid (GA) was utilized as a standard for the calibration curve, and the results were expressed as gallic acid equivalents (mg GAE/g DM, mg gallic acid/g of sample). Samples included DOP extracts, DOP dry pastes and freeze-dried powder.

#### 2.5.2. 2,2′-Diphenyl-1-picrylhydrazyl Radical (DPPH) Assay

The total antioxidant activity by DPPH assay was implemented based on the slightly modified method of Zhao et al. [16]. In total, 195 µL methanolic solution of DPPH (MW 394.32 Da, 2.108 mg/100 mL methanol, Abs = 0.589) was added to 5 µL of the 0.2 mg/mL sample in 96 cuvettes microplate (Corning In., Corning, NY, USA). The absorbance of the remaining DPPH was determined at 515 nm after 30 min dark incubation at room temperature (RT). The percentage inhibition of DPPH of the test sample and known solutions of Trolox were calculated by the following formula: % Inhibition = $100 \times (A0 - A)/A0$, where A0 is the absorbance of DPPH methanolic solution without any inhibition, which was equal to the beginning absorbance at 515 nm, acquired by measuring the same volume of solvents of both extracted sample and the methanol solution of DPPH and A is the final absorbance of the test sample at 515 nm in a Synergy H1 microplate reader (BioTek, Winooski, VT, USA). Blank was made with 5 µL 70% methanol and 195 µL methanol without DPPH. The calibration curve between % Inhibition and known solutions of Trolox was then established. The radical scavenging activities of the test samples were expressed as Trolox equivalent (TrE) antioxidant capacity (µmol TrE/g of sample, MW of Trolox: 250.29 g/mol) on their percentage inhibitions. Trolox standard solutions were prepared at a concentration ranging from 0.1 to 1.0 µmol/mL. Samples included DOP dry pastes and freeze-dried powder.

#### 2.5.3. Ferric Reducing Antioxidant Power (FRAP)

The FRAP assay was implemented according to the method of D'Amato et al. [17]. The fresh working FRAP solution was prepared by mixing: (1) 10 mM 2,4,6-tripirydyl-Striazine (TPTZ, MW 312.33 Da) dissolved in 40 mM HCl; (2) 20 mM ferric chloride; and (3) 300 mM acetate buffer at pH 3.6, with a ratio of 1:1:10, respectively. The FRAP was kept at 37 °C before use. In total, 50 µL of 0.2 mg/mL antioxidant solution was added to 950 µL of FRAP solution, then the mixture was incubated at 37 °C in the dark for 30 min. In addition, the absorbance was measured at 595 nm in a Synergy H1 microplate reader (BioTek, Winooski, VT, USA). Ethanol and Trolox solution (0.1–1.0 µmol/mL) were used for positive control and the standard curve, respectively. The results were expressed as μmol TrE/g of sample which included DOP dry pastes and freeze-dried powder.

*2.6. Statistical Analysis*

Triplicate data were analyzed by multiple comparison tests with Fisher's least significant difference (LSD, $p < 0.05$) method by R software. Heatmap cluster analysis was implemented in MATLAB 2020a (The Mathworks Inc., Natick, MA, USA).

## 3. Results and Discussion

*3.1. Chemical Composition in Olive Pomace*

The basic chemical composition of pitted drum-dried olive pomace (OP) is reported in Table 1. The total carbohydrates estimated by difference was the major component which was up to 66.15%. In addition, the olive contained 11.72% residual oils. The results were in line with other works that reported total carbohydrates as the major component, including OP from Portugal containing 84.9% total carbohydrates [18], although it most likely contained pits and skins that increased their total carbohydrates. Our pitted drum-dried OP, obtained from OP pulp, also indicated that soluble saccharides could be the major impurity of the olive pomace extracts for phenolics extraction.

**Table 1.** Basic chemical composition analysis of Arbequina olive pomace.

| Nutritional Component | Contents % |
| --- | --- |
| Protein | 10.28 ± 0.11 |
| Moisture | 2.83 ± 0.08 |
| Fat | 11.72 ± 0.07 |
| Ash | 9.02 ± 0.04 |
| Total carbohydrates | 66.15 |

*3.2. Total Phenolic Contents in Olive Pomace and Extracts*

As shown in Figure 2, extractable TPC in DOP was 36.49 ± 0.29 mg GAE/g (DOP solids) by water extraction, 41.68 ± 0.95 mg/g by 70% methanol and 43.25 ± 2.08 mg/g by 70% ethanol; 70% ethanol was selected to be followed by resin purification for its highest phenolic extraction and nontoxicity food-grade potential. A similar TPC range (30–40 mg GAE/g) in exhausted olive pomace (EOP) in Spain has been reported using a variety of combinations of water-solvents extraction, and the TPC extracted by solvents was normally higher than that by water and acidified water [2]. In addition, another similar TPC range in caffeic acid equivalent (CAE) has been reported in an Italy OP, which was 38–52 mg CAE/g [19]. However, to the best of our knowledge, this is the first report on the extractable TPC range (36–43 mg GAE/g DOP solids) in the Arbequina olive pomace in the US. Moreover, 70% methanol is a common standard solvent for the analytical purpose of phenol extractions [20,21]. Water and 70% ethanol were also selected in this study and are considered as green, safe and nontoxic media for extraction compared to other organic solvents [22,23]. We are currently studying other olive cultivars in the US for a more comprehensive investigation, as Arbequina cultivars tend to have lower TPC compared to many other cultivars.

The yields of dried crude extracts from water, 70% methanol and 70% ethanol were 45.75%, 45.58% and 47.58% (g dry paste/g DOP), respectively. While almost half of the soluble matters were extracted, the TPC increased to 52.42 ± 2.21 mg GAE/g dry paste solids, 69.17 ± 2.10 mg/g and 65.83 ± 2.43 mg/g, by the three extraction solvents, respectively, which increased to about 1.5 times compared to the TPC in DOP (Figure 2).

To concentrate the TPC and eliminate free sugar impurities, Amberlite® XAD7HP macroporous resin chromatography on pilot-scale was implemented for the purification. Because the XAD7HP has been reported to have 91% adsorption and 97% desorption of olive phenols [5], we selected it for the purification of the OP extract. TPC in the resin purified freeze-dried powders increased 4.6 times to 303.03 ± 6.74 mg GAE/g freeze-dried solids from 43.25 ± 2.08 mg/g of the 70% ethanol extracted dry paste (Figure 2). Although several studies [5,24,25] have reported the kinetics of macroporous resin absorbing OP phenols based on an analytical scale of several grams of resin, limited studies obtained freeze-dried powders of resin purified extract and reported the TPC. This study used preparative and pilot-scale resin column by filling hundreds of grams of resin, demonstrating that it may be feasible to scale-up the purification process to an industrial level.

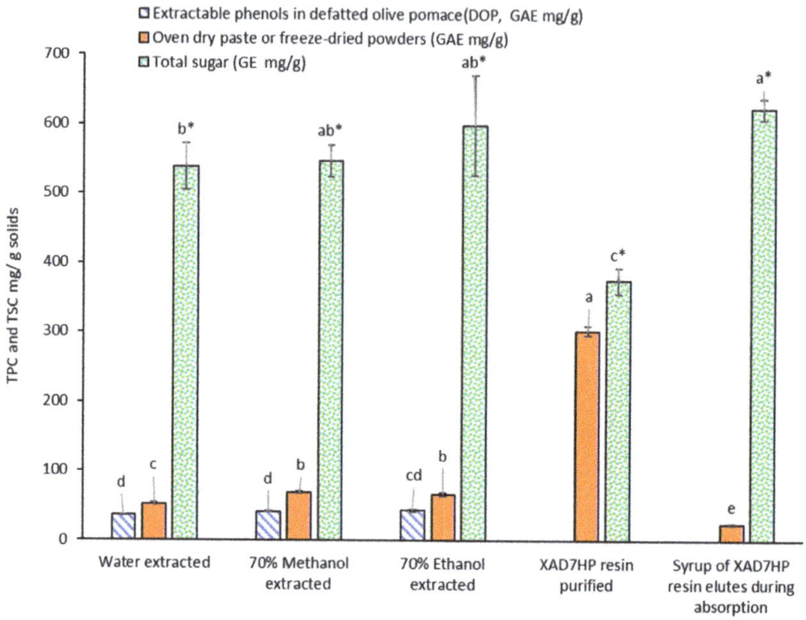


**Figure 2.** Comparison of total phenolic content (TPC) expressed by gallic acid equivalents (GAE) and total sugar content (TSC) expressed by glucose equivalents (GE) from different extraction methods and purified steps, different letters indicate significantly different, $p < 0.05$. Note: 0.2 mg/mL gallic acid standard was used as the control of the determination of total sugar content, and no interference from gallic acid was observed. Significant differences were only compared in either GAE without "*" on significant markers or GE with "*" on significant markers. Syrup of the resin elutes during sample absorption after freeze-drying still had 36.60 ± 0.17% moisture, and data were reported to dry matters (DM).

### 3.3. Individual Phenolic Contents in Olive Pomace and Extracts

Figure 3a shows 19 phenolic compounds were identified by phenol standards, and these 19 phenolic compounds and 5 phenolic sugar derivatives and others were further putatively identified by mass spectrometry in Table 2. The tentatively identified phenols included unique compounds in olive and olive pomaces, such as hydroxytyrosol, tyrosol, 4-hydroxyphenyl acetic acid (4-HPA), verbascoside, oleuropein and their sugar derivatives, as well as several compounds commonly found in other plants, such as gallic acid, rutin and luteolin, etc. About ten unknown peaks have yet to be elucidated by HPLC-ESI-QToF-$MS^n$.

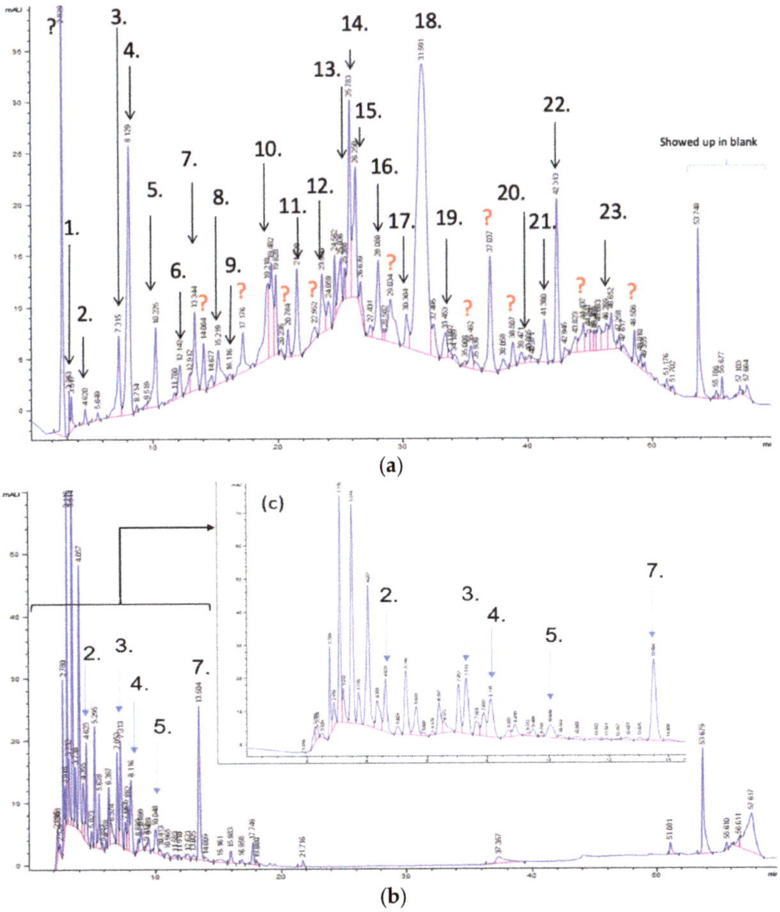


**Figure 3.** (**a**) Chromatography of XAD7HP resin purified freeze-dried powder at 280 nm and tentatively identified compounds by HPLC–DAD and HPLC-ESI-QToF-MS$^n$; (**b**) Chromatography of the syrup of resin elutes during absorption; (**c**) 0–14 min of (**b**). Note: 1. Vanillin–glucoside-pentoseside, 2. Gallic acid, 3. Hydroxytyrosol-glucoside, 4. Hydroxytyrosol, 5. Tyrosol-glucoside, 6. Tyrosol, 7. 4-Hydroxyphenylacetate (4-HPA), 8. Vanillic acid, 9. Caffeic acid, 10. Vanillin, 11. P-coumaric acid, 12. Ferulic acid, 13. Verbascoside 14. Rutin, 15. Luteolin-7-glucoside, 16. O-coumaric acid, 17. Apigenin-7-glucoside, 18. 3,4-DHPEA-EDA, 19. Oleuropein, 20. Pinoresinol, 21. Cinnamic acid, 22. Luteolin, and 23. Apigenin. Mass spectrometric data refer to the Table 3. Spectra at 320, 365 nm and total ion chromatogram (TIC) refer Figure S1.

**Table 2.** Retention times and mass spectrometric data of tentatively identified compounds by HPLC-ESI-QToF-$MS^n$.

| No. | Tentatively Identified Compounds | Retention Time (min) | MW/Da | MS (*m/z*) $[M-H]^-$ | MS/MS (*m/z*) | Molecular Structure |
|---|---|---|---|---|---|---|
| 01 | Vanillin-glucoside | 3.25 | 448.15 | 447.15 | 447.15, 315.11, 169.05, 153.06, 123.05 | NA |
| 02 | Gallic acid | 4.62 | 170.05 | 169.05 | 141.02, 123.01 | 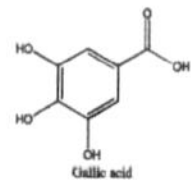 |
| 03 | Hydroxytyrosol-glucoside | 7.32 | 316.11 | 315.11 | 153.06, 123.05, 101.02 | NA |
| 04 | Hydroxytyrosol | 8.13 | 154.16 | 153.06 | 123.05, 109.03, 101.02 | 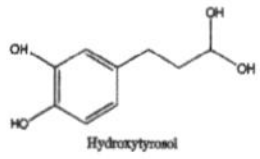 |
| 05 | Tyrosol-glucoside | 10.23 | 300.14 | 299.14 | 183.06, 139.07, 119.04, 101.02 | NA |
| 06 | Tyrosol | 12.14 | 138.16 | 137.16 | 119.04, 101.03 | 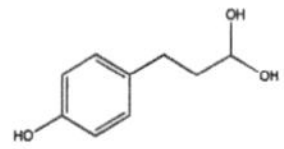 |
| 07 | 4-Hydroxyphenylacetate (4-HPA) | 13.34 | 152.15 | 151.04 | 163.04, 123.05 | 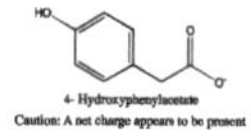 |
| 08 | Vanillic acid | 15.22 | 168.03 | 167.03 | ND | 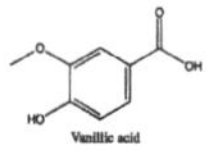 |
| 09 | Caffeic acid | 16.12 | 180.06 | 179.06 | ND | 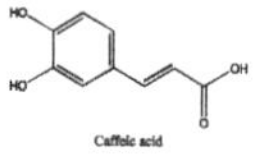 |
| 10 | Vanillin | 19.22 | 152.15 | 151.04 | 123.05, 139.00 | 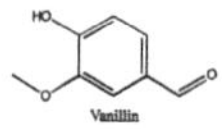 |
| 11 | p-coumaric acid | 21.62 | 164.04 | 163.04 | 119.04, 101.02 | 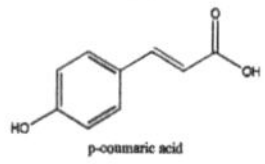 |
| 12 | Ferulic acid | 23.38 | 195.06 | 194.06 | 151.04, 135.04 | 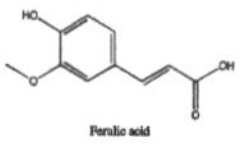 |
| 13 | Verbascoside | 25.78 | 624.59 | 623.20 | 458.20, 461.15, 151.04 | 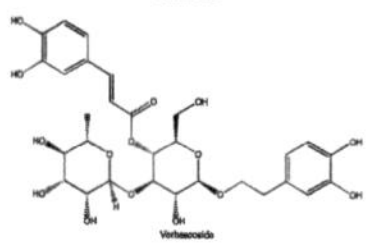 |
| 14 | Rutin | 25.25 | 610.52 | 609.52 | 609.15, 301.03 | 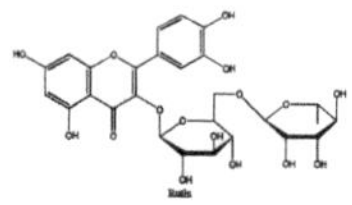 |
| 15 | Luteolin-7-glucoside | 26.25 | 448.09 | 447.09 | 285.04 | 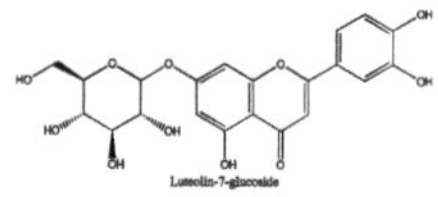 |
| 16 | o-coumaric acid | 28.08 | 165.05 | 164.05 | ND | 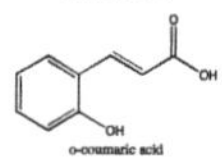 |

**Table 2.** *Cont.*

| No. | Tentatively Identified Compounds | Retention Time (min) | MW/Da | MS (*m*/*z*) [M-H]$^-$ | MS/MS (*m*/*z*) | Molecular Structure |
|---|---|---|---|---|---|---|
| 17 | Apigenin-7-glucoside | 30.30 | 432.38 | 431.38 | ND | 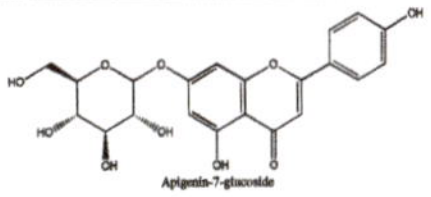 |
| 18 | 3,4-DHPEA-EDA | 31.59 | 320.12 | 319.12 | 139.08, 123.05 | NA |
| 19 | Oleuropein | 33.45 | 540.18 | 539.18 | 377.12, 307.08, 275.09 | 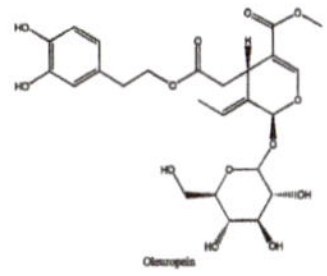 |
| 20 | Pinoresinol | 39.47 | 358.10 | 357.10 | 341.12, 151.04 | 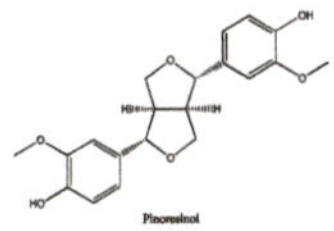 |
| 21 | Cinnamic acid | 41.30 | 148.16 | 147.16 | ND | 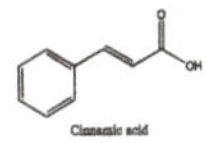 |
| 22 | Luteolin | 42.31 | 286.04 | 285.04 | 217.00 | 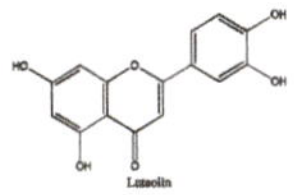 |
| 23 | Apigenin | 46.27 | 270.04 | 269.04 | 241.07, 141.02 | 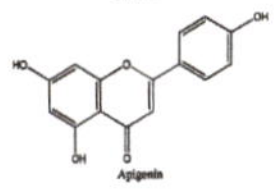 |

Note: ND means ms/ms was not detected by HPLC-QToF-MS$^n$, but the compound was identified by HPLC-DAD with its standard chemical by retention time. NA means not available. Retention time were matched to HPLC-DAD in Figure 3a.

The concentration of individual phenolic compounds is in Table 3 and visualized in Figure 4a. The hydroxytyrosol, 4-HPA, rutin and many others of three extracted dry paste were about 2 times the extractable individual phenolics in DOP. In addition, those major individual phenolic compounds of the resin purified freeze-dried powders increased about 4–5 times as that of the 70% ethanol extracted dry paste. The results of individual phenolic compounds were in line with TPC discussed in the previous section. Although resin purification captured most of the OP phenolic compounds as shown in Figure 3a, some polar molecules in Figure 3 b and c (0–14 min) escaped from the resin column, which explained the TPC detected in syrup of resin elutes during absorption (Figure 2).

Figure 4b shows the heatmap cluster analysis where we transformed absolute values of individual phenolic contents to standardized values in each row for comparing phenolic compound profiles. By observing the 'red hot zone' of the heatmap, major compounds included hydroxytyrosol, oleuropein, rutin, verbascoside, 4-HPA, hydroxytyrosol-glucoside and tyrosol-glucoside. 3,4-DHPEA-EDA at the peak 18 in Figure 3a was found to possibly be the dominantly abundant phenol in the OP if the concentration was calculated to oleuropein equivalents. However, due to the lack of a standard of 3,4-DHPEA-EDA, the large area of the peak 18 could be attributed to the intensive response of the chemical to the UV at 280 nm, instead of an actual high concentration. Although 3,4-DHPEA-DEDA has been reported to be the most abundant phenol in a Spanish OP [26], we lack the confirmation of standard chemical and the standard curve to report the compound as the abundant phenol in the US OP.

**Table 3.** Individual phenolic contents (mg/g of sample) of different extracted methods and purified steps of olive pomace.

| Name | Water Extractable Phenols in DOP | 70% Methanol Extractable Phenols in DOP | 70% Ethanol Extractable Phenols in DOP | Water Extracted Dry Paste | 70% Methanol Extracted Dry Paste | 70% Ethanol Extracted Dry Paste | XAD7HP Resin Purified Freeze-Dried Powders |
|---|---|---|---|---|---|---|---|
| 01. Vanillin-glucoside | 0.165 ± 0.004 | 0.054 ± 0.002 | 0.225 ± 0.021 | 0.151 ± 0.001 | 0.155 ± 0.001 | 0.152 ± 0.000 | 0.187 ± 0.002 |
| 02. Gallic acid | 0.223 ± 0.006 | 0.045 ± 0.004 | 0.000 ± 0.009 | 0.007 ± 0.002 | 0.008 ± 0.011 | 0.010 ± 0.003 | −0.018 ± 0.026 |
| 03. Hydroxytyrosol-glucoside | 1.407 ± 0.034 | 0.657 ± 0.024 | 0.250 ± 0.006 | 1.423 ± 0.006 | 1.475 ± 0.034 | 1.480 ± 0.003 | 4.423 ± 0.278 |
| 04. Hydroxytyrosol | 1.978 ± 0.039 | 2.017 ± 0.089 | 1.356 ± 0.054 | 3.508 ± 0.500 | 3.880 ± 0.027 | 4.219 ± 0.459 | 17.298 ± 0.363 |
| 05. Tyrosol-glucoside | 1.096 ± 0.022 | 0.679 ± 0.026 | 0.384 ± 0.018 | 1.555 ± 0.031 | 1.581 ± 0.006 | 1.639 ± 0.085 | 6.519 ± 0.421 |
| 06. Tyrosol | 0.460 ± 0.008 | 0.365 ± 0.008 | 0.162 ± 0.012 | 0.624 ± 0.014 | 0.811 ± 0.091 | 0.666 ± 0.002 | 3.514 ± 0.060 |
| 07. 4-HPA | 1.700 ± 0.044 | 0.800 ± 0.027 | 0.660 ± 0.023 | 1.691 ± 0.005 | 1.755 ± 0.013 | 1.742 ± 0.010 | 4.450 ± 0.021 |
| 08. Vanillic acid | 0.203 ± 0.004 | 0.208 ± 0.007 | 0.223 ± 0.004 | 0.509 ± 0.044 | 0.609 ± 0.009 | 0.585 ± 0.001 | 2.530 ± 0.018 |
| 09. Caffeic acid | 0.050 ± 0.001 | 0.044 ± 0.002 | 0.039 ± 0.002 | 0.073 ± 0.011 | 0.102 ± 0.009 | 0.091 ± 0.001 | 0.420 ± 0.001 |
| 10. Vanillin | 0.371 ± 0.005 | 0.329 ± 0.014 | 0.375 ± 0.011 | 0.285 ± 0.044 | 0.385 ± 0.006 | 0.269 ± 0.001 | 2.439 ± 0.027 |
| 11. p-coumaric acid | 0.084 ± 0.006 | 0.097 ± 0.003 | 0.086 ± 0.004 | 0.131 ± 0.001 | 0.168 ± 0.001 | 0.157 ± 0.000 | 0.884 ± 0.002 |
| 12. Ferulic acid | 0.047 ± 0.001 | 0.023 ± 0.002 | 0.029 ± 0.001 | 0.043 ± 0.003 | 0.046 ± 0.000 | 0.047 ± 0.000 | 0.326 ± 0.012 |
| 13. Verbascoside | 0.833 ± 0.007 | 1.074 ± 0.035 | 1.232 ± 0.037 | 1.135 ± 0.003 | 1.858 ± 0.016 | 1.507 ± 0.000 | 10.159 ± 0.052 |
| 14. Rutin | 0.770 ± 0.011 | 1.360 ± 0.050 | 1.031 ± 0.035 | 0.791 ± 0.019 | 2.409 ± 0.216 | 2.108 ± 0.120 | 11.048 ± 0.003 |
| 15. Luteolin-7-glucoside | 0.042 ± 0.000 | 0.042 ± 0.000 | 0.042 ± 0.000 | 0.312 ± 0.236 | 0.175 ± 0.000 | 0.785 ± 0.529 | 4.086 ± 0.022 |
| 16. o-coumaric acid | 0.101 ± 0.000 | 0.070 ± 0.000 | 0.070 ± 0.000 | 0.352 ± 0.041 | 0.416 ± 0.004 | 0.369 ± 0.001 | 1.562 ± 0.005 |
| 17. Apigenin-7-glucoside | 0.055 ± 0.000 | 0.121 ± 0.003 | 0.088 ± 0.001 | 0.293 ± 0.012 | 0.341 ± 0.002 | 0.336 ± 0.001 | 1.345 ± 0.034 |
| 18. Oleuropein | 0.811 ± 0.012 | 1.270 ± 0.324 | 0.930 ± 0.093 | 1.298 ± 0.188 | 2.609 ± 0.073 | 2.393 ± 0.052 | 12.231 ± 0.066 |
| 19. Pinoresinol | 0.084 ± 0.001 | 0.257 ± 0.005 | 0.175 ± 0.012 | 0.300 ± 0.087 | 0.478 ± 0.136 | 0.461 ± 0.025 | 2.775 ± 0.495 |
| 20. Cinnamic acid | 0.027 ± 0.002 | 0.019 ± 0.000 | 0.013 ± 0.004 | 0.012 ± 0.000 | 0.043 ± 0.024 | 0.063 ± 0.004 | 0.205 ± 0.111 |
| 21. Luteolin | 0.010 ± 0.000 | 0.487 ± 0.016 | 0.515 ± 0.034 | 0.041 ± 0.000 | 0.714 ± 0.003 | 0.678 ± 0.001 | 3.515 ± 0.003 |
| 22. Apigenin | 0.007 ± 0.000 | 0.062 ± 0.004 | 0.066 ± 0.004 | 0.030 ± 0.000 | 0.111 ± 0.006 | 0.107 ± 0.007 | 0.469 ± 0.031 |

Note: Vanillin-glucoside, hydroxytyrosol-glucoside and tyrosol-glucoside are expressed by equivalents of vanillin, hydroxytyrosol and tyrosol, respectively.

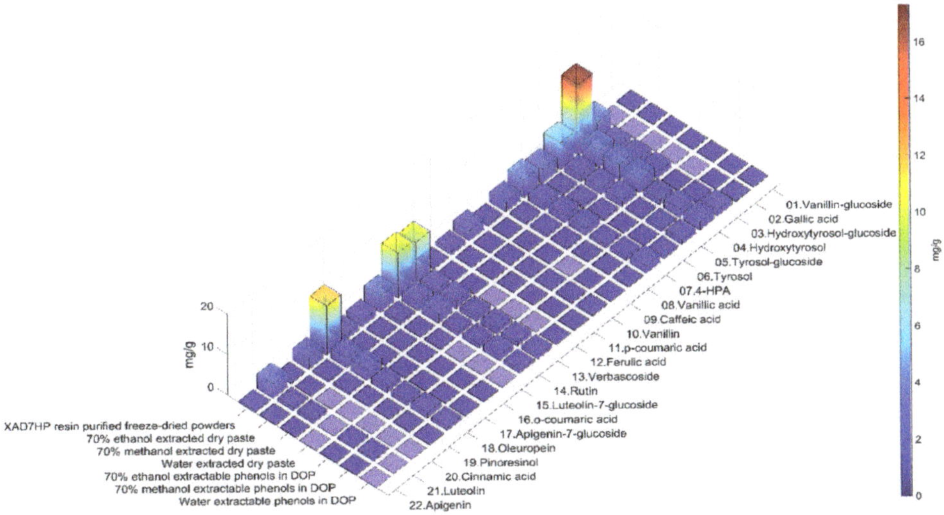


(a)

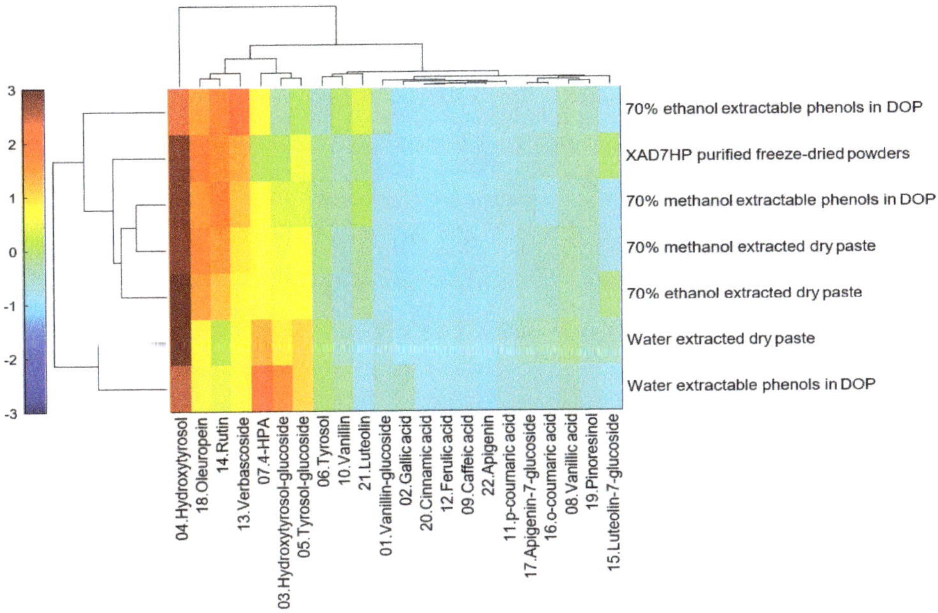


(b)

**Figure 4.** (**a**) Heatmap of individual phenolic contents (mg/g) of different extracted methods and purified steps; average and standard deviation can be found in Table 3; (**b**) heatmap cluster analysis of individual phenolic content of different extracted methods and purified steps; data were standardized along each row. Note: XAD7HP resin purified freeze-dried powder was made from 70% ethanol extracts, without oven-drying.

The first tier of the vertical cluster generally classified samples to the water extracts group and alcoholic extracts group, respectively. Meanwhile, the orange-to-red 'hot zone' of the third–fourth tiers of the horizontal clusters showed the major water-extracted compounds were 4-HPA, hydroxytyrosol and tyrosol-glucoside which were relatively higher polar and water-soluble molecules, whereas the primary alcohol-extracted compounds were oleuropein, verbascoside and rutin which were relatively less polar and less water soluble. Furthermore, methanol and ethanol extracts were grouped to their subvertical clusters; however, the grouping trend was not entirely true for the ethanol-extracted dry paste because of phenolic compound profile difference. Compared with the fresh 70% extractable ethanol in DOP, the verbascoside content in ethanol-extracted dry paste decreased while the hydroxytyrosol increased, which indicated that the verbascoside decomposed and released its hydroxytyrosol sidechain during the oven drying. Similarly indicated compound shifts were also observed by the changes from 4-HPA and hydroxytyrosol-glucoside to hydroxytyrosol in water-extracted dry paste and from verbascoside to hydroxytyrosol in methanol extracted dry paste, respectively. The precursors of hydroxytyrosol and tyrosol, such as oleuropein tend to decompose into hydroxytyrosol under high temperatures [27]. Previously, our group [8] reported that hydroxytyrosol, tyrosol, caffeic acid, verbascoside and luteolin-7-glucoside were well preserved by drum-drying at 135 °C while oleuropein decomposed, which was in agreement with this study. Individual phenols in olive pomace from the production stream of virgin olive oil (VOO) in Span have been extracted by superheated liquid extraction at 160 °C for presenting 5 min [15], and they found many aglycon derivatives, such as *p*-HPEA-EDA derived from ligstroside and 3,4-DHPEA-EA derived from oleuropein, as well as decarboxymethylated aglycone derivatives, such as 3,4-DHPEA-EDA derived from oleuropein and p-HPEAEDA derived from ligstroside. However, not all those derivatives were found in this study, which may be due to the drum-drying process that decomposed those derivatives to more stable compounds such as hydroxytyrosol and tyrosol as well as their glucoside derivatives. In addition, although major olive phenols have been identified in this study, further study is still necessary to identify those unknown peaks in Figure 3a.

### 3.4. Removal of Sugar Impurity

Figure 2 shows sugar impurity accounted for 538.37 ± 33.39 mg glucose equivalents (GE)/g dry paste, 546.95 ± 22.73 mg/g and 598.69 ± 71.34 mg/g in water, 70% methanol, 70% ethanol extracts, respectively, which were the major components of the three dry pastes of the crude extracts. Although resin purified freeze-dried powders increased 4.6 times from that of the 70% ethanol-extracted dry paste, the total sugar content decreased by 37.33%. However, the total sugar content was still up to 375.20 ± 19.00 mg GE/g which may be primarily attributed to those phenol sugar derivates in Figure 3 instead of free sugars, because free sugars in water cannot be adsorbed by macroporous resin [28], theoretically, and should be removed during the DI water rinsing step [29]. While many previous studies [5,24,30] evaluated the purification of olive phenols by resin, few studies have revealed the removal of sugar impurity alone with the increment of olive phenols. This study provided a more comprehensive evaluation of the resin purification step, but further study on the individual free sugar molecules should be conducted to reveal the mechanism of the separations of sugars from phenols in olive pomace extracts.

### 3.5. In Vitro Antioxidant Activities of Olive Pomace Extracts

It can be seen from Figure 5, while the three crude extracts performed less than 0.5 TrE of both 2,2-diphenyl-1-picrylhydrazyl (DPPH) and ferric reducing antioxidant power (FRAP), XAD7HP resin purified freeze-dried powders increased 3.7 times TrE for DPPH and 4.7 times TrE for FRAP compared to that of the 70% ethanol extracted dry pasted. The results generally correlated with the 4.6 times increment in TPC, although the increments of DPPH were not proportionally correlated with the TPC increase. It has been reported that DPPH was not always consistent with other antioxidant activities [31] or proportional to TPC, because some phenolic compounds may be inactive to participate in scavenging DPPH radicals [32]. Very recently, three selected food models (flour, whole-wheat flour and sugar) were fortified by hydroxytyrosol from olive oil, while the oil and fat were removed by supercritical fluid carbon dioxide, leaving olive phenols in the foodstuffs and increasing the antioxidant activities of the food models [33]. This study purified phenols from olive pomace (OP) instead of olive oil, which would be a more cost-effective way for nutrition fortification by directly adding and distributing OP phenols into foodstuffs and their premix.

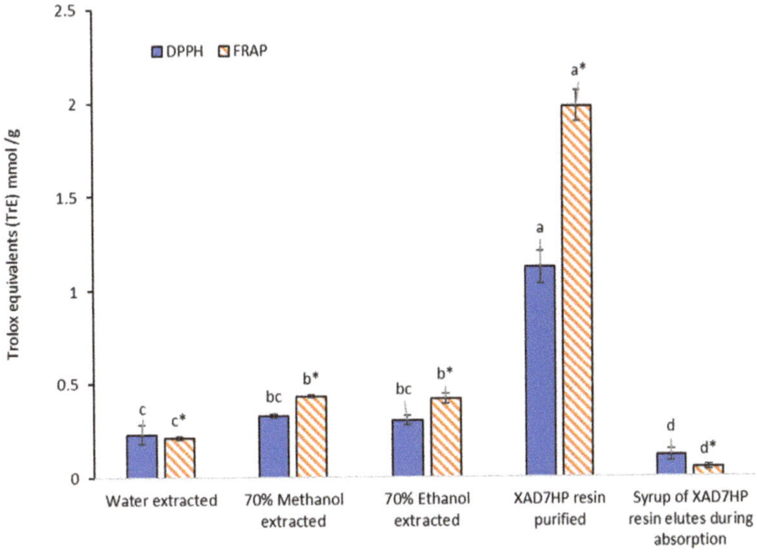


**Figure 5.** DPPH scavenging and FRAP activity among crude extracts, resin-purified extracts, different letters indicate significantly different, $p < 0.05$. Note: significant differences were only compared in either DPPH without"*" on significant markers or FRAP with "*" on significant markers. Data of syrup of the resin elutes were reported to dry matters (DM).

## 4. Conclusions

Our study demonstrated that major phenols in the Aberquina olive pomace from California are hydroxytyrosol, oleuropein, rutin, verbascoside, 4-HPA, hydroxytyrosol-glucoside and tyrosol-glucoside. Macroporous resin purification increased the total phenolic content (TPC) by 4.6 times the ethanol crude extracts of DOP while removing 37.33% of total sugar, while the extractable TPC was 36–43 mg GAE/g DOP. Meanwhile, the antioxidant activities increased 3.7 times TrE for DPPH and 4.7 times TrE for FRAP compared to the ethanol crude extracts. This new data on the chemical compositions of the selected US OP provides important and practical knowledge for the valorization and industrial food applications of the US olive wastes. Extracting bioactive compounds from the US OP not only could directly produce value-added products from food byproducts and increase consumers' health via the delivery of natural antioxidants but also reduce negative environmental impact. The chemical and phenolic composition of more olive cultivars in the US should be determined and compared for better assessment of the valorization of the US OP.

**Supplementary Materials:** The following supporting information can be downloaded at: https://www.mdpi.com/article/10.3390/foods11020174/s1, Figure S1: Chromatography of XAD7HP resin purified freeze-dried powder (**a**) at 320 and (**b**) at 365 nm and (**c**) total ion chromatogram (TIC).

**Author Contributions:** Conceptualization, H.Z. and S.C.W.; methodology, H.Z. and S.C.W.; software, H.Z.; validation, H.Z. and S.C.W..; formal analysis, H.Z.; investigation, H.Z., R.J.A.-B., and S.C.W.; resources, S.C.W.; data curation, H.Z.; writing—original draft preparation, H.Z.; writing—review and editing, H.Z., R.J.A.-B., and S.C.W.; visualization, H.Z.; supervision, S.C.W.; project administration, S.C.W.; funding acquisition, S.C.W. and R.J.A.-B. All authors have read and agreed to the published version of the manuscript.

**Funding:** California Department of Food and Agriculture, 2020 Specialty Crop Block Grant Program (20-0001-033-SF). Pitted drum-dried olive pomace used in this study was obtained from research supported by USDA/ARS/WRRC/HPFR CRADA Agreement No. 58-2030-9-005 signed with California Olive Ranch.

**Institutional Review Board Statement:** Not applicable.

**Data Availability Statement:** The data presented in this study are available in within the article and Supplementary Materials.

**Conflicts of Interest:** The authors declare no competing financial interest.

## References

1. Conte, P.; Fadda, C.; Del Caro, A.; Urgeghe, P.P.; Piga, A. Table Olives: An Overview on Effects of Processing on Nutritional and Sensory Quality. *Foods* **2020**, *9*, 514. [CrossRef] [PubMed]
2. Gómez-Cruz, I.; Cara, C.; Romero, I.; Castro, E.; Gullón, B. Valorisation of Exhausted Olive Pomace by an Eco-Friendly Solvent Extraction Process of Natural Antioxidants. *Antioxidants* **2020**, *9*, 1010. [CrossRef]
3. Antónia Nunes, M.; Pawlowski, S.; Costa, A.S.G.; Alves, R.C.; Oliveira, M.B.P.P.; Velizarov, S. Valorization of olive pomace by a green integrated approach applying sustainable extraction and membrane-assisted concentration. *Sci. Total Environ.* **2019**, *652*, 40–47. [CrossRef]
4. Tundis, R.; Conidi, C.; Loizzo, M.R.; Sicari, V.; Cassano, A. Olive Mill Wastewater Polyphenol-Enriched Fractions by Integrated Membrane Process: A Promising Source of Antioxidant, Hypolipidemic and Hypoglycaemic Compounds. *Antioxidants* **2020**, *9*, 602. [CrossRef]
5. Şahin, S.; Bilgin, M. Selective adsorption of oleuropein from olive (*Olea europaea*) leaf extract using macroporous resin. *Chem. Eng. Commun.* **2017**, *204*, 1391–1400. [CrossRef]
6. Kammerer, D.R.; Carle, R.; Stanley, R.A.; Saleh, Z.S. Pilot-Scale Resin Adsorption as a Means To Recover and Fractionate Apple Polyphenols. *J. Agric. Food Chem.* **2010**, *58*, 6787–6796. [CrossRef]
7. Tierney, M.; Gallagher, A.M.; Giotis, E.S.; Pentieva, K. An Online Survey on Consumer Knowledge and Understanding of Added Sugars. *Nutrients* **2017**, *9*, 37. [CrossRef] [PubMed]
8. Sinrod, A.J.G.; Avena-Bustillos, R.J.; Olson, D.A.; Crawford, L.M.; Wang, S.C.; McHugh, T.H. Phenolics and Antioxidant Capacity of Pitted Olive Pomace Affected by Three Drying Technologies. *J. Food Sci.* **2019**, *84*, 412–420. [CrossRef] [PubMed]
9. Zhao, H.; Shen, C.; Wu, Z.; Zhang, Z.; Xu, C. Comparison of wheat, soybean, rice, and pea protein properties for effective applications in food products. *J. Food Biochem.* **2020**, *44*, e13157. [CrossRef]
10. Zhang, W.-H.; Wu, J.; Weng, L.; Zhang, H.; Zhang, J.; Wu, A. An improved phenol-sulfuric acid method for the determination of carbohydrates in the presence of persulfate. *Carbohydr. Polym.* **2020**, *227*, 115332. [CrossRef]
11. De Bruno, A.; Romeo, R.; Fedele, F.L.; Sicari, A.; Piscopo, A.; Poiana, M. Antioxidant activity shown by olive pomace extracts. *J. Environ. Sci. Health Part B* **2018**, *53*, 526–533. [CrossRef]
12. Xu, C.; Yagiz, Y.; Zhao, L.; Simonne, A.; Lu, J.; Marshall, M.R. Fruit quality, nutraceutical and antimicrobial properties of 58 muscadine grape varieties (*Vitis rotundifolia Michx.*) grown in United States. *Food Chem.* **2017**, *215*, 149–156. [CrossRef]
13. Jiang, L.; Shen, X.; Shoji, T.; Kanda, T.; Zhou, J.; Zhao, L. Characterization and Activity of Anthocyanins in Zijuan Tea (*Camellia sinensis var. kitamura*). *J. Agric. Food Chem.* **2013**, *61*, 3306–3310. [CrossRef]
14. Li, C.; Zheng, Y.; Wang, X.; Feng, S.; Di, D. Simultaneous separation and purification of flavonoids and oleuropein from *Olea europaea* L.(*olive*) leaves using macroporous resin. *J. Sci. Food Agric.* **2011**, *91*, 2826–2834. [CrossRef] [PubMed]
15. Peralbo-Molina, Á.; Priego-Capote, F.; Luque de Castro, M.D. Tentative Identification of Phenolic Compounds in Olive Pomace Extracts Using Liquid Chromatography–Tandem Mass Spectrometry with a Quadrupole–Quadrupole-Time-of-Flight Mass Detector. *J. Agric. Food Chem.* **2012**, *60*, 11542–11550. [CrossRef]
16. Zhao, H.; Xie, X.; Read, P.; Loseke, B.; Gamet, S.; Li, W.; Xu, C. Biofortification with selenium and lithium improves nutraceutical properties of major winery grapes in the Midwestern United States. *Int. J. Food Sci. Technol.* **2021**, *56*, 825–837. [CrossRef]
17. D'Amato, R.; De Feudis, M.; Guiducci, M.; Businelli, D. *Zea mays* L. Grain: Increase in Nutraceutical and Antioxidant Properties Due to Se Fortification in Low and High Water Regimes. *J. Agric. Food Chem.* **2019**, *67*, 7050–7059. [CrossRef]
18. Antónia Nunes, M.; Costa, A.S.G.; Bessada, S.; Santos, J.; Puga, H.; Alves, R.C.; Freitas, V.; Oliveira, M.B.P.P. Olive pomace as a valuable source of bioactive compounds: A study regarding its lipid- and water-soluble components. *Sci. Total Environ.* **2018**, *644*, 229–236. [CrossRef]
19. Aliakbarian, B.; Sampaio, F.C.; de Faria, J.T.; Pitangui, C.G.; Lovaglio, F.; Casazza, A.A.; Converti, A.; Perego, P. Optimization of spray drying microencapsulation of olive pomace polyphenols using Response Surface Methodology and Artificial Neural Network. *LWT* **2018**, *93*, 220–228. [CrossRef]

20. Baiano, A.; Gambacorta, G.; Terracone, C.; Previtali, M.A.; Lamacchia, C.; La Notte, E. Changes in Phenolic Content and Antioxidant Activity of Italian Extra-Virgin Olive Oils during Storage. *J. Food Sci.* **2009**, *74*, C177–C183. [CrossRef] [PubMed]
21. Xu, C.; Yagiz, Y.; Hsu, W.-Y.; Simonne, A.; Lu, J.; Marshall, M.R. Antioxidant, Antibacterial, and Antibiofilm Properties of Polyphenols from Muscadine Grape (*Vitis rotundifolia Michx.*) Pomace against Selected Foodborne Pathogens. *J. Agric. Food Chem.* **2014**, *62*, 6640–6649. [CrossRef]
22. Parniakov, O.; Apicella, E.; Koubaa, M.; Barba, F.J.; Grimi, N.; Lebovka, N.; Pataro, G.; Ferrari, G.; Vorobiev, E. Ultrasound-assisted green solvent extraction of high-added value compounds from microalgae *Nannochloropsis* spp. *Bioresour. Technol.* **2015**, *198*, 262–267. [CrossRef] [PubMed]
23. Rodríguez-Rojo, S.; Visentin, A.; Maestri, D.; Cocero, M.J. Assisted extraction of rosemary antioxidants with green solvents. *J. Food Eng.* **2012**, *109*, 98–103. [CrossRef]
24. Pinelli, D.; Molina Bacca, A.E.; Kaushik, A.; Basu, S.; Nocentini, M.; Bertin, L.; Frascari, D. Batch and Continuous Flow Adsorption of Phenolic Compounds from Olive Mill Wastewater: A Comparison between Nonionic and Ion Exchange Resins. *Int. J. Chem. Eng.* **2016**, *2016*, 9349627. [CrossRef]
25. Niknam, S.M.; Kashaninejad, M.; Escudero, I.; Sanz, M.T.; Beltrán, S.; Benito, J.M. Valorization of olive mill solid residue through ultrasound-assisted extraction and phenolics recovery by adsorption process. *J. Clean. Prod.* **2021**, *316*, 128340. [CrossRef]
26. Katsinas, N.; Bento da Silva, A.; Enríquez-de-Salamanca, A.; Fernández, N.; Bronze, M.R.; Rodríguez-Rojo, S. Pressurized Liquid Extraction Optimization from Supercritical Defatted Olive Pomace: A Green and Selective Phenolic Extraction Process. *ACS Sustain. Chem. Eng.* **2021**, *9*, 5590–5602. [CrossRef]
27. Yuan, J.-J.; Wang, C.-Z.; Ye, J.-Z.; Tao, R.; Zhang, Y.-S. Enzymatic Hydrolysis of Oleuropein from Olea europea (*Olive*) Leaf Extract and Antioxidant Activities. *Molecules* **2015**, *20*, 2903–2921. [CrossRef] [PubMed]
28. Heinonen, J.; Farahmandazad, H.; Vuorinen, A.; Kallio, H.; Yang, B.; Sainio, T. Extraction and purification of anthocyanins from purple-fleshed potato. *Food Bioprod. Process.* **2016**, *99*, 136–146. [CrossRef]
29. Sandhu, A.K.; Gu, L. Adsorption/Desorption Characteristics and Separation of Anthocyanins from Muscadine (*Vitis rotundifolia*) Juice Pomace by Use of Macroporous Adsorbent Resins. *J. Agric. Food Chem.* **2013**, *61*, 1441–1448. [CrossRef]
30. Vavouraki, A.I.; Dareioti, M.A.; Kornaros, M. Olive Mill Wastewater (OMW) polyphenols adsorption onto polymeric resins: Part i—batch anaerobic digestion of OMW. *Waste Biomass Valorization* **2021**, *12*, 2271–2281. [CrossRef]
31. Li, G.; Zhan, J.; Hu, L.; Yuan, C.; Takaki, K.; Ying, X.; Hu, Y. Identification of a new antioxidant peptide from porcine plasma by in vitro digestion and its cytoprotective effect on $H_2O_2$ induced HepG2 model. *J. Funct. Foods* **2021**, *86*, 104679. [CrossRef]
32. Lewoyehu, M.; Amare, M. Comparative evaluation of analytical methods for determining the antioxidant activities of honey: A review. *Cogent Food Agric.* **2019**, *5*, 1685059. [CrossRef]
33. Bartella, L.; Mazzotti, F.; Talarico, I.R.; Santoro, I.; Di Donna, L. Hydroxytyrosol-Fortified Foods Obtained by Supercritical Fluid Extraction of Olive Oil. *Antioxidants* **2021**, *10*, 1619. [CrossRef] [PubMed]

foods

MDPI

Article

# Efficacy of L-Arabinose in Lowering Glycemic and Insulinemic Responses: The Modifying Effect of Starch and Fat

**Korrie Pol [1], Marie-Luise Puhlmann [1,2] and Monica Mars [1,*]**

1 Division of Human Nutrition and Health, Wageningen University, P.O. Box 17, NL-6700 AA Wageningen, The Netherlands; korrie.pol@wur.nl (K.P.); marie-luise.puhlmann@wur.nl (M.-L.P.)
2 Lab of Microbiology, Wageningen University, P.O. Box 17, NL-6700 AA Wageningen, The Netherlands
* Correspondence: monica.mars@wur.nl; Tel.: +31-317-485-340

**Abstract:** L-arabinose is a bio-active compound derived from the side-streams of plant food processing. L-arabinose lowers glycemic and insulinemic responses when added to simple water-based sugary liquids. However, the effect in more complex foods, including fat and starch, is inconsistent. This study assessed the effect of fat or starch in a sugary drink on the efficacy of L-arabinose. Twenty-three healthy volunteers (12 female/11 male; aged 24 ± 3 years; BMI 23 ± 3 kg/m$^2$) participated in a randomised cross-over trial with six drinks: control: 50 g sucrose in water; fat: control + 22 g oil; starch: control + 50 g starch; and all three with and without the addition of 5 g L-arabinose. The addition of L-arabinose to the control drink lowered glucose and insulin peaks by 15% and 52%; for the fat drink by 8% and 45%; and for the starch drink by 7% and 29%. For all three drinks, adding L-arabinose increased glucagon-like peptide 1 (GLP-1) responses and lowered Glucose-dependent insulinotropic polypeptide (GIP) responses. Despite adding large quantities of starch and fat to sugary drinks, L-arabinose significantly lowered postprandial glycemic and insulinemic responses in healthy subjects. These findings suggest that L-arabinose can be functional in more complex foods; however, the factors affecting its efficacy in solid food matrices need to be studied in more detail.

**Keywords:** L-arabinose; sucrose; fat; starch; glucose; insulin; GLP-1; GIP; healthy adults

**Citation:** Pol, K.; Puhlmann, M.-L.; Mars, M. Efficacy of L-Arabinose in Lowering Glycemic and Insulinemic Responses: The Modifying Effect of Starch and Fat. *Foods* **2022**, *11*, 157. https://doi.org/10.3390/foods11020157

Academic Editors: Michela Verni and Federico Casanova

Received: 19 November 2021
Accepted: 30 December 2021
Published: 8 January 2022

**Publisher's Note:** MDPI stays neutral with regard to jurisdictional claims in published maps and institutional affiliations.



## 1. Introduction

Worldwide, more than 463 million people are currently diagnosed with diabetes and 374 million with impaired glucose tolerance [1]. Postprandial glycemic control is of high importance in these populations. Systematic reviews have shown that diets containing mainly foods with a low potency to increase blood glucose levels, so-called low glycemic index (GI) diets, improve biomarkers of glycemic control [2,3]. In addition, several clinical guidelines advise diabetic patients to especially incorporate foods with a low GI into their diets [4–6].

Simple sugars, such as sucrose, are often added to foods for their sweet taste. However, they are also known for their fast and high postprandial glycemic responses. The food industry has a high interest in novel functional ingredients that can lower the glycemic potency of sugar-rich foods without compromising the food's sweet taste and other food properties such as texture. L-arabinose is such a functional ingredient. L-arabinose is an aldopentose molecule with a relatively sweet taste [7]. L-arabinose can be derived from hemicellulose by enzymatic hydrolysis [8]. Hemicellulose is abundantly present in the side-streams of plant food processing, such as sugar production [8]. L-arabinose has been shown to slow down glucose absorption by inhibiting the intestinal enzyme sucrase in an uncompetitive manner [9]. Sucrase is present in the brush border of the intestine and critical for hydrolysing sucrose into a glucose and fructose molecule [9].

Several randomised clinical trials with healthy subjects show that adding L-arabinose to simple drinks or gels containing sucrose reduces postprandial glycemic responses significantly [10–12]. When 0.9 g L-arabinose was added to a tropical fruit-flavoured drink sweetened with 10 g sucrose, glycemic and insulinemic responses were lowered by 9% and 35%, respectively [13]. In addition, the replacement of 30% sucrose by L-arabinose lowered glycemic and insulinemic responses by 22% and 67% [14]. However, the efficacy of L-arabinose is less evident for more complex foods. For example, no effect was found on glucose or insulin responses when L-arabinose was incorporated into a mixed breakfast meal [15], cereals [14], and when sucrose was replaced with L-arabinose in muffins [13]. Although in the latter two studies no effect was found on glycemic response, trends towards lower insulinemic responses were observed [13,14]. These findings suggest that L-arabinose may have had an inhibiting effect on sucrose breakdown, but that the net effects could not be observed in the circulating glucose levels of healthy subjects. In addition, it may very well be that the other components present in the foods, such as fat and starch, may have modified the effect. For example, fat is known to slow down gastric emptying, and may therefore also slow down glucose absorption [16,17]. At the same time, the abundance of starch present in the foods may overshadow the effect of an inhibited sucrose breakdown. Therefore, the impact that individual food components can have on the efficacy of L-arabinose needs to be disentangled before L-arabinose can be implemented into real life and complex foods.

This manuscript describes the results of a model food experiment that was conducted in order to explore the modifying effect of isolated food components on the efficacy of L-arabinose in sucrose-rich foods. The model foods were sucrose drinks containing large quantities of either starch or fat with or without the addition of L-arabinose. With this model food experiment, the opportunities and bandwidth of food applications are illustrated. It is demonstrated that L-arabinose still lowers the postprandial glycemic and insulinemic response in healthy subjects, despite the presence of starch and fat. Moreover, effects are shown on associated physiological markers—that is, the incretin hormones glucagon-like peptide 1 (GLP-1) and glucose-dependent insulinotropic polypeptide (GIP). These findings suggest that L-arabinose maintains its functionality in the presence of starch and fat, two important compounds present in more complex foods and real-life foods.

## 2. Materials and Methods

### 2.1. Subjects

Healthy adults (both sexes, age 18–35 years) were recruited in Wageningen and surrounding areas. For the recruitment, emails were sent to persons in a database of volunteers who had previously expressed their interest in participating in nutrition studies. Inclusion criteria were as follows: apparently healthy, as judged by the subject and measured by a health and lifestyle questionnaire (i.e., no known diseases, no medicine use, and no allergy or intolerance to the food products under study); normal fasting glucose (<6.1 mmol/L); normal haemoglobin concentration (>8.5 mmol/L for males and >7.5 mmol/L for females); and able to communicate in spoken and written Dutch. Subjects who were pregnant or lactating or reported weight fluctuations of >5 kg in the two months prior to the screening or consumed excessive alcoholic beverages (≥21 glasses/week) were excluded.

A total of 87 subjects completed the health and lifestyle questionnaire to determine their eligibility. In addition, the Dutch Eating Behaviour Questionnaire (DEBQ) was completed to identify eating restraint, emotional eaters, and external eaters [18]. However, none of these personality traits affected the glycemic or insulinemic responses, hence they will not be further discussed. Eligible subjects came after fasting to a physical examination, during which weight and height were measured, blood glucose and haemoglobin were measured with a finger prick, and a nurse checked the suitability of the participants' antecubital veins for inserting a cannula several times.

In total, 25 subjects were included in the trial, of which 23 subjects completed all treatments (Table 1). Two subjects dropped out: one of them was not able to consume the total amount of the product (Fat + Ara), while the other subject dropped out due to side effects of the blood drawings (painful elbow cavities due to a combination of the cannula and playing volleyball). The data of these 2 subjects were not analysed.

**Table 1.** Baseline characteristics of the study subjects (mean ± SD).

| | Total ($n$ = 23) | Men ($n$ = 12) | Women ($n$ = 11) |
|---|---|---|---|
| Age (y) | 23.8 ± 3.3 | 24.4 ± 3.3 | 23.2 ± 3.3 |
| Body weight (kg) | 74.5 ± 11.2 | 77.6 ± 9.9 | 71.2 ± 12.6 |
| Height (m) | 1.79 ± 0.09 | 1.84 ± 0.08 | 1.74 ± 0.07 |
| BMI (kg/m$^2$) | 23.1 ± 2.7 | 22.8 ± 1.6 | 23.5 ± 3.7 |

Subjects received financial compensation after participation. The study was performed in accordance with Dutch law and the principles laid down in the current version of the Declaration of Helsinki. The medical ethical committee of Wageningen University gave ethical approval (ABR nr: NL61428.081.17). All eligible subjects provided oral and written informed consent. The trial and its primary outcomes were pre-registered in the Dutch Trial Register as NTR6636 (www.trialregister.nl Trial NL6458).

The sample size was based on the effect of L-arabinose on glycemic response peaks. Sample size estimations were performed with the software program G-power (University of Kiel, Kiel, Germany). For all calculations, a paired Student's $t$-test was assumed, with a two-tailed $t$-test with $\alpha$ = 0.05, power = 0.8, and SD of 0.75 mmol/L (obtained from previous experiments [13,14]). Furthermore, an effect of 0.5 mmol/L was assumed to be relevant in this population, which means an effect size (dz) of 0.67. This resulted in an estimated sample size of minimal 20 participants. Previous studies from our lab [13,14] showed dropout rates of 17% and 11%. Taking into account a safe dropout rate of 20%, it was decided to enrol 24 subjects.

### 2.2. Design

The study was a 3 × 2 factorial randomised cross-over trial. The study was carried out between August and October 2017 at the division of Human Nutrition and Health, Wageningen University, The Netherlands. The treatments consisted of three drinks (control drink, fat drink, or starch drink) with or without 0 g or 5 g L-arabinose (see Section 2.3 Test Products).

The researchers and subjects were blinded for the addition of L-arabinose, but not for the type of drink (control, fat, starch), due to the difference in mouthfeel and taste. Subjects were randomised to all six treatments. An independent research assistant generated a list with randomised sequences, using a Williams design. Another independent research assistant was responsible for the allocation of the randomisation codes to the products. The researchers allocated the subjects in consecutive order to the randomized sequences. To minimise carry-over effects, a washout period of a minimum of three days was used.

### 2.3. Test Products

Ingredients, energy content, and liking of the test products are described in Table 2. All test drinks contained 50 g of sucrose and 500 g of water. The L-arabinose-containing drinks contained 5 g of L-arabinose (10% $w/w$ sucrose). The preparation protocol of the drinks was based on extensive in vitro experimenting (see Supplemental Data S1). The drinks were freshly prepared every morning according to the preparation protocol and were distributed in identical opaque cups wrapped in aluminium foil. The subjects received the following instructions: "Drink the full amount gradually within 5 min with a straw". As the test products were clearly different in taste and mouthfeel, it was only possible to blind the subjects to the presence or absence of L-arabinose, but not to the presence of starch or fat.

**Table 2.** Nutrient content, energy content, weight, and liking of the test products.

| | Water | Sucrose | L-Arabinose | Fat | Starch | Energy | Weight | Liking [1] |
|---|---|---|---|---|---|---|---|---|
| | (g) | (g) | (g) | (g) | (g) | (kcal) | (g) | |
| Control | 500 | 50 | 0 | 0 | 0 | 200 | 550 | 5.2 ± 1.2 [a] |
| Control + Ara | 500 | 50 | 5 | 0 | 0 | 200 | 555 | 5.0 ± 1.3 [ad] |
| Fat | 500 | 50 | 0 | 22 | 0 | 398 | 572 | 4.2 ± 1.0 [bd] |
| Fat + Ara | 500 | 50 | 5 | 22 | 0 | 398 | 577 | 4.1 ± 1.3 [b] |
| Starch | 500 | 50 | 0 | 0 | 50 | 400 | 600 | 3.1 ± 1.4 [c] |
| Starch + Ara | 500 | 50 | 5 | 0 | 50 | 400 | 605 | 2.7 ± 1.3 [c] |

[1] Liking was measured on a 7-point Likert scale, anchored from "not liked at all" to "very much liked". Mean ± SD are presented. Different letters indicate significant differences in liking between test products ($p < 0.05$). Nutrient content was estimated based on the ingredients and energy content was calculated by using the Atwater factors—that is, 4 kcal for carbohydrates and 9 kcal for fat. The energetic value of L-arabinose was ignored, as its metabolic energetic content is unknown.

The subjects scored the drinks on a liking scale directly after they finished the drinks. A 7-point Likert scale ranging from "not liked at all" to "very much liked" was used. On average, the control drinks were scored as 5.1 ± 1.3, the fat drinks as 4.2 ± 1.2, and the starch drinks as 2.9 ± 1.4 (Table 2). The addition of L-arabinose did not change the liking score of the drinks (all $p$-values > 0.05).

#### 2.3.1. Test Drink Preparation

Control Drinks

The control drinks consisted of sugar solutions with 9% ($w/w$) sucrose, and in the case of Control + Ara, also 0.9% ($w/w$) L-arabinose (Cosun, Breda, The Netherlands). Sugar (and L-arabinose) were mixed with cold tap water in a Waring blender twice for 10 s (CB102T, Waring, Torrington, CT, USA). These sugar solutions were then also used as a base for the fat and starch drinks.

Fat Drinks

The fat drinks were a 3.8% ($w/w$) oil-in-water emulsion using sunflower oil and 0.4% ($w/w$) Tween 80 (Lamesorb SMO 20, BASF, Basel, Germany). The emulsifier was added to prevent the fat from phase separating beforehand or in the stomach. The sunflower oil and the emulsifier were blended into the pre-made sugar solutions using a Waring blender for 90 s at full speed (HGB500, Waring, Torrington, CT, USA). Both fat treatments had the same average droplet sizes, which were $d_{3,2}$ = 1.63 µm and $d_{4,3}$ = 4.54 µm (Supplemental Data S1).

Starch Drinks

The starch drinks were a dispersion of a pregelatinised waxy maize starch (C Gel–Instant 12410, Cargill, Minneapolis, MN, USA) that can be solubilised in water without heating. The starch drinks were prepared by quickly and carefully incorporating the pregelatinised starch into the premade sugar solutions using a hand blender set to full speed for at least 30 s (MSM87110, Bosch, BSH group, Nazarje, Slovenia).

### 2.4. *Study Procedures*

#### 2.4.1. Before the Test Day

To standardise the fasting state, subjects were asked to maintain their habitual diet and other lifestyle habits throughout the whole study period. Subjects were not allowed to drink alcohol nor to perform heavy exercise two days before test days. Subjects reported their evening meal before the first test session in their study diary—that is, the ingredients, preparation methods, and the amounts they ate in household measures. Additionally, subjects registered the amount of physical activity they did the evening before the test day. They were then instructed to eat and exercise in the same way in the evenings prior to the test day.

#### 2.4.2. Test Days

The subjects arrived at the study site between 7.30 and 8.00 a.m. after an overnight fast; subjects did not eat or drink anything other than water, coffee, or tea between 20:00 and 22:00 h and were only allowed to drink water from 22:00 h the evening before the test. The study personnel checked the diaries for deviations from the study protocol and the subjects then completed the well-being and gastro-intestinal comfort questionnaire. Then, a trained and experienced nurse inserted an intravenous cannula. After at least 30 min rest, the nurse drew the baseline blood samples and the subjects completed an appetite questionnaire. After that ($t = 0$ min), the subjects drank the test product within 5 min. After the drink was consumed, subjects drank an additional 135 mL of water and scored the test product on liking (see Section 2.3 Test Products). Blood samples and appetite questionnaires were conducted at $t = 15, 30, 45, 60, 90, 120, 180$ min after the subjects started to consume the test products. After 120 min, subjects again drank 135 mL of water. At $t = 180$ min, a gastro-intestinal comfort questionnaire was completed. Subjects remained seated in the living room of the study facility and could do sedentary activities such as quietly reading, watching television, or listening to music. After the last blood sample was drawn, the intravenous cannula was removed. Subjects were then directed to the eating lab and received an *ad libitum* meal (see Section 2.6). The morning after the test day, the subjects received an online evaluation questionnaire and were asked which experimental product they thought they had had. Subjects were instructed to fill out a study diary throughout the whole study period in which they could report any complaints and deviations from the study protocol. In addition, a well-being questionnaire was filled out to register any adverse effects(see Section 2.7 Gastro-Intestinal Tolerance).

### 2.5. Biochemical Measures

Venous blood samples for plasma glucose analysis were collected into NaF vacutainers (Becton Dickinson, Franklin Lakes, NJ, USA), and for insulin analyses, blood samples were collected into EDTA vacutainers (Becton Dickinson, Franklin Lakes, NJ, USA) and kept in ice water for a maximum of 15 min before being centrifuged. For GLP-1 and GIP analyses, EDTA + Aprotinine vacutainers were manually pre-treated by injecting 50 μL DPP-IV inhibitor (Catalog no. DPP4-010, Millipore, Burlington, MA, United States USA) to prevent proteolytic cleavage. These tubes were kept in ice water before use and venous blood samples were collected into these vacutainers and put in ice water for a maximum of 15 min before being centrifuged. All tubes were centrifuged for 10 min at $1200\times g$ at 4 °C. Plasma was then aliquoted and stored at −80 °C until analysis.

Plasma glucose samples were measured with the hexokinase assay using the Siemens Dimension Vista System (Siemens Healthcare, The Hague, The Netherlands). Plasma insulin samples were measured using an enzyme-linked immunosorbent assay (ELISA) (catalogue no. 10-1113-10, Mercodia Insulin ELISA, Uppsala, Sweden) (the lowest detectable level for insulin was 1.0 mU/L; intra-assay-coefficient of variation (CV): 4% and inter-assay CV: 4%). Plasma total GLP-1 and GIP were measured with a custom-made multiplex assay (Meso Scale Discovery, Rockville, MD, USA). All samples of one subject were analysed within one run or plate, which also contained a positive and negative control.

### 2.6. Subjective Appetite Ratings and Ad Libitum Food Intake

Hunger, fullness, desire to eat, prospective food consumption, and thirst were measured with 100 mm visual analogue scales (VAS) with the anchors: "not at all" (left) to "very much" (right) and filled out at the same time points as the blood drawings [19].

An *ad libitum meal* was consumed after 3 h. Subjects received 18 small wheat buns (~22 g each) and several toppings, being 100 g of low-fat margarine, 8 slices of cheese, 100 g of strawberry jam, and 100 g of chocolate sprinkles. Further, 500 mL water, 500 mL hot water with tea bags, 500 mL coffee, and sachets of creamer powder and sugar were offered. Subjects were instructed to eat as much as they wanted until they felt comfortably full. Left-overs were covertly weighed. Food intake was calculated by subtracting the weight of the leftovers from the food that was provided. Energy and macronutrient intake was estimated using the Dutch Food Composition Database [20]. None of the subjects finished all buns or asked for a second portion.

### 2.7. Gastro-Intestinal Tolerance

Potential impact on gastro-intestinal tolerance was measured with questions on bloating, regurgitation, flatulence, and nausea, giving a grade being no (0), little (1), moderate (2), or severe (3) at the study site at $t = 0$ and $t = 180$ min. Further, similar questions were completed three times in the diary: on days preceding a study day, on every study day, and the day after a study day.

### 2.8. Calculations and Statistical Analysis

Several parameters were extracted from the individual time curves. For glucose, insulin, GLP-1, and GIP responses these were: the incremental area under the curve above the baseline ($iAUC_{0-180min}$) by the trapezoidal rule, peak-value ($C_{max}$), and time-to-peak ($T_{peak}$). In addition, the area under the curve (AUC) for the appetite ratings ($AUC_{0-180min}$) was calculated by the trapezoidal rule.

Outcomes are shown as least square means and 95% confidence intervals, unless stated otherwise. Statistical analyses were performed two-sided, and $p$-values < 0.05 were considered statistically significant. All data were analysed using SAS for Windows software v9.4 (SAS Institute Inc., Cary, NC, USA). All continuous variables were checked for normal distribution by visual inspection of histograms and QQ-plots of the Studentised residuals. There were ten missing values for GLP-1 and these were left missing in the mixed model analyses.

All statistical analyses were performed separately for control, fat, and starch drinks. Comparisons were made between arabinose and non-arabinose containing drinks. The responses over time (glucose, insulin, GLP-1, GIP, appetite score) were analysed using a linear mixed model ANOVA (PROC MIXED; SAS Institute Inc., Cary, NC, USA). The fixed factors were treatment (with or without L-arabinose), time, and the interaction of treatment x time, with time being a repeated factor and ID number as a random factor. The compound symmetry covariance structure was used, as this was the best-fitting covariance structure based on the lowest AIC. The contrasts between time points were assessed by treatment x time effects sliced by time. Post hoc tests were performed with Tukey correction. These results are presented in Figure 1. Moreover, $iAUC_{0-180min}$ (total response), $C_{max}$ (peak), and $T_{peak}$ (time-to-peak) were compared between treatments with and without L-arabinose by means of a paired $t$-test. Mean ad libitum energy intake and the change in gastro-intestinal tolerance was compared between treatments with the mixed model ANOVA described above.

Before the actual statistical analyses, all endpoints were tested in a separate model to check for order and baseline effects; in these analyses, no statistically significant effects were found.

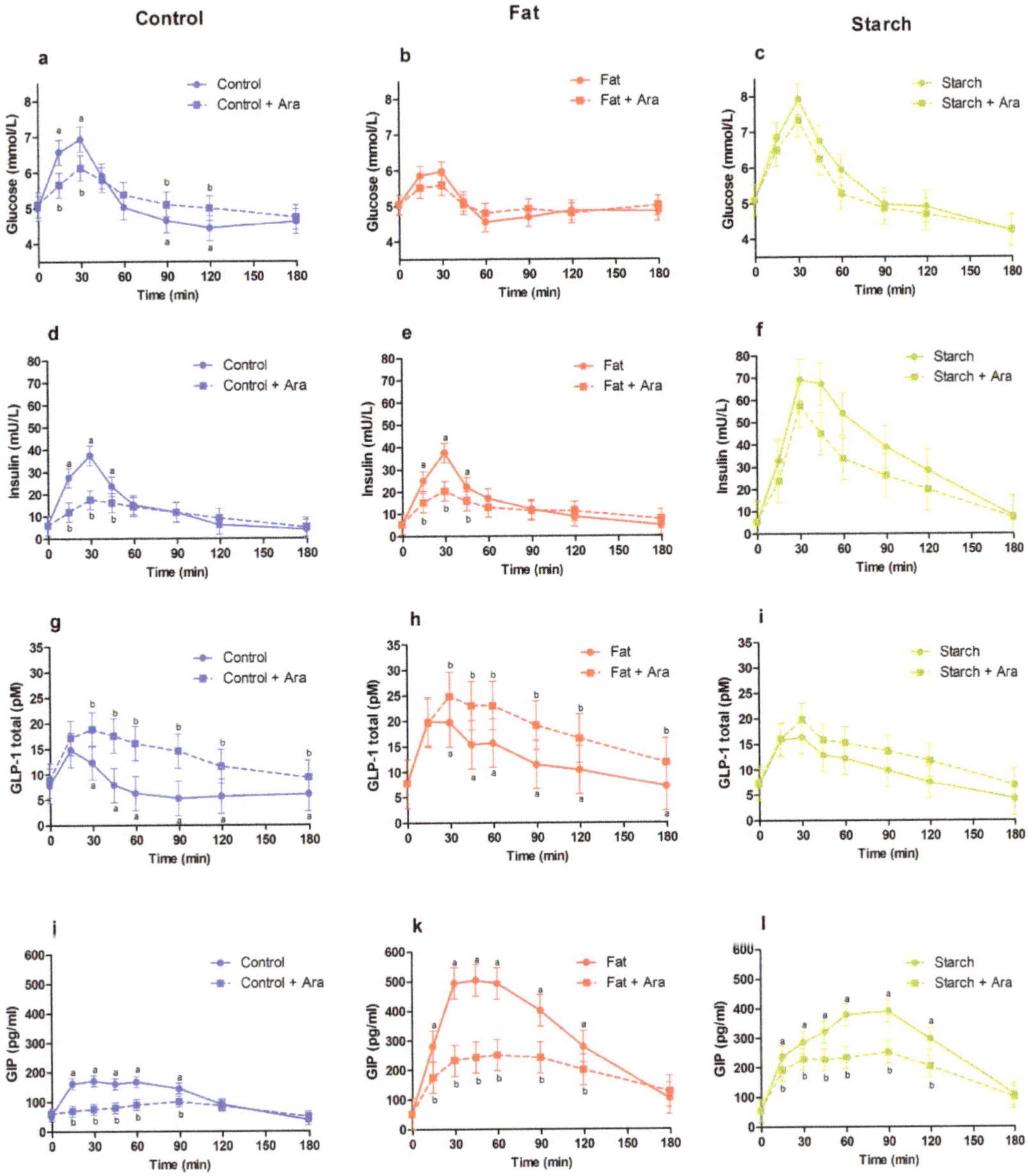


**Figure 1.** Glycemic and hormone concentrations (mean and SEM) before and after the consumption of each of the 6 drinks: control drink = blue, control with added fat = red, control with added starch = green. Dotted lines represent drinks with added arabinose, solid lines represent drinks without arabinose. (**a–c**) Glucose curves, (**d–f**) insulin curves, (**g–i**) GLP-1, (**j–l**) GIP. Different letters represent statistical differences ($p < 0.05$) between treatments at a specific time point (mixed-model ANOVA, post hoc contrasts with Tukey correction).

## 3. Results

### 3.1. Control Drink

#### 3.1.1. Glycaemic and Insulinemic Responses

The plasma glucose and insulin curves rose sharply and declined quickly after consumption of the control drink. The rises and declines in glucose and insulin were less steep when L-arabinose was added to the drink (Figure 1). The glucose peak was, on average, 1.1 mmol/L (15%) lower and the total glucose response (iAUC) was 32% lower when L-arabinose was added (Table 3). Glucose levels were significantly lower at 15 and 30 min, and L-arabinose seemed to prevent a drop in plasma glucose at 90 and 120 min (Figure 1a). Total insulin responses were 38% lower; the peak in insulin was 21 mU/L (52%) lower and occurred, on average, 12 min later (Table 3). A significantly lower insulin concentration was observed at 15, 30, and 45 min (Figure 1d).

#### 3.1.2. GLP-1 and GIP Responses

The control drink showed a sharp increase in plasma GLP-1, which declined quickly. This rise in GLP-1 was prolonged and the decline was delayed when L-arabinose was added to the drink (Figure 1g). The total GLP-1 response was 300% higher, the peak was 16 min later, and 6 pM (40%) higher (Table 3).

The plasma GIP curve rose immediately and declined after 90 min for the control drink (Figure 1j). When L-arabinose was added to the drink, the GIP response did not change from the baseline until 90 min (time effect = 0.99). From 15 to 90 min, the GIP response was significantly lower with L-arabinose compared to the control drink. L-arabinose addition reduced the peak ($p < 0.001$) for GIP with 88 pg/mL (43%) and prolonged the time to peak by 31 min ($p < 0.001$) (Table 3).

### 3.2. Fat Drink

#### 3.2.1. Glycaemic and Insulinemic Responses

Compared to the control drink, the fat drink showed a less sharp increase in plasma glucose, but the decline in glucose was almost as quick as the control drink (Figure 1b). Adding L-arabinose to the fat drink reduced the peak in plasma glucose by 0.5 mmol/L (8%). No effects of L-arabinose on the total response or the time of the peak were observed (Table 3).

The fat drink showed a similar insulin response compared to the control drink. Moreover, adding L-arabinose to the fat drinks resulted in a reduced total insulin response of 26%, and a 20 mU/L (45%) lower peak (Table 3). Moreover, insulin levels were significantly lower at 15, 30, and 45 min after consumption when L-arabinose was added (Figure 1e).

#### 3.2.2. GLP-1 and GIP Responses

The fat drink had a higher GLP-1 response compared to the control drink, which then decreased more slowly. Adding L-arabinose to the fat drink resulted in an increased overall GLP-1 response of 83% (Table 3). GLP-1 levels were higher at all time points from 30 to 120 min (Figure 1h).

The fat drink showed a higher GIP response than the control drink (Figure 1). Adding L-arabinose to the fat drink resulted in a 65% reduced total response, an on-average 18 min later peak, and an 88 pg/mL (48%) lower peak (Table 3). GIP levels were lower at all time points from 15 to 120 min (Figure 1k).

### 3.3. Starch Drink

#### 3.3.1. Glycemic and Insulinemic Responses

Overall, the starch drink had a more rapid and greater increase in glucose and insulin responses compared to the control and the fat drinks (Figure 1c,f). There were no statistical differences at individual time points in glucose or insulin levels between the starch drinks with or without L-arabinose. However, the glucose peak was lower (0.6 mmol/L; 7%) and the total glucose response was 32% lower. Moreover, the insulin peak was 25 mU/L (29%) lower and the overall insulin response was 33% lower (Table 3).

**Table 3.** Postprandial responses ($iAUC_{0-180min}$), peak ($C_{max}$) and $T_{peak}$ (time to peak) of glucose, insulin, GLP-1 and GIP after a single dose of the sucrose solution without or with L-arabinose and without (control) or with fat or starch.

| | | Without L-Arabinose | 5 g L-Arabinose | *p*-Value [2] |
|---|---|---|---|---|
| **Glucose** | | | | |
| Peak ($C_{max}$) (mmol/L) | Control | 7.4 [7.0; 7.7] [1] | 6.3 [5.9; 6.6] | <0.001 |
| | Fat | 6.4 [6.1; 6.6] | 5.9 [5.6; 6.1] | <0.01 |
| | Starch | 8.2 [7.7; 8.6] | 7.6 [7.1; 8.0] | <0.05 |
| Time to peak ($T_{peak}$) (min) | Control | 27 [22; 32] | 33 [28; 38] | NS |
| | Fat | 28 [17; 39] | 32 [21; 43] | NS |
| | Starch | 34 [27; 41] | 29 [22; 36] | NS |
| Response (AUC) [3] (mmol/Lxmin) | Control | 91.7 [61.2; 122.3] | 61.9 [31.4; 92.5] | <0.01 |
| | Fat | 46.8 [29.3; 64.3] | 41.4 [24.0; 58.9] | NS |
| | Starch | 151.1 [104.8; 197.3] | 103.5 [57.2; 149.7] | <0.05 |
| **Insulin** | | | | |
| Peak ($C_{max}$) (mU/L) | Control | 41.0 [33.4; 48.6] | 19.7 [12.1; 27.3] | <0.001 |
| | Fat | 44.0 [35.6; 52.4] | 24.1 [15.7; 32.4] | <0.001 |
| | Starch | 85.1 [69.1; 101.1] | 60.6 [44.6; 76.6] | <0.01 |
| Time to peak ($T_{peak}$) (min) | Control | 32 [23; 40] | 44 [36; 53] | <0.01 |
| | Fat | 31 [20; 41] | 38 [27; 48] | NS |
| | Starch | 50 [39; 60] | 42 [32; 52] | NS |
| Response (AUC) (mU/Lxmin) | Control | 1538 [1124; 1952] | 952 [538; 1366] | <0.001 |
| | Fat | 1654 [1321; 1987} | 1228 [896; 1561] | <0.01 |
| | Starch | 5501 [4358; 6644] | 3662 [2519; 4805] | <0.01 |
| **GLP-1** | | | | |
| Peak ($C_{max}$) (pM) | Control | 15.4 [10.4; 20.5] | 21.7 [16.7; 26.6] | <0.001 |
| | Fat | 22.3 [15.4; 29.2] | 27.5 [20.6; 34.4] | NS |
| | Starch | 19.0 [14.1; 23.8] | 22.1 [17.3; 26.9] | NS |
| Time to peak ($T_{peak}$) (min) | Control | 24 [9; 39] | 40 [25; 55] | <0.001 |
| | Fat | 30 [13; 47] | 52 [35; 69] | 0.05 |
| | Starch | 38 [22; 54] | 47 [31; 63] | NS |
| Response (AUC) (pM × min) | Control | 268 [45; 491] | 803 [581; 1026] | <0.001 |
| | Fat | 987 [394; 1581] | 1805 [1212; 2399] | <0.01 |
| | Starch | 627 [418; 836] | 962 [752; 1171] | <0.001 |
| **GIP** | | | | |
| Peak ($C_{max}$) (pg/mL) | Control | 204 [179; 230] | 116 [90; 142] | <0.001 |
| | Fat | 578 [507; 649] | 278 [207; 348] | <0.001 |
| | Starch | 421 [370; 472] | 282 [231; 333] | <0.001 |
| Time to peak ($T_{peak}$) (min) | Control | 42 [27; 57] | 73 [58; 88] | <0.001 |
| | Fat | 45 [33; 57] | 63 [51; 76] | <0.05 |
| | Starch | 83 [71; 96] | 77 [65; 89] | NS |
| Response (AUC) (pg/mL × min) | Control | 6375 [3985; 8765] | 5161 [2771; 7551] | NS |
| | Fat | 35,750 [28,255; 43,244] | 23,121 [15,627; 30,616] | <0.01 |
| | Starch | 28,383 [22,081; 34,684] | 21,755 [15,453; 28,057] | NS |

[1] Least square means (LSMEANS) with 95% confidence intervals between brackets. [2] *p*-value mixed model ANOVA contrast between without and with arabinose addition. *p*-values were categorised into <0.001, <0.01, <0.05, and NS (non-significant). [3] AUC = Area Under the Curve.

3.3.2. GLP-1 and GIP Responses

The starch drink had a similar GLP-1 response compared to the control drink, declining more gradually (Figure 1i). L-arabinose addition to the starch drink led to a significant 53% increased GLP-1 response ($p < 0.001$) (Table 3).

Starch addition to the control drinks induced a higher and slower increase in GIP concentration compared to the control drink (Figure 1l). This was also visible in the 139 pg/mL (33%) lower peak ($p < 0.001$) (Table 3). L-arabinose addition to the starch drink reduced GIP immediately and significantly from 15 to 120 min postprandial.

### 3.4. *Subjective Appetite Parameters, Thirst and Comfort*

All subjective appetite parameters changed over time (all time effects $p < 0.001$) (see Supplemental Data S2), and no treatment×time effects were observed (all $p > 0.05$). No effects on thirst and comfort feelings were observed (all $p > 0.05$).

The subjects consumed on average 1019 ± 421 kcal (4.3 ± 1.8 MJ) from the lunch that was offered *ad libitum*. This intake was similar for all three test drinks and no effect of L-arabinose addition was observed (all $p$-values > 0.05) (see Supplemental Data S3).

### 3.5. *Gastro-Intestinal Comfort*

None of the gastrointestinal comfort markers—these being bloating, regurgitation, nausea, and flatulence—changed during the test morning after the consumption of the test products, and there were no differences between the test products whether L-arabinose was added or not (all $p$-values > 0.05) (see Supplemental Data S4).

## 4. Discussion

In the current study, the modifying effects of fat and starch on the efficacy of L-arabinose for lowering glycemic and insulinemic responses were assessed. It was confirmed that adding L-arabinose to sucrose-containing drinks lowers the glycemic and insulinemic responses in healthy subjects, by 15% and 52%, respectively. As hypothesised, adding fat or starch modified and reduced this effect. However, despite the effects being reduced, the glycemic peaks were still significantly lower than without the addition of L-arabinose. In the fat drink, L-arabinose reduced the glycemic peak by 8% and the insulin peak by 45%, while in the starch drink, the glucose peak was reduced by 7% and the insulin peak by 29%. Moreover, it was observed that, for all three model drinks, the addition of L-arabinose affected the post-prandial response of other physiological markers associated with glucose homeostasis, i.e., GLP-1 responses were increased and GIP responses were reduced. These findings suggest that L-arabinose affects postprandial glucose homeostasis in response to sucrose consumption, also when added to foods containing fat or starch.

To our knowledge, this is the first study that has assessed the effects of isolated food components on the functionality of L-arabinose in glucose homeostasis regulation. Although L-arabinose has been investigated often in simple single-nutrient (sucrose) liquid foods [10–12], only a few studies have investigated the effects of L-arabinose in more complex solid meals. For example, Halschou-Jensen investigated the effects of the addition of 5%, 10%, and 20% L-arabinose to sucrose- and/or starch-rich breakfast meals [15]. These meals were buns and muffins served with strawberry jam, buns with cheese, or liquified buns eaten as porridge. In this study, no effects were observed on glucose, insulin, or c-peptide concentrations. Our lab recently performed two studies on the addition of L-arabinose to ground cereals [14] and the replacement of sucrose with L-arabinose in muffins [13]. In both these studies, no effects were found on glucose peaks, but trends towards lower insulinemic responses were found (e.g., 10 mU/L lower peak after 8% L-arabinose replacement cereal [14]). However, these responses were not as clear and consistent as the effects in the simple liquid foods (fruit-flavoured drinks and water with added sucrose). Moreover, it was not clear which factor hindered the functionality of the L-arabinose, as the foods contained different nutrients; the texture, fat content, or starch content could have modified the functionality. Therefore, model drinks were used for the current study, carefully developed and only manipulated to be rich in one single nutrient, that being fat or starch. These model foods allowed for a better disentanglement of these effects.

Adding fat to the model drink resulted in a lower glucose peak of 1 mmol/L compared to the control drink with only sucrose. Yet, the addition of L-arabinose showed an additional effect on the plasma glucose peak—that is, 0.5 mmol/L (8%)—and on the insulin peak—that is, 20 mU/L (45%). It is known that the addition of fat lowers the gastric emptying rate and may therefore slow down the absorption of glucose [16,17]. These results suggest that even if L-arabinose and sucrose enter the duodenum at a slower pace, L-arabinose still binds to the sucrase enzyme and inhibits the absorption of glucose and fructose. The increase in GLP-1 and decrease in insulin response suggests that the GLP-1-producing L-cells in the proximal intestine detect the sucrose molecules despite L-arabinose being present.

Next to the modifying effect of fat, the modifying effect of starch was assessed. On top of the 50 g of sucrose which was present in the control drink, the starch drink contained an additional 50 g of available carbohydrates. Due to the higher level of available carbohydrates, it was hypothesised that the effect of L-arabinose could be overshadowed by the additional glucose from the starch. Indeed, the addition of starch led to a 0.8 mmol/L higher peak in glucose and also a 44 mU/L higher insulin peak after the consumption of the starch-rich drink compared to the control drink. However, when L-arabinose was added to the starch drink, there was still a 7% decrease in the glucose peak and a 29% lower insulin peak. So, L-arabinose was functional despite the additional glucose coming from starch in this model drink.

It has been speculated previously that L-arabinose not only inhibits sucrase but also inhibits maltase [11,15,21,22]. Maltase (α-glucosidase) is needed for the hydrolysis of maltose and is important for starch breakdown in the lumen of the duodenum. It is suggested that the inhibition of maltase by L-arabinose is not as pronounced as that of sucrase, but taking the amount of starch that is consumed every day in our diets, this may be an important additional strategy to lower blood glycemic responses [11,15]. Animal studies have shown inconsistent results in this respect. For example, Jurgonski did find a significant decrease in blood glucose when adding L-arabinose to starch challenges in rats [22], while Preuss did not find such effects [21]. In addition, Halschou-Jensen et al. did not find inhibiting effects when L-arabinose was added to solid mixed meals containing sucrose and starch from wheat flour [15]. From the current data, it is difficult to conclude whether L-arabinose also affected maltase, as it is impossible to disentangle the effects from the glucose coming from sucrose and the glucose coming from the starch present in the model drink. Moreover, for the present study, pregelatinised starch was chosen, as this starch is water-soluble at room temperature, which made it practically suitable for a model food and resembles the starch typically present in common foods such as bread and pasta. However, it cannot be ruled out that other factors such as physicochemical properties or the chain length of the starch molecules may affect the functionality of L-arabinose in starch digestion.

In order to better understand the underlying physiological response, not only postprandial glycemic and insulinemic responses were measured, but also GLP-1 and GIP responses. These are incretin hormones that are rapidly released after food consumption, and both stimulate insulin secretion and thus decrease glucose fluctuations after food intake [23]. GLP-1 responses were higher after all three drinks with added L-arabinose, ranging from 53% to 300%, while GIP responses were decreased by 33% to 48% and also showed later peaks. These higher GLP-1 responses and lower GIP responses are in line with earlier results [11,14]. The increase and prolonged rise in GLP-1 is presumably the result of higher and longer stimulation in the first part of the gut, while the lower GIP responses may be the result of a lower stimulation of k-cells in the duodenum and proximal jejunum. This is in line with the proposed mechanism of L-arabinose—that is, inhibiting the hydrolysis of sucrose in fructose and glucose, and thus the absorption of nutrients.

As for the simple sucrose-containing foods, the findings of earlier studies were reproduced [13,14]. Compared to previous studies, the current study used relatively high doses of L-arabinose: 5 g of L-arabinose per 50 g of sugar (10%). A sugar concentration of 10 g of sucrose/100 mL was chosen as this is equal to the amount of sucrose in sugar-sweetened drinks in the Netherlands [20]. A larger effect of L-arabinose may be accomplished by replacing sucrose with L-arabinose, which was demonstrated in one of the previous studies, where 30% of sucrose was replaced by L-arabinose and the glucose and insulin responses were even reduced by 22% and 67% [14]. These results show that L-arabinose can be used as a functional ingredient in sugar-rich drinks to prevent large fluctuations in glucose after the consumption of sugar-sweetened beverages. This may be especially relevant for (pre-)diabetic patients that are more prone to fluctuations in glucose.

None of the drinks showed consistent effects on subjective appetite responses or subsequent food intake, despite large effects on insulin and GLP-1. These findings are in line with other studies that did not see any effects on appetite or energy intake by L-arabinose [11,14,24].

The high doses of L-arabinose did not lead to any side effects or gastro-intestinal complaints, which is consistent with other studies [10,11,13,14]. If the hydrolysis of sucrose is inhibited, sucrose can flow from the small intestine into the colon, and be metabolised by microbiota, resulting in gastrointestinal symptoms such as flatulence and diarrhoea. Medicines such as acarbose, voglibose, and miglitol also inhibit digestive enzymes. These medicines are known for these side effects, due to the dumping of the undigested sugar fractions into the colon [25,26]. The current study shows that up to 1 g of L-arabinose per 10 g of available carbohydrates is a well-tolerated dose in liquids with a fast gastro-intestinal passage.

It is not clear to what extent L-arabinose is metabolised by the body. One of the previous studies has shown that L-arabinose is to some extent absorbed in the intestine, as up to 15% of supplemented L-arabinose was recovered in urine collected 24 h after consumption [14]. Moreover, it is likely that L-arabinose is also fermented by microbiota in the gut. This evidence comes from animal studies in pigs and rats [27,28], and, recently, in vitro and functional genomic studies showed that arabinose-containing fibres can be utilised by members of the Bacteroidetes phylum, which are abundantly present in the human colon [29,30]. However, more fundamental research is needed to further unravel the digestive and metabolic route of L-arabinose in the human body.

The advantage of adding L-arabinose to a sucrose-rich product is that the sweet taste is maintained and that other sensory properties, such as texture and mouthfeel, are not affected. The participants of the study were not able to tell the difference between the versions of the model drinks with and without L-arabinose. This study shows that even in model drinks rich in fat and starch, adding L-arabinose can affect glycemic control. The model drinks were manipulated beyond the nutrient content of existing foods, which makes them valid model drinks. The model drink rich in fat contained 3.8% *w*/*w* fat, which is a bit higher than full-fat milk in the Netherlands with 3.5% *w*/*w* fat [20]. The starch model drink contained 8.3% *w*/*w* pregelatinised starch and a total of 16.7% *w*/*w* available carbohydrates, which is twice as high as commercially available vanilla custard, which contains 4.6% *w*/*w* starch and 10.0% *w*/*w* total available carbohydrates [20]. Hence, this study illustrates the opportunities and bandwidth of foods to which L-arabinose can be added. It needs to be further investigated what the other food matrix related factors are that affect the functionality of L-arabinose.

Lastly, as the study population consisted of healthy, lean, young individuals, their glucose levels were relatively well-controlled. The range of foods and their properties, such as nutrient content and matrix, that may be supplemented with L-arabinose need to be investigated further in populations with impaired glucose tolerance and/or diabetes. This study illustrates how novel functional ingredients can be derived from plant food processing side-streams.

## 5. Conclusions

In conclusion, this study shows that L-arabinose addition to sugar-sweetened drinks, even with other nutrients present, such as fat or starch, flattens the blood glucose, insulin, and GIP response curves and increases GLP-1 concentrations in circulation in healthy subjects. L-arabinose addition to sugar-sweetened drinks may be beneficial for people that want to dampen their glycemic and insulinemic responses, such as people with (or at risk of) type 2 diabetes. Moreover, their glucose tolerance may be improved by reducing GIP secretion and increasing GLP-1 secretion. The advantage of adding L-arabinose to a sucrose-rich product is that the sweet taste is maintained. Therefore, L-arabinose addition to real sugar-sweetened foods with some starch or fat seems to be a promising strategy to affect acute postprandial glycemic responses. This may be a strategy to target consumers that are keen on drinks or foods that are high in sucrose content and do not want to use other (artificial) sweeteners. Further research should further focus on applications in solid foods and patient populations that would benefit from better glucose control, such as (pre)diabetics.

**Supplementary Materials:** The following are available online at: https://www.mdpi.com/article/10.3390/foods11020157/s1, Data S1: Results of in vitro experimentation of the preparation protocol of the drinks, Data S2. Results of appetite ratings, Data S3. Results of food intake, and Data S4. Results of gastro-intestinal comfort.

**Author Contributions:** Conceptualization, K.P. and M.M.; methodology, K.P. and M.M.; formal analysis, K.P.; investigation, M.-L.P.; writing—original draft preparation, K.P. and M.M.; writing—review and editing, K.P., M.-L.P. and M.M.; visualization, M.M.; supervision, M.M.; project administration, K.P. and M.M.; funding acquisition, M.M. All authors have read and agreed to the published version of the manuscript.

**Funding:** This work was supported by the Pulp2Value project from the BioBased Industries Joint Undertaking under the European Union's Horizon 2020 Research and Innovation Programme (under grant agreement No 669105).

**Institutional Review Board Statement:** The study was conducted in accordance with the Declaration of Helsinki, and approved by the Institutional Review Board (or Ethics Committee) of Wageningen University (protocol code NL61428.081.17 on 29 June 2017).

**Informed Consent Statement:** Informed consent was obtained from all subjects involved in the study.

**Data Availability Statement:** The data presented in this study are available on request from the corresponding author. The data are not publicly available due to privacy and ethical reasons.

**Acknowledgments:** The authors are grateful to the participants for their valuable time and effort. The authors are grateful to Royal Cosun for the L-arabinose supplied, and the consortium members of Pulp2Value for their constructive input. Furthermore, the authors would like to thank Henriette Fick-Brinkhof, Els Siebelink, Corine Perenboom, Nhien Ly, and Ineke Klöpping, for their help and advice during the preparations of the study. We would like to thank Anja Janssen, Maurice Strubel, and Jos Sewalt for their assistance during the in vitro digestion assessment of the drinks.

**Conflicts of Interest:** The authors declare no conflict of interest. The funding source had no role in the design of the study; in the collection, analyses, or interpretation of data; in the writing of the manuscript; or in the decision to publish the results.

## References

1. International Diabetes Federation. *Diabetes Atlas*, 9th ed.; International Diabetes Federation: Brussels, Belgium. Available online: https://www.diabetesatlas.org (accessed on 10 October 2021).
2. Wang, Q.; Xia, W.; Zhao, Z.; Zhang, H. Effects comparison between low glycemic index diets and high glycemic index diets on hba1c and fructosamine for patients with diabetes: A systematic review and meta-analysis. *Prim. Care Diabetes* **2015**, *9*, 362–369. [CrossRef] [PubMed]
3. Ojo, O.; Ojo, O.O.; Adebowale, F.; Wang, X.-H. The effect of dietary glycaemic index on glycaemia in patients with type 2 diabetes: A systematic review and meta-analysis of randomized controlled trials. *Nutrients* **2018**, *10*, 373. [CrossRef] [PubMed]

4. Nederlandse Diabetes Federatie. Voedingsrichtlijn Diabetes. 2020. Available online: https://diabetesfederatie.Nl/ndf-toolkit-persoonsgerichte-diabeteszorg/persoonsgerichte-voedingszorg (accessed on 10 October 2021).
5. Dyson, P.A.; Twenefour, D.; Breen, C.; Duncan, A.; Elvin, E.; Goff, L.; Hill, A.; Kalsi, P.; Marsland, N.; McArdle, P.; et al. Diabetes uk evidence-based nutrition guidelines for the prevention and management of diabetes. *Diabet. Med.* **2018**, *35*, 541–547. [CrossRef] [PubMed]
6. Canada Diabetes. Clinical Practice Guidelines. 2018. Available online: http://guidelines.Diabetes.Ca/cpg/chapter11#sec5 (accessed on 10 October 2021).
7. Birch, G.G.; Shamil, S. Structure, sweetness and solution properties of small carbohydrate molecules. *J. Chem. Soc. Faraday Trans. 1 Phys. Chem. Condens. Phases* **1988**, *84*, 2635–2640. [CrossRef]
8. Maki-Arvela, P.; Salmi, T.; Holmbom, B.; Willfor, S.; Murzin, D.Y. Synthesis of sugars by hydrolysis of hemicelluloses—A review. *Chem. Rev.* **2011**, *111*, 5638–5666. [CrossRef]
9. Seri, K.; Sanai, K.; Matsuo, N.; Kawakubo, K.; Xue, C.; Inoue, S. L-arabinose selectively inhibits intestinal sucrase in an uncompetitive manner and suppresses glycemic response after sucrose ingestion in animals. *Metabolism* **1996**, *45*, 1368–1374. [CrossRef]
10. Inoue, S.; Sanai, K.; Seri, K. Effect of l-arabinose on blood glucose level after ingestion of sucrose-containing food in human (translated from Japanese). *Nippon. Eiyo Shokuryo Gakkaishi* **2000**, *53*, 243–247. [CrossRef]
11. Krog-Mikkelsen, I.; Hels, O.; Tetens, I.; Holst, J.J.; Andersen, J.R.; Bukhave, K. The effects of l-arabinose on intestinal sucrase activity: Dose-response studies in vitro and in humans. *Am. J. Clin. Nutr.* **2011**, *94*, 472–478. [CrossRef] [PubMed]
12. Shibanuma, K.; Degawa, Y.; Houda, K. Determination of the transient period of the eis complex and investigation of the suppression of blood glucose levels by l-arabinose in healthy adults. *Eur. J. Nutr.* **2011**, *50*, 447–453. [CrossRef] [PubMed]
13. Pol, K.; Mars, M. L-arabinose and d-xylose: Sweet pentoses that may reduce postprandial glucose and insulin responses. *Food Nutr. Res.* **2021**, *65*. [CrossRef] [PubMed]
14. Pol, K.; de Graaf, K.; Diepeveen-de Bruin, M.; Balvers, M.; Mars, M. The effect of replacing sucrose with l-arabinose in drinks and cereal foods on blood glucose and plasma insulin responses in healthy adults. *J. Funct. Foods* **2020**, *73*, 104114. [CrossRef]
15. Halschou-Jensen, K.; Bach Knudsen, K.E.; Nielsen, S.; Bukhave, K.; Andersen, J.R. A mixed diet supplemented with l-arabinose does not alter glycaemic or insulinaemic responses in healthy human subjects. *Br. J. Nutr.* **2015**, *113*, 82–88. [CrossRef] [PubMed]
16. Cunningham, K.M.; Read, N.W. The effect of incorporating fat into different components of a meal on gastric emptying and postprandial blood glucose and insulin responses. *Br. J. Nutr.* **1989**, *61*, 285–290. [CrossRef]
17. Gentilcore, D.; Chaikomin, R.; Jones, K.L.; Russo, A.; Feinle-Bisset, C.; Wishart, J.M.; Rayner, C.K.; Horowitz, M. Effects of fat on gastric emptying of and the glycemic, insulin, and incretin responses to a carbohydrate meal in type 2 diabetes. *J. Clin. Endocrinol. Metab.* **2006**, *91*, 2062–2067. [CrossRef] [PubMed]
18. Van Strien, T.; Frijters, J.E.; Bergers, G.; Defares, P.B. The dutch eating behavior questionnaire (debq) for assessment of restrained, emotional, and external eating behavior. *Int. J. Eat. Disord.* **1986**, *5*, 295–315. [CrossRef]
19. Blundell, J.; de Graaf, C.; Hulshof, T.; Jebb, S.; Livingstone, B.; Lluch, A.; Mela, D.; Salah, S.; Schuring, E.; van der Knaap, H.; et al. Appetite control: Methodological aspects of the evaluation of foods. *Obes Rev.* **2010**, *11*, 251–270. [CrossRef] [PubMed]
20. Stichting Nederlands Voedingsstoffenbestand. *Dutch Food Composition Data Base "Nevo Tabel: Nederlands Voedingsstoffenbestand"*; Stichting Nederlands Voedingstoffenbestand: Zeist, The Nederlands, 2011.
21. Preuss, H.G.; Echard, B.; Bagchi, D.; Stohs, S. Inhibition by natural dietary substances of gastrointestinal absorption of starch and sucrose in rats and pigs: 1. Acute studies. *Int. J. Med. Sci.* **2007**, *4*, 196–202. [CrossRef] [PubMed]
22. Jurgoński, A.; Krotkiewski, M.; Juśkiewicz, J.; Billing-Marczak, K.; Oulai, P.D.; Zoue, L.T.; Niamke, S.L.; Abiose, S.H.; Abiodun, V.I.; Aboderin, F.I. Suppression of postprandial glycaemia by l-arabinose in rats is more 9 associated with starch than sucrose ingestion-short report. *Pol. J. Food Nutr. Sci.* **2015**, *65*, 57–60. [CrossRef]
23. Eelderink, C.; Noort, M.W.J.; Sozer, N.; Koehorst, M.; Holst, J.J.; Deacon, C.F.; Rehfeld, J.F.; Poutanen, K.; Vonk, R.J.; Oudhuis, L.; et al. Difference in postprandial glp-1 response despite similar glucose kinetics after consumption of wheat breads with different particle size in healthy men. *Eur. J. Nutr.* **2017**, *56*, 1063–1076. [CrossRef] [PubMed]
24. Pol, K.; de Graaf, C.; Meyer, D.; Mars, M. The efficacy of daily snack replacement with oligofructose-enriched granola bars in overweight and obese adults: A 12-week randomised controlled trial. *Br. J. Nutr.* **2018**, *119*, 1076–1086. [CrossRef] [PubMed]
25. Usman, B.; Sharma, N.; Satija, S.; Mehta, M.; Vyas, M.; Khatik, L.G.; Khurana, N.; Hansbro, M.P.; Williams, K.; Dua, K. Recent developments in alpha-glucosidase inhibitors for management of type-2 diabetes: An update. *Curr. Pharm. Des.* **2019**, *25*, 2510–2525. [CrossRef] [PubMed]
26. Chiasson, J.-L.; Josse, R.G.; Gomis, R.; Hanefeld, M.; Karasik, A.; Laakso, M. Acarbose for prevention of type 2 diabetes mellitus: The stop-niddm randomised trial. *Lancet* **2002**, *359*, 2072–2077. [CrossRef]
27. Schutte, J.B.; de Jong, J.; van Weerden, E.J.; Tamminga, S. Nutritional implications of l-arabinose in pigs. *Br. J. Nutr.* **1992**, *68*, 195–207. [CrossRef]
28. Osaki, S.; Kimura, T.; Sugimoto, T.; Hizukuri, S.; Iritani, N. L-arabinose feeding prevents increases due to dietary sucrose in lipogenic enzymes and triacylglycerol levels in rats. *J. Nutr.* **2001**, *131*, 796–799. [CrossRef] [PubMed]

29. Chang, C.; Tesar, C.; Li, X.; Kim, Y.; Rodionov, D.A.; Joachimiak, A. A novel transcriptional regulator of l-arabinose utilization in human gut bacteria. *Nucleic Acids Res.* **2015**, *43*, 10546–10559. [CrossRef] [PubMed]
30. Pereira, G.V.; Abdel-Hamid, A.M.; Dutta, S.; D'Alessandro-Gabazza, C.N.; Wefers, D.; Farris, J.A.; Bajaj, S.; Wawrzak, Z.; Atomi, H.; Mackie, R.I.; et al. Degradation of complex arabinoxylans by human colonic bacteroidetes. *Nat. Commun.* **2021**, *12*, 459. [CrossRef] [PubMed]

*foods*

MDPI

*Article*

# Evaluation of the Structural, Physicochemical and Functional Properties of Dietary Fiber Extracted from Newhall Navel Orange By-Products

**Jiaqi Sang [1,2], Lu Li [1], Jing Wen [1,*], Qingqing Gu [2], Jijun Wu [1], Yuanshan Yu [1], Yujuan Xu [1], Manqin Fu [1] and Xian Lin [1]**

1 Sericulture and Agri-Food Research Institute, Guangdong Academy of Agricultural Sciences/Key Laboratory of Functional Foods, Ministry of Agriculture/Guangdong Key Laboratory of Agricultural Products Processing, Guangzhou 510610, China; sjq05176729@163.com (J.S.); lilu045@163.com (L.L.); wujijun@gdaas.cn (J.W.); yuyuanshan@gdaas.cn (Y.Y.); guoshuxuyujuan@163.com (Y.X.); fumanqin@gdaas.cn (M.F.); linxian@gdaas.cn (X.L.)

2 College of Agronomy, Jiangxi Agricultural University, Nanchang 330045, China; qingqinggu2006@126.com

* Correspondence: jingw988@163.com; Tel.: +86-13694258705

**Citation:** Sang, J.; Li, L.; Wen, J.; Gu, Q.; Wu, J.; Yu, Y.; Xu, Y.; Fu, M.; Lin, X. Evaluation of the Structural, Physicochemical and Functional Properties of Dietary Fiber Extracted from Newhall Navel Orange By-Products. *Foods* **2021**, *10*, 2772. https://doi.org/10.3390/foods10112772

Academic Editors: Federico Casanova and Michela Verni

Received: 10 October 2021
Accepted: 6 November 2021
Published: 11 November 2021



**Abstract:** Ultrasound-assisted enzymatic treatment was used to treat Newhall navel orange peel and residue, and then the structural, physicochemical and functional properties of extracted soluble dietary fibers (SDF) and insoluble dietary fibers (IDF) were investigated. The structural properties were determined using scanning electron microscopy, X-ray diffraction, FT-IR and monosaccharide composition. Among these dietary fibers, residue-SDF showed a more complex structure, while peel-IDF exhibited a looser structure. Four samples showed representative infrared spectral features of polysaccharides, typical cellulose crystalline structure and diverse monosaccharide composition. Furthermore, residue-IDF exhibited higher oil-holding capacity (2.08 g/g), water-holding capacity (13.43 g/g) and nitrite ion adsorption capacity (NIAC) than other three samples, and residue-SDF showed the highest swelling capacity (23.33 mL/g), cation exchange capacity (0.89 mmol/g) and cholesterol adsorption capacity (CAC) among these dietary fibers. In summary, this study suggests that the residue-SDF and residue-IDF could be used as the ideal dietary fibers for application in the functional food industry.

**Keywords:** Newhall navel orange; dietary fiber; structural properties; functional properties

## 1. Introduction

Orange (*Citrus* L.) fruit is one of the most important fruits in the world. Newhall navel is the main variety of navel orange, which is famous for its large seedless fruit, thin skin, bright color and fragrant, sweet taste. Because of its high vitamin content and rich aroma characteristics, it is well-received by consumers. In addition to a small number of oranges for processing, most oranges are used for marketing fresh. Moreover, a large number of by-products are formed during processing and daily consumption [1], such as peel and residue, accounting for 20~35% of the whole fruit mass. The by-products of orange have been mostly utilized as molasses for animal feed, pectin and fuel production [2,3]. Peel and residue are main by-products of orange, which are rich in dietary fiber (DF). In order to better use the by-products of Newhall navel orange, it is necessary to investigate the structural, physicochemical and functional properties of dietary fiber exacted from Newhall navel orange peel and residue.

Nowadays, DF has attracted wide attention from researchers, which due to its functions of improving human nutritional status, regulating bodily functions, removing heavy metal ions and inhibiting antioxidant activity [4]. Total dietary fiber (TDF) is usually divided into insoluble (IDF) and soluble (SDF) dietary fiber based on their solubility. Meanwhile, IDF and SDF have different physiological functions. IDF, which accounts for 2/3 of natural fiber, has the effect of accelerating fecal excretion and high viscosity. SDF can be used as a therapeutic agent to reduce blood fat and blood sugar levels [5]. A previous study showed that the water-soluble fraction contained 30–50% of the total dietary fiber, which could not only balance dietary fiber intake but also replenish the essential nutrient element of the human body [6]. Compared with cereals, the DF of orange peel contains higher SDF, which has attracted the attention of researchers. Furthermore, the insoluble fibers of Liucheng orange peel had water-holding capacities and oil-holding capacities [2]. Moreover, the water swelling capacity of citrus fiber treated with chemo-mechanical was significantly higher than that of commercial fiber [7], and citrus fiber could also improve the water-binding ability of low-fat frankfurters [8]. Although the effects on the dietary fiber of [9] citrus have been widely studied, the dietary fiber and its functional characteristics of Newhall orange by-products have few reports.

At present, the methods of extracting DF can be divided into enzymatic method, chemical method, ultrasound, wet milling, microbiological methods and microwave treatment [9,10]. The enzymatic method uses two or more enzymes to destroy the cell wall, thereby removing impurities, such as proteins, starches and soluble sugars [9,11], which has the advantage of security, high efficiency and low consumption of energy. Ultrasonic treatment can result in plant cell wall cavitation in a liquid medium, which can increase the yield or rate of extraction as well as reduce extraction time [12]. The chemical method can destroy the cellulose and hemicellulose in the cell wall, thereby altering the morphology of the cell wall [13]. In previous studies, the combination of two different extraction methods is better than the single extraction method. Ultrasound-assisted enzymatic hydrolysis has the advantages of mild extraction conditions, low investment cost and low energy requirements [14]. Yang found the pectin yield attained with the combined enzymatic and ultrasonic extraction was higher than enzymatic extraction, ultrasonic extraction and acidic extraction [15]. Moczkowska found enzymatic-ultrasonic had the highest SDF content and apparent viscosity of flaxseed [16]. However, the preparation of DF extracted from Newhall navel orange by-products by ultrasonic-assisted enzymatic hydrolysis has rarely been reported. Thus, the purpose of this study was to evaluate the effects of an ultrasound-assisted complex enzyme method on the structural, physicochemical and functional properties of DFs extracted from Newhall navel orange peel and residue, which could provide insight into studies on Newhall orange DF and offer a theoretical basis for comprehensive utilization.

## 2. Materials and Methods

### 2.1. Materials and Chemicals

Fresh oranges (*Citrus sinensis* (L.) *Osbeck*) were harvested and processed in Yichang city of Hubei province of China in July 2020. Glycosylase (100 KU/g), neutral protease (100 KU/g) and α-amylase (4000 U/g, Termamyl) were provided by Shanghai Yuanye biotechnology Co., Ltd. (Shanghai, China). Sodium hydroxide (NaOH), disodium hydrogen phosphate dodecahydrate ($Na_2HPO_4 \cdot 12H_2O$) and methanol were HPLC grade. Other chemicals utilized were of analytical grade.

### 2.2. Preparation of Citrus Dietary Fiber

#### 2.2.1. Preliminary Treatment of Orange

The orange fruits were divided into the peel and the residue after extraction of juice. Then, the peel and residue were dried in an air-oven at 50 °C for 24 h. Finally, the dried samples were ground and sieved (80 μm) to obtain peel and residue powders.

#### 2.2.2. Extraction of IDFs and SDFs

The extraction method of IDF and SDF was according to a method with slight modification [17]. An amount of 10 g of peel and residue powders was suspended in water with a material-liquid ratio of 1:20 (*w*/*v*), and their pH was adjusted to 7.0; then, the mixture was treated with ultrasound for 30 min, 40 °C, 400 W. Hence, 1.5% mixed enzyme (α-amylase-glycosylase enzyme 1:1, m/m) was put into the above mixture, followed by 1 h incubation on the 70 °C water bath. The resulting slurries were heated at 100 °C for 10 min to terminate the enzymatic reaction and allowed to cool to room temperature. After that, the suspension was mixed with protease (500 μL) in a 40 °C water bath for 1 h. The resulting slurries were heated at 100 °C for 5 min to terminate the enzymatic reaction, allowed to cool to room temperature and centrifuged at 5000 rpm for 10 min. The resulting precipitate was freeze-dried to obtain the P-IDF and R-IDF. Meanwhile, while the supernatants were collected and added to 95% ethanol (four-fold volumes) for 24 h to collect residues. The residues were freeze-dried to obtain the P-SDF and R-SDF.

### 2.3. Molecular Weight of IDFs and SDFs

According to a previous study [18], the molecular weight of DF was determined by high performance liquid chromatography equipped with a refractive index detector (RI-10A; Shimadzu Corp., Tokyo, Japan) and a BRT105-104-102 column (8 mm × 300 mm). The mobile phase was 0.05 M NaCl solution and was eluted at 0.6 mL/min for 60 min. The column temperature was 40 °C, and the injection volume was 20 μL. The SDFs and IDFs sample was formulated into aqueous solution (5 mg/mL) and filtered through a 0.22 μm filter. A standard dextran curve was prepared, and the molecular weight of SDFs and IDFs was calculated.

### 2.4. Monosaccharide Compositions of IDFs and SDFs

The monosaccharide composition of the samples was detected by HPLC [5]. Each 0.01 g IDF or SDF sample was added into 2 mL 2 mol/L trifluoroacetic acid, hydrolyzed at 100 °C for 8 h, dried trifluoroacetic acid, washed with 1 mL methanol and then added to 1 mL distilled water to dissolve. Then, the hydrolytes of SDF and IDF were derivatized with 0.5 mol/L PMP-methanol solution and 0.3 mol/L NaOH for 1 h at 70 °C. After cooling at ambient temperature, the reacted product was neutralized with 300 μL 0.3 mol/L HCl and 1 mL chloroform. Then, it underwent 10 min of centrifugation at 4800 rpm. Once the supernatant was absorbed, then 1 mL chloroform was added, and the extraction was repeated three times, with the last supernatant filtered through a 0.22 μm membrane. The injection volume was 20 μL. Acetonitrile and 0.1 mol/L phosphate buffer (pH = 6.7) were used as mobile phases A and B at a ratio of 18:82. The flow rate was 1 mL/min.

### 2.5. Scanning Electron Microscopy (SEM)

The micrograph of IDFs and SDFs were observed by scanning electron microscopy SEM (EVO 18, ZEISS, Karlsruhe, Germany). The IDFs and SDFs were placed on a sample holder with double-sided scotch tape and sputter with gold. Subsequently, each sample was transferred to the scanning electron microscope with 800× and 2000× magnification and an acceleration voltage of 15.0 kV.

### 2.6. Fourier Transfer-Infrared Spectrometry (FT-IR)

The FT-IR spectrum of samples was performed in a total reflection Fourier Transform Infrared (ATR-FTIR) instrument (VERTEX 33, Bruker Co. Ltd., Karlsruhe, Germany). The dry powder sample was mixed with potassium bromide (KBr) powder (1:100, $v/v$) and pressed into particles for spectrometric measurement. The spectra were measured in the frequency range of 4500~500 $cm^{-1}$ with a resolution of 4 $cm^{-1}$.

### 2.7. X-ray Diffraction (XRD)

The crystalline structure of IDFs and SDFs was analyzed by X-ray polycrystalline (D8 Advance, Bruker Corp., Karlsruhe, Germany) diffractometer at the working voltage of 40 kV and an incident current of 40 mA. The scanning region of the diffraction Angle (2θ) was 3–55°, and the scanning velocity was 2°/min [19]. The relative degree of crystalline

(%) was calculated using MDI Jade 6.5 software (Materials Data, Inc., Livermore, CA, USA) using the following equation:

$$DC(\%) = A_c \times 100/(A_c + A_a)$$

where DC is the degree of crystallinity, and $A_c$ and $A_a$ represent the crystalline and the amorphous area on the X-ray diffractogram, respectively.

### 2.8. Physicochemical and Functional Properties of Orange DF

#### 2.8.1. Water-Holding Capacity (WHC)

Water-holding capacity (WHC) was conducted according to the method described in [19]. First, we added 0.5 g of IDF or SDF sample to 25 mL of distilled water at room temperature (25 °C) for 1 h. After centrifugation at 4800 rpm for 10 min, the residues were extracted weighed. Finally, WHC was determined by the following equation:

$$WHC(g/g) = \frac{W_1 - W_2}{W_2}$$

where $W_1$ and $W_2$ are the weights of wet and dry sample, respectively.

#### 2.8.2. Oil-Holding Capacity (OHC)

The OHC was determined using a method modified from Liu et al. [20], adding 1.0 g of IDF or SDF sample to 25 mL of soybean oil for 2 h. After centrifugation at 4800 rpm for 10 min, the precipitate was extracted and then weighed. The OHC was determined according to the following equation.

$$OHC(g/g) = \frac{W_1 - W_2}{W_2}$$

where $W_1$ and $W_2$ are the weights of wet and dry sample, respectively.

#### 2.8.3. The Swelling Capacity (SC)

The SC was determined using a method modified from Zhang et al. [21]. An accurately weighed dry sample of 0.3 g was placed in a test tube, the original volume was recorded, and 5 mL of water was added. It was hydrated for 24 h at 25 °C, and the final volume of the swollen fiber was recorded. The SC was calculated using the following equation:

$$SC(mL/g) = \frac{V_1 - V_2}{M}$$

where $V_2$ is the volume of the dried sample, $V_1$ is the volume of the hydrated sample and M is the weight of the sample.

#### 2.8.4. Cation Exchange Capacity (CEC)

The CEC was determined using a method modified from He et al. [22] with slight modifications. It began with 0.50 g of the sample and 15 mL of 0.1 mol/L HCl, stirred thoroughly and standing at room temperature for 24 h, then filtered with filter paper and repeatedly washed with distilled water to move acid. The residue was added to a conical flask with 100 mL of 15% (*w*/*v*) NaCl with constant stirring. Distilled water was used as a blank sample, titrated with a 0.01 mol/L NaOH solution, and the volume of consumed NaOH solution was recorded. The cation exchange capacity was calculated as follows:

$$CEC(mmol/g) = \frac{(V_1 - V_0) \times 0.1}{m}$$

where $V_1$ is the titrated volume of NaOH of the sample, $V_0$ is the titrated volume of NaOH of the blank, 0.1 is the concentration of NaOH, and m is the dry weight of the sample.

2.8.5. Cholesterol Adsorption Capacity (CAC)

CAC was determined using a method modified from Zhang et al. [23]. The yolk was diluted with 9 volumes of distilled water. A total of 1.0 g of the IDF or SDF sample was mixed with 50 mL diluted yolk. The mixture was adjusted to pH 7.0 or 2.0 and incubated in a shaking water bath at 37 °C for 10 min, and the 0.02 mL supernatant was absorbed. The color was developed by adding 0.1 mL of glacial acetic acid reagent and 2 mL of $H_2SO_4$. The absorbance of the sample reagent was measured at 550 nm. According to the standard curve, the corresponding cholesterol content and cholesterol adsorption were determined.

$$\mathrm{CAC(mg/g)} = \frac{\mathrm{M_1 - M_2}}{\mathrm{W}}$$

where $M_1$—cholesterol content before adsorption; $M_2$—cholesterol content after adsorption; W—IDF and SDF quality.

2.8.6. Nitrite Ion Absorption Capacity (NIAC)

NIAC was determined using a method modified from Zhu et al. [5] with slight modification. A dried sample (1.0 g) was mixed with 100 mL 250 μmol/L $NaNO_2$ solution in a conical flask. After centrifugation at 4800 rpm for 5 min, 5 mL of the supernatant was transferred to a 25 mL glass tube and 2 mL of p-aminobenzenesulfonic acid (4 μg/mL) and 1 mL of naphthalenediamine hydrochloride (2 μg/mL) were added to the mixture. The concentration of $NaNO_2$ was measured with a UV-1800 spectrophotometer (Shimadzu Corp., Tokyo, Japan) at 538 nm and quantified according to the standard curve.

$$\mathrm{NIAC(\mu g/g)} = \frac{\mathrm{M_1 - M_2}}{\mathrm{W}}$$

where $M_1$—nitrite content before adsorption; $M_2$—nitrite content after adsorption; W—IDF and SDF quality.

### *2.9. Statistical Analysis*

All analyses were conducted in triplicate, and the results are expressed as mean standard deviation. One-way analysis of variance (ANOVA) was performed using SPSS 24.0 statistical software (SPSS Inc., Chicago, IL, USA).

## 3. Results

### *3.1. The Extraction Yield of SDFs and IDFs*

The SDF and IDF of Newhall navel orange peel were 9.29% and 70.91%, respectively. Compared with SDF and IDF of Newhall navel orange peel, the Newhall navel orange residue had the higher SDF content (15.05%) and IDF content (79.10%), which may be attributed to the disruption of the glycosidic linkages in dietary fiber under ultrasound, thus causing in orange peel a loss of SDF, IDF, hemicellulose and cellulose [24].

### *3.2. Molecular Weight*

The HPGPC results showed that there were many small peaks in R-SDF with the main peak having 49.79% peak area, while there were only two small peaks and one main peak of P-SDF with 89.47% peak area. Moreover, there were many small peaks with one main peak in R-IDF with the main peak having 81.87% peak area, while there were only three small peaks of P-SDF with one main peak of 61.66% peak area. According to the retention time and the dextran standard carve, the Mw of four samples was R-SDF ($1.07 \times 10^6$ Da), R-IDF ($1.57 \times 10^6$ Da), P-SDF ($0.57 \times 10^6$ Da) and P-IDF ($0.56 \times 10^6$ Da), respectively. It could be found that DF of residue possessed higher molecular weight than DF of peel, which indicated that the molecular weight of DF in orange peel was significantly affected by ultrasound-assisted enzymatic method. It might be explained by the fact that ultrasound and enzymatic hydrolysis could destroy the skeleton of DF of peel and convert large molecules into small molecules [18].

### 3.3. The Monosaccharide Composition of SDFs and IDFs

In order to facilitate better exploration of the properties of dietary fiber, the monosaccharide components of four samples were analyzed. Table 1 showed the monosaccharide composition of different DF samples. All the DF samples contained seven monosaccharides, including mannose, rhamnose, galacturonic acid, glucose, xylose, galactose and arabinose. Glucose was the major monosaccharide, followed by galactose and arabinose. This result indicated that the DFs were mainly composed of glucose in cellulose, xylose and arabinose in hemicellulose. Glucose is mainly derived from starch and cellulose, and the glucose contents of P-SDF and R-SDF were relatively high, which might be because ultrasound promoted the hydrolysis of cellulose in the cell wall [25]. Moreover, galacturonic acid and rhamnose are constituents of pectin [25], and galacturonic acid contents of R-SDF and P-SDF were at high levels along with the relatively high content of rhamnose in P-IDF and P-SDF. Therefore, our analysis shows that pectin is the main component of SDF extracted from orange peel. These were the relatively high contents of arabinose and galactose in R-IDF, which indicated that most of the cellulose has been degraded, and part of the hemicellulose has been released. The high content of xylose and galactose in P-IDF indicated that these sugars were important components of orange cell wall. The contents of mannose in these four samples were very low, but mannose can be used to synthesize glycoprotein, which can participate in the immune regulation of the body and has a good effect on the health of the body.

**Table 1.** The monosaccharide composition of R-SDF, P-SDF, R-IDF and P-IDF.

| Sample | Mannose (mg/g) | Rhamnose (mg/g) | Galacturonic Acid (mg/g) | Glucose (mg/g) | Galactose (mg/g) | Xylose (mg/g) | Arabinose (mg/g) |
|---|---|---|---|---|---|---|---|
| R-SDF | $0.44 \pm 0.12$ [a] | $0.58 \pm 0.28$ [d] | $1.41 \pm 0.11$ [b] | $20.51 \pm 0.16$ [a] | $3.72 \pm 0.14$ [b] | $0.16 \pm 0.20$ [d] | $3.18 \pm 0.27$ [d] |
| P-SDF | $0.75 \pm 0.23$ [b] | $1.71 \pm 0.1$ [b] | $1.57 \pm 0.3$ [a] | $20.16 \pm 0.22$ [b] | $1.79 \pm 0.07$ [d] | $0.26 \pm 0.17$ [b] | $3.52 \pm 0.29$ [c] |
| R-IDF | $0.33 \pm 0.10$ [c] | $1.40 \pm 0.14$ [c] | $1.19 \pm 0.29$ [c] | $8.18 \pm 0.28$ [c] | $5.12 \pm 0.41$ [a] | $0.22 \pm 0.04$ [c] | $11.31 \pm 0.50$ [a] |
| P-IDF | $0.36 \pm 0.12$ [c] | $2.27 \pm 0.12$ [a] | $1.58 \pm 0.15$ [a] | $5.94 \pm 0.23$ [d] | $3.53 \pm 0.22$ [c] | $1.28 \pm 0.11$ [a] | $8.19 \pm 0.38$ [b] |

R-SDF represents the residue soluble fiber; P-SDF represents the residue soluble fiber; R-IDF represents the residue insoluble fiber; P-IDF represents the residue insoluble fiber; data are expressed as the mean ± standard deviation (SD). Different letters (a, b, c, d) indicate significantly different means ($p < 0.05$).

### 3.4. SEM Analyses of SDFs and IDFs

In this study, SEM was carried out to investigate the microstructures of IDFs and SDFs extracted from Newhall navel orange peel and residue. As shown in Figure 1, R-SDF had a block shape and a wrinkled surface with a compact texture. In addition, the surface of the R-SDF presented a dense structure, covered with many spherical particles and massive materials, most of which were starch granules and protein wedges [20,26]. P-IDF had looser and more complicated structures, while R-IDF had more wave drapes; this was possibly due to ultrasound treatment having potent oxidation during the extraction process, which could enlarge its surface area. Many voids were formed on the surface of SDFs, which was attributed to enzymes working deep into the fiber molecules, loosening the tightly packed structure of cellulose molecules. However, the spatial network structure of the IDFs was very loose, with large pores and more prominences on the surface. DF with a looser spatial structure had a higher specific surface area, which might affect its adsorption capacities of water, oil and nitrite ion [25].

### 3.5. FT-IR

FT-IR was carried out to analyze the spectroscopic features of different DF samples. As shown in Figure 2, there was a large absorption peak at 3200–3600 $cm^{-1}$, which was the absorption of O-H stretching vibration band of cellulose and semi-fiber [27]. Compared with P-SDF and R-SDF, blue shifts were observed in both P-IDF and R-IDF, which may be due to the destruction of the hydrogen bonds formed in the hydroxyl groups of the polysaccharide during the ultrasonic treatment [28]. The breaking of hydrogen bonding could increase the hydrogen bonds between the citrus fiber and water, which might lead to increase the water-holding capacity of citrus fibers.

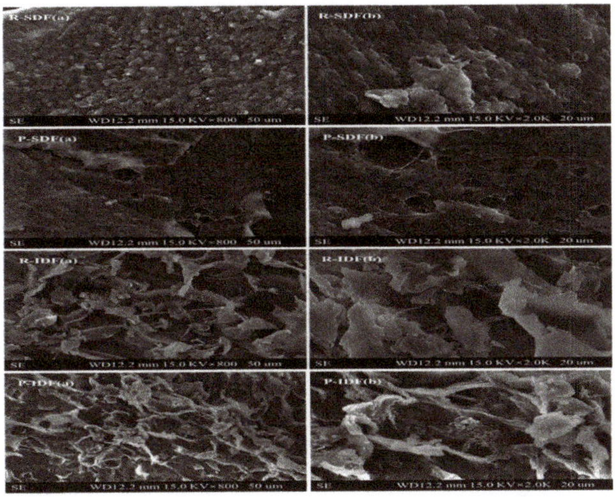

**Figure 1.** SEM images for R-SDF (**a**,**b**), P-SDF (**a**,**b**), R-IDF (**a**,**b**) and P-IDF (**a**,**b**); (**a**,**b**) represent the micrographs taken at the small and large magnification, respectively. R-SDF represents the residue soluble fiber; P-SDF represents the peel soluble fiber; R-IDF represents the residue insoluble fiber; P-IDF represents the peel insoluble fiber.

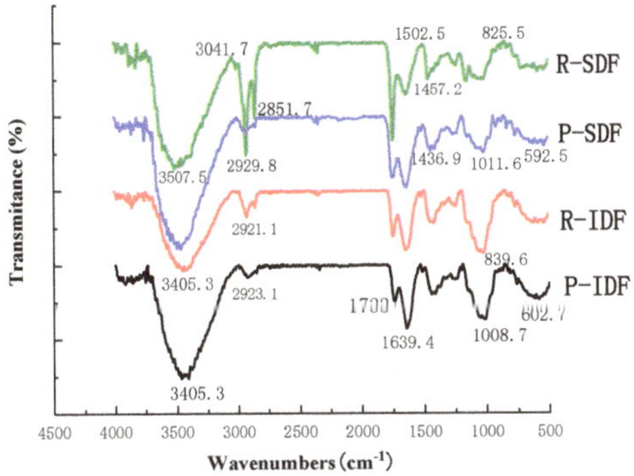


**Figure 2.** FT-IR spectra for IDFs and SDFs. R-SDF represents the residue soluble fiber; P-SDF represents the peel soluble fiber; R-IDF represents the residue insoluble fiber; P-IDF represents the peel insoluble fiber.

The broad peak appearing at 2800–3000 cm$^{-1}$ was the stretching vibration of the C-H bond, which was a characteristic absorption peak of the polysaccharide methyl, in dictating that all four substances were polysaccharides [29]. All orange fibers had C-H stretching vibrations from methyl and methylene groups of polysaccharides at approximately 2927 cm$^{-1}$. All samples had a characteristic peak at 1740–1760 cm$^{-1}$ (C-O stretching of COOH), indicating that the ultrasound-assisted enzymatic method caused the hydrolysis of hemicellulose. In addition, the absorption peaks near 1646 cm$^{-1}$ and 1419 cm$^{-1}$ contained carbonyl C=O and carboxyl COOH, exhibiting that uranic acid was presented in DF. The strength of R-SDF intermolecular hydrogen bonds at the band of 1640 cm$^{-1}$ was lower than that of the other three samples, which showed that ultrasonic cavitation might lead to the destruction of partial hydrogen bonds between R-SDF cellulose chains. This change could help citrus fiber hydrogen bonding loosen its internal structure and more fibers water hydrogen bonds when exposed to water.

In P-SDF and R-SDF, the peak intensity of 1016–1032 $cm^{-1}$ was observed to decrease, indicating that the hydrogen bond between cellulose was broken after ultrasonic treatment, which is beneficial to the combination of pectin and cellulose. The absorption peak was at 867 $cm^{-1}$ of carbohydrate characteristics, that is, the stretching vibration peak of β-glycosidic bond, and the peak values of R-IDF and P-IDF were more noticeable. The absorption peak of R-IDF at 1747 $cm^{-1}$ was slightly stronger, which might contain more lignin. P-IDF had a strong absorption peak at 619 $cm^{-1}$, which can probably be attributed to the presence of free alcohol or phenol. In addition, the four samples showed red shifts at some peak positions, indicating that the ultrasonic-assisted enzyme treatment destroyed the structure of organic molecules.

### 3.6. X-ray Diffractometry

DF can be divided into ordered crystalline region and amorphous region [11]. The characteristic sharp front diffraction reflects the crystalline structure, and the diffusion diffraction reflects the amorphous structure. As shown in Figure 3, other samples had a crystalline peak at 15.53°2θ except the R-SDF sample. All the samples had the characteristic crystalline peaks at 21.14°2θ. For three kinds of DFs except R-SDF, the peak appeared at 15.53°2θ. Among them, the signal peak of 15.53° came from the crystal plane 101, the diffraction peak of 21.14°2θ corresponded to the crystal plane 002 [30]. The crystal plane diffraction peak showed that the four DFs had the coexistence of crystalline and amorphous states, which was a typical cellulose type I structure. The peak patterns of R-IDF and P-IDF were sharper, indicating higher crystallinity, which might be attributed to the decreased protein and lignin content [11]. The characteristic peaks of R-SDF and P-SDF were relatively gentle, and there was no very sharp and obvious crystallization peak [19], which might be because of SDF having a higher pectin content. There were no significant differences among the crystallinity values of P-IDF (42%) and R-IDF (44%), although the values of R-SDF (17%) and P-SDF (29%) were decreased using ultrasonic treatments. This latter observation was attributed to the fact that ultrasonic treatment leads to the disruption of cellulose chain and affects the molecular tissue of SDF [10]; however, the treatment might also have an effective on improving the swelling, water-holding and oil-holding capacities.

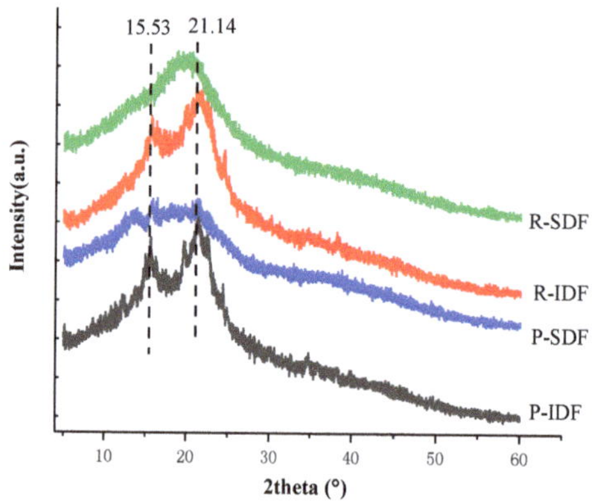


**Figure 3.** X-ray diffraction pattern of dietary fiber from R-SDF, P-IDF, P-SDF, R-IDF. R-SDF represents the residue soluble fiber; P-SDF represents the peel soluble fiber; R-IDF represents the residue insoluble fiber; P-IDF represents the peel insoluble fiber.

### 3.7. Physicochemical and Functional Properties of SDFs and IDFs

#### 3.7.1. WHC, OHC, SC and CEC

Water-holding capacity (WHC), oil-holding capacity (OHC), swelling capacity (SC) and cation exchange capacity (CEC) are important indexes to evaluate the quality and physiological function of DF.

The WHC stands for the water retention capacity of the material; it can reduce dehydration and contraction of the product and promote digestion in the human body. As shown in Table 2, the WHC values of four samples ranged from 8.81 to 13.43 g/g. R-IDF (13.43 ± 0.20 g/g) had the best WHC, followed by R-SDF (11.97 ± 0.49 g/g), P-IDF (11.75 ± 0.35 g/g) and P-SDF (8.81 ± 0.36 g/g). The WHC values of DF samples extracted were higher than lemon (6.46 ± 0.79 g/g), orange (7.76 ± 0.81 g/g) and grapefruit seeds (7.06 ± 0.16 g/g) [31], which maybe because the special functional properties of ultrasound could result in a loose porous structure of DF samples. Furthermore, the WHC of extruded Liucheng sweet orange pomace of 6.73 g/g of orange dry sample was less than P-SDF (8.81 ± 0.36 g/g). The possible reason for this result may be that the high temperature in barrel destructed the water-holding capacity of the extrudate, which could not be well retained water within the DF [32]. This value of WHC indicated that ultrasound retains moisture better than extrusion.

**Table 2.** Physiochemical properties of R-SDF, P-SDF, R-IDF and P-IDF.

| | R-SDF | P-SDF | R-IDF | P-IDF |
|---|---|---|---|---|
| WHC (g/g) | 11.97 ± 0.49 $^{b}$ | 8.81 ± 0.36 $^{c}$ | 13.43 ± 0.20 $^{a}$ | 11.75 ± 0.35 $^{b}$ |
| OHC (g/g) | 1.34 ± 0.05 $^{b}$ | 1.26 ± 0.05 $^{b}$ | 2.08 ± 0.21 $^{a}$ | 1.36 ± 0.03 $^{b}$ |
| SC (mL/g) | 23.33 ± 1.15 $^{a}$ | 19.33 ± 1.53 $^{b}$ | 18.44 ± 0.51 $^{bc}$ | 16.89 ± 0.38 $^{c}$ |
| CEC (mmol/g) | 0.89 ± 0.01 $^{a}$ | 0.76 ± 0.02 $^{c}$ | 0.82 ± 0.02 $^{b}$ | 0.73 ± 0.03 $^{c}$ |

WHC, water-holding capacity; OHC, oil-holding capacity; SC, swelling capacity; CEC, Cation exchange capacity. R-SDF represents the residue soluble fiber; P-SDF represents the peel soluble fiber; R-IDF represents the residue insoluble fiber; P-IDF represents the peel insoluble fiber; data are expressed as the mean ± standard deviation (SD). Different letters (a, b, c) indicate significantly different means ($p < 0.05$).

In contrast to WHC, SC always depends on many factors such as network density, molecular size and the cellulose components of the fiber [19]. Table 2 shows that the SC of R-SDF (23.33 ± 1.15 mL/g) were significantly greater ($p < 0.05$) than those of P-SDF (19.33 ± 1.53 mL/g), R-IDF (18.44 ± 0.51 mL/g) and P-IDF (16.89 ± 0.38 mL/g). Soluble orange peel fibers had 4.83 and 6.28 mL/g SC values by steam explosion and dilute acid soaking, which were less than P-SDF and R-SDF [33]. This result may be because the ultrasound treatment can loosen the orange fibers' inner structure and break hydrogen bonds, and the enzymatic treatment can further disrupt cellulose chains and remove some cellulose, which can be beneficial to the increase in swelling capacity.

Table 2 showed the OHC values of Newhall navel orange peel and residue DFs ranged between 1.26 and 2.08 g oil/g fiber. R-IDF (2.08 ± 0.21 g/g) was 65.08% greater than that of P-SDF (1.26 ± 0.05). In another study, the OHC values of various orange fibers ranged from 0.9 to 1.3 g/g [34]. Therefore, the OHC of R-SDF is within the reported ranges for the OHC values. The other three samples have higher OHC values, which may be related to their better structure and surface area. Previous studies have shown that fiber particles can absorb and combine oily components [35], which could make food have better palatability [27]. Therefore, four dietary fibers can become effective raw materials for the functional food industry.

There are amino and hydroxyl groups in the molecular structure of dietary fiber. When these groups are exchanged with $Zn^{2+}$ and $Cu^{2+}$ plasma, the pH and osmotic pressure of human digestive tract can be changed, and the blood pressure can be lowered. As shown in Table 2, the CEC of R-SDF was the highest, followed by R-IDF and P-SDF, and P-IDF was the least. This might be due to the decomposition of some citrus pulp fibers caused by ultrasonic treatment, exposing more side groups such as hydroxyl, carboxyl and amino groups. DFs with high CEC will reduce the diffusion and absorption of lipids and cholesterol [2]. Besides, CEC can decrease the utilization of cholesterol in vivo [22].

3.7.2. Cholesterol Adsorption Capacity (CAC)

Cholesterol metabolism has a close relationship with the incidence of cardiovascular and cerebrovascular diseases, while DF can reduce the digestion and absorption of cholesterol and triglyceride by combining with food. Therefore, the CAC of Newhall navel orange peel and residue DFs was studied. The CACs of the DF samples are shown in Figure 4. The CACs of all the DF samples in simulated small intestinal pH value environment (pH = 7.0) were significantly ($p < 0.05$) stronger than those of under the condition of simulated stomach acidity (pH = 2.0), which is consistent with the behavior of bamboo shoot dietary fibers [36]. This may be because there is more $H^+$ in an acidic environment, and the cholesterol in dietary fiber has a partial positive charge which produces repulsive force, thus reducing the adsorption force. Furthermore, the R-SDF showed the highest CAC in both pH = 2.0 (8.07 ± 0.13 mg/g) and pH = 7.0 (13.88 ± 0.10 mg/g), which might be due to the dietary fiber's surface area. In addition, SDF exhibited a better healthy effect on reducing cholesterol than IDF in the cholesterol absorption, which is similar to previous studies [37,38].

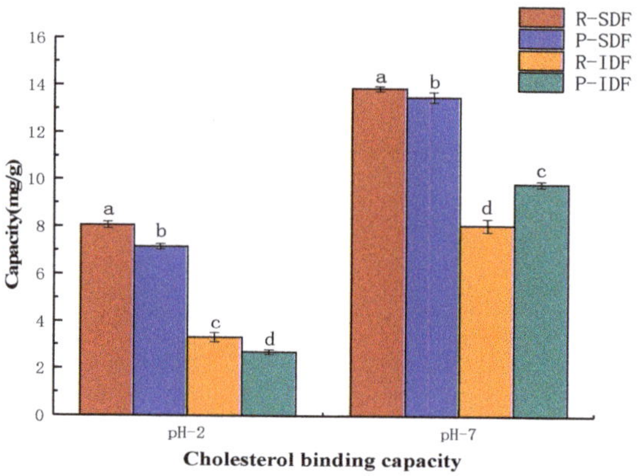


**Figure 4.** The CAC of R-IDF, P-IDF, R-SDF, P-SDF. R-SDF represents the residue soluble fiber; P-SDF represents the peel soluble fiber; R-IDF represents the residue insoluble fiber; P-IDF represents the peel insoluble fiber; values with different letters (a, b, c, d) for IDF or SDF are significantly different means ($p < 0.05$).

3.7.3. Nitrite Ion Absorption Capacity (NIAC)

Nitrite is commonly used as a color protector for meat products. Nitrite reacts with secondary amines, tertiary amines and amines in food to form a strong carcinogen, N-nitrosamine [9]. The result of this study was presented in Figure 5. For all the dietary fiber samples had increased NIAC in pH = 2.0 environment compared with that at pH = 7.0, and IDFs had a higher NIAC than SDFs [9], which is owing to $NO_2^-$ combining with $H^+$ to form $HNO_2$, forming more nitrogen oxides. Among these DF samples, R-IDF revealed the highest NIAC in both pH = 2.0 (159.16 ± 0.16 µg/g) and pH = 7.0 (123.31 ± 0.18 µg/g), which might be due to the R-IDF sample having a better structure interaction with more nitrite ion [9]. In addition, R-SDF exhibited significantly higher NIAC than P-SDF in both pH = 2.0 and pH = 7.0 (146.99 ± 0.60 µg/g, 141.50 ± 0.99 µg/g, respectively) and pH = 7.0 (80.08 ± 1.97 µg/g, 66.49 ± 0.39 µg/g, respectively). The above results indicated that R-SDF had more functional groups that could interact with $NO_2^-$.

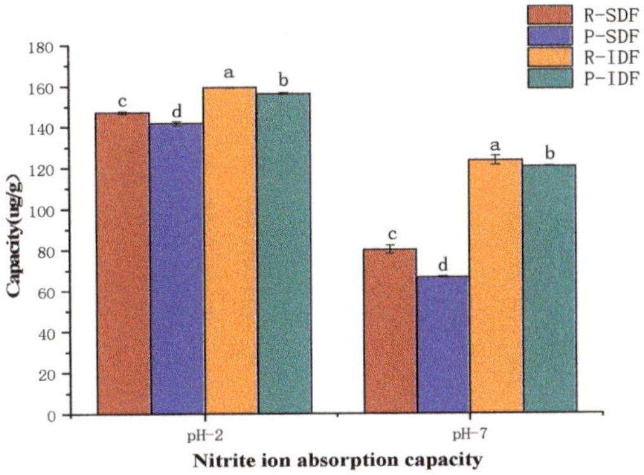


**Figure 5.** The NIAC of R-IDF, P-IDF, R-SDF, P-SDF. R-SDF represents the residue soluble fiber; P-SDF represents the peel soluble fiber; R-IDF represents the residue insoluble fiber; P-IDF represents the peel insoluble fiber; values with different letters (a, b, c, d) for IDF or SDF are significantly different ($p < 0.05$).

## 4. Discussion

In this study, the structural, physicochemical and functional properties of dietary fiber exacted from Newhall navel orange by-products were investigated. We accordingly found that SDF and IDF of orange residue in by-products extracted by ultrasound-assisted enzymatic method had the highest yield. Glucose and arabinose were the dominant monosaccharides of all DF samples in this study, and the content of monosaccharide in navel orange residue DFs was higher than navel peel DFs. Besides, the surface of P-IDF more than R-IDF has looser and more complicated structures, and R-SDF has a more complete surface than P-SDF. Among these four DF samples, R-IDF has the highest WHC ($13.43 \pm 0.20$ g/g) and OHC ($2.08 \pm 0.21$ g/g) values, whereas R-SDF has the highest SC ($23.33 \pm 1.15$ mL/g) value and CEC ($0.89 \pm 0.01$ mmol/g). The result of functional properties revealed that the R-SDF extracted from residue and peel had the highest CAC and NIAC. Considering the extraction efficiency and good quality retention, the R-SDF and R-IDF have the broadest market prospect for potential application, which could improve the economic efficiency and utilization rate of Newhall navel orange by-products.

**Author Contributions:** J.W. (Jing Wen): Conceptualization, Methodology; J.S.: Data Curation, Writing; L.L.: Reviewing and Editing; Y.X.: Visualization, Investigation; Q.G. and J.W. (Jijun Wu): Supervision; M.F. and Y.Y.: Software, Validation; X.L.: Project Administration. All authors have read and agreed to the published version of the manuscript.

**Funding:** We are thankful for the financial support of Guangdong Provincial Key Research Project of China (2020B020226010), the National Natural Science Foundation of China (Grant Nos. 32060649), the National Natural Science Foundation of China (Grant No. 31901713), the Program for Guangdong YangFan Introducing Innovative and Enterpreneurial Teams (No. 2017YT05H045), Science and Technology Project of Guangzhou (201909020001) and the Talent Project of Guangdong Academy of Agricultural Sciences (R2020PY-JX011).

**Conflicts of Interest:** The authors declare no conflict of interest.

## References

1. Huang, J.Y.; Liao, J.S.; Qi, J.R.; Jiang, W.X.; Yang, X.Q. Structural and physicochemical properties of pectin-rich dietary fiber prepared from citrus peel. *Food Hydrocoll.* **2021**, *110*, 106140. [CrossRef]
2. Chau, C.F.; Huang, Y.L. Comparison of the chemical composition and physicochemical properties of different fibers prepared from the peel of *Citrus sinensis* L. Cv. Liucheng. *J. Agric. Food Chem.* **2003**, *51*, 2615–2618. [CrossRef] [PubMed]
3. Llorach, R.; Espín, J.C.; Tomas-Barberan, F.A.; Ferreres, F. Valorization of Cauliflower (*Brassica oleracea* L. var. botrytis) by-products as a source of antioxidant phenolics. *J. Agric. Food Chem.* **2003**, *51*, 2181–2187. [CrossRef] [PubMed]

4. Rodríguez, R.; Jiménez, A.; Fernández-Bolaños, J.; Guillén, R.; Heredia, A. Dietary fibre from vegetable products as source of functional ingredients. *Trends Food Sci. Technol.* **2006**, *17*, 3–15. [CrossRef]
5. Zhu, Y.; Chu, J.X.; Lu, Z.X.; Lv, F.X.; Bie, X.M.; Zhang, C.; Zhao, H.Z. Physicochemical and functional properties of dietary fiber from foxtail millet (Setaria italic) bran. *J. Cereal Sci.* **2018**, *79*, 456–461. [CrossRef]
6. Jing, Y.; Chi, Y.J. Effects of twin-screw extrusion on soluble dietary fibre and physicochemical properties of soybean residue. *Food Chem.* **2013**, *138*, 884–889. [CrossRef] [PubMed]
7. Zhang, Y.; Liao, J.; Qi, J.R. Functional and structural properties of dietary fiber from citrus peel affected by the alkali combined with high-speed homogenization treatment. *Food Sci. Technol.* **2020**, *128*, 109397. [CrossRef]
8. Song, J.H.; Pan, T.; Wu, J.P.; Ren, F.Z. The improvement effect and mechanism of citrus fiber on the water-binding ability of low-fat frankfurters. *J. Food Sci. Technol.* **2016**, *53*, 4197–4204. [CrossRef] [PubMed]
9. Gan, J.; Huang, Z.Y.; Yu, Q.; Peng, G.Y.; Xie, J.H.; Nie, S.P.; Xie, M.Y. Microwave assisted extraction with three modifications on structural and functional properties of soluble dietary fibers from grapefruit peel. *Food Hydrocoll.* **2020**, *101*, 105549. [CrossRef]
10. Dong, W.; Wang, D.; Hu, R.S.; Long, Y.Z.; Lv, L.S. Chemical composition, structural and functional properties of soluble dietary fiber obtained from coffee peel using different extraction methods. *Food Res. Search Int.* **2020**, *136*, 109497. [CrossRef]
11. Ma, M.M.; Mu, T.H. Effects of extraction methods and particle size distribution on the structural, physicochemical, and functional properties of dietary fiber from deoiled cumin. *Food Chem.* **2016**, *194*, 237–246. [CrossRef] [PubMed]
12. Bagherian, H.; Ashtiani, F.Z.; Fouladitajar, A.; Mohtashamy, M. Comparisons between conventional, microwave- and ultrasound-assisted methods for extraction of pectin from grapefruit. *Chem. Eng. Process.* **2011**, *50*, 1237–1243. [CrossRef]
13. Nyman, E.M.G.L.; Svanberg, S.J.M. Modification of physicochemical properties of dietary fibre in carrots by mono- and divalent cations. *Food Chem.* **2002**, *76*, 273–280. [CrossRef]
14. Chen, S.; Chen, H.; Tian, J.; Wang, J.; Wang, Y.; Xing, L. Enzymolysis-ultrasonic assisted extraction, chemical characteristics and bioactivities of polysaccharides from corn silk. *Carbohydr. Polym.* **2014**, *101*, 332–341. [CrossRef]
15. Yang, Y.S.; Wang, Z.M.; Hu, D.; Xiao, K.J.; Wu, J.Y. Efficient extraction of pectin from sisal waste by combined enzymatic and ultrasonic process. *Food Hydrocoll.* **2018**, *79*, 189–196. [CrossRef]
16. Moczkowska, M.; Karp, S.; Niu, Y.; Kurek, M.A. Enzymatic, enzymatic-ultrasonic and alkaline extraction of soluble dietary fibre from flaxseed—A physicochemical approach. *Food Hydrocoll.* **2019**, *90*, 105–112. [CrossRef]
17. Zhao, G.H.; Zhang, R.F.; Dong, L.H.; Huang, F.; Tang, X.J.; Wei, Z.C.; Zhang, M.W. Particle size of insoluble dietary fiber from rice bran affects its phenolic profile, bioaccessibility and functional properties. *Food Sci. Technol.* **2018**, *87*, 450–456. [CrossRef]
18. Gu, M.; Fang, H.C.; Gao, Y.H.; Su, T.; Niu, Y.G.; Yu, L.I. Characterization of enzymatic modified soluble dietary fiber from tomato peels with high release of lycopene. *Food Hydrocoll.* **2020**, *99*, 105321. [CrossRef]
19. Zhang, W.M.; Zeng, G.L.; Pan, Y.G.; Chen, W.X.; Huang, W.Y.; Chen, H.M.; Li, Y.S. Properties of soluble dietary fiber-polysaccharide from papaya peel obtained through alkaline or ultrasound-assisted alkaline extraction. *Carbohydr. Polym.* **2017**, *172*, 102–112. [CrossRef] [PubMed]
20. Liu, Y.L.; Zhang, H.B.; Yi, C.P.; Quan, K.; Lin, B.P. Chemical composition, structure, physicochemical and functional properties of rice bran dietary fiber modified by cellulase treatment. *Food Chem.* **2021**, *342*, 128352. [CrossRef]
21. Zhang, M.; Bai, X.; Zhang, Z.S. Extrusion process improves the functionality of soluble dietary fiber in oat bran. *J. Cereal Sci.* **2011**, *54*, 98–103. [CrossRef]
22. He, Y.Y.; Li, W.; Zhang, X.Y.; Li, T.T.; Ren, D.F.; Lu, J. Physicochemical, functional, and microstructural properties of modified insoluble dietary fiber extracted from rose pomace. *J. Food Sci. Technol.* **2020**, *57*, 1421–1429. [CrossRef] [PubMed]
23. Zhang, N.; Huang, C.H.; Ou, S.Y. In vitro binding capacities of three dietary fibers and their mixture for four toxic elements, cholesterol, and bile acid. *J. Hazard. Mater.* **2011**, *186*, 236–239. [CrossRef] [PubMed]
24. Khan, G.M.; Khan, N.M.; Khan, Z.U.; Ali, F.; Jan, A.K.; Muhammad, N.; Elahi, R. Effect of extraction methods on structural, physiochemical and functional properties of dietary fiber from defatted walnut flour. *Food Sci. Biotechnol.* **2018**, *27*, 1015–1022. [CrossRef] [PubMed]
25. Wang, K.; Li, M.; Wang, Y.X.; Liu, Z.H.; Ni, Y.Y. Effects of extraction methods on the structural characteristics and functional properties of dietary fiber extracted from kiwifruit (*Actinidia deliciosa*). *Food Hydrocoll.* **2021**, *110*, 106162. [CrossRef]
26. Xie, F.Y.; Zhao, T.; Wan, H.C.; Li, M.; Sun, L.N.; Wang, Z.J.; Zhang, S. Structural and physicochemical characteristics of rice bran dietary fiber by cellulase and high-pressure homogenization. *Appl. Sci.* **2019**, *9*, 1270. [CrossRef]
27. Mahloko, L.M.; Silungwe, H.; Mashau, M.E.; Kgatla, T.E. Bioactive compounds, antioxidant activity and physical characteristics of wheat-prickly pear and banana biscuits. *Heliyon* **2019**, *5*, e02479. [CrossRef] [PubMed]
28. Zhang, Y.; Qi, J.R.; Zeng, W.Q.; Huang, Y.X.; Yang, X.Q. Properties of dietary fiber from citrus obtained through alkaline hydrogen peroxide treatment and homogenization treatment. *Food Chem.* **2020**, *311*, 125873. [CrossRef] [PubMed]
29. Su, Y.; Li, L. Structural characterization and antioxidant activity of polysaccharide from four auriculariales. *Carbohydr. Polym.* **2020**, *229*, 115407. [CrossRef]
30. Wen, Y.; Niu, M.; Zhang, B.J.; Zhao, S.M.; Xiong, S.B. Structural characteristics and functional properties of rice bran dietary fiber modified by enzymatic and enzyme-micronization treatments. *Food Sci. Technol.* **2017**, *75*, 344–351. [CrossRef]
31. Karaman, E.; Yılmaz, E.; Tuncel, N.B. Physicochemical, microstructural and functional characterization of dietary fibers extracted from lemon, orange and grapefruit seeds press meals. *Bioact. Carbohydr. Diet. Fibre* **2017**, *11*, 9–17. [CrossRef]

32. Huang, Y.L.; Ma, Y.S. The effect of extrusion processing on the physiochemical properties of extruded orange pomace. *Food Chem.* **2016**, *192*, 363–369. [CrossRef] [PubMed]
33. Wang, L.; Xu, H.; Yuan, F.; Fan, R.; Gao, Y. Preparation and physicochemical properties of soluble dietary fiber from orange peel assisted by steam explosion and dilute acid soaking. *Food Chem.* **2015**, *185*, 90–98. [CrossRef]
34. Tejada-Ortigoza, V.; Garcia-Amezquita, L.E.; Serna-Saldívar, S.O.; Welti-Chanes, J. Advances in the functional characterization and extraction processes of dietary fiber. *Food Eng. Rev.* **2016**, *8*, 251–271. [CrossRef]
35. Almeida, J.D.S.O.; Dias, C.O.; Arriola, N.D.A.; Freitas, B.S.M.; Francisco, M.; Petkowicz, C.L.O.; Araujo, L.; Guerra, M.P.; Nodari, R.O.; Amboni, R.D.M.C. Feijoa (*Acca sellowiana*) peel flours: A source of dietary fibers and bioactive compounds. *Food Biosci.* **2020**, *38*, 100789. [CrossRef]
36. Wu, W.J.; Hu, J.; Gao, H.Y.; Chen, H.J.; Fang, X.J.; Mu, H.L.; Han, Y.C.; Liu, R.L. The potential cholesterol-lowering and prebiotic effects of bamboo shoot dietary fibers and their structural characteristics. *Food Chem.* **2020**, *332*, 127372. [CrossRef]
37. Xu, H.G.; Jiao, Q.; Yuan, F.; Gao, Y.X. In vitro binding capacities and physicochemical properties of soluble fiber prepared by microfluidization pretreatment and cellulase hydrolysis of peach pomace. *Food Sci. Technol.* **2015**, *63*, 677–684. [CrossRef]
38. Jia, M.Y.; Chen, J.J.; Liu, X.Z.; Xie, M.Y.; Nie, S.P.; Chen, Y.; Xie, J.H.; Yu, Q. Structural characteristics and functional properties of soluble dietary fiber from defatted rice bran obtained through Trichoderma viride fermentation. *Food Hydrocoll.* **2019**, *94*, 468–474. [CrossRef]

*foods*

MDPI

*Review*

# Recent Advances and Applications in Starch for Intelligent Active Food Packaging: A Review

**Dandan Liu [1], Pei Zhao [1,*], Jinyu Chen [2], Yali Yan [1] and Zijian Wu [2,*]**

1 School of Biotechnology and Food Science, Tianjin University of Commerce, Tianjin 300134, China
2 Tianjin Key Laboratory of Food Biotechnology, Tianjin University of Commerce, Tianjin 300134, China
* Correspondence: zhaopei@tjcu.edu.cn (P.Z.); wzjian@tjcu.edu.cn (Z.W.)

**Abstract:** At present, the research and innovation of packaging materials are in a period of rapid development. Starch, a sustainable, low-cost, and abundant polymer, can develop environmentally friendly packaging alternatives, and it possesses outstanding degradability and reproducibility in terms of improving environmental issues and reducing oil resources. However, performance limitations, such as less mechanical strength and lower barrier properties, limit the application of starch in the packaging industry. The properties of starch-based films can be improved by modifying starch, adding reinforcing groups, or blending with other polymers. It is of significance to study starch as an active and intelligent packaging option for prolonging shelf life and monitoring the extent of food deterioration. This paper reviews the development of starch-based films, the current methods to enhance the mechanical and barrier properties of starch-based films, and the latest progress in starch-based activity, intelligent packaging, and food applications. The potential challenges and future development directions of starch-based films in the food industry are also discussed.

**Keywords:** starch-based film; degradability; active packaging; intelligent packaging; mechanical properties; barrier properties; freshness indicator; time-temperature sensor

**Citation:** Liu, D.; Zhao, P.; Chen, J.; Yan, Y.; Wu, Z. Recent Advances and Applications in Starch for Intelligent Active Food Packaging: A Review. *Foods* **2022**, *11*, 2879. https://doi.org/10.3390/foods11182879

Academic Editors: Federico Casanova and Michela Verni

Received: 30 August 2022
Accepted: 11 September 2022
Published: 16 September 2022



## 1. Introduction

From farm to table, food ingredients or products are susceptible to external damage or contamination, predominantly by pathogenic bacteria and oxidation, resulting in deterioration quality [1]. Therefore, adequate measures must be taken to maintain the original quality of the food. Among them, proper food packaging can maintain quality, avoid spoilage, extend shelf life, and reduce waste. In existing packaging materials, plastics are commonly used in food packaging due to their low price, good mechanical properties, and moderate barrier properties [2]. Each year, approximately 300 million tons of plastic are manufactured worldwide, and 40% are used in packaging materials [3]. Nonetheless, most plastic packaging is non-degradable [4] and causes severe environmental pollution [5]. Accordingly, the research and developing of environment-friendly food packaging to replace plastic packaging has gained increasing attention [6].

In recent years, degradable food packaging has developed rapidly. Diverse degradation technologies emerged with endless successions, such as photodegradable and biodegradable [7]. However, some degradable materials are toxic and hazardous to humans, and not all of them are suitable for food packaging [8]. Biodegradable materials, by contrast, are optimal for food packaging [9]. The function of biodegradable packaging film is the same as that of conventional packaging: to protect food quality, promote food circulation, and increase added value. At the end of the previous century, the focus of research switched from biodegradable film to completely degradable biofilms. To date, molecules used to assemble completely biodegradable films include polysaccharides (such as starch, chitosan, and cellulose) [10,11], protein (such as whey protein, soy protein, and silk protein) [12,13], lipids (such as beeswax and lauric acid) [14,15], and so on. The high cost of production and use limits the application of biodegradable materials in food packaging. Among the various natural polymers, starch is possible for food packaging because of its low price, abundant reserves, edibility, and degradability [16].

Starch is the primary carbohydrate storage form in plant tubers and seed endosperm, generally found in maize, potato, cassava, and cereals [17]. Starch is divided into amylose and amylopectin. The former is composed for α-1,4-glycosidic bonds connected end to end, which is a non-branching helical structure. The latter consists of α-1,4-glycosidic bonds, α-1,6-glycosidic bonds, which forms highly branched polymers with 24 to 30 glucose residues. Amylose content varies with different plant sources [18]. Due to the characteristics of organic starch itself (such as insolubility in cold water, hygrometry, poor structure, degradation, etc.), the application of natural starch in the industry is constrained [19]. In contrast, the mechanical properties of the pure starch-based film are considerably lower than those of conventional ordinary plastics [20]. In addition, starch molecules contain many hydroxyl groups, making them highly hydrophilic, resulting in poor water resistance and hydrophobicity and poor mechanical properties in wet environments [21,22]. Therefore, it is necessary to strengthen the mechanical and barrier properties of starch-based packaging materials.

Starch modification or blending with different materials is primarily used to resolve the above problems [23,24]. With the large-scale development of integrated technology in food processing, transportation, and storage, the food needs to have a long shelf life and maintain the quality of fresh food. The requirements for food packaging are also relatively increased [25]. The emerging active, intelligent food packaging can extend the shelf life of food, ensure food safety, and show information about food and its current status in the food supply chain to processors, retailers, and consumers [26,27]. Active packaging refers to the packaging system containing certain active substances (such as organic acids, enzymes, bacteriocins, natural plant extracts, etc.) [28], which can be released into the packaged food or the surrounding environment, thus extending the shelf life of food and retaining their quality, safety, and sensory properties [29]. Intelligent packaging is a system (including pH indicators and time or temperature sensors, etc.) that can monitor the storage status/cycle of packaged food or inform consumers about the quality of the food [30]. These intelligent packaging materials can broadly be pasted as labels or direct film formation [31].

In recent years, most of the studies on starch-based biodegradable films have focused on the sources of starch films [32], their processing methods [33], and the challenges and opportunities of starch-based materials [34]. Nevertheless, there is no current review on the status and application of starch-based biodegradable films in active, intelligent food packaging; this review first proposes the degradability of starch-based materials and methods to improve the mechanical and barrier properties of starch-based films. Then, the starch-based activity, the preparation of intelligent packaging for monitoring food quality, maintaining safety, and extending shelf life were evaluated, and the application of starch-based films as packaging materials in food preservation are analyzed. Finally, the existing problems are discussed, and the future research direction is adopted. It is of great significance and practical value to develop green, safe, and functional starched-based food packaging materials and their application in food preservation.

## 2. Starch-Based Biodegradable Film Materials

Biodegradable plastics refer to those that can be degraded by indigenous microorganisms under natural or unique conditions and eventually convert into environment-friendly biomass or small nontoxic molecule (such as $CO_2$, $CH_4$, or $H_2O$) [35] (Figure 1). Among these, starch-based biodegradable materials have become one of the favored materials [36,37]. Starch is a kind of natural polymer compound belonging to the polysaccharide, which mainly exists in seeds, tubers, roots, or fruits of plants [38]. The raw materials used in starch manufacturing should first have the characteristics of high starch content and then have the characteristics of easy extraction, low processing cost, easy storage, and by-product production. Therefore, the main raw materials that meet the requirements are cereals, tubers, and legumes, which all contain a large amount of starch [39]. At present, starch extraction methods mainly include the alkali method, enzymatic method, and surfactant method. The Alkali method and enzymatic method use the action of alkali or enzymes to hydrolyze starch in combination with components, such as proteins and cellulose, and then release starch, improving the yield of starch; the surfactant method is to use sodium alkyl benzene sulfonate and other surfactants combined with protein so that the modified protein and starch form complex separation to achieve the purpose of starch extraction [40–42]. Starch properties vary by grain variety, growing climate, soil quality, and other growing conditions, ranging from more than 70% starch in cereals to 36% to 47% starch in dried beans. The content of vegetable starch is different. Potatoes account for approximately 14.7% and green leafy vegetables less than 0.2% [43,44].

Starch-based degradable packaging research and development began in the 1980s, and it experienced rapid growth during this period. After approximately 10 years of development, it demonstrated that only the starch component in those so-called degradable films could be degraded, while other film materials were just broken into fragments, which continued to exist in soil and water, still causing harm to the environment [45]. Hence, completely degraded materials were introduced at that time. Starch-based biodegradable materials are developed in three stages: Starch-filled plastics stage, Blended starch plastics stage, and All starch plastics stage. The increase in starch content has accompanied the evolution of starch-based plastics. Among them, all starch plastics can be completely biodegraded and have a comprehensive source of raw materials and low prices, which are the focus of starch-based material development.

1. Starch-filled plastics are made by mixing a small amount of original starch or modified starch with polyethylene or other thermoplastics and adding other applicable additives. Its purpose is to enhance the biodegradability of traditional petroleum-based starch materials. Nonetheless, its degradation still needs several years and cannot be thoroughly degraded [46];
2. Blended starch plastics are made of starch mixed with synthetic resin or other natural polymer materials. They are generally blends of starch/modified starch (30–60%) and synthetic biodegradable materials, which can be completely biodegradable and do not pollute the environment [47]. Compared with purely synthetic polymers, the blends degrade quickly and have better mechanical properties. Nonetheless, the added synthetic resins or other natural polymer materials are primarily polar compounds with hydrophilicity, and long-term exposure or contact with water will considerably degrade the properties of the plastic [48]. In addition, the compatibility between starch and additives, such as synthetic resins or other natural polymers is likewise problematic [49];
3. All starch plastics, also known as thermoplastic starch plastic, are a natural polymer biodegradable material. They are prepared by adding degradable plasticizers and other additives [50] through processes, such as extrusion, injection moulding, blow moulding, and calendering, which result in a "disordered" arrangement of starch molecule. The starch content of all starch plastics is above 90%, and a small number of different substances added as additives are nontoxic and can be completely degraded. Thus, all starchy plastics are genuinely and completely biodegradable. In addition, almost all plastic processing methods can be applied to all starch plastics [51].

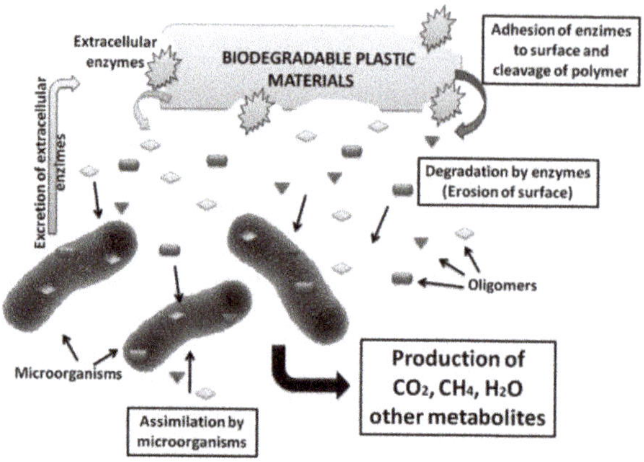


**Figure 1.** General mechanism of plastic biodegradation ref. [52]. 2019 Hesham Moustafa, Ahmed M. Youssef, Nabila A. Darwish, Ahmed I. Abou-Kandil.

The increasing use of starch-based biodegradable materials in food packaging has many advantages. These include reduced use of petroleum products, environmental friendliness and safety and reliability of food packaging [21]. Despite these advantages and benefits, biodegradable plastics currently account for less than 1% of total plastic production. Compared with traditional packaging materials, biodegradable materials have high production costs and poor mechanical and barrier properties, which are the main reasons for their limited application [53]. With the improvement of film material requirements, it will be very promising to develop antibacterial, antioxidant, and other multifunctional films based on improving the properties of the starch-based films.

## 3. Mechanical Properties of Starch-Based Films

Good mechanical properties are essential for the actual use of food packaging materials [54]. However, compared with similar applications of traditional plastics, the starch-based degradable packaging still has obvious disadvantages because of its poor mechanical properties [20], which limits its application in food and product packaging [22]. Excellent mechanical properties can be used to an impressive advantage in traditional packaging. Therefore, it is essential to strengthen the mechanical properties of starch-based films. Table 1 lists the mechanical properties of some starch-based films.

**Table 1.** Mechanical properties of starch-based films.

| Films | Additives | Thickness (mm) | Moisture Content (%) | Tensile Strength (MPa) | Elongation(%) | References |
|---|---|---|---|---|---|---|
| Cassava starch | glycerol | 0.103 | 19.22 | 2.85 | 18.82 | [55] |
| Mungbean starch | | 0.098 | 19.66 | 9.34 | 21.37 | |
| Cassava: Mungbean (50:50) | | 0.090 | 22.11 | 7.93 | 21.32 | |
| Cassava starch | sorbitol | 0.101 | 9.43 | 6.77 | 14.86 | |
| Mungbean starch | | 0.113 | 9.16 | 19.20 | 12.89 | |
| Cassava: Mungbean (50:50) | | 0.105 | 8.84 | 15.87 | 10.84 | |
| Wheat | | 0.074 | 44.5 | 3.29 | 15.21 | [56] |
| Corn | | 0.112 | 36.7 | 3.72 | 19.13 | |
| Potato | | 0.055 | 31.6 | 6.56 | 5.67 | |
| PV:PB (50:50, 60:40, 70:30, 80:20, 90:10, 100:0) | glycerol | 0.061~0.070 | - | 27.5~52.6 | 108.1~241.8 | [57] |
| NF | | 0.064 | 13.06 | 3.49 | 19.21 | [58] |
| ACT (4%, 8%) | | 0.071, 0.072 | 14.47, 13.43 | 3.69, 2.86 | 31.4, 19.5 | |
| HPS (10%, 30%) | | 0.070, 0.067 | 16.49, 18.82 | 3.10, 2.54 | 57.17, 64.81 | |
| Cassava starch | | | | 5.5 | 45.5 | [59] |
| | (5~15) wt% metakaolin+ glycerol | | | 5.7~8.1 | 23.1~33.2 | |

Table 1. *Cont.*

| Films | Additives | Thickness (mm) | Moisture Content (%) | Tensile Strength (MPa) | Elongation(%) | References |
|---|---|---|---|---|---|---|
| Rice starch | sorbitol | | | 10.75 | 7.56 | [60] |
| | (10~50) % NaOH+ sorbitol | | | 2.75~9.87 | 11.36~53.03 | |
| Octenyl succinate starch | glycerol | 0.087 | 29.54 | 9.60 | 32.41 | [61] |
| | (0.025~0.100) % PSE+ glycerol | 0.090~0.091 | 29.22~29.62 | 7.56~8.62 | 23.99~30.98 | |
| | (0.025~0.100) % HSE+ glycerol | 0.090~0.091 | 29.74~30.03 | 7.31~10.58 | 29.58~31.65 | |

PB: Pinto Bean Starch, PV: Polyvinyl Alcohol. NF: native starch; PSE: pecan nutshell extract, HSE: hazelnut skin extract.

Studies have shown that the elongation of the starch film is negatively correlated with amylopectin content when starch is 6.3% to 25.0%. In contrast, the tensile strength of starch films was positively correlated with the amylose content, which was increased to 40% [62,63]. Hoang and Nguyen [64] studied cassava starch and mung bean starch with 26% and 33% amylose content, respectively, and found that the peak elongation and breaking elongation of mung bean starch film were 11% and 50% higher than those of cassava starch film, respectively. Paulina and Izabela [65] prepared films from starches isolated from pumpkin fruits, lentils, and quinoa seeds and compared them with potato starch films. The results showed that the tensile strength and elongation at the break of films are 8.98~13.85 MPa and 3.35~4.44%, respectively. All thin films are continuous elastic and have unlimited elastic behavior.

Although the pure starch-based film has mechanical properties that can be selected by researching the properties of various starches, it does not meet food packaging requirements. Currently, chemical and physical modification of starch is predominantly used to enhance the mechanical properties of the starch film [66,67]. Dai and Zhang [68] researched the impact of modification methods on the mechanical properties of starch-based films. The results indicated that the comprehensive properties of cross-linked cassava starch films were superior to that of other modified cassava starch films. The cross-linked modified starch molecules in the film increased the molecular weight of starch and expanded intermolecular interactions, resulting in better tensile strength. Adding cellulose nanofibers also improves the tensile strength and Young's modulus of starch-based films [69]. The surface of cellulose particles contains many hydroxyl groups, which interact with starch molecules to form a dense network structure [70]. Ana and Buddhi [71] added lignocellulose into cassava starch film. Compared with the control group, the tensile stress of the film was 6.6 MPa (37.5% increased). The elongation at break was 44.43%, which was lower than that of the control group (54.92%).

In addition, starch can be acidified and hydrolyzed into starch nanocrystals to prepare enhanced mechanical properties of the material [72]. It has been reported that the natural crystals of starch do not improve the hardness and tensile strength of extruded film, while the plasticized starch obtained by recrystallization of V- and B-starch uses its compounding effect to increase tensile strength but reduce elongation at break [73]. Ren and Fu [74] hydrolyzed waxy corn starch with H6-sodium metaphosphate or glutaric acid to prepare cross-linked modified starch nanocrystals. The results showed that compared with the control group, the cross-linked starch nanocrystals improved the tensile strength and elongation at the break of starch film, but Young's modulus remained unchanged.

## 4. Hydrophobic and Barrier Properties of Starch-Based Films

One of the main functions of food packaging is to maintain food's stability and extend its storage period. In contrast, the barrier performance of packaging materials is a decisive factor affecting the shelf life of food [75]. Starch-based materials can permeate small molecules, such as gas and water vapor [76,77]. Accordingly, the barrier properties of starch-based degradable materials directly determine whether they can be used in food packaging.

### 4.1. Water Vapor Barrier

Spoilage of food is closely related to environmental humidity, and the water vapor barrier is critical for retaining or extending food shelf life [78]. Different foods have different requirements for a water vapor barrier. Dehydration should be avoided for fresh foods, but water vapor should be prevented for bread or cooked food [23]. Since starch is rich in hydroxyl groups and hydrophilic, the starch-based material is susceptible to water, and when it is combined with glycerol, the expansion of the network maintains a large amount of water. This expansion destroys the structural integrity of the matrix, leads to poor barrier performance, and cannot meet the packaging requirements of protective products, especially limiting the application scope of the film [79]. Researchers have added hydrophobic groups through starch modification to strengthen the hydrophobic properties of starch-based packaging materials. Hydrophilic hydroxyl groups in starch are esterified, etherified, crosslinked, and grafted with other substances to reduce the number of hydrophilic hydroxyl groups, thus enhancing the hydrophobic properties of starch [80,81]. Wongphan and Panrong [82] synthesized blended film with polybutylene adipate terephthalate (PBAT) by extrusion blending of protostars (NS), acetylated starch (AS), octenyl succinate starch (OS), and hydroxypropyl starch (HS). The results showed that the hydrophobic starch improved the compatibility and interaction with PBAT and greatly improved the barrier performance (82~89%). Nevertheless, modification of starch alone is insufficient to reinforce material properties.

In contrast, synergistic composite modification can increase the effectiveness of starch modification, thereby increasing the water resistance of starch-based materials. Cheng and Cui [30] researched the effects of natural and distinct binary changes [hydroxypropyl starch oxide, distarch acetyl phosphate, and oxidized acetyl starch] on the hydrophobic properties of intelligent films. The results demonstrated that the film of double-modified cassava starch had better water and water vapor resistance than the film of natural cassava starch.

In addition to directly modifying starch, it can be blended with some hydrophobic materials to improve the water-resistance of starch-based materials. Hydrophobic substances, such as lipids, hydrophobic nanoparticles, biomass materials, and their derivatives, are often added to starch films to strengthen their hydrophobicity. On the one hand lipids in the films can form a double layer by stacking hydrophobic lipids on a preformed starch-based film. On the other hand, lipid substances can also be added to starch film solutions in the form of emulsions to obtain starch-lipid composite films [83]. Bedroom [84] researched the effects of various lipids (oleic acid, palm oil, and margarine) and their concentrations on the water vapor permeability (WVP) of rice starch matrix film. The results showed that the addition of lipids reduced the WVP of rice starch matrix film. Compared with margarine and palm oil, the WVP value of oleic acid-doped films decreased with the oil content increase: Katiany and Adriana [85] mixed carboxymethyl cellulose with corn starch and cassava starch, respectively. The results showed that the WVP of the sample films was 48 and 40%, respectively. The interaction between starch and glycerol hydroxy and carboxymethyl cellulose carboxyl groups is supported by the high content of straight chain starch in the corn starch films and the high hydrophobicity. Mehran and Nahal [86] mixed Zataria multiflora Boiss (ZEO) or Mentha Pulegium (MEO) into a starch film by solution flow diffusion method. ZEO or MEO have enhanced the barrier properties of the starch film. It reduced the WVP by 50% compared to the control group. This may be due to the hydrogen and covalent interaction between the starch network and polyphenols, which lower the availability of hydrophilic groups and the affinity of the starch matrix for water molecules, leading to a decrease in the relationship of the film for water [87]. Some nanoparticles, such as ZnO nanoparticles (ZnO NPs), $TiO_2$ NPs, $SiO_2$ NPs, and nano-clay, have been demonstrated to be potent in enhancing water repellency. Zhang and Wang [88] studied the impact of $SiO_2$ NPs with different particle sizes on the properties of potato starch film. The results showed that the addition of $SiO_2$ NPs improved the water resistance of the film. Ni and Zhang [89] studied the adding of ZnO NPs into starch solution under an ultrasonic and magnetic angle and then flowed to form thin films. The contact Angle of the film increases from 85.73° to 121.45°.

### 4.2. Oxygen Barrier

Due to the growth of aerobic microorganisms and the chemical deterioration of food ingredients, the infiltration of oxygen dramatically affects the quality of packaged food [90]. To improve the barrier properties of starch-based materials to oxygen small molecules, appropriate fillers are usually added in the preparation process to form the composite material composed of starch matrix and stuffing to prepare the starch-based composites with high barrier properties. Compared with starch molecules, the structure of filler is more stable and denser. Small mass-transferring molecules, such as oxygen molecules, cannot directly transfer through, enhancing the barrier properties of composite films to small mass transfer molecules.

The combination of nanoparticles and starch has been proven to be an effective method to improve the oxygen barrier properties of starch-based materials. For example, Wattinee and Phatthtanit [91] used thermoplastic starch (TPS) as raw material to prepare composite starch films by extrusion and blow moulding using nirite and polybutylene diacrylate (PBAT). The results showed that nitrite modified the carbonyl group bond of PBAT, improved its compatibility with the TPS network, compacted its microstructure, and reduced the permeability of oxygen. Adding 5% nitrite to the PBAT/TPS blend system can effectively produce oxygen permeability similar to high-barrier fossil-based plastics. Pramod and Kalyani [92] prepared nanoclay-reinforced starch-polyacrylic acid hybrid nanocomposite films in an aqueous solution by in-situ polymerization technology. The results show that the oxygen resistance of starch-co-polyacrylic acid/clay film was significantly fortified during the loading process. Wang and Zhang [93] starch/polyvinyl alcohol/clay nanocomposite films were prepared by extrusion blow molding. Compared with starch/clay film, the oxygen permeability of starch /PVA/ clay film with 50% PVA content was reduced by approximately 210%. They also proposed a continuous phase change mechanism to explain the improved film properties. It is evident that starch/polyvinyl alcohol/clay nanocomposite films are a promising food packaging material with high barrier properties.

## 5. Starch-Based Active Films

Oxidation and microbial contamination are the leading causes of food spoilage. Starch packaging is similar to traditional packaging based on the problem of how to prolong the shelf life of products. It does not delay food spoilage by itself [94]. Food preservation is predominantly accomplished by adding antioxidants or antimicrobial substances to starch substrates to release active substances [95]. It can improve food quality and prolong food shelf life and has become a research focus [96]. In general, the antioxidant activity of film mainly depends on the potential release of active substances from the film matrix, which is closely related to the source of active substances, extraction conditions, additional amount, and the interaction between active substances and film matrix, and the microstructure of the film [97,98].

### 5.1. Antioxidant Active Starch-Based Films

The antioxidant active film can achieve an antioxidant effect by adding antioxidants to the packaging materials that can delay or prohibit food oxidation. When food is packaged in this packaging material, antioxidants are released into the interior of the packaging to extend the shelf life of the food and, to some extent, maintain the quality of the food [99]. Various antioxidants are used in food packaging and can be classified as naturally derived and synthetic sources. It is generally believed that chemically synthesized antioxidants have potential safety risks. Accordingly, naturally extracted antioxidants, such as essential oils and spices, have become the mainstream of current research [29]. Among them, absorption packaging by adding antioxidants to the packaging, absorption packaging of $O_2$ to prevent food oxidation rancidity, packaging through the diffusion of antioxidants to the food surface or released into the packaging environment can inhibit its oxidation rancidity, so that the food is in a safer state. The fixed type can only keep the parts directly in contact with the package.

Oxidative degradation of foods is one of the significant non-microbial causes of food spoilage. Prevention of oxidation is most important for maintaining nutritional quality in foods, such as fresh produce, processed foods, and fresh meat [100]. The use of active film with antioxidant properties can inhibit the oxidative deterioration of food by adding active antioxidant substances during storage [101]. Various types of antioxidants can be classified as synthetic and natural antioxidants according to their sources.

Synthetic antioxidants, such as butyl-hydroxytoluene and butyl-hydroxyanisole (BHA) starch substrates to prevent lipid oxidation, have been commonly used in food packaging [102]. Nonetheless, the demand for natural antioxidants has lately expanded due to synthetic compounds' potential toxicity and carcinogenicity [103]. Natural antioxidants, such as polyphenols, tocopherols, plant extracts, and essential oils, are preferred to be added to active packaging materials [104]. Kumar and Akhila [105] added 20% pineapple peel extract to polyvinyl alcohol (vinyl alcohol)-corn starch film, and the results showed that the control film had no antioxidant activity, the film containing 20% pineapple peel extracts had a DPPH scavenging activity of approximately 42% in the film. Dariusz and Waldemar [106] prepared oxidized potato films loaded with sodium ascorbate (SA); at 100 mM ascorbate ion concentration, the sample's oxidation resistance activity and anti-free radical activity films were 7 times and 20 times higher than those of the control.

Due to their apparent volatility, volatile antioxidants in packaging materials delay food oxidation more effectively than non-volatile compounds. Essential oil, as a volatile antioxidant, is widely used in starch-based films to enhance their antioxidant properties. Elham and Majid [107] analyzed the antioxidant capacity of corn starch film supplemented with multi-flower corn essential oil and cinnamaldehyde in conventional, nano, and enhanced nanoemulsions. The results showed that the starch film containing nanoemulsion had higher antioxidant activity than the traditional sample film. In starch-based film materials, essential oil vapor diffuses into the internal atmosphere. It directly interacts with food, producing antioxidant protection [108]. These findings can be applied to the food packaging industry, especially meat and meat products susceptible to spoilage. Of course, the antioxidant effect of the starch films with essential oil nanoparticles in different food systems needs to be further research.

### 5.2. *Antibacterial Active Starch-Based Films*

The growth of microorganisms in food can lead to food spoilage, which reduces food's nutritional value and safety [109]. As with traditional petroleum-based packaging, although modified starch-based films have good properties when applied to food packaging, they have weak antibacterial activity [110]. To resolve this problem, the antibacterial properties of the film can be accomplished through a hydrogen bond, electrostatic interaction, and other interactions between starch functional groups and antibacterial substances [111]. Compared to adding antimicrobial agents directly to food, making antimicrobial starch-based film prevents food spoilage by interacting with the active substance inside the package and the food. During this process, the active substance is slowly released around the food, effectively inhibiting the growth of bacteria. Depending on the user's requirements, starch-based packaging materials can be satisfied by adding the appropriate active substance inside. The antibacterial agents added to the packaging film are generally classified natural antibacterial agents, inorganic antibacterial agents, and organic antibacterial agents.

Natural antibacterial materials mainly come from animals and plants, as well as microorganisms and their derivatives [112]. Common materials include essential plant oil, chitosan, antimicrobial peptides, etc. all of which have a bactericidal role: releasing bactericidal substances to change cell permeability and antagonize microorganisms [113,114]. Paola and Daniela [115] studied the fresh-keeping effect of rice starch film containing essential oregano oil on frozen fish. The results showed that the composite film containing oregano leaf essential oil had antibacterial activity compared with the control group. The shelf life of fish fillets packaged by the active film was prolonged. Cristina and Lorena [116] obtained cassava starch-chitosan films by melt bending and compression molding. The results showed that film could reduce the coliform group and total oxygen demand of frozen pork slices and prolong the shelf life of pork slices. Although there are more and more studies on starch-based natural antibacterial films due to their green, safe, broad-spectrum antibacterial properties and good biocompatibility, the problems of poor chemical stability and high extraction cost of raw antibacterial materials should also be considered.

Organic antimicrobial with some application are quaternary ammonium salts, polyphenols, pyridine, etc. They kill bacteria by electrostatic adsorption, have a powerful antibacterial effect, and are inexpensive [117]. However, organic antimicrobial agents have not received much attention because of their high toxicity and the tendency to produce drug-resistant bacteria in excessive use. Inorganic antibacterial agents primarily refer to metal antimicrobial agents. It is favored for its broad antimicrobial spectrum and excellent antimicrobial properties, which have become a leading research direction [118].

Inorganic antimicrobial agents generally refer to nano-metallic materials, such as Ag and Cu, or photocatalytic antimicrobial agents, such as $TiO_2$ and ZnO, which have a solid binding ability with the active enzyme center of bacteria. Metal ions released by inorganic antibacterial materials in specific environments will either compound with nitrogen and oxygen in proteins after entering bacteria, or destroy the spatial conformation of protein molecules, inhibit DNA replication of cells, hinder the normal physiological functions of bacteria, and lead to bacterial death [119]. Hu and Jia [120] prepared composite films by incorporating chitosan nanoparticles in a modified starch matrix. The antibacterial activity of starch-based film was positively correlated with the loading of nanoparticles, and the antibacterial activity against Gram-positive *Staphylococcus aureus* was stronger than Gram-negative *Escherichia coli*. Chen and Li [121] prepared composite films by in-situ reduction using carboxymethyl cellulose (CMC) and starch as reductants and stabilizers. The results showed that ACS film had apparent antibacterial activity against *S. aureus* and *E. coli*. With the increase of $AgNO_3$ solution concentration, the inhibitory effect of ACS film was greatly amplified.

### *5.3. Controlled Release Starch-Based Active Films*

Commonly used antioxidant and antibacterial materials have good antioxidant and antibacterial effects [122]. However, if the release of the active substance is slow, resulting in insufficient concentration of the active substance, the food is prone to spoilage; when the release rate is accelerated, the concentration of the active substance is too high, leading to degradation or interaction with the food components. Therefore, current and future research is focused on regulating the release of active ingredients from films.

Controlled release packaging (CRP) extends product shelf life by controlling the release of active substances in food storage [123]. There are many ways to design CRP, generally including chemical modification, multilayer preparation, and cross-linkers. In recent years, microencapsulating active compounds and utilizing a metal-organic framework (MOF) in an active starch film matrix are the latest techniques for preparing CRP systems [124]. According to the mechanism of action, the controlled release system can be divided into a release system and an absorption system. In the release system, antimicrobial agents to the food surface as active agents to prevent food spoilage and quality loss [125]. Surfactants combine with starch matrix to form composite films, which are induced to release active substances through expansion, disintegration, diffusion, or disintegration [126]. The starch-based film belongs to the bottom-induced release type. Due to its moderate diffusion coefficient in the starch-based film system, the added active agent cannot diffuse in the starch matrix. Since most foods are humid and contain a lot of water, when the starch matrix is placed in a consonant liquid medium, the starch expands into the matrix through the water. In the expansion state, the diffusion coefficient of the active agent increases and then diffuses outwards [127]. In other words, the higher the moisture content of food, the higher the spoilage rate, and the higher the release rate of active substances in starch-based film, indicating that film can prolong the shelf life of products.

Zhang and Zhao [128] prepared active films using corn starch (CS) and zein rutin composite nanoparticles (RNs) as raw materials. The experimental results showed that the initial release rate was fast, and the cumulative release amount reached 19.8~27.1% after 2 h due to the weak binding or adsorption between RNs and CS. Rutin is released from CS film in the sustained release stage. After 12 h, the release of rutin was only 27.1~36.9% of the total rutin due to the migration of rutin from nanoparticles to solution. RNs dispersed in CS film can be controlled and released in aqueous food packaging. Farrag and Ide [129] prepared a starch film containing doughnut-like starch-quercetin particles, using pea and corn starch as raw materials. The in vitro release of quercetin film in aqueous ethanol was researched. The quercetin release of grain starch film reached equilibrium within 1 to 4 days, and that of legume starch film reached equilibrium over 1 week.

## 6. Starch-Based Intelligent Films

In addition to extending the shelf life of food through active substances, a new type of intelligent food packaging can be developed by giving new functions to starch-based packaging materials [130]. Intelligent packaging can monitor the quality of internally packaged food or detect the surrounding environment of food [131]. Natural active substances used in intelligent films usually have antibacterial and antioxidant activities. Consequently, in most cases, intelligent packaging is simultaneously active, but it is rarely studied to evaluate the two functions simultaneously [132].

Compared with other degradable polymers, the most significant advantage of starch-based film is that it is colorless and transparent, and the color change of food packaging film will not be affected by the sample matrix. Starch-based intelligent food packaging mainly combines indicators and provides intuitive, quantitative, or semi-quantitative information about packaged food through visual changes, such as color [133]. It includes explicitly freshness indicator (indicating the remaining shelf life of food by reacting with some characteristic gases generated in the storage process), time-temperature indicator (showing the remaining shelf life of food by time-temperature accumulation effect), etc. [134].

### 6.1. Freshness Indicator

The film-forming ability of starch makes the biopolymer an ideal proppant for preparing intelligent colorimetric films. Recently, interest in developing intelligent pH-sensitive films using starch has increased. pH changes are the primary food freshness and standard conditions. As food rots under the action of microorganisms, the pH value around food changes, so the relationship between food freshness or quality and pH value can be verified [135]. Organic pH indicators are not harmful to the human body or the environment. They come from a wide range of sources [136]. It is currently a popular topic of research in intelligent food packaging, including anthocyanins (ATH), curcumin (CR), and carotenoids, etc. [137]. Compared with other natural pH indicators, anthocyanins have a more comprehensive color range and a more significant color difference. The main methods of pH determination are colorimetry and electrochemical process.

Starch as a film matrix, mixed with indicators, can respond by sensing changes in food. When food rots, ammonia, dimethylamine, triathlon, and other gases will be produced. These gases will change the pH value around the food, and the natural pigment in the packaging film will change the color through its mechanism. For example, bok choy anthocyanins generally increase with pH, ranging from mauve to blue-purple to blue-green. This distinct color change allows the consumer to clearly identify whether the food is at the fresh, medium fresh, or spoiled stage and bring reference to consumers' consumption. Choi and Lee [138] designed a colorimetric pH indicator film based on agar/potato starch/anthocyanins extracted from sweet potatoes. When the film is used as a pork package, the color shifted from red to green with the change of pH value and the deterioration of the sample. Mayra and José [133] studied the pH monitoring system of chitosan/corn starch/purple cabbage extract. They used it as a visual indicator of fish decay.

Electrochemical methods in pH sensors are used to convert chemical information into electrical signals for analytical experiments. The sensor receives chemical information and converts it into usable energy, which it converts into electrical signals [139]. In food packaging, chemical byproducts of spoiled food interact with the electrodes and begin to produce chemical changes. Compared with electrochemical sensors, the sensitivity of pH sensors is more strongly correlated with colorimetric sensors [140]. With colorimetric sensors, many visual perceptions come from the color intensity, sensitivity, or pH range. Increasing the proportion of pH-sensitive substances and decreasing the number of binding substrates can promote this increase in color intensity. The higher the anthocyanin content, the higher the color intensity; the more porous the starch content, the higher the sensitivity; the higher the cellulose binder content, the greater the mechanical strength of the sensor.

### 6.2. Time-Temperature Sensor

Perishable food is sensitive to temperature, and low-temperature storage can effectively prolong its shelf life. High temperatures accelerate the deterioration of food quality and lead to food reaching the end of shelf life in advance. Time and temperature are key factors affecting the quality of most foods. The time–temperature indicator (TTIs) can record and indicate the temperature change of the remaining shelf life of food during its circulation [141]. Through product time and temperature information, the temperature change of the product in each link can be monitored to ensure the quality and safety of food. This provides irreversible, visible color changes associated with temperature changes [142], which are caused by chemical changes [72], microbial changes [143], enzyme changes [144], or physical changes [145]. The range from activation to termination is usually reflected by color changes, corresponding to the shelf life of accompanying foods [146].

Compared to other areas, research on time and temperature indicators began late, and there is little relevant research. For example, Carolina and Pricila [147] added myoglobin extract and nitrite to the thermoplastic sensor film of cassava starch as an alternative to traditional electronic time-temperature sensors. They developed a natural, non-toxic, biodegradable thermochromic protein-based sensor. To investigate the color changes of myoglobin and myoglobin nitrite proteins under temperature. The sample film's visual and instrument color changes in different environments demonstrate its feasibility as a time-temperature sensor for packaging or labels. Nogueira and Fakhouri [148] subjected starch-based edible films containing freeze-dried blackberry particles to sterilization at 127 °C for 15 min. The films underwent a significant change from red to brown color.

Starch films have made great progress in mechanical and barrier properties, and a large amount of research work on this subject shows the promise of starch films as an alternative to petroleum-based polymers as food packaging materials. Starch modification and additives have proved successful in producing films with similar properties to conventional packaging materials. In addition, starch-based films are used as carriers of functional ingredients to prepare active and smart packaging by combining antibacterial, antioxidant, and indicator agents to improve shelf life and quality, while facilitating the observation of food spoilage levels.

## 7. Starch-Based Active and Intelligent Films Application in the Food Industry

In response to consumer demand, there is a need to extend the shelf life of food products in the food industry. In the overall food circulation link, it is essential to maintain the high level of food quality. Certain starch-based biodegradable films have been used in food packaging. Table 2 lists their applications to some highly perishable, semi-perishable, and highly durable foods.

**Table 2.** Highlights for applications of starch-based films on food products.

| Starch | Additives | Product | Finding | References |
|---|---|---|---|---|
| Yam starch | eugenol | pork preservation | with 3% eugenol can extend the shelf-life of pork beyond 50% | [149] |
| Job's tears starch | clove bud essential oil | pork belly | with 0.5% CBEO can reduce Lipid oxidation | [150] |
| Potato starch | carrot anthocyanins | row milk | used as an indicator to monitor freshness/spoilage of milk | [151] |
| Cassava starch | gelatin and casein | guavas | increased the guavas shelf-life by 2 days | [152] |
| Maize starch | grape juice | chicken breast fillets | delayed the lipid oxidation and microbiological growth of chicken breast fillets. | [153] |
| Brazilian pine seed starch | citric pectin and functionalized | grapes and bread | maintained the quality for 30 days of storage | [154] |
| Hydroxypropyl distarch phosphate | ε-polylysine and gelatin | fresh bread | delayed microbial spoilage | [95] |
| Corn starch | carboxymethyl cellulose | food simulant | excellent antimicrobial activity towards *E. coli.* | [155] |
| Potato starch | betacyanin | fish | visual change from pink to yellow color of the package label paralleled the increase in total volatile base nitrogen (TVB-N) | [156] |
| Corn starch | curcumin-loaded Pickering emulsion | fish | the color of films changed from yellow to red | [157] |
| Cassava starch | lycium ruthenicum anthocyanins-loaded nano-complexes | micropterus salmoides | when the fillet of perch deteriorates, the film shows significant color change | [158] |

### 7.1. Active Packaging

The application of starch-based active film in food packaging can effectively inhibit the growth of microorganisms and lipid oxidation in food, thus extending the shelf life of food [159]. This effect is mainly achieved through the antibacterial and antioxidant agent action of the film. When the film is used for food packaging, the active substances in the film can reach the food surface or the upper space of the packaging through diffusion, thus inhibiting the deterioration of food quality. Its preservation mechanism is ultimately the result of the interaction between active substances and food [160]. Currently, there is a growing number of studies on applying starch-based active film in food preservation. Cheng and Wang [149] prepared the starch-based antibacterial film for pork preservation using yam starch as a matrix and eugenol (YDE) (Figure 2A). The results showed that the antibacterial activity of YDE's antibacterial activity against *E. coli* was superior to that of *Listeria monocytogenes* and *S. aureus*. The antibacterial activity of YDE3 film was better, which could prolong the shelf life of pork by more than 50% (Figure 2B). Studies have shown that the hydrophobic active substance of EO can attach to the cell surface of microorganisms and enter through plasma, plasma-binding enzyme, and other targets, resulting in cell wall rupture and leakage of intracellular substances [161]. In active packaging, EOs are embedded in a starch-based film, which allows for the bacteriostatic active compound to be released from the package longer, prolonging the time of food transportation and storage [162]. Thermoplastic starch/montmorillonite films containing EO components, such as thymol and carvacrol, were prepared and placed in PET containers to release EO as water vapor for the preservation of strawberries [163].

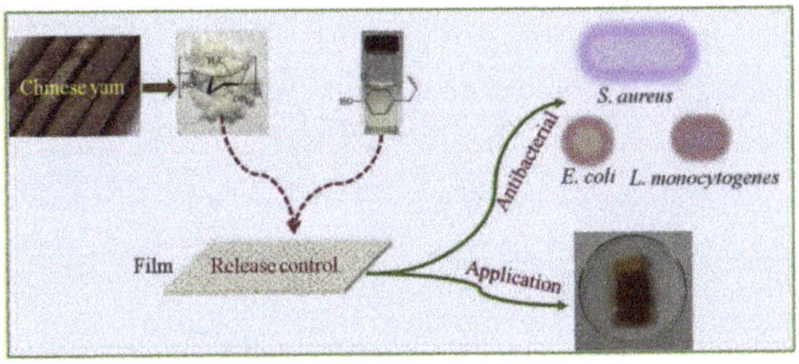


**(A)**

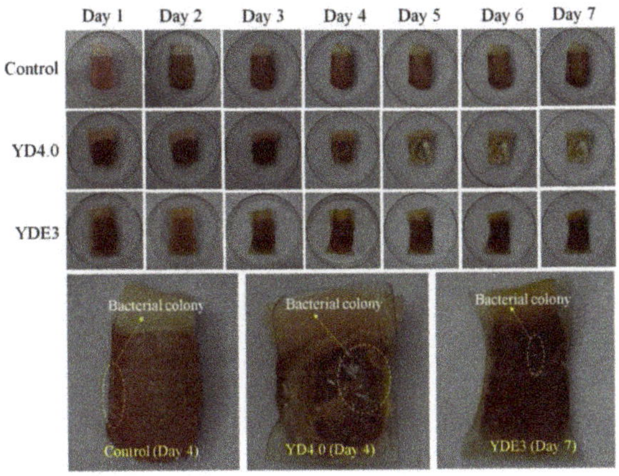


**(B)**

**Figure 2.** Application of active starch-based film (**A**) active packaging film based on yam starch with eugenol (**B**) application of antibacterial films to pork preservation. ref. [149]. 2019 Junfeng Cheng, Hualin Wang, Shaolei Kang, etc.

However, research into the application of starch-based active film in food packaging is still in the preliminary stage. There are still many problems worthy of further development, such as the antibacterial/antioxidant mechanism between starch film and food, the effect on food flavor, and other quality issues [164].

### 7.2. Intelligent Packaging

In addition to applied research on active packaging, scholars have also developed intelligent packaging film materials based on starch. At present, they are primarily focus on indicator intelligent packaging materials. Indicative intelligent packaging combines intelligent functions with standard packaging technology and provides consumers information through external color changes. Chen and Zhang [165] prepared visual pH-sensitive films containing (CR) and (ATH) as packaging indication labels for real-time non-destructive detection of fish freshness (Figure 3A). The results showed that starch film mixed with CR and ATH could provide three different colors: indicators of freshness, medium freshness, and spoilage of packaged fish. Using potato starch (PS), chitosan (CH), and floss lonicera anthocyanins (LCA) as raw materials, pH and $NH_3$ response coloristic film (PS-CH-LCA) were prepared by controlling the pH value of the film-forming solution and applied to real-time monitoring of shrimp freshness (Figure 3B). The results showed that PS-CH-LCA (pH = 2.5) film was sensitive to color changes and highly correlated with spoilage indicators, indicating that the film could well reflect the freshness, sub-freshness, and spoilage degree of shrimp [166].

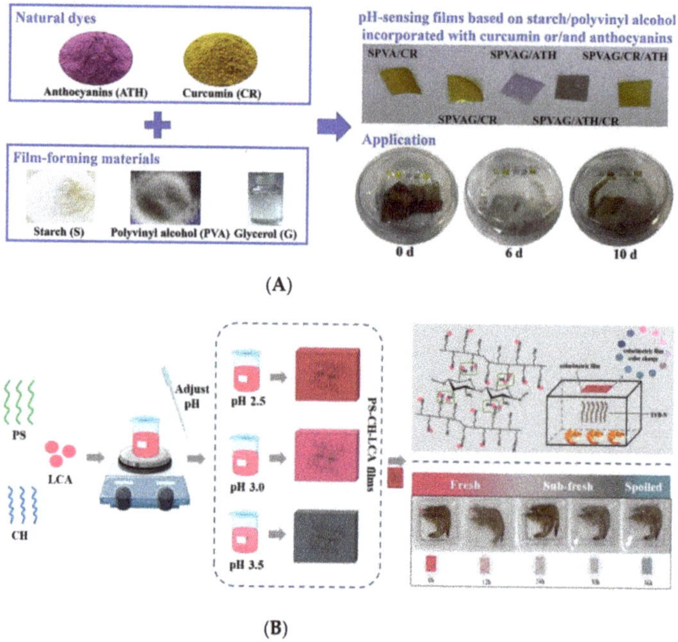


**Figure 3.** Application of Intelligent starch-based film (**A**) novel pH-sensitive films containing curcumin and anthocyanins to monitor fish freshness. (**B**) preparation and application of chitosan/starch based colorimetric film for sub-freshness monitoring. Ref. [164]. 2022 Bin Li, Yiwen Bao, etc. Ref. [165]. 2020 Hui-zhi Chen, Min Zhang, etc.

Based on the existing studies, intelligent packaging materials are mainly prepared using the color principle of CR and ATH, which has a good indicator effect on food with a significant change in pH value during food spoilage. Still, the precise relationship between pH value change and quality change remains to be further studied. In addition, the film material with antibacterial function has been prepared based on intelligent packaging, so developing intelligent, active packaging material is a research direction in the future.

## 8. Conclusions and Future Perspective

Starch-based biodegradable materials may play an important role in the future development of sustainable food packaging materials, reducing the energy and environmental stress of petroleum-based packaging materials. The current limitation is mainly due to the poor mechanical properties and barrier properties of starch-based packaging materials due to the properties of starch itself. Researchers can solve this problem by physically or chemically modifying starch or mixing it with other biopolymers and functional additives. After meeting the basic conditions of food packaging, starch-based active packaging can be prepared by adding antioxidant or anti-bacterial substances, which can extend the shelf life of food and reduce food waste. Starch time and temperature indicator films can detect food freshness in real-time by the color reaction. Starch-based films are widely used in food packaging and have a good protective effect on fresh food. In the future, it will be possible to prepare active or smart packaging using starch-based materials.

Although some of these methods have improved the properties of starch-based films, more research is still needed to create starch-based films with similar mechanical and barrier properties to traditional plastic packaging. Starch-based biodegradable materials can be affected by starch type, preparation techniques, storage conditions and other factors during preparation, resulting in a lack of homogeneity and stability of the product. Currently, most research on food packaging applications is completed at the laboratory level. Industrial manufacturing, safety regulations, environmental issues, and consumer acceptance have also limited the commercialization of starch-based films. In addition, the combination of starch and other materials needs further study to meet practical needs. Many additives have great potential in developing antibacterial and antioxidant films. There is still quite a lot of interesting work to be done in terms of developing antimicrobial agents that can be used in starch matrices. There is a lack of appropriate criteria for assessing and quantifying antimicrobial and antioxidant activity. For example, in "food mimics" solutions, the mimics corresponding to different foods should be expanded rather than limited to water. The field of starch-based smart films has great potential, although so far its functionality has been relatively limited. At present, pH indicators are mostly used in intelligent packaging. While changing the environment around the food in package, active substances will also change the pH change of the system, which may lead to the irreversibility or timeliness of indicators in packaging film and cause consumers to make a wrong judgment on product quality. Further studies are needed to investigate the combination of functional food components with multireactive starch membranes. By mixing with heat-sensitive, moisture-sensitive, gas-sensitive, and other multifunctional packaging materials to prepare new intelligent packaging, the film can have a variety of different functions, can prepare the corresponding stimulus response packaging according to the food needs, and have a slow/controlled release effect on the active substances in the film.

In summary, although starch-based active and intelligent packaging still has shortcomings in terms of material selection, preparation, and its role, there is no doubt that starch has become the most likely substrate to replace petroleum-based traditional packaging due to its structural properties and other advantages, and it has the possibility of substrates leading the development of intelligent food packaging in the future.

**Author Contributions:** D.L.: Collected the literature sources and wrote the manuscript. P.Z.: Conceptualized the idea and critically reviewed the manuscript. J.C. and Y.Y.: Formal analysis and supervision. Z.W.: Review and editing and funding acquisition. All authors have read and agreed to the published version of the manuscript.

**Funding:** This research was funded by Tianjin Key Research and Development Program and Tianjin Applied Basic and Frontier Technology Research Program, grant number 19YFLHSN00080 and 19JCTPJC54600.

**Data Availability Statement:** Data is contained within the article.

**Conflicts of Interest:** The authors declare no conflict of interest.

## Abbreviations

| | |
|---|---|
| PB | Pinto Bean Starch |
| PV | Polyvinyl Alcohol |
| NF | native starch |
| PSE | pecan nutshell extract |
| HSE | hazelnut skin extract |
| WVP | water vapor permeability |
| CR | curcumin |
| ATH | anthocyanin |

## References

1. Koketso, N.L.; Uchenna, U.A.; Nifise, O.E.; Rozli, Z.; Nongwe, B.I. An Overview of Plastic Waste Generation and Management in Food Packaging Industries. *Recycling* **2021**, *6*, 12.
2. Kumar, G.M.; Irshad, A.; Raghunath, B.; Rajarajan, G. Waste management in food packaging industry. In *Integrated Waste Management in India*; Springer: Heidelberg/Berlin, Germany, 2016; pp. 265–277.

3. Ellen, P.; Jacob, H.; Karl, H.; Dan, N.T. Narrating plastics governance: Policy narratives in the European plastics strategy. *Environ. Politics* **2022**, *31*, 365–385.
4. Biron, M. *Thermoplastics and Thermoplastic Composites*; William Andrew: Norwich, NY, USA, 2018.
5. Williams, K.S. Plastic Packaging: Not a Throw-away Resource. In *Issues in Environmental Science and Technology: Waste as a Resource*; The Royal Society for Chemistry: Cambridge, UK, 2013; pp. 83–109.
6. Robertson, G.L. Definitions, Functions, Attributes and Environments of Food Packaging. In *Reference Module in Food Science*; Elsevier: Amsterdam, The Netherlands, 2018.
7. Shen, M.; Song, B.; Zeng, G.; Zhang, Y.; Huang, W.; Wen, X.; Tang, W. Are biodegradable plastics a promising solution to solve the global plastic pollution? *Environ. Pollut.* **2020**, *263 Pt A*, 114469. [CrossRef]
8. Lokesh, K.; Dakuri, R.; Konala, A.; Gaikwad, K.K. Edible films and coatings for food packaging applications: A review. *Environ. Chem. Lett.* **2021**, *52*, 533–552.
9. Moshood, T.D.; Gusman, N.; Fatimah, M.; Fazeeda, M.; Hanafiah, A.M.; Airin, A. Sustainability of biodegradable plastics: New problem or solution to solve the global plastic pollution? *Curr. Res. Green Sustain. Chem.* **2022**, *5*, 100273. [CrossRef]
10. Fan, Z. Polysaccharide based films and coatings for food packaging: Effect of added polyphenols. *Food Chem.* **2021**, *359*, 129871.
11. Anna, K.; Katarzyna, K.; Katarzyna, P.; Mariola, S.; Ewa, S.; Paulina, H. Polysaccharides as Edible Films and Coatings: Characteristics and Influence on Fruit and Vegetable Quality—A Review. *Agronomy* **2021**, *11*, 813.
12. Sergio, J.C.-E.; Jiménez-Fernández, M.; Lugo-Cervantes, E. Protein-Based Films: Advances in the Development of Biomaterials Applicable to Food Packaging. *Food Eng. Rev.* **2019**, *11*, 78–92.
13. Yunpeng, X.; Ying, W.; Tao, Z.; Guangqing, M.; Shujuan, J.; Xuemei, Z.; Yanfeng, T.; Fang, Q. Evaluation of the properties of whey protein films with modifications. *J. Food Sci.* **2021**, *86*, 923–931.
14. Ochoa, T.A.; Almendárez, B.E.G.; Reyes, A.A.; Dulce, M.; Rivera, P.; Gustavo, F.; Gutiérrez, L.; Olga Martín, B.; Carlos, R.-G. Design and Characterization of Corn Starch Edible Films Including Beeswax and Natural Antimicrobials. *Food Bioprocess Technol.* **2017**, *10*, 103–114. [CrossRef]
15. Usman, A.; Usman, K.M.; Yaqoob, M.; Maksim, R.; Mars, K.; Elena, B.; Ali, S.M.; Min, C.I.; Muthu, T. Potentials of polysaccharides, lipids and proteins in biodegradable food packaging applications. *Int. J. Biol. Macromol.* **2021**, *183*, 2184–2198.
16. Shilpi, A. Major factors affecting the characteristics of starch based biopolymer films. *Eur. Polym. J.* **2021**, *160*, 110788.
17. Guo, B.; Wang, Y.; Pang, M.; Wu, J.; Hu, X.; Huang, Z.; Wang, H.; Xu, S.; Luo, S.; Liu, C. Annealing treatment of amylose and amylopectin extracted from rice starch. *Int. J. Biol. Macromol.* **2020**, *164*, 3496–3500. [CrossRef]
18. Ettelaie, R.; Holmes, M.; Chen, J.; Farshchi, A. Steric stabilising properties of hydrophobically modified starch: Amylose vs. amylopectin. *Food Hydrocoll.* **2016**, *58*, 364–377. [CrossRef]
19. Punia, B.S.; Omodunbi, A.A.; Arashdeep, S.; Vandana, C.; Scott, W.W. Enzymatic modification of starch: A green approach for starch applications. *Carbohydr. Polym.* **2022**, *287*, 119265. [CrossRef]
20. Khan, B.; Niazi, M.B.K.; Samin, G.; Jahan, Z. Thermoplastic Starch: A Possible Biodegradable Food Packaging Material—A Review. *J. Food Process Eng.* **2017**, *40*, e12447. [CrossRef]
21. Liu, P.; Li, Y.; Shang, X.; Xie, F. Starch–zinc complex and its reinforcement effect on starch-based materials. *Carbohydr. Polym.* **2018**, *206*, 528–538. [CrossRef]
22. Congli, C.; Na, J.; Yanfei, W.; Liu, X.; Qingjie, S. Bioactive and intelligent starch-based films: A review. *Trends Food Sci. Technol.* **2021**, *116*, 854–869.
23. Weerapoprasit, C.; Prachayawarakorn, J. Effects of Polymethacrylamide-Grafted Branch on Mechanical Performances, Hydrophilicity, and Biodegradability of Thermoplastic Starch Film. *Starch-Stärke* **2019**, *71*, 11–12. [CrossRef]
24. Kwaśniewska, A.; Chocyk, D.; Gładyszewski, G.; Borc, J.; Świetlicki, M.; Gładyszewska, B. The Influence of Kaolin Clay on the Mechanical Properties and Structure of Thermoplastic Starch Films. *Polymers* **2020**, *12*, 73. [CrossRef]
25. Alizadeh, S.M.; Maryam, A.; Milad, T.; Keyhan, M.; Julian, M.D. Recent Advances in the Development of Smart and Active Biodegradable Packaging Materials. *Nanomaterials* **2021**, *11*, 1331.
26. Vilas, C.; Mauricio-Iglesias, M.; García, M.R. Model-based design of smart active packaging systems with antimicrobial activity. *Food Packag. Shelf Life* **2020**, *24*, 100446. [CrossRef]
27. Kanatt, S.R. Development of active/intelligent food packaging film containing Amaranthus leaf extract for shelf life extension of chicken/fish during chilled storage. *Food Packag. Shelf Life* **2020**, *24*, 100506.
28. Tarsila, R.A.; Patrícia, C.B.; Allan, R.F.e.M.; Nilda, d.F.F.S. Natural bioactives in perspective: The future of active packaging based on essential oils and plant extracts themselves and those complexed by cyclodextrins. *Food Res. Int.* **2022**, *156*, 111160.
29. Menzel, C.; González-Martínez, C.; Vilaplana, F.; Diretto, G.; Chiralt, A. Incorporation of natural antioxidants from rice straw into renewable starch films. *Int. J. Biol. Macromol.* **2020**, *146*, 976–986. [CrossRef]
30. Meng, C.; Yingjun, C.; Xiaoran, Y.; Rongfei, Z.; Juan, W.; Xiangyou, W. Effect of dual-modified cassava starches on intelligent packaging films containing red cabbage extracts. *Food Hydrocoll.* **2022**, *124*, 107225.
31. Luman, Z.; Liming, L.; Jiahao, Y.; Ping, S. Novel trends and applications of natural pH-responsive indicator film in food packaging for improved quality monitoring. *Food Control* **2022**, *134*, 108769.
32. Helen, O.; KeChrist, O.; Golden, M.; Nwabunwanne, N. Current Research and Applications of Starch-Based Biodegradable Films for Food Packaging. *Polymers* **2022**, *14*, 1126.

33. Liu, W.; Wang, Z.; Liu, J.; Dai, B.; Hu, S.; Hong, R.; Xie, H.; Li, Z.; Chen, Y.; Zeng, G. Preparation, reinforcement and properties of thermoplastic starch film by film blowing. *Food Hydrocoll.* **2020**, *108*, 106006. [CrossRef]
34. Larissa, d.V.S.; la Fuente, A.C.I.; Chieregato, M.B.; Cecília, T.C. Starch-based biodegradable plastics: Methods of production, challenges and future perspectives. *Curr. Opin. Food Sci.* **2020**, *38*, 122–130.
35. Chuanyan, G.; Hongge, G. Progress in the Degradability of Biodegradable Film Materials for Packaging. *Membranes* **2022**, *12*, 500.
36. Hao, C.; Long, C.; Julian, M.D.; Tianyi, Y.; Zipei, Z.; Fei, R.; Ming, M.; Yaoqi, T.; Zhengyu, J. Starch-based biodegradable packaging materials: A review of their preparation, characterization and diverse applications in the food industry. *Trends Food Sci. Technol.* **2021**, *114*, 70–82.
37. Jiménez, A.; Fabra, M.J.; Talens, P.; Chiralt, A. Edible and biodegradable starch films: A review. *Food Bioprocess Technol.* **2012**, *5*, 2058–2076. [CrossRef]
38. da Silva, L.R.; de Carvalho, C.W.P.; Velasco, J.I.; Fakhouri, F.M. Extraction and characterization of starches from pigmented rice. *Int. J. Biol. Macromol.* **2020**, *156*, 485–493. [CrossRef]
39. Oyeyinka, S.A.; Adeloye, A.A.; Olaomo, O.O.; Kayitesi, E. Effect of fermentation time on physicochemical properties of starch extracted from cassava root. *Food Biosci.* **2020**, *33*, 100485. [CrossRef]
40. Madsar, H.; Ahmad, S.W.; Sajjad, A.; Qutab, H.G.; Muhammad, D.; Muhammad, I. Enzymatic extraction of potato starch: A parametric optimization study using response surface methodology. *Pol. J. Chem. Technol.* **2020**, *22*, 48–54.
41. Hernández-Carmona, F.; Morales-Matos, Y.; Lambis-Miranda, H.; Pasqualino, J. Starch extraction potential from plantain peel wastes. *J. Environ. Chem. Eng.* **2017**, *5*, 4980–4985. [CrossRef]
42. Lu, Z.; Donner, E.; Liu, Q. The Effect of Various Extracting Agents on the Physicochemical and Nutritional Properties of Pea Starch. *Starch-Stärke* **2019**, *71*, 11–12. [CrossRef]
43. Kringel, D.H.; Dias, A.R.G.; Zavareze, E.d.; Gandra, E.A. Fruit Wastes as Promising Sources of Starch: Extraction, Properties, and Applications. *Starch-Stärke* **2020**, *72*, 3–4. [CrossRef]
44. Kim, S.; Kang, J.; Song, K.B. Development of a Sword Bean (Canavalia gladiata) Starch Film Containing Goji Berry Extract. *Food Bioprocess Technol.* **2020**, *13*, 911–921. [CrossRef]
45. Zuzanna, Ż.; Alicja, K. The Influence of Starch Origin on the Properties of Starch Films: Packaging Performance. *Materials* **2021**, *14*, 1146.
46. Abioye, A.A.; Obuekwe, C.C. Investigation of the biodegradation of low-density polyethylene-starch Bi-polymer blends. *Results Eng.* **2020**, *5*, 100090. [CrossRef]
47. Dammak, M.; Fourati, Y.; Tarrés, Q.; Delgado-Aguilar, M.; Mutjé, P.; Boufi, S. Blends of PBAT with plasticized starch for packaging applications: Mechanical properties, rheological behaviour and biodegradability. *Ind. Crops Prod.* **2020**, *144*, 112061. [CrossRef]
48. Castaño, J.; Rodríguez-Llamazares, S.; Sepúlveda, E.; Giraldo, D.; Bouza, R.; Pozo, C. Morphological and structural changes of starch during processing by melt blending. *Starch-Stärke* **2017**, *69*, 9–10. [CrossRef]
49. Huan, H.; Ang, X.; Dianfeng, Z.; Weiyi, Z.; Shaoxian, P.; Xipo, Z. High-Toughness Poly(Lactic Acid)/Starch Blends Prepared through Reactive Blending Plasticization and Compatibilization. *Molecules* **2020**, *25*, 5951.
50. Hirpara, N.J.; Dabhi, M.N. Development of Potato Starch Based Biodegradable Packaging Film. *J. Food Process. Technol.* **2021**, *12*, 529–541.
51. Gülay, B.; Faik, D. Investigation and preparation of biodegradable starch-based nanofilms for potential use of curcumin and garlic in food packaging applications. *J. Biomater. Sci. Polym. Ed.* **2020**, *31*, 1127–1143.
52. Moustafa, H.; Youssef, A.M.; Darwish, N.A.; Abou-Kandil, A.I. Eco-friendly polymer composites for green packaging: Future vision and challenges. *Compos. Part B* **2019**, *172*, 16–25. [CrossRef]
53. Kuz, P.; Ateş, M. Starch-Based Bioplastic Materials for Packaging Industry. *J. Sustain. Constr. Mater. Technol.* **2020**, *5*, 399–406.
54. Lauer, M.K.; Smith, R.C. Recent advances in starch-based films toward food packaging applications: Physicochemical, mechanical, and functional properties. *Compr. Rev. Food Sci. Food Saf.* **2020**, *19*, 3031–3083. [CrossRef]
55. Vu, H.P.N.; Lumdubwong, N. Starch behaviors and mechanical properties of starch blend films with different plasticizers. *Carbohydr. Polym.* **2016**, *154*, 112–120.
56. Basiak, E.; Lenart, A.; Debeaufort, F. Effect of starch type on the physico-chemical properties of edible films. *Int. J. Biol. Macromol.* **2017**, *98*, 348–356. [CrossRef] [PubMed]
57. Ali, K. Evaluation of Physical, Mechanical and Antibacterial Properties of Pinto Bean Starch-Polyvinyl Alcohol Biodegradable Films Reinforced with Cinnamon Essential Oil. *Polymers* **2021**, *13*, 2778.
58. Shaikh, M.; Haider, S.; Ali, T.M.; Hasnain, A. Physical, thermal, mechanical and barrier properties of pearl millet starch films as affected by levels of acetylation and hydroxypropylation. *Int. J. Biol. Macromol.* **2018**, *124*, 209–219. [CrossRef] [PubMed]
59. Namory, M.; Koffi, K.L.; Tohoué, T.M.; Samuel, O. Effect of metakaolin content on mechanical and water barrier properties of cassava starch films. *S. Afr. J. Chem. Eng.* **2022**, *40*, s186–s194.
60. Pornchai, R.; Sarinthip, T.; Auras, A.R.; Nareekan, C.; Kittisak, J.; Pensak, J.; Yuthana, P.; Phisit, S.; Noppol, L.; Thanongsak, C.; et al. Morphology, Mechanical, and Water Barrier Properties of Carboxymethyl Rice Starch Films: Sodium Hydroxide Effect. *Molecules* **2022**, *27*, 331.
61. Leon-Bejarano, M.; Durmus, Y.; Ovando-Martínez, M.; Simsek, S. Physical, Barrier, Mechanical, and Biodegradability Properties of Modified Starch Films with Nut By-Products Extracts. *Foods* **2020**, *9*, 226. [CrossRef]
62. Cano, A.; Jiménez, A.; Cháfer, M.; Gónzalez, C.; Chiralt, A. Effect of amylose:amylopectin ratio and rice bran addition on starch films properties. *Carbohydr. Polym.* **2014**, *111*, 543–555. [CrossRef]

63. Yuyue, Z.; Lingyu, T.; Andreas, B.; Li, D.; Klaus, H.; Jianzhou, Q.; Anzhou, X.; Dongwei, G.; Henrik, H.K.; Xingxun, L. High-amylose starch: Structure, functionality and applications. *Crit. Rev. Food Sci. Nutr.* **2022**, 1–23. [CrossRef]
64. Vu, H.P.N.; Lumdubwong, N. Fabrication of starch blend films with different matrices and their mechanical properties. *Polym. Test.* **2020**, *90*, 106694.
65. Pająk, P.; Przetaczek-Rożnowska, I.; Juszczak, L. Development and physicochemical, thermal and mechanical properties of edible films based on pumpkin, lentil and quinoa starches. *Int. J. Biol. Macromol.* **2019**, *138*, 441–449. [CrossRef] [PubMed]
66. Totosaus, A.; Godoy, I.A.; Ariza, O.T.J. Structural and mechanical properties of edible films from composite mixtures of starch, dextrin and different types of chemically modified starch. *Int. J. Polym. Anal. Charact.* **2020**, *25*, 517–528. [CrossRef]
67. Peng, Y.; Chunhao, C.; Hongpeng, M.; Huijuan, G.; Bin, G.; Panxin, L. Surface cross-linked thermoplastic starch with different UV wavelengths: Mechanical, wettability, hygroscopic and degradation properties. *RSC Adv.* **2020**, *10*, 44815–44823.
68. Dai, L.; Zhang, J.; Cheng, F. Effects of starches from different botanical sources and modification methods on physicochemical properties of starch-based edible films. *Int. J. Biol. Macromol.* **2019**, *132*, 897–905. [CrossRef] [PubMed]
69. Pelissari, F.M.; Andrade-Mahecha, M.M.; Sobral, P.J.d.A.; Menegalli, F.C. Nanocomposites based on banana starch reinforced with cellulose nanofibers isolated from banana peels. *J. Colloid Interface Sci.* **2017**, *505*, 154–167. [CrossRef] [PubMed]
70. Jiang, S.; Liu, C.; Wang, X.; Xiong, L.; Sun, Q. Physicochemical properties of starch nanocomposite films enhanced by self-assembled potato starch nanoparticles. *LWT-Food Sci. Technol.* **2016**, *69*, 251–257. [CrossRef]
71. Travalini, A.P.; Lamsal, B.; Magalhães, W.L.E.; Demiate, I.M. Cassava starch films reinforced with lignocellulose nanofibers from cassava bagasse. *Int. J. Biol. Macromol.* **2019**, *139*, 1151–1161. [CrossRef] [PubMed]
72. Antonia, A.; Rolf, I.; Verena, R.; Werner, R.; Dietrich, H.; Judith, K. Implementation of Time Temperature Indicators to Improve Temperature Monitoring and Support Dynamic Shelf Life in Meat Supply Chains. *J. Packag. Technol. Res.* **2020**, *4*, 23–32.
73. van Soest, J.J.G.; Hulleman, S.H.D.; de Wit, D.; Vliegenthart, J.F.G. Changes in the mechanical properties of thermoplastic potato starch in relation with changes in Btype crystallinity. *Carbohydr. Polym.* **1996**, *29*, 225–232. [CrossRef]
74. Ren, L.; Fu, Y.; Chang, Y.; Jiang, M.; Tong, J.; Zhou, J. Performance improvement of starch films reinforced with starch nanocrystals (SNCs) modified by cross-linking. *Starch-Stärke* **2017**, *69*. [CrossRef]
75. Thakur, R.; Pristijono, P.; Golding, J.B.; Stathopoulos, C.E.; Scarlett, C.J.; Bowyer, M.; Singh, S.P.; Vuong, Q.V. Amylose-lipid complex as a measure of variations in physical, mechanical and barrier attributes of rice starch- ι -carrageenan biodegradable edible film. *Food Packag. Shelf Life* **2017**, *14*, 108–115. [CrossRef]
76. Ni, S.; Zhang, H.; Dai, H.; Xiao, H. Starch-Based Flexible Coating for Food Packaging Paper with Exceptional Hydrophobicity and Antimicrobial Activity. *Polymers* **2018**, *10*, 1260. [CrossRef] [PubMed]
77. Wang, X.; Huang, L.; Zhang, C.; Deng, Y.; Xie, P.; Liu, L.; Cheng, J. Research advances in chemical modifications of starch for hydrophobicity and its applications: A review. *Carbohydr. Polym.* **2020**, *240*, 116292. [CrossRef] [PubMed]
78. Siracusa, V.; Rocculi, P.; Romani, S.; Rosa, M.D. Biodegradable polymers for food packaging: A review. *Trends Food Sci. Technol.* **2008**, *19*, 634–643. [CrossRef]
79. Perdomo, J.; Cova, A.; Sandoval, A.J.; García, L.; Laredo, E.; Müller, A.J. Glass transition temperatures and water sorption isotherms of cassava starch. *Carbohydr. Polym.* **2009**, *76*, 305–313. [CrossRef]
80. Chenglong, L.; Bin, Y.; Haiteng, T.; Pengfei, L.; Haibo, Z.; Congping, T.; Bo, C. Effects of soy protein isolate on mechanical and hydrophobic properties of oxidized corn starch film. *LWT* **2021**, *147*, 111529.
81. Frida, I.; Krister, H.; Romain, B. Surface Treatment by Hydrophobic Particles: Influence of Starch and Ionic Strength. *ACS Sustain. Chem. Eng.* **2017**, *5*, 6107–6115.
82. Phanwipa, W.; Theeraphorn, P.; Nathdanai, H. Effect of different modified starches on physical, morphological, thermomechanical, barrier and biodegradation properties of cassava starch and polybutylene adipate terephthalate blend film. *Food Packag. Shelf Life* **2022**, *32*, 100844.
83. Milani, M.J.; Nemati, A. Lipid-Based Edible Films and Coatings: A Review of Recent Advances and Applications. *J. Packag. Technol. Res.* **2022**, *6*, 11–22. [CrossRef]
84. Bourtoom, T. Improvement of Water Barrier Property of Rice Starch-chitosan Composite Film Incorporated with Lipids. *Food Sci. Technol. Int.* **2009**, *15*, 149–158. [CrossRef]
85. Tavares, K.M.; de Campos, A.; Mitsuyuki, M.C.; Luchesi, B.R.; Marconcini, J.M. Corn and cassava starch with carboxymethyl cellulose films and its mechanical and hydrophobic properties. *Carbohydr. Polym.* **2019**, *223*, 115055. [CrossRef] [PubMed]
86. Ghasemlou, M.; Aliheidari, N.; Fahmi, R.; Shojaee-Aliabadi, S.; Keshavarz, B.; Cran, M.J.; Khaksar, R. Physical, mechanical and barrier properties of corn starch films incorporated with plant essential oils. *Carbohydr. Polym.* **2013**, *98*, 1117–1126. [CrossRef] [PubMed]
87. Shen, X.L.; Wu, J.M.; Chen, Y.; Zhao, G. Antimicrobial and physical properties of sweet potato starch films incorporated with potassium sorbate or chitosan. *Food Hydrocoll.* **2009**, *24*, 285.0–290.0. [CrossRef]
88. Zhang, R.; Wang, X.; Cheng, M. Preparation and Characterization of Potato Starch Film with Various Size of Nano-SiO2. *Polymers* **2018**, *10*, 1172. [CrossRef] [PubMed]
89. Ni, S.; Zhang, H.; Godwin, P.M.; Dai, H.; Xiao, H. ZnO nanoparticles enhanced hydrophobicity for starch film and paper. *Mater. Lett.* **2018**, *230*, 207–210. [CrossRef]
90. Onwulata, C.I. *Food Packaging Principles and Practice*, 3rd ed.; Robertson, G.L., Ed.; CRC Press Taylor & Francis Group: Boca Rotan, FL, USA; p. 2055.

91. Wattinee, K.; Phanwipa, W.; Phatthranit, K.; Nathdanai, H. Thermoplastic starch blown films functionalized by plasticized nitrite blended with PBAT for superior oxygen barrier and active biodegradable meat packaging. *Food Chem.* **2022**, *374*, 131709.
92. Sethy, P.K.; Prusty, K.; Mohapatra, P.; Swain, S.K. Nanoclay decorated polyacrylic acid/starch hybrid nanocomposite thin films as packaging materials. *Polym. Compos.* **2019**, *40*, 229–239. [CrossRef]
93. Wang, W.; Zhang, H.; Jia, R.; Dai, Y.; Dong, H.; Hou, H.; Guo, Q. High performance extrusion blown starch/polyvinyl alcohol/clay nanocomposite films. *Food Hydrocoll.* **2018**, *79*, 534–543. [CrossRef]
94. Salman, K.; Saqib, A.; Mohsin, A.T.; Mehwish, I.H.; Marium, S.; Haris, K.; QurratulAin, A.; Shahid, Y. Effect of Silver Nanoparticles Prepared from Saraca asoca Leaf Extract on Morphological, Functional, Mechanical, and Antibacterial Properties of Rice Starch Films. *Starch-Stärke* **2022**, *74*, 2100228.
95. Yue, C.; Shan, G.; Wen-tao, W.; Han-xue, H.; Loong-Tak, L. Low temperature extrusion blown ε-polylysine hydrochloride-loaded starch/gelatin edible antimicrobial films. *Carbohydr. Polym.* **2022**, *278*, 118990.
96. Nattinee, B.; Jungwook, C.; Seonghyuk, K. Applications of Nanomaterials in Food Packaging. *J. Nanosci. Nanotechnol.* **2015**, *15*, 6357–6372.
97. Zheng, K.; Li, W.; Fu, B.; Fu, M.; Ren, Q.; Yang, F.; Qin, C. Physical, antibacterial and antioxidant properties of chitosan films containing hardleaf oatchestnut starch and Litsea cubeba oil. *Int. J. Biol. Macromol.* **2018**, *118*, 707–715. [CrossRef]
98. Rungsima, C.; Siraprapa, P.; Watthana, B.; Nattaporn, K.; Kunat, K.; Rungsinee, S.; Prakit, S.; Udomlak, S.; Nathdanai, H. Antioxidant and antibacterial activities of cassava starch and whey protein blend films containing rambutan peel extract and cinnamon oil for active packaging. *LWT* **2020**, *130*, 109573.
99. Pardeep, K.; Rohit, T.; Vidhi, G.; Aakash, U.; Anil, K.; Kirtiraj, G.K. Pineapple peel extract incorporated poly(vinyl alcohol)-corn starch film for active food packaging: Preparation, characterization and antioxidant activity. *Int. J. Biol. Macromol.* **2021**, *187*, 223–231.
100. Gaikwad, K.K.; Suman, S.; Suk, L.Y. A new pyrogallol coated oxygen scavenging film and their effect on oxidative stability of soybean oil under different storage conditions. *Food Sci. Biotechnol.* **2017**, *26*, 1535–1543. [CrossRef]
101. Nisa, I.U.; Ashwar, B.A.; Shah, A.; Gani, A.; Gani, A.; Masoodi, F.A. Development of potato starch based active packaging films loaded with antioxidants and its effect on shelf life of beef. *J. Food Sci. Technol.* **2015**, *52*, 7245–7253. [CrossRef]
102. Barbosa-Pereira, L.; Aurrekoetxea, G.P.; Angulo, I.; Paseiro-Losada, P.; Cruz, J.M. Development of new active packaging films coated with natural phenolic compounds to improve the oxidative stability of beef. *Meat Sci.* **2014**, *97*, 249–254. [CrossRef]
103. Cláudia, S.A.A.; Francielly, C.A.; Maria, d.S.F.e.S.L.; Aparecida, C.F.L.; Emília, d.S.G.M.; José, P.C. Effect of natural and synthetic antioxidants on oxidation and storage stability of mechanically separated tilapia meat. *LWT* **2022**, *154*, 112679.
104. Tampau, A.; González-Martínez, C.; Chiralt, A. Biodegradability and disintegration of multilayer starch films with electrospun PCL fibres encapsulating carvacrol. *Polym. Degrad. Stab.* **2020**, *173*, 109100. [CrossRef]
105. Kumar, J.; Akhila, K.; Gaikwad, K.K. Recent Developments in Intelligent Packaging Systems for Food Processing Industry: A Review. *J. Food Process. Technol.* **2021**, *12*, 895.
106. Kowalczyk, D.; Kazimierczak, W.; Zięba, E.; Mężyńska, M.; Basiura-Cembala, M.; Lisiecki, S.; Karaś, M.; Baraniak, B. Ascorbic acid- and sodium ascorbate-loaded oxidized potato starch films: Comparative evaluation of physicochemical and antioxidant properties. *Carbohydr. Polym.* **2018**, *181*, 317–326. [CrossRef] [PubMed]
107. Amiri, E.; Aminzare, M.; Azar, H.H.; Mehrasbi, M.R. Combined antioxidant and sensory effects of corn starch films with nanoemulsion of Zataria multiflora essential oil fortified with cinnamaldehyde on fresh ground beef patties. *Meat Sci.* **2019**, *153*, 66–74. [CrossRef] [PubMed]
108. Ângelo, L.; Ana, R.; Fernanda, D. Pullulan Films Containing Rockrose Essential Oil for Potential Food Packaging Applications. *Antibiotics* **2020**, *9*, 681.
109. William, O.; Zhong, Z.; Yang, B.; Reza, T. Application of starch-based coatings incorporated with antimicrobial agents for preservation of fruits and vegetables: A review. *Prog. Org. Coat.* **2022**, *166*, 106800.
110. Muller, J.; Quesada, A.C.; González-Martínez, C.; Chiralt, A. Antimicrobial properties and release of cinnamaldehyde in bilayer films based on polylactic acid (PLA) and starch. *Eur. Polym. J.* **2017**, *96*, 316–325. [CrossRef]
111. Anwar, M.; Istiqomah, L.; Ekaningrum, M.; Yembise, D. Antibacterial activities of biocomposite plastic-based phenolic acids-grafted chitosan and sugar palm starch (Arenga pinata). *IOP Conf. Ser. Earth Environ. Sci.* **2020**, *462*, 012046. [CrossRef]
112. Saberi, B.; Chockchaisawasdee, S.; Golding, J.B.; Scarlett, C.J.; Stathopoulos, C.E. Characterization of pea starch-guar gum biocomposite edible films enriched by natural antimicrobial agents for active food packaging. *Food Bioprod. Process.* **2017**, *105*, 51–63. [CrossRef]
113. Evangelho, J.A.d.; Dannenberg, G.d.; Biduski, B.; el Halal, S.L.M.; Kringel, D.H.; Gularte, M.A.; Fiorentini, A.M.; Zavareze, E.D. Antibacterial activity, optical, mechanical, and barrier properties of corn starch films containing orange essential oil. *Carbohydr. Polym.* **2019**, *222*, 114981. [CrossRef]
114. Shapi'i, R.A.; Othman, S.H.; Nordin, N.; Basha, R.K.; Naim, M.N. Antimicrobial properties of starch films incorporated with chitosan nanoparticles: In vitro and in vivo evaluation. *Carbohydr. Polym.* **2020**, *230*, 115602. [CrossRef]
115. Martins, P.C.; Bagatini, D.C.; Martins, V.G. Oregano essential oil addition in rice starch films and its effects on the chilled fish storage. *J. Food Sci. Technol.* **2020**, *58*, 1562–1573. [CrossRef]
116. Valencia-Sullca, C.; Atarés, L.; Vargas, M.; Chiralt, A. Physical and Antimicrobial Properties of Compression-Molded Cassava Starch-Chitosan Films for Meat Preservation. *Food Bioprocess Technol.* **2018**, *11*, 1339–1349. [CrossRef]

117. Yasa, S.R.; Kaki, S.S.; Poornachandra, Y.; Kumar, C.G.; Penumarthy, V. Synthesis, characterization, antimicrobial and biofilm inhibitory studies of new esterquats. *Bioorganic Med. Chem. Lett.* **2016**, *26*, 1978–1982. [CrossRef] [PubMed]
118. Zexin, W. Talking about the New Uses of Several Inorganic Antibacterial Materials. *J. Phys. Conf. Ser.* **2021**, *1965*, 012076.
119. Chen, R.; Han, Z.; Huang, Z.; Karki, J.; Wang, C.; Zhu, B.; Zhang, X. Antibacterial activity, cytotoxicity and mechanical behavior of nano-enhanced denture base resin with different kinds of inorganic antibacterial agents. *Dent. Mater. J.* **2017**, *36*, 693–699. [CrossRef]
120. Hu, X.; Jia, X.; Zhi, C.; Jin, Z.; Miao, M. Improving the properties of starch-based antimicrobial composite films using ZnO-chitosan nanoparticles. *Carbohydr. Polym.* **2019**, *210*, 204–209. [CrossRef]
121. Chenwei, C.; Chenxi, L.; Shaohua, Y.; Qinjun, Z.; Fuxin, Y.; Zhipeng, T.; Jing, X. Development of New Multilayer Active Packaging Films with Controlled Release Property Based on Polypropylene/Poly(Vinyl Alcohol)/Polypropylene Incorporated with Tea Polyphenols. *J. Food Sci.* **2019**, *84*, 1836–1843. [CrossRef]
122. Chenwei, C.; Zhewei, X.; Yarui, M.; Jinliang, L.; Qinjun, Z.; Zhipeng, T.; Kaijia, F.; Fuxin, Y.; Jing, X. Properties, vapour-phase antimicrobial and antioxidant activities of active poly(vinyl alcohol) packaging films incorporated with clove oil. *Food Control* **2018**, *88*, 105–112. [CrossRef]
123. Hadi, A.; Mahsa, J.O.; Ayda, S. A review on techniques utilized for design of controlled release food active packaging. *Crit. Rev. Food Sci. Nutr.* **2020**, *61*, 1–21.
124. Cornelia, V.; Mihaela, B. Progresses in Food Packaging, Food Quality, and Safety—Controlled-Release Antioxidant and/or Antimicrobial Packaging. *Molecules* **2021**, *26*, 1263.
125. Kumar, Y.; Yadav, D.N.; Ahmad, T.; Narsaiah, K. Recent Trends in the Use of Natural Antioxidants for Meat and Meat Products. *Compr. Rev. Food Sci. Food Saf.* **2015**, *14*, 796.0–812.0. [CrossRef]
126. Mastromatteo, M.; Mastromatteo, M.; Conte, A.; del Nobile, M.A. Advances in controlled release devices for food packaging applications. *Trends Food Sci. Technol.* **2010**, *21*, 591–598. [CrossRef]
127. Fernando, M.Y.D.; Walter, M.A.; Arturo, J.R.Á.; Jairo, M.A.J.; Esteban, A.P.Á. Encapsulation of phenols of gulupa seed extract using acylated rice starch: Effect on the release and antioxidant activity. *J. Funct. Foods* **2021**, *87*, 104788.
128. Zhang, S.; Zhao, H. Preparation and properties of zein–rutin composite nanoparticle/corn starch films. *Carbohydr. Polym.* **2017**, *169*, 385–392. [CrossRef] [PubMed]
129. Farrag, Y.; Ide, W.; Montero, B.; Rico, M.; Rodríguez-Llamazares, S.; Barral, L.; Bouza, R. Starch films loaded with donut-shaped starch-quercetin microparticles: Characterization and release kinetics. *Int. J. Biol. Macromol.* **2018**, *118*, 2201–2207. [CrossRef] [PubMed]
130. Qin, Y.; Liu, Y.; Zhang, X.; Liu, J. Development of active and intelligent packaging by incorporating betalains from red pitaya (Hylocereus polyrhizus) peel into starch/polyvinyl alcohol films. *Food Hydrocoll.* **2020**, *100*, 105410. [CrossRef]
131. Sajad, P.; Karimi, S.I.; Sadat, M.S. Nano-biocomposite based color sensors: Investigation of structure, function, and applications in intelligent food packaging. *Food Packag. Shelf Life* **2022**, *31*, 100789.
132. Chen, W.; Ma, S.; Wang, Q.; Julian, M.D.; Liu, X.; Ngai, T.; Liu, F. Fortification of edible films with bioactive agents: A review of their formation, properties, and application in food preservation. *Crit. Rev. Food Sci. Nutr.* **2021**, *62*, 5029–5055. [CrossRef]
133. Silva-Pereira, M.C.; Teixeira, J.A.; Pereira-Júnior, V.A.; Stefani, R. Chitosan/corn starch blend films with extract from Brassica oleraceae (red cabbage) as a visual indicator of fish deterioration. *LWT-Food Sci. Technol.* **2015**, *61*, 258–262. [CrossRef]
134. Nitya, B.; Sharanagat, V.S.; Mor, R.S.; Kumar, K. Active and intelligent biodegradable packaging films using food and food waste-derived bioactive compounds: A review. *Trends Food Sci. Technol.* **2020**, *105*, 385–401.
135. Swarup, R.; Jong-Whan, R. Anthocyanin food colorant and its application in pH-responsive color change indicator films. *Crit. Rev. Food Sci. Nutr.* **2020**, *61*, 1–29.
136. Golasz, L.B.; da Silva, J.; da Silva, S.B. Film with anthocyanins as an indicator of chilled pork deterioration. *Food Sci. Technol.* **2013**, *33*, 155–162. [CrossRef]
137. Etxabide, A.; Kilmartin, P.A.; Maté, J.I. Color stability and pH-indicator ability of curcumin, anthocyanin and betanin containing colorants under different storage conditions for intelligent packaging development. *Food Control* **2021**, *121*, 107645. [CrossRef]
138. Choi, I.; Lee, J.Y.; Lacroix, M.; Han, J. Intelligent pH indicator film composed of agar/potato starch and anthocyanin extracts from purple sweet potato. *Food Chem.* **2017**, *218*, 122–128. [CrossRef] [PubMed]
139. Anabel, V.; Alfredo, S.; Beatriz, M.; Julio, R.; Reynaldo, V. Electrochemical biosensors for food bioprocess monitoring. *Curr. Opin. Food Sci.* **2022**, *43*, 18–26.
140. Boyu, M.; Guoqing, C.; Luwei, Z.; Yu, Z.; Xinqing, X. Flexible wireless pH sensor system for fish monitoring. *Sens. Bio-Sens. Res.* **2021**, *34*, 100465.
141. Wang, S.; Liu, X.; Yang, M.; Zhang, Y.; Xiang, K.; Tang, R. Review of Time Temperature Indicators as Quality Monitors in Food Packaging. *Packag. Technol. Sci.* **2015**, *28*, 839.0–867.0. [CrossRef]
142. Hao, C.; Hao, X.; David, J.M.; Long, C.; Aiquan, J.; Yaoqi, T.; Ming, M.; Zhengyu, J. Recent advances in intelligent food packaging materials: Principles, preparation and applications. *Food Chem.* **2021**, *375*, 131738.
143. Zhang, X.; Sun, G.; Xiao, X.; Liu, Y.; Zheng, X. Application of microbial TTIs as smart label for food quality: Response mechanism, application and research trends. *Trends Food Sci. Technol.* **2016**, *51*, 12–23. [CrossRef]
144. Preetam, A.; Dominic, S.; Anastasia, E.L. Time-Temperature Indicator Based on Enzymatic Degradation of Dye-Loaded Polyhydroxybutyrate. *Biotechnol. J.* **2017**, *12*, 1700050.

145. Jafry, A.T.; Lim, H.; Sung, W.K.; Lee, J. Flexible time–temperature indicator: A versatile platform for laminated paper-based analytical devices. *Microfluid. Nanofluid* **2017**, *21*, 57. [CrossRef]
146. Kim, J.U.; Ghafoor, K.; Ahn, J.; Shin, S.; Lee, S.H.; Shahbaz, H.M.; Shin, H.; Kim, S.; Park, J. Kinetic modeling and characterization of a diffusion-based time-temperature indicator (TTI) for monitoring microbial quality of non-pasteurized angelica juice. *LWT-Food Sci. Technol.* **2016**, *67*, 143–150. [CrossRef]
147. Santos, C.T.; Veiga-Santos, P.; Sestari, P.; Sorrini, N.C.; Roça, R.O. Protein time–temperature sensor for intelligent starch polymers. *J. Food Process. Preserv.* **2020**, *44*, e14428. [CrossRef]
148. Nogueira, G.F.; Fakhouri, F.M.; de Oliveira, R.A. Incorporation of spray dried and freeze dried blackberry particles in edible films: Morphology, stability to pH, sterilization and biodegradation. *Food Packag. Shelf Life* **2019**, *20*, 100313. [CrossRef]
149. Cheng, J.; Wang, H.; Kang, S.; Xia, L.; Jiang, S.; Chen, M.; Jiang, S. An active packaging film based on yam starch with eugenol and its application for pork preservation. *Food Hydrocoll.* **2019**, *96*, 546.0–554.0. [CrossRef]
150. Kang, J.; Song, K.B. Characterization of Job's tears (Coix lachryma-jobi L.) starch films incorporated with clove bud essential oil and their antioxidant effects on pork belly during storage. *LWT* **2019**, *111*, 711–718. [CrossRef]
151. Goodarzi, M.M.; Moradi, M.; Tajik, H.; Forough, M.; Ezati, P.; Kuswandi, B. Development of an easy-to-use colorimetric pH label with starch and carrot anthocyanins for milk shelf life assessment. *Int. J. Biol. Macromol.* **2020**, *153*, 240–247. [CrossRef]
152. Pellá, M.C.G.; Silva, O.A.; Pellá, M.G.; Beneton, A.G.; Caetano, J.; Simões, M.R.; Dragunski, D.C. Effect of gelatin and casein additions on starch edible biodegradable films for fruit surface coating. *Food Chem.* **2020**, *309*, 125764. [CrossRef]
153. Meral, Y.; Hasan, S.; Mahmut, Ş. Characterization of edible film based on grape juice and cross-linked maize starch and its effects on the storage quality of chicken breast fillets. *LWT* **2021**, *142*, 111012. [CrossRef]
154. Gustavo, S.W.; Gonçalves, d.C.; Gomes, d.A.P.; Peruzzo, F.J.; Stremel, A.M.; Tânia, F.C.; Ramos, N.M.; Paula, d.V.A. Application in situ of biodegradable films produced with starch, citric pectin and functionalized with feijoa (Acca sellowiana (Berg) Burret) extracts: An effective proposal for food conservation. *Int. J. Biol. Macromol.* **2021**, *189*, 544–553.
155. Ni, S.; Jiao, L.; Zhang, H.; Zhang, Y.; Fang, G.; Xiao, H.; Dai, H. Enhancing hydrophobicity, strength and UV shielding capacity of starch film via novel co-cross-linking in neutral conditions. *R. Soc. Open Sci.* **2018**, *5*, 181206.
156. Shahab, N.; Masoud, R.; Mehdi, A. A starch-based pH-sensing and ammonia detector film containing betacyanin of paperflower for application in intelligent packaging of fish. *Int. J. Biol. Macromol.* **2021**, *191*, 161–170.
157. Xinxin, L.; Di, L.; Shuai, D.; Ling, Z.; Kang, M. Corn starch/polyvinyl alcohol based films incorporated with curcumin-loaded Pickering emulsion for application in intelligent packaging. *Int. J. Biol. Macromol.* **2021**, *188*, 974–982.
158. Yan, Q.; Dawei, Y.; Fengfeng, X.; Dan, C.; Juan, K.; Jun, L. Smart packaging films based on starch/polyvinyl alcohol and Lycium ruthenicum anthocyanins-loaded nano-complexes: Functionality, stability and application. *Food Hydrocoll.* **2021**, *119*, 106850.
159. Goudar, H.V.D.; Gasti, N.; Khanapure, T.; Vanjeri, S.; Sataraddi, V.N.; D'souza, S.; Oshin, V.J.; Kumar, S.; Saraswati, M.P.; Malabadi, R.B.; et al. Exploration of Multifunctional Properties of Piper betel Leaves Extract Incorporated Polyvinyl Alcohol-Oxidized Maize Starch Blend Films for Active Packaging Applications. *J. Polym. Environ.* **2022**, *30*, 1314–1329.
160. Zhikun, M.; Yanfei, Z.; Panfang, L. Novel active starch films incorporating tea polyphenols-loaded porous starch as food packaging materials. *Int. J. Biol. Macromol.* **2021**, *192*, 1123–1133.
161. Shi, C.; Zhang, X.; Guo, N. The antimicrobial activities and action-mechanism of tea tree oil against food-borne bacteria in fresh cucumber juice. *Microb. Pathog.* **2018**, *125*, 262–271. [CrossRef]
162. Quintavalla, S.; Vicini, L. Antimicrobial food packaging in meat industry. *Meat Sci.* **2002**, *62*, 373–380. [CrossRef]
163. Campos-Requena, V.H.; Rivas, B.L.; Pérez, M.A.; Figueroa, C.R.; Sanfuentes, E.A. The synergistic antimicrobial effect of carvacrol and thymol in clay/polymer nanocomposite films over strawberry gray mold. *LWT-Food Sci. Technol.* **2015**, *64*, 390–396. [CrossRef]
164. Bin, L.; Yiwen, B.; Jiaxin, L.; Jinfeng, B.; Qinqin, C.; Huijun, C.; Yuxuan, W.; Jinlong, T.; Chi, S.; Yuehua, W.; et al. A sub-freshness monitoring chitosan/starch-based colorimetric film for improving color recognition accuracy via controlling the pH value of the film-forming solution. *Food Chem.* **2022**, *388*, 132975.
165. Chen, H.; Zhang, M.; Bhandari, B.; Yang, C. Novel pH-sensitive films containing curcumin and anthocyanins to monitor fish freshness. *Food Hydrocoll.* **2020**, *100*, 105438. [CrossRef]
166. Mohammadalinejhad, S.; Almasi, H.; Moradi, M. Immobilization of Echium amoenum anthocyanins into bacterial cellulose film: A novel colorimetric pH indicator for freshness/spoilage monitoring of shrimp. *Food Control* **2020**, *113*, 107169. [CrossRef]

*foods*

*Review*

# Pulsed Electric Field: Fundamentals and Effects on the Structural and Techno-Functional Properties of Dairy and Plant Proteins

**Ahmed Taha [1,2], Federico Casanova [3,*], Povilas Šimonis [1], Voitech Stankevič [1], Mohamed A. E. Gomaa [2] and Arūnas Stirkė [1,4,*]**

1 Department of Functional Materials and Electronics, Center for Physical Sciences and Technology, Saulėtekio al. 3, LT-10257 Vilnius, Lithuania; ahmed-taha@alexu.edu.eg (A.T.); povilas.simonis@ftmc.lt (P.Š.); voitech.stankevic@ftmc.lt (V.S.)
2 Department of Food Science, Faculty of Agriculture (Saba Basha), Alexandria University, Alexandria 21531, Egypt; m.gomaa@alexu.edu.eg
3 Food Production Engineering, National Food Institute, Technical University of Denmark, 2800 Kongens Lyngby, Denmark
4 Micro and Nanodevices Laboratory, Institute of Solid State Physics, University of Latvia, Kengaraga Str. 8, LV-1063 Riga, Latvia
* Correspondence: feca@food.dtu.dk (F.C.); arunas.stirke@ftmc.lt (A.S.)

**Abstract:** Dairy and plant-based proteins are widely utilized in various food applications. Several techniques have been employed to improve the techno-functional properties of these proteins. Among them, pulsed electric field (PEF) technology has recently attracted considerable attention as a green technology to enhance the functional properties of food proteins. In this review, we briefly explain the fundamentals of PEF devices, their components, and pulse generation and discuss the impacts of PEF treatment on the structure of dairy and plant proteins. In addition, we cover the PEF-induced changes in the techno-functional properties of proteins (including solubility, gelling, emulsifying, and foaming properties). In this work, we also discuss the main challenges and the possible future trends of PEF applications in the food proteins industry. PEF treatments at high strengths could change the structure of proteins. The PEF treatment conditions markedly affect the treatment results with respect to proteins' structure and techno-functional properties. Moreover, increasing the electric field strength could enhance the emulsifying properties of proteins and protein-polysaccharide complexes. However, more research and academia–industry collaboration are recommended to build highly effective PEF devices with controlled processing conditions.

**Keywords:** pulsed electric field; pulse generation; milk proteins; plant proteins; functional properties; protein structure

**Citation:** Taha, A.; Casanova, F.; Šimonis, P.; Stankevič, V.; Gomaa, M.A.E.; Stirkė, A. Pulsed Electric Field: Fundamentals and Effects on the Structural and Techno-Functional Properties of Dairy and Plant Proteins. *Foods* **2022**, *11*, 1556. https://doi.org/10.3390/foods11111556

Academic Editor: Urszula Tylewicz

Received: 19 April 2022
Accepted: 23 May 2022
Published: 25 May 2022



## 1. Introduction

Food proteins play vital roles in human nutrition, food production, and nutraceutical industries. Dairy proteins have excellent functional properties and high nutritional values [1,2]. The dairy industry also significantly contributes to the world economy, particularly in the European Union (EU). For example, according to the milk market observatory of the European Commission, the EU exported more than 500 thousand tons of whey powders in 2020 [3]. As a raw material in many food products, dairy proteins are required to possess excellent functional properties, including high solubility and improved emulsifying, foaming, and gelling properties [4]. To improve the functional properties of dairy proteins, several changes in the structural and conformational state of such proteins are needed. For instance, the solubility of proteins depends mainly on the hydrophilic and hydrophobic residues on the proteins' surfaces, as well as the content of hydrogen bonds. Moreover, the gelling properties could be altered if the contents of sulfhydryl (SH) groups and disulfide bonds are changed. Additionally, the emulsifying properties are closely related to the surface activity of proteins [5]. Plant proteins are increasingly being utilized as alternative sources of animal proteins due to their importance in developing sustainable food systems [6,7]. Moreover, plant proteins have been used as emulsifiers, foam stabilizers, and for other applications in the food industry due to their affordable price and amphiphilic structures [8,9]. The weak electrostatic repulsion, low solubility, and high molecular weight of proteins can limit the applications of these proteins [10]. Therefore, emerging green technologies are needed to alter the structural and techno-functional properties of plant and dairy proteins with minimal effects on the nutritional value and flavor of these proteins and their products.

Electric field devices are classified into two major categories based on the electric field strength. Pulsed electric field (PEF) devices have electric field strengths of 20–100 kV/cm, whereas moderate electric field (MEF) devices have electric field strengths of <1000 V/cm [11]. PEF, as an emerging ecofriendly technology, has been used in the food industry for the inactivation of enzymes and microorganisms [12–16]. During PEF treatment, pulses of high-voltage electric fields for a short time (from nanoseconds to milliseconds) were applied to the material between two electrodes [17,18]. PEF technology offers several advantages over traditional pasteurization methods in the food industry, such as shelf-life extension, nutrient retention, quality preservation, and cost effectiveness [19,20]. Therefore, PEF is being adopted increasing rapidly in many industrial sectors, including for the processing of plant and dairy products [13,21]. Some studies have shown that PEF could be used to alter the structural and techno-functional properties of food proteins [5,22–34].

The main mechanism behind the effects of electric fields on proteins is not very clear. However, some researchers have proposed that polar groups of proteins absorb energy during PEF treatment and generate free radicals. The produced free radicals can affect intramolecular interactions within protein molecules, including hydrophobic and electrostatic interactions, disulfide bridges, hydrogen bonds, salt bridges, and Van der Waals forces [5]. Moreover, PEF treatments could change the apparent charge of proteins due to the alteration of their ionic interactions [4,35]. Therefore, the structural and functional properties of proteins can be altered. Several review articles have discussed the effects and food applications of PEF technology [4,5,13,14,18,20,26,36–47]. However, there is a lack of systematic reviews discussing PEF fundamentals and their applications in the food industry. Moreover, to the best of our knowledge, few studies have been conducted on the effects of PEF on the structure and techno-functional properties of food proteins; thus, systematic reviews discussing this point are scarce. Therefore, a review of the fundamentals of PEF and the effects of PEF treatments on the structural and techno-functional properties of food proteins is needed. Accordingly, with this review, we aim to (1) briefly compare PEF with other processing techniques; (2) discuss the theory and fundamentals of PEF technology; and (3) discuss the effects of PEF treatments on the structural and techno-functional properties (including solubility, gelling, emulsifying, and foaming properties) of dairy and plant proteins. This review could guide both researchers and industry leaders to develop new applications of PEF as a green and sustainable technology in the food industry.

## 2. PEF vs. Other Processing Technologies

Besides PEF, some other emerging technologies have been used to alter the structure and techno-functional properties of food proteins, such as ultrasound [48–51], high-pressure processing [52–55], microwave treatment [56–59], and cold plasma processing [60–62]. The main mechanisms, processing parameters, and effects of these technologies on proteins are summarized in Table 1. However, there is currently a growing interest in PEF because of its sustainable approach and wide range of applications in the food and biotechnology sectors [63]. For example, PEF can be used for:

- Microbial inactivation [64];
- Enhancing the quality of freeze-dried fruits [65] and improving the quality of potato chip products [66];
- Improving drying and extraction kinetics [67];
- Winemaking, biogas production, and extraction of protein from algae [63]; and
- Improving the functional properties of proteins [5].

Moreover, PEF offers many advantages over other techniques [47], including:

- Short processing times;
- Waste-free process;
- Low energy consumption;
- Environmentally-friendly technique;
- Better retention of nutrients, flavors, and colors; and
- Suitability for processing heat-sensitive foods.

The main disadvantage of PEF technology is the existence of bubbles during treatment, which could result in operational problems and non-uniform treatment. Additionally, commercial PEF units not widely available in many regions worldwide [47].

**Table 1.** Brief comparison of different emerging processing techniques and their effects on protein structure and techno-functional properties.

| Processing Technology | Processing Parameters | Mechanism | Effects on Protein Structure | Effects on Proteins' Techno-Functional Properties |
|---|---|---|---|---|
| PEF | Pulse-wave shape, pulse duration, electric field strength, frequency, temperature, and treatment duration [68]. | Polarization of protein molecules and release of free radicals can induce changes in protein structures and functionalities [4]. | Depends on the electric field strength and the type of proteins.<br>Main changes occurred in the secondary structure and exposure of hydrophobic groups to the surface of protein molecules [33]. | Different waveforms and protein types can have different effects on protein solubility.<br>Emulsifying and foaming properties improved [69]. |
| Ultrasound | Amplitude, frequency, acoustic energy, intensity, energy density (J/mL), time, and temperature [70]. | Acoustic cavitation (the formation and collapse of air bubbles) induces chemical reactions and physical effects, which influence the structure and techno-functional properties of proteins [71]. | Changes in the secondary and tertiary structures.<br>Increases in surface hydrophobicity and free sulfhydryl groups [51] | Ultrasound improved the emulsifying and gelling properties of proteins [72]. |
| High pressure processing | Pressure, temperature, and time [53] | Protein unfolding can occur due to the penetration of water into the protein matrix [38]. | Depends on the applied conditions and protein system.<br>Mainly protein denaturation and aggregation occurred [73]. | Depends on the applied pressure.<br>Emulsifying and foaming capabilities enhanced.<br>Solubility of proteins improved [74]. |
| Microwave | Power, frequency, time, and temperature [56]. | At the molecular level, exposed proteins interact with electromagnetic energy. Then, heat is generated from the electromagnetic energy through the motion of molecules during treatment [71]. | Changes in the secondary structure.<br>Protein aggregation [75]. | Gelling properties improved [76]. |
| Cold-plasma processing | Voltage, frequency, time, and temperature [77]. | Several high-energy radicals, such as nitric oxide, atomic oxygen, superoxide, and hydroxyl radicals to break the covalent bonds and promote several chemical reactions [78]. | The high-energy reaction could break peptide bonds and oxidize the side chains of amino acids.<br>They may also facilitate the formation of protein–protein interactions.<br>Changes in the secondary structures were observed [61]. | Water- and oil-holding capacities enhanced, reflecting the improvement of emulsifying and gelling properties of proteins [62]. |

## 3. Fundamentals of PEF Technology: Device Components and Pulse Generation

An electric circuit simply means a closed-loop that carries electricity. The electric current ($I$) is the flow of electrons in the circuit, measured in amperes (A) and can be calculated according to Equation (1). The voltage ($V$, in volts) is the electric pressure or source that causes the flow of the current. The resistance ($R$, in ohms) in the electric circuit opposes the current flow [79].

$$I = \frac{V}{R} \quad (1)$$

Contemporary PEF is based on a direct application of power pulses to the food material placed between two electrodes for micro- to nanoseconds at an intensity range of 10–80 kV/cm [80]. The PEF processing time can be calculated by multiplying the effective pulse duration by the number of pulses. The magnitude and time course of PEF are controlled by a voltage generator and electrode geometry.

PEF devices basically consist of a treatment chamber with a suitable cuvette, a high-voltage pulse generator, and necessary controlling and monitoring devices. Figure 1 shows a diagram of a continuous PEF device used to treat food samples. The electric field depends on the applied electric voltage, the distance between the two electrodes, and pulse width and waveform (Equation (2)). $E$ represents the electric field strength (V/m), $u(t)$ represents the applied voltage over time ($V$), and $d$ represents the distance between electrodes (m) [81].

$$E(t) = \frac{1}{d} * \int_{0}^{t} u(t)\,dt \quad (2)$$

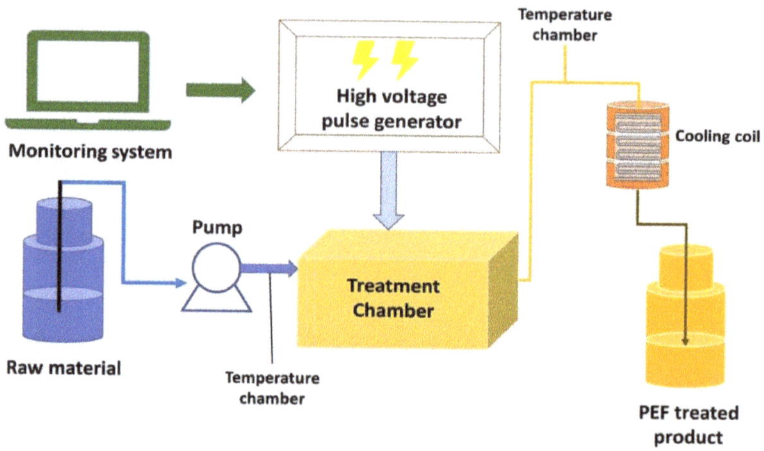


**Figure 1.** Diagram of a possible continuous PEF device used to treat food samples.

Many types of circuits (pulse generators) in PEF devices have different circuit components to fulfill the required functions. Figure 2 is a schematic diagram that represents an example of an electric circuit in a PEF device used for food processing. Electrical pulses are acquired by charging a capacitor, and the discharging of the capacitor is controlled by a trigger or switcher that controls the decay in an electronic circuit [82]. Table 2 summarizes the components of PEF systems and their functions. As shown in Table 2, the electric circuit of PEF devices employed for food processing has several electric elements. A high-voltage pulse generator is used to charge the capacitors and can be a direct current (DC) or alternating current (AC) switched to DC using a rectifier. The high-voltage pulse generator is also used to discharge the high voltage from capacitors in the form of a pulse with a specific pulse shape and width through a pulse-forming network (PFN) (Figure 2).

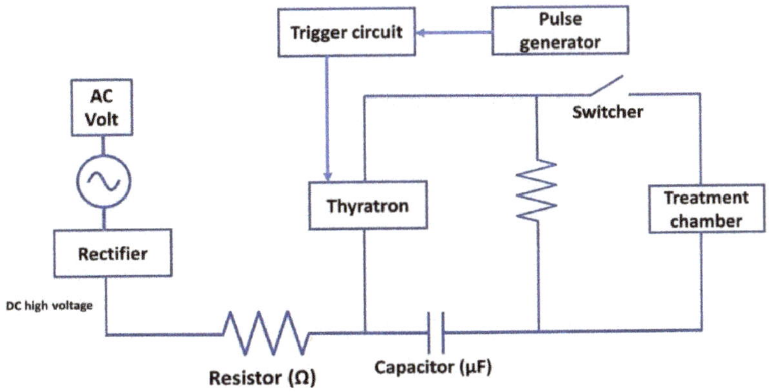


**Figure 2.** Schematic diagram of an electric circuit in a PEF device used for food processing.

**Table 2.** Functions of main PEF devices components.

| Component | Description and Function(s) | References |
|---|---|---|
| High-voltage pulse generator | • To generate high-voltage direct current (DC) at a specific intensity by a power supply.<br>• To discharge high voltage in the form of pulses with specific shapes and widths through a pulse-forming network (PFN). | [80] |
| Resistors | Delay the current flow and impose a voltage reduction. | [83] |
| Capacitors | Energy (voltage) storage. | [84] |
| Switchers | Connect or disconnect the electric current and control the discharge of the stored energy. | [20] |
| Treatment chambers | Specific containers are used to carry food samples during exposure to PEF. | [81,85] |

A capacitor is an electrical device used to store energy that used to generate electric pulses in an electric circuit (PEF systems). Several types of capacitors are available in the market, such as electrolytic, ceramic, paper, film, mica, and non-polarized capacitors [86]. Electrochemical capacitors (ECs), also known as supercapacitors or electrochemical double-layer capacitors (EDLC), have recently been used in many electronic applications on a large scale [84]. Generally, capacitors consist of two parallel conductive (metal) electrodes isolated using non-conducting materials (dielectrics), such as ceramic, waxed paper, plastic, mica, or a liquid gel, as utilized in electrolytic capacitors. Due to the presence of dielectric materials between two conductive materials, the direct current cannot flow through the capacitor. Thus, a voltage is stored in the conductive metal plates as an electrical charge [84]. Several factors determine the power needed to charge the capacitor, such as the size and number of capacitors, as well as the resistance of the charging resistor [87]. It has been concluded that a larger capacitor requires more power and time to be charged than a smaller one. The capacitance, $C_0$ ($F$), of a capacitor can be calculated with Equation (3), where $R$ (Ω) is the resistance, $A$ (m$^2$) is the area of the electrode surface, $\sigma$ (S/m) is the conductivity of the food, $\tau$ (s) is the pulse duration, and $d$ (m) is the distance between electrodes [80].

$$C_0 = \frac{\tau}{R} = \frac{\tau \sigma A}{d} \quad (3)$$

The energy stored ($Q$) in a capacitor is calculated based on the values of capacitance ($C_0$) and charge voltage ($V$), as shown in Equation (4) [80].

$$Q = 0.5\, C_0\, V^2 \quad (4)$$

The switching device (switcher) is a critical device for the efficiency of PEF systems. It is needed to connect or disconnect the electric current and discharge the stored energy in a capacitor through the PFN. There are two main types of switchers: ON (semi-controlled) and ON/OFF (fully controlled) switchers. The suitable switcher in a circuit is selected based on its repetition rate and potential to operate at high voltage. Capacitors should be fully discharged to turn the switcher off in semi-controlled switches (including thyratron, trigatron, gas spark gap, and ignitron). These switches can control high voltages at a lower cost. However, low repetition rate and short life are the main disadvantages of such switches. Fully controlled switches (including symmetrical gate commutated thyristors (SGCT), insulated gate bipolar transistors (IGBT), and the gate turn-off (GTO) thyristors) can control the pulse generation process and can be switched on and off with full or partial discharge of the capacitors. The development of fully controlled semiconductor switches increased life spans and improved switch performance [20,80]. The relative electrical value of each component of PFN systems influences the pulse shape. For instance, an exponentially decaying pulse shape is formed in a simple resistance–capacitance (*RC*) circuit. On the other hand, complex PFN systems can produce instantaneously reversal, as well as bipolar and square pulses (Figure 3). To generate exponential decay pules, the generation of an exponential decay pulse only needs semi-controlled (ON) switchers in which the capacitor is completely discharged. Square wave pulses can be generated by the partial discharge of a capacitor with fully controlled (ON/OFF) switchers [85].

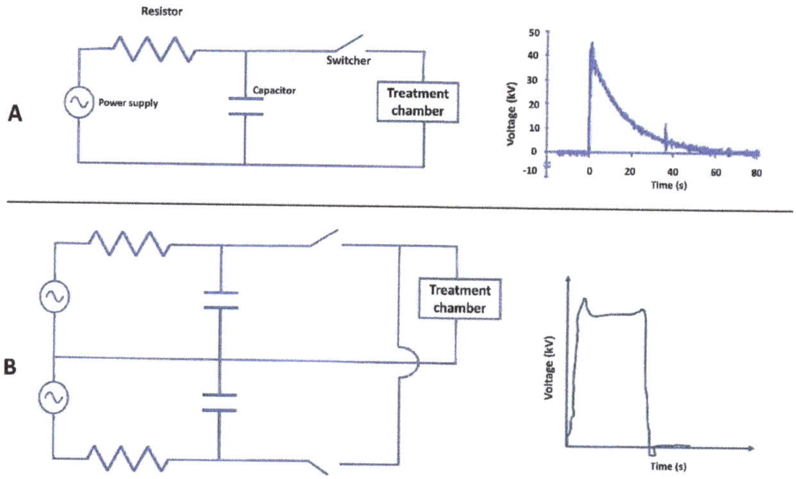


**Figure 3.** The two main types of electrical circuits and their possible pulses shapes: (**A**) simple resistance–capacitance (*RC*) circuit and its exponentially decaying pulse shape; (**B**) complex electric circuit and its square pulse shape.

Resistors are one of the main components of electric circuits. They are used to regulate the current flow and force voltage reduction. The theory of electric resistance is similar to the water flow in pipes; the resistor can be considered a thin pipe (wire in the case of an electric circuit) that reduces the water flow [83]. In reducing the current flow, the electrical energy is absorbed by the resistor and dissipated as heat. Capacitors are classified based on their production materials: wire-wound, film, or cermet (made of metal or metal-oxide); carbon composition; and semiconductor capacitors. Ohm's law of resistance specifies the relationship between the resistance (*R*), voltage (*V*), and current (*I*) (Equation (1)); an ohm ($\Omega$) equals a volt per ampere [83,88].

Various treatment chambers are designed to expose food samples to the electric field. Static chambers are used in batch processing and laboratories, whereas continuous chambers can meet the requirements of industrial-scale applications. Batch chambers offer several advantages at a laboratory scale, such as treating small-volume samples, efficiently controlling temperature by cooling the electrodes, and slowing the repetition rates. However, continuous chambers are essential to reach high-volume capacity; they are also easily integrated into existing food processing lines [81,85]. The material used for building treatment chambers should be washable or autoclavable. Currently, three main types of treatment chambers are designed based on the arrangement of electrodes in different geometric configurations, including parallel plates, as well as coaxial and colinear chambers [41,85]. Parallel and coaxial plates are commonly used for batch processing, whereas colinear chambers are utilized in continuous processing devices (Figure 4).

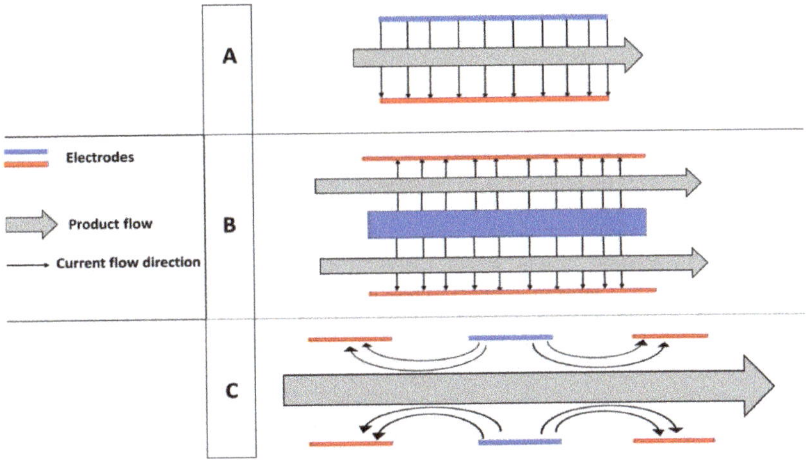


**Figure 4.** Schematic diagrams of the three main types of treatment chambers in PEF devices. (**A**) parallel plate chambers; (**B**) coaxial plate chambers; (**C**) colinear plate chambers.

Recently, several companies have developed emerging PEF systems for industrial applications. Current large-scale PEF devices are based on Marx generators or transformers, and electric pulses are applied continuously. In Marx generators, a stack of capacitors is used, charged in parallel, and discharged in a sequence, providing a high-power conversion rate. For transformers, a pulse transformer is used with a low-voltage switch. Most PEF units have an average power ranging from 20 to 400 kW [89]. Treatment chambers are designed based on the scope of application. Two major types of treatment chamber are belt systems, which are used for processing of solid products, such as potatoes or seafood, and pipe systems are used for processing of liquid products [90].

The outcome of PEF treatment is influenced by many product and process factors. Product factors include chemical composition, pH, rheological properties, temperature, and electrical conductivity. Process parameters include electric field strength, pulse number, pulse frequency, the shape of the pulse wave, pulse width, type of treatment chamber, flow conditions, and flow rate [91]. It is worth mentioning that it is difficult to compare the data from different research groups due to many factors that affect PEF treatment results, which will be discussed in this review. In the following sections, we will discuss the effects of PEF on the structural and techno-functional properties of milk proteins.

## 4. Effects of PEF on the Structure of Dairy and Plant Proteins

As shown in Figure 5, caseins consist of four major subunits, including αs1, αs2, β, and kappa caseins. Whey proteins have several subunits, including β-lactoglobulin (β-LG), α-lactalbumin (α-LA), bovine serum albumin (BSA), lactoferrin, and traces of some other components, such as immunoglobulins and glycomacropeptide [92,93]. In general, upon thermal treatment of milk proteins, proteins unfold because of covalent bonds breaking, and sulfhydryl (-SH) groups are exposed to the protein surface; then, aggregates are produced due to the formation of disulfide bonds.

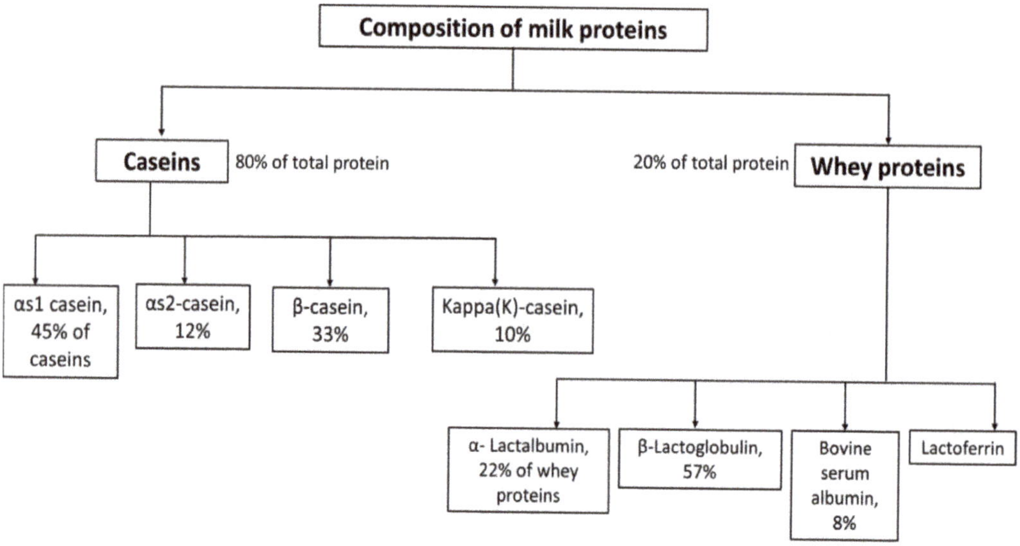


**Figure 5.** Flow chart of the composition of bovine milk (Data collected from Abd El-Salam et al. [92] and Onwulata et al. [93]).

Moreover, as free thiol groups are not available in α-LA, it is less sensitive to thermal treatment than β-LG [4]. Studies have shown that PEF can change the structure of dairy proteins, especially at higher electric strengths at a wide range of temperatures [18]. The energy generated by PEF devices could expose amino acid and/or free-SH groups to the protein molecules' surface. Moreover, non-covalent interactions, such as hydrophobic and hydrogen bonds, may be disrupted [4]. Furthermore, it was found that PEF can change the charge density around amino acids (at the -COOH and -$NH3^+$ moieties), influencing the catalytic activity of peptides [94]. Whey proteins have recently attracted attention due to their nutritional benefits and industrial applications.

As summarized in Table 3, the available results in the literature about PEF effects on the structure of whey proteins are somehow contradictory. Sui et al. [95] investigated the effects of PEF and heat treatments (30–35 kV/cm, 19.2–211 μs, 30–75 °C) on the physico-chemical and functional properties of whey protein isolates (WPI). They concluded that PEF treatment did not influence protein unfolding, surface hydrophobicity, of free-SH group content [95]. Using a different treatment chamber with a different distance between electrodes, Xiang et al. [22] found that PEF increased the surface hydrophobicity and the extrinsic fluorescence intensity of WPI. Similarly, Perez et al. [31] noticed that PEF treatment (12.5 kV/cm) with up to 10 pulses changed the native structure of β-LG and induced protein aggregation. The differences between the results may occur due to the use of various experimental conditions, such as treatment chamber, electric field intensity, frequency, and temperature [20]. Bovine lactoferrin was treated using PEF at different temperatures (30–70 °C) and compared with non-PEF-treated samples at the same temperatures [24]. The results showed that the lactoferrin concentration was not changed by the PEF treatment (35 kV/cm, 19.2 μs, 30–70 °C). Moreover, SDS-PAGE results indicated no significant difference in the gel profile of PEF and non-PEF-treated lactoferrin. The surface hydrophobicity increased with increased temperature. There were no significant differences in surface hydrophobicity values between PEF- and non-PEF-treated lactoferrin [24]. Bekard et al. [96] studied the effects of a low-intensity electric field on the conformational state of BSA using circular dichroism (CD) spectroscopy. They concluded that a low-intensity electric field (500 V/m, 3 h at 22.7–24.2 °C) changed the tertiary structure of BSA, probably due to perturbation in the hydrogen bonds that stabilized the native structure of BSA [96]. Sharma et al. [97] preheated milk samples to 55 °C for 24 s and then applied PEF at 20–26 kV/cm for 34 μs. The results indicated that the surface hydrophobicity of milk proteins considerably increased with increased electric field intensity. Thermal treatment at 30–55 °C can dissociate β-LG dimers into monomers [98]. Thermal pre-treatment associated with PEF might facilitate the dissociation of β-LG dimers of milk samples and expose hydrophobic groups and free-SH groups to the protein molecules' surfaces [97]. Rodrigues et al. [99] compared conventional heat treatment with moderate electric field (MEF, 20–80 V/cm) heating at 50–90 °C. They found that with 70 °C and 80 °C treatments, moderate electric field treatment exhibited higher content of α-helix and random coils and lower content of β-sheet compared to conventional heat treatment at the same temperature. These structural changes probably occurred due to the effects of both heat treatment and electric field on the conformational state of β-LG.

**Table 3.** Effects of pulsed electric field (PEF) on the structure of dairy proteins.

| Dairy Protein | PEF Conditions | Structural Changes | References |
| --- | --- | --- | --- |
| Whey protein | 35.5 kV/cm for 300 or 1000 μs, pulse duration of 7 μs, and pulse repetition rate set at 111 Hz. | Significant differences in the concentration of α-LA, β-LG, and serum albumin between PEF-treated samples for 300 μs and 1000 μs. | [100] |
| Whey protein isolate (WPI) | 12, 16, and 20 kV/cm; number of pulses (10, 20, and 30) | • More hydrophobic groups exposed.<br>• Partial denaturation of WPI fractions. | [22] |
| WPI | 30–35 kV/cm, 19.2–211 μs, 30–75 °C, flow rate of 60 mL/min | • No obvious changes in the gel pattern of SDS-PAGE analysis between PEF and non-PEF control samples.<br>• No significant change in surface hydrophobicity after PEF treatment. | [95] |
| Lactoferrin | Intensity of 35 kV/cm, pulse width of 2 μs, and pulse frequency of 200 or 100 Hz.; flow rate of 60 mL/min. | • No significant differences in surface hydrophobicity values between PEF- and non-PEF-treated lactoferrin.<br>• No significant change in surface hydrophobicity. | [24] |
| β-lactoglobulin | Intensity of 12.5 kV/cm with 40 μF of capacitance. 1–10 pulses, with 15 s between pulses. | PEF partially denatured β-lactoglobulin. | [31] |
| Whole milk | Intensity of 20 or 26 kV/cm for 34 μs, bipolar square wave pulses, pulse width of 20 μs for 34 μs. | The surface hydrophobicity of milk proteins increased with increased electric field intensity. | [97] |
| Sodium caseinate | 10–150 V/cm for 5 s—2 h using a 60 Hz sine wave alternating current. | Moderate electric field altered the secondary structure of sodium caseinate and unfolded the protein molecules. | [33] |
| β-lactoglobulin | 20 V/cm during holding and 80 V/cm during heating at a frequency of 20 kHz for 5–7 min. | Changes in the secondary structure of β-lactoglobulin. | [99] |
| Bovine serum albumin (BSA) | Strengths of 78, 150, 300, and 500 V/m for 3 h. | Low-intensity electric field changed the tertiary structure of BSA. | [96] |

Caseins are the major proteins in milk (80% of total milk protein) and one of the main protein sources in human nutrition. Studies on the effects of PEF on the structure of caseins are scarce. Subaşı et al. [33] studied the impact of MEF (230 V/cm) on the structural changes of sodium caseinate compared to sunflower protein. FTIR data revealed that MEF can change the secondary structure of sodium caseinate and unfold the protein molecules.

This is probably because MEF treatment can polarize the surface of protein molecules, facilitating the exposure of hydrophobic regions to the surface of protein molecules [31,33].

The mechanism of PEF effects on milk protein structures can be proposed based on the available information in the literature. Generally, PEF treatments at low electric field intensities have no apparent effects on the structure of milk proteins. In contrast, PEF treatments at high electric field intensities can considerably change protein structures, especially in whey proteins.

As summarized in Figure 6, some polar groups of milk proteins absorb energy and produce free radicals when exposed to intensive electric fields. These free radicals can disrupt the several interactions among protein molecules, including disulfide and hydrogen bonds, as well as hydrophobic, electrostatic, and Van der Waals interactions. Moreover, the electric field can affect the strong dipole moment of the polypeptide chains, increasing the dielectric constant of proteins. These changes may facilitate the unfolding of protein molecules and the exposure of hydrophobic and -SH groups to the surface of protein molecules. Increasing the duration of PEF treatment could result in the formation of aggregates, as covalent and hydrophobic interaction may occur to crosslink unfolded protein molecules [4,5,31]. It is worth mentioning that an increase in temperature during PEF treatment could facilitate the denaturation of protein molecules. Thus, further study of the effects of PEF on protein structures under controlled temperatures is recommended.

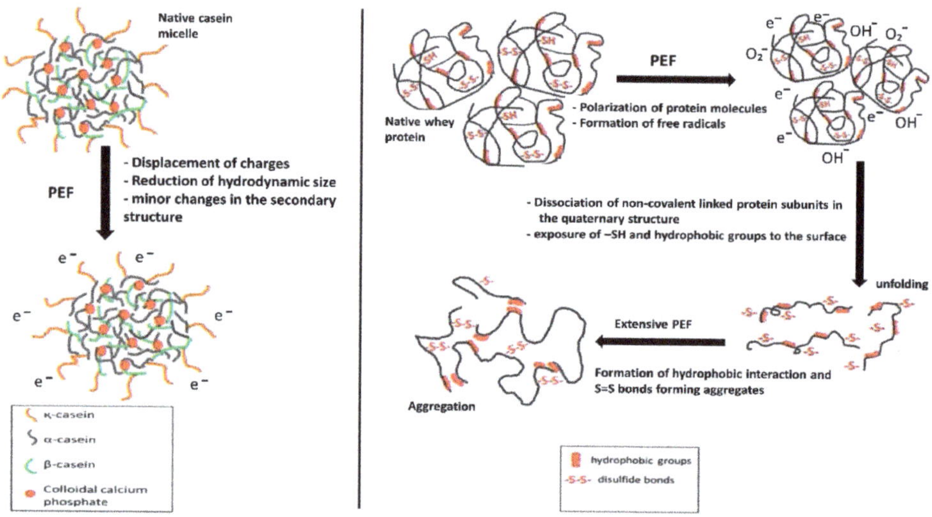


**Figure 6.** Mechanism of PEF effects on the structural properties of milk proteins (caseins and whey proteins); adapted with permission (License number: 5001830472118) from [4]. Copyright (2019) Elsevier.

As shown in Table 4, PEF treatment changed the structures of plant proteins. The secondary structure of soy protein isolate (SPI) changed after PEF treatment at 30–50 kV/cm. PEF caused denaturation and aggregation to SPI, probably due to the formation of hydrophobic interactions and S–S bonds [29,101]. Exposure of sunflower protein to moderate electric field strength (150 V for 20 s at a temperature < 45 °C) resulted in secondary and tertiary structural changes. PEF treatment broke the hydrophobic bonds and facilitated the crosslinking of amino acid side chains [33]. Similar results were also reported with pea and canola proteins [69,102]. Generally, PEF treatment is able to alter the structure of plant proteins. These changes could also affect the techno-functional properties of such proteins.

**Table 4.** Effects of pulsed electric field (PEF) on the structure of plant proteins.

| Plant Protein | PEF Conditions | Structural Changes | References |
|---|---|---|---|
| Soy protein isolate (SPI) | 0–40 kV/cm for 0–547 μs, 2 ms pulse width, and 500 pulse per second (pps) pulse frequency. | - PEF caused slight changes in the secondary structures.<br>- PEF treatment caused denaturation and aggregation of SPI. | [29] |
| SPI | 0 to 50 kV/cm, 40 μs pulse width, 1.0 kHz frequency, and 10 mL/min flow speed. | - PEF changed the vibration of polar groups and reduced the strength of hydrogen bonding, leading to a decrease in the β-turns and an increase in the antiparallel β-sheets. | [101] |
| Sunflower protein | 10–150 V/cm for 5 s-2 h at 25–45 °C. | - Moderate electric field at 150 V for 20 s altered the secondary and tertiary structures of sunflower protein. | [33] |
| Canola protein | 10 to 35 kV, pulse frequency of 600 Hz, and pulse width of 8 μs. | - PEF caused protein molecule aggregation.<br>- PEF reduced β-turns and random coils and increased α-helices and β-sheets. | [69] |
| Pea protein isolate | 5, 10, and 20 V/cm and frequencies of 50 Hz and 20 kHz. | - Moderate electric field treatment (50 Hz and 20 V/cm) unfolded the α-helix into a β-sheet structure.<br>- Aromatic amino acids were exposed to the solvent. | [102] |

## 5. Effects of PEF on the Techno-Functional Properties of Dairy and Plant Proteins

The functionality of milk proteins is determined by physicochemical properties that affect the behavior of proteins during their utilization in food systems [103]. The modifications of protein structures can alter their functional properties [10]. Techno-functional properties, including solubility, gelling, emulsifying, and foaming properties, are of considerable interest in the food industry [104]. Therefore, in this section, we will discuss the effects of electric field treatment on the techno-functional properties of milk proteins. Table 5 presents the main studies on the effects of PEF on the techno-functional properties of dairy proteins, whereas Table 6 summarizes studies on the effects of PEF on the techno-functional properties of plant proteins.

**Table 5.** Effects of pulsed electric field (PEF) on the techno-functional properties of dairy proteins.

| Dairy Protein | PEF Conditions | Changes in Protein Functionality | References |
|---|---|---|---|
| Raw milk | Intensity of 30 kV/cm, outlet temperature of 50 ± 1 °C; pulse number of 80 and 120 pulses, pulse width of 2 μs, and pulse frequency of 2 Hz. | Rennet coagulation time (RCT) higher than that of raw milk but lower than that of pasteurized milk. | [105] |
| Whey protein isolate (WPI) | 15–22 V/cm heating phase and 4 to 8 V/cm holding phase, frequency of 25 kHz. | Moderate electric field treatment resulted in a weaker gel structure than conventional heat treatment. | [106] |

**Table 5.** *Cont.*

| Dairy Protein | PEF Conditions | Changes in Protein Functionality | References |
|---|---|---|---|
| β-lactoglobulin | 20 V/cm during holding, 80 V/cm during heating, and frequency of 20 kHz. | At pH 7, moderate electric field and thermal treatment (up to 60 °C) had similar effects on the free SH group relativity. At higher temperatures, conventional heat-treated samples had higher free-SH-group relativity than moderate electric field-treated samples. | [99] |
| WPI | 30–35 kV/cm, 19.2–211 μs, 30–75 °C. | • Emulsions stabilized by PEF-treated and heat-treated (72 °C for 15 s) WPI showed similar droplet sizes and similar emulsifying properties. Increasing the duration of heat treatment to 10 min caused a significant increase in the droplet size of emulsions stabilized by heat-treated WPI. • PEF-treated WPI showed lower gel strength than untreated samples. Increasing the duration of PEF further decreased the gel strength. | [95] |
| β-lactoglobulin | Intensity of 12.5 kV/cm with 40 μF of capacitance. | PEF improved the gelling rate of β-lactoglobulin (at 72 °C) when the number of pulses was less than six. | [31] |
| WPI | 15 to 55 kV/cm, 2 to 8 and 50 to 90 °C. | The gelling properties of WPI increased when treated at 35 kV/cm but decreased after treatment at 45 kV/cm. | [107] |

**Table 6.** Effects of pulsed electric field (PEF) on the techno-functional properties of plant proteins.

| Plant Protein | PEF Conditions | Changes in Protein Functionality | References |
|---|---|---|---|
| Soy protein isolate (SPI) | 0–40 kV/cm for 0–547 μs, 2 ms pulse width, and 500 pulse per second (pps) pulse frequency. | - PEF decreased the solubility and surface hydrophobicity. | [29] |
| Canola protein | 10 to 35 kV, pulse frequency of 600 Hz, and pulse width of 8 μs. | - PEF treatment improved several functional properties of canola protein, including solubility, foaming, and emulsifying properties. | [69] |
| Sunflower protein | 10–150 V/cm for 5 s–2 h at 25–45 °C. | - Moderate electric field treatment at 20 V reduced the interfacial tension at the sunflower protein solution/water interface. | [33] |
| Pea protein isolate | 5, 10, and 20 V/cm and frequencies of 50 Hz and 20 kHz. | - Moderate electric field treatment (50 Hz and 20 V/cm) increased the surface hydrophobicity and improved the gelling properties of pea protein. | [102] |

### 5.1. Protein Solubility

Protein solubility is commonly determined by measuring the concentration of soluble proteins after the centrifugation of protein samples and relating it to the total protein concentration before centrifugation [108]. Protein solubility is influenced by several intrinsic factors, such as amino acid composition, protein molecular weight, the content of hydrophilic and hydrophobic groups on proteins molecules' surfaces, and the content of hydrogen bonds [53,109]. Several extrinsic factors can also affect the protein solubility, including temperature, ionic strength, pH, and the presence of solvents [110]. Protein solubility is important for several protein applications, such as emulsions and foams. Therefore, it is recommended to use highly soluble proteins to form well-dispersed colloidal systems [111]. The effects of electric field treatments on the solubility of several proteins were investigated. There was a decrease in the solubility of pea (from 23.2 to 17.2%), rice (from 16.4 to 9.2%), and gluten (from 25 to 22.4%) concentrates after treatment with moderate electric field strength (1.65 kV/cm, square pulse system) [112]. Similarly, the content of soluble egg white proteins decreased (7.84%) after PEF treatment using a PEF system with square-wave pulses (at 25 kV/cm) [113]. The authors also observed that the average particle size of egg white proteins increased (36.9%) after PEF treatment. PEF unfolded protein molecules and formed insoluble protein molecules. Moreover, intermolecular interactions, such as S-S bonds could occur, resulting in reduced protein solubility. However, with soy protein isolates, Li et al. [29] found that PEF treatment of up to 30 kV/cm using a PEF system with bipolar waveforms improved solubility, whereas PEF at strengths higher than 30 kV/cm resulted in a slight decrease in protein solubility. Additionally, PEF treatment (35 kV/cm for 8 μs) increased the solubility (50.07%) of canola protein compared to that of control samples (43.25%) [69]. Therefore, we conclude that different waveforms and protein types can affect protein solubility differently. However, there is a lack of available knowledge about factors behind the desired solubility of milk proteins after PEF treatment, probably due to the confirmed higher solubility of milk proteins.

### 5.2. Gelling Properties

The gelling properties of proteins are closely associated with the content of -SH groups and disulfide bonds. In the dairy industry, the gelation of milk proteins is an essential factor influencing the quality of many dairy products, including cheese, yogurt, and dairy-based desserts [114]. Perez et al. [31] found that PEF improved the gelling rate of β-lactoglobulin (at 72 °C) when samples were exposed to fewer than six pulses. Yu et al. [105] studied the effects of PEF (20 and 30 kV/cm) at different outlet temperatures on the rennet coagulation characteristics of raw milk. They found that PEF (at 20 °C)-treated milk had higher curd firmness than pasteurized milk samples. Moreover, PEF-treated milk samples had a lower rennet coagulation time (RCT) than pasteurized milk samples. It is known that lower RCT values result in better gelling properties [105]. Jin et al. [107] concluded that the gelling properties of WPI increased when treated at 35 kV/cm but decreased after PEF treatment at 45 kV/cm. As proposed in Figure 7, during PEF treatment, the unfolding of milk proteins and the exposure of -SH groups, followed by the formation of S-S bonds, could be the reason behind the improved gelling properties of milk. Another reason for the improved gelling properties reported in these studies could be the polarization of protein molecules during treatment. Polarized molecules can attract each other through electrostatic forces [31]. However, Sui et al. [24] found that PEF (30 kV/cm)-treated WPI showed lower gel strength than untreated samples; increasing the PEF duration decreased the gel strength of WPI samples. Rodrigues et al. [106] concluded that conventional heat-treated WPI samples had higher gel strength than those subjected to moderate electric field treatment (15–22 V/cm). At pH 7, electrostatic repulsion among protein molecules may reduce the size of protein aggregates [115]. Moreover, applying an electric field could destroy some of the non-covalent bonds between proteins [106]. The water-holding capacity of PEF-treated canola protein increased at lower electric field strength (25 kV/cm) and decreased at higher electric field strength (35 kV/cm) [69]. For pea protein isolate, lower electric field treatment resulted in cohesive, more elastic, and weaker gels with higher water-holding capacity [102]. The inconsistency of gelling properties reported in different studies could be due to the use of different PEF conditions, such as voltage, the shape of the pulse wave, and the type of treatment chamber used.

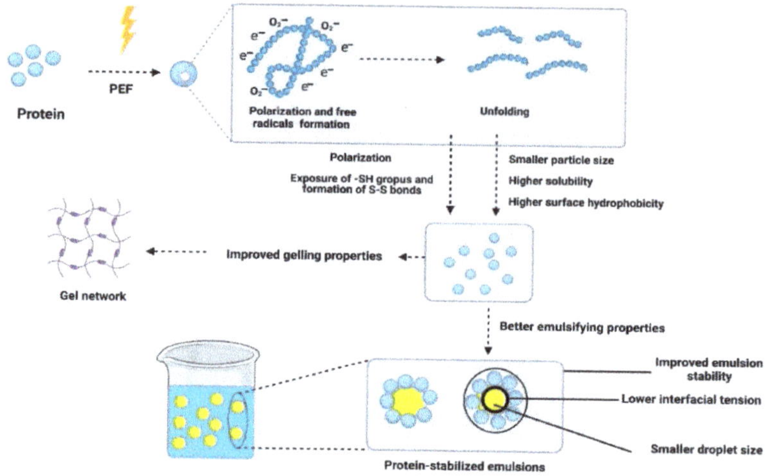


**Figure 7.** Proposed mechanism of PEF effects on the emulsifying and gelling properties of proteins. This figure was created with BioRender.com, (accessed on 18 May 2022) with publication permission.

### 5.3. *Emulsifying and Foaming Properties*

The stability of emulsions is vital for improving the shelf life of emulsion-based food products, such as mayonnaise, ice cream, butter, milk, and margarine. Therefore, several emulsifiers are used to reduce interfacial tension, improving the stability of emulsions [70]. Among them, proteins are widely used as natural emulsifiers due to their surface-active properties [116,117]. Several processing technologies, such as high-pressure treatment [53], ultrasound [70,118], cold plasma treatment [61], and microwaves [56], have been used to improve the emulsifying properties of proteins. Studies on the effects of PEF on the emulsifying foaming properties of milk proteins are scarce. Sui et al. [95] compared the effects of heat treatment and PEF on the emulsifying properties of WPI. They observed that emulsions stabilized by PEF-treated (30 kV/cm) and heat-treated (72 °C for 15 s) samples had similar droplet sizes (~4 µm), whereas emulsions stabilized by WPI heated for 10 min had significantly larger droplet sizes (18.3 µm). Sun et al. [119] studied the effects of PEF treatment (15 and 30 kV/cm) on the emulsifying properties of a WPI–dextran mixture. They found that the PEF-treated mixture had a higher emulsifying activity index (EAI) than the untreated mixture [119]. PEF could facilitate the glycosylation reaction between WPI and dextran. The combination of protein and polysaccharides was confirmed to improve the stability of emulsions. This could be because the hydrophobic regions of proteins can be adsorbed at the surface of oil droplets, and the hydrophilic part of polysaccharides can be oriented towards the water phase, preventing the coalescence of oil droplets through steric stabilization [70]. Zhang et al. [69] found that PEF pre-treatment of canola seeds prior oil and protein extraction improved the emulsifying and foaming properties of the resulting canola proteins. PEF could improve the solubility of plant proteins and promote the exposure of their hydrophobic groups to the surface, thus improving their emulsifying and foaming properties. More studies are needed to understand the effects of PEF treatment on the emulsifying and foaming properties of plant and milk proteins. The changes in the protein structures induced by PEF treatment could improve the techno-functional properties of proteins. As proposed in Figure 7, PEF could polarize and unfold protein molecules, exposing the hydrophobic groups to the surface of molecules [5]. Additionally, PEF can increase solubility and reduce the particle size of protein molecules at a certain electric field strength. These changes could reduce the interfacial tension at the oil/water interface, improve the emulsifying properties of food proteins, and enhance the stability of protein-stabilized emulsions [11]. However, there is still a lack of detailed information on the mechanism of action of PEF and its effects on protein functionality due to the limited number of studies conducted in this area. Thus, more fundamental research at the molecular scale is required to establish a clear mechanism of PEF effects on protein functionality.

## 6. Conclusions and Future Perspectives

Pulsed electric field is a promising green technology that can be utilized in many food applications. With an increase in sustainable development needs, the utilization of PEF in the food industry is expected to increase in the coming years. The conclusions of the available studies that investigated the effects of PEF on the structure and techno-functional properties of milk and plant proteins can be summarized in the following points:

- In general, PEF treatment at low electric strength (<10 kV/cm) cannot change the structure of proteins;
- PEF treatment conditions, such as electric strength, pulse shape, pulse duration, and the type of treatment chamber, have a significant impact on the effects of PEF on the structure and techno-functional properties of proteins'
- The effects of PEF on structure and techno-functional properties are vary from one protein type to another.

As a limited number of studies have been conducted to investigate the effects of PEF on food proteins, several aspects need to be investigated in the future. The impact of different electric field strengths on the structure and techno-functional properties of proteins must be studied to define the optimum PEF conditions to improve the techno-functional properties of such proteins. Investigation of the impacts of using nanosecond PEF treatment on the structure and techno-functional properties of food proteins is recommended. Moreover, PEF, as a promising green technology, can be introduced at a large scale to produce highly effective emulsifiers. Research could be conducted to determine the possibility of using PEF technology as an emulsification technique to produce food-based stable emulsions. The main challenge of PEF applications is that many factors (such as PEF device parameters and external factors, such as conductivity, pH, and concentration of treated solutions) can affect the treatment results. Consequently, studies focusing on thermal, chemical, and biophysical components of PEF effects on protein structures should be conducted until clear mechanisms are elucidated. It is also extremely important that authors provide all the necessary details about treatment conditions so the analog studies can be implemented, and results can be compared between those studies. We recommend referring to the guidelines and recommendations proposed by Cemazar et al. [68] for reporting on PEF applications. Moreover, collaborations between the food industry and academic institutions are needed to design and build more effective and energy-efficient PEF devices with controlled treatment conditions.

**Author Contributions:** A.T.: literature search, data analysis, writing—original draft preparation, reviewing and editing; F.C.: data interpretation, writing—reviewing and editing; P.Š.: preparation of figures and tables preparation, original draft preparation, reviewing and editing; V.S.: reviewing and editing; M.A.E.G.: data collection and data analysis; A.S.: data interpretation, supervision, reviewing and editing. All authors have read and agreed to the published version of the manuscript.

**Funding:** This research received no external funding.

**Institutional Review Board Statement:** Not applicable.

**Informed Consent Statement:** Not applicable.

**Data Availability Statement:** Data is contained within the article.

**Conflicts of Interest:** The authors declare no conflict of interest.

## References


1. Jambrak, A.R.; Mason, T.J.; Lelas, V.; Herceg, Z.; Herceg, I.L. Effect of Ultrasound Treatment on Solubility and Foaming Properties of Whey Protein Suspensions. *J. Food Eng.* **2008**, *86*, 281–287. [CrossRef]
2. Wagoner, T.; Vardhanabhuti, B.; Foegeding, E.A. Designing Whey Protein-Polysaccharide Particles for Colloidal Stability. *Annu. Rev. Food Sci. Technol.* **2016**, *7*, 93–116. [CrossRef] [PubMed]
3. European Comission Milk Market Observatory. 2020. Available online: https://ec.europa.eu/info/food-farming-fisheries/farming/facts-and-figures/markets/trade/trade-sector/animal-products/milk-and-dairy-products_en (accessed on 21 December 2020).
4. Nunes, L.; Tavares, G.M. Thermal Treatments and Emerging Technologies: Impacts on the Structure and Techno-Functional Properties of Milk Proteins. *Trends Food Sci. Technol.* **2019**, *90*, 88–99. [CrossRef]
5. Han, Z.; Cai, M.J.; Cheng, J.H.; Sun, D.W. Effects of Electric Fields and Electromagnetic Wave on Food Protein Structure and Functionality: A Review. *Trends Food Sci. Technol.* **2018**, *75*, 1–9. [CrossRef]
6. Sá, A.G.A.; Moreno, Y.M.F.; Carciofi, B.A.M. Plant Proteins as High-Quality Nutritional Source for Human Diet. *Trends Food Sci. Technol.* **2020**, *97*, 170–184. [CrossRef]
7. Geng, M.; Wang, Z.; Qin, L.; Taha, A.; Du, L.; Xu, X.; Pan, S.; Hu, H. Effect of Ultrasound and Coagulant Types on Properties of β-Carotene Bulk Emulsion Gels Stabilized by Soy Protein. *Food Hydrocoll.* **2022**, *123*, 107146. [CrossRef]
8. Akharume, F.U.; Aluko, R.E.; Adedeji, A.A. Modification of Plant Proteins for Improved Functionality: A Review. *Compr. Rev. Food Sci. Food Saf.* **2021**, *20*, 198–224. [CrossRef]
9. Geng, M.; Hu, T.; Zhou, Q.; Taha, A.; Qin, L.; Lv, W.; Xu, X.; Pan, S.; Hu, H. Effects of Different Nut Oils on the Structures and Properties of Gel-like Emulsions Induced by Ultrasound Using Soy Protein as an Emulsifier. *Int. J. Food Sci. Technol.* **2021**, *56*, 1649–1660. [CrossRef]
10. Nishinari, K.; Fang, Y.; Guo, S.; Phillips, G.O. Soy Proteins: A Review on Composition, Aggregation and Emulsification. *Food Hydrocoll.* **2014**, *39*, 301–318. [CrossRef]
11. Rodrigues, R.M.; Avelar, Z.; Machado, L.; Pereira, R.N.; Vicente, A.A. Electric Field Effects on Proteins—Novel Perspectives on Food and Potential Health Implications. *Food Res. Int.* **2020**, *137*, 109709. [CrossRef]
12. Jaeger, H.; Meneses, N.; Knorr, D. Impact of PEF Treatment Inhomogeneity Such as Electric Field Distribution, Flow Characteristics and Temperature Effects on the Inactivation of *E. Coli* and Milk Alkaline Phosphatase. *Innov. Food Sci. Emerg. Technol.* **2009**, *10*, 470–480. [CrossRef]
13. Alirezalu, K.; Munekata, P.E.S.; Parniakov, O.; Barba, F.J.; Witt, J.; Toepfl, S.; Wiktor, A.; Lorenzo, J.M. Pulsed Electric Field and Mild Heating for Milk Processing: A Review on Recent Advances. *J. Sci. Food Agric.* **2020**, *100*, 16–24. [CrossRef] [PubMed]
14. Soltanzadeh, M.; Peighambardoust, S.H.; Gullon, P.; Hesari, J.; Gullón, B.; Alirezalu, K.; Lorenzo, J. Quality Aspects and Safety of Pulsed Electric Field (PEF) Processing on Dairy Products: A Comprehensive Review. *Food Rev. Int.* **2020**, 1–22. [CrossRef]
15. Simonis, P.; Kersulis, S.; Stankevich, V.; Sinkevic, K.; Striguniene, K.; Ragoza, G.; Stirke, A. Pulsed Electric Field Effects on Inactivation of Microorganisms in Acid Whey. *Int. J. Food Microbiol.* **2019**, *291*, 128–134. [CrossRef] [PubMed]
16. Knorr, D.; Froehling, A.; Jaeger, H.; Reineke, K.; Schlueter, O.; Schoessler, K. Emerging Technologies in Food Processing. *Annu. Rev. Food Sci. Technol.* **2011**, *2*, 203–235. [CrossRef]
17. Jin, W.; Wang, Z.; Peng, D.; Shen, W.; Zhu, Z.; Cheng, S.; Li, B.; Huang, Q. Effect of Pulsed Electric Field on Assembly Structure of α-Amylase and Pectin Electrostatic Complexes. *Food Hydrocoll.* **2020**, *101*, 105547. [CrossRef]
18. Giteru, S.G.; Oey, I.; Ali, M.A. Feasibility of Using Pulsed Electric Fields to Modify Biomacromolecules: A Review. *Trends Food Sci. Technol.* **2018**, *72*, 91–113. [CrossRef]
19. Toepfl, S.; Heinz, V.; Knorr, D. High Intensity Pulsed Electric Fields Applied for Food Preservation. *Chem. Eng. Processing Process. Intensif.* **2007**, *46*, 537–546. [CrossRef]
20. Arshad, R.N.; Abdul-Malek, Z.; Munir, A.; Buntat, Z.; Ahmad, M.H.; Jusoh, Y.M.M.; Bekhit, A.E.D.; Roobab, U.; Manzoor, M.F.; Aadil, R.M. Electrical Systems for Pulsed Electric Field Applications in the Food Industry: An Engineering Perspective. *Trends Food Sci. Technol.* **2020**, *104*, 1–13. [CrossRef]
21. Jaeger, H.; Balasa, A.; Knorr, D. Food Industry Applications for Pulsed Electric Fields. In *Electrotechnologies for Extraction from Food Plants and Biomaterials*; Food Engineering Series; Springer: New York, NY, USA, 2008; pp. 181–216.
22. Xiang, B.Y.; Ngadi, M.O.; Ochoa-Martinez, L.A.; Simpson, M.V. Pulsed Electric Field-Induced Structural Modification of Whey Protein Isolate. *Food Bioprocess. Technol.* **2011**, *4*, 1341–1348. [CrossRef]
23. Zhang, S.; Sun, L.; Ju, H.; Bao, Z.; Zeng, X.; Lin, S. Research Advances and Application of Pulsed Electric Field on Proteins and Peptides in Food. *Food Res. Int.* **2020**, *139*, 109914. [CrossRef] [PubMed]
24. Sui, Q.; Roginski, H.; Williams, R.P.W.; Versteeg, C.; Wan, J. Effect of Pulsed Electric Field and Thermal Treatment on the Physicochemical Properties of Lactoferrin with Different Iron Saturation Levels. *Int. Dairy J.* **2010**, *20*, 707–714. [CrossRef]
25. Fernandez-Diaz, M.D.; Barsotti, L.; Dumay, E.; Cheftel, J.C. Effects of Pulsed Electric Fields on Ovalbumin Solutions and Dialyzed Egg White. *J. Agric. Food Chem.* **2000**, *48*, 2332–2339. [CrossRef] [PubMed]
26. Sharma, P.; Oey, I.; Everett, D.W. Effect of Pulsed Electric Field Processing on the Functional Properties of Bovine Milk. *Trends Food Sci. Technol.* **2014**, *35*, 87–101. [CrossRef]
27. Zhao, W.; Yang, R. Pulsed Electric Field Induced Aggregation of Food Proteins: Ovalbumin and Bovine Serum Albumin. *Food Bioprocess. Technol.* **2012**, *5*, 1706–1714. [CrossRef]

28. Franco, I.; Pérez, M.D.; Conesa, C.; Calvo, M.; Sánchez, L. Effect of Technological Treatments on Bovine Lactoferrin: An Overview. *Food Res. Int.* **2018**, *106*, 173–182. [CrossRef]
29. Li, Y.; Chen, Z.; Mo, H. Effects of Pulsed Electric Fields on Physicochemical Properties of Soybean Protein Isolates. *LWT—Food Sci. Technol.* **2007**, *40*, 1167–1175. [CrossRef]
30. Li, Y.-Q. Structure Changes of Soybean Protein Isolates by Pulsed Electric Fields. *Phys. Procedia* **2012**, *33*, 132–137. [CrossRef]
31. Perez, O.E.; Pilosof, A.M.R. Pulsed Electric Fields Effects on the Molecular Structure and Gelation of β-Lactoglobulin Concentrate and Egg White. *Food Res. Int.* **2004**, *37*, 102–110. [CrossRef]
32. Wei, J.N.; Zeng, X.A.; Tang, T.; Jiang, Z.; Liu, Y.Y. Unfolding and Nanotube Formation of Ovalbumin Induced by Pulsed Electric Field. *Innov. Food Sci. Emerg. Technol.* **2018**, *45*, 249–254. [CrossRef]
33. Subaşı, B.G.; Jahromi, M.; Casanova, F.; Capanoglu, E.; Ajalloueian, F.; Mohammadifar, M.A. Effect of Moderate Electric Field on Structural and Thermo-Physical Properties of Sunflower Protein and Sodium Caseinate. *Innov. Food Sci. Emerg. Technol.* **2021**, *67*, 102593. [CrossRef]
34. Tobajas, A.P.; Agulló-García, A.; Cubero, J.L.; Colás, C.; Segura-Gil, I.; Sánchez, L.; Calvo, M.; Pérez, M.D. Effect of High Pressure and Pulsed Electric Field on Denaturation and Allergenicity of Pru p 3 Protein from Peach. *Food Chem.* **2020**, *321*, 126745. [CrossRef] [PubMed]
35. Jaeger, H.; Meneses, N.; Knorr, D. Food Technologies: Pulsed Electric Field Technology. In *Encyclopedia of Food Safety*; Motarjemi, Y., Ed.; Academic Press: Waltham, MA, USA, 2014; pp. 239–244. ISBN 978-0-12-378613-5.
36. Knorr, D.; Geulen, M.; Grahl, T.; Sitzmann, W. Food Application of High Electric Field Pulses. *Trends Food Sci. Technol.* **1994**, *5*, 71–75. [CrossRef]
37. Soliva-Fortuny, R.; Balasa, A.; Knorr, D.; Martín-Belloso, O. Effects of Pulsed Electric Fields on Bioactive Compounds in Foods: A Review. *Trends Food Sci. Technol.* **2009**, *20*, 544–556. [CrossRef]
38. Liu, Y.-F.; Oey, I.; Bremer, P.; Carne, A.; Silcock, P. Modifying the Functional Properties of Egg Proteins Using Novel Processing Techniques: A Review. *Compr. Rev. Food Sci. Food Saf.* **2019**, *18*, 986–1002. [CrossRef]
39. Gerlach, D.; Alleborn, N.; Baars, A.; Delgado, A.; Moritz, J.; Knorr, D. Numerical Simulations of Pulsed Electric Fields for Food Preservation: A Review. *Innov. Food Sci. Emerg. Technol.* **2008**, *9*, 408–417. [CrossRef]
40. Mosqueda-Melgar, J.; Elez-Martínez, P.; Raybaudi-Massilia, R.M.; Martín-Belloso, O. Effects of Pulsed Electric Fields on Pathogenic Microorganisms of Major Concern in Fluid Foods: A Review. *Crit. Rev. Food Sci. Nutr.* **2008**, *48*, 747–759. [CrossRef]
41. Huang, K.; Wang, J. Designs of Pulsed Electric Fields Treatment Chambers for Liquid Foods Pasteurization Process: A Review. *J. Food Eng.* **2009**, *95*, 227–239. [CrossRef]
42. Gabrić, D.; Barba, F.; Roohinejad, S.; Gharibzahedi, S.M.T.; Radojčin, M.; Putnik, P.; Bursać Kovačević, D. Pulsed Electric Fields as an Alternative to Thermal Processing for Preservation of Nutritive and Physicochemical Properties of Beverages: A Review. *J. Food Process. Eng.* **2018**, *41*, e12638. [CrossRef]
43. Huang, K.; Tian, H.; Gai, L.; Wang, J. A Review of Kinetic Models for Inactivating Microorganisms and Enzymes by Pul sed Electric Field Processing. *J. Food Eng.* **2012**, *111*, 191–207. [CrossRef]
44. Buckow, R.; Ng, S.; Toepfl, S. Pulsed Electric Field Processing of Orange Juice: A Review on Microbial, Enzymatic, Nutritional, and Sensory Quality and Stability. *Compr. Rev. Food Sci. Food Saf.* **2013**, *12*, 455–467. [CrossRef] [PubMed]
45. Niu, D.; Zeng, X.A.; Ren, E.F.; Xu, F.Y.; Li, J.; Wang, M.S.; Wang, R. Review of the Application of Pulsed Electric Fields (PEF) Technology for Food Processing in China. *Food Res. Int.* **2020**, *137*, 109715. [CrossRef] [PubMed]
46. Yogesh, K. Pulsed Electric Field Processing of Egg Products: A Review. *J. Food Sci. Technol.* **2016**, *53*, 934–945. [CrossRef] [PubMed]
47. Gómez, B.; Munekata, P.E.S.; Gavahian, M.; Barba, F.J.; Martí-Quijal, F.J.; Bolumar, T.; Campagnol, P.C.B.; Tomasevic, I.; Lorenzo, J.M. Application of Pulsed Electric Fields in Meat and Fish Processing Industries: An Overview. *Food Res. Int.* **2019**, *123*, 95–105. [CrossRef] [PubMed]
48. Flores-Jiménez, N.T.; Ulloa, J.A.; Silvas, J.E.U.; Ramírez, J.C.R.; Ulloa, P.R.; Rosales, P.U.B.; Carrillo, Y.S.; Leyva, R.G. Effect of High-Intensity Ultrasound on the Compositional, Physicochemical, Biochemical, Functional and Structural Properties of Canola (*Brassica Napus* L.) Protein Isolate. *Food Res. Int.* **2019**, *121*, 947–956. [CrossRef]
49. Geng, M.; Liu, J.; Hu, H.; Qin, L.; Taha, A.; Zhang, Z. A Comprehensive Study on Structures and Characterizations of 7S Protein Treated by High Intensity Ultrasound at Different PH and Ionic Strengths. *Food Chem.* **2021**, *373*, 131378. [CrossRef]
50. Zheng, T.; Li, X.; Taha, A.; Wei, Y.; Hu, T.; Fatamorgana, P.B.; Zhang, Z.; Liu, F.; Xu, X.; Pan, S.; et al. Effect of High Intensity Ultrasound on the Structure and Physicochemical Properties of Soy Protein Isolates Produced by Different Denaturation Methods. *Food Hydrocoll.* **2019**, *97*, 105216. [CrossRef]
51. Hu, H.; Wu, J.; Li-Chan, E.C.Y.; Zhu, L.; Zhang, F.; Xu, X.; Fan, G.; Wang, L.; Huang, X.; Pan, S. Effects of Ultrasound on Structural and Physical Properties of Soy Protein Isolate (SPI) Dispersions. *Food Hydrocoll.* **2013**, *30*, 647–655. [CrossRef]
52. Xue, S.; Xu, X.; Shan, H.; Wang, H.; Yang, J.; Zhou, G. Effects of High-Intensity Ultrasound, High-Pressure Processing, and High-Pressure Homogenization on the Physicochemical and Functional Properties of Myofibrillar Proteins. *Innov. Food Sci. Emerg. Technol.* **2018**, *45*, 354–360. [CrossRef]
53. Zhu, S.M.; Lin, S.L.; Ramaswamy, H.S.; Yu, Y.; Zhang, Q.T. Enhancement of Functional Properties of Rice Bran Proteins by High Pressure Treatment and Their Correlation with Surface Hydrophobicity. *Food Bioprocess. Technol.* **2017**, *10*, 317–327. [CrossRef]
54. Tang, C.H.; Ma, C.Y. Effect of High Pressure Treatment on Aggregation and Structural Properties of Soy Protein Isolate. *LWT—Food Sci. Technol.* **2009**, *42*, 606–611. [CrossRef]

55. He, X.H.; Liu, H.Z.; Liu, L.; Zhao, G.L.; Wang, Q.; Chen, Q.L. Effects of High Pressure on the Physicochemical and Functional Properties of Peanut Protein Isolates. *Food Hydrocoll.* **2014**, *36*, 123–129. [CrossRef]
56. Zheng, Y.; Li, Z.; Zhang, C.; Zheng, B.; Tian, Y. Effects of Microwave-Vacuum Pre-Treatment with Different Power Levels on the Structural and Emulsifying Properties of Lotus Seed Protein Isolates. *Food Chem.* **2020**, *311*, 125932. [CrossRef] [PubMed]
57. Wei, S.; Yang, Y.; Feng, X.; Li, S.; Zhou, L.; Wang, J.; Tang, X. Structures and Properties of Chicken Myofibrillar Protein Gel Induced by Microwave Heating. *Int. J. Food Sci. Technol.* **2020**, *55*, 2691–2699. [CrossRef]
58. Xiang, S.; Zou, H.; Liu, Y.; Ruan, R. Effects of Microwave Heating on the Protein Structure, Digestion Properties and Maillard Products of Gluten. *J. Food Sci. Technol.* **2020**, *57*, 2139–2149. [CrossRef]
59. Sun, X.; Ohanenye, I.C.; Ahmed, T.; Udenigwe, C.C. Microwave Treatment Increased Protein Digestibility of Pigeon Pea (*Cajanus Cajan*) Flour: Elucidation of Underlying Mechanisms. *Food Chem.* **2020**, *329*, 127196. [CrossRef]
60. Moosavi, M.H.; Khani, M.R.; Shokri, B.; Hosseini, S.M.; Shojaee-Aliabadi, S.; Mirmoghtadaie, L. Modifications of Protein-Based Films Using Cold Plasma. *Int. J. Biol. Macromol.* **2020**, *142*, 769–777. [CrossRef]
61. Sharma, S.; Singh, R.K. Cold Plasma Treatment of Dairy Proteins in Relation to Functionality Enhancement. *Trends Food Sci. Technol.* **2020**, *102*, 30–36. [CrossRef]
62. Pérez-Andrés, J.M.; Álvarez, C.; Cullen, P.J.; Tiwari, B.K. Effect of Cold Plasma on the Techno-Functional Properties of Animal Protein Food Ingredients. *Innov. Food Sci. Emerg. Technol.* **2019**, *58*, 102205. [CrossRef]
63. Parniakov, O.; Wiktor, A.; Toepfl, S. Application Concepts for PEF in Food and Biotechnology. *Innov. Food Processing Technol.: A Compr. Rev.* **2021**, 160–172. [CrossRef]
64. Wan, J.; Coventry, J.; Swiergon, P.; Sanguansri, P.; Versteeg, C. Advances in Innovative Processing Technologies for Microbial Inactivation and Enhancement of Food Safety—Pulsed Electric Field and Low-Temperature Plasma. *Trends Food Sci. Technol.* **2009**, *20*, 414–424. [CrossRef]
65. Ammelt, D.; Lammerskitten, A.; Wiktor, A.; Barba, F.J.; Toepfl, S.; Parniakov, O. The Impact of Pulsed Electric Fields on Quality Parameters of Freeze-Dried Red Beets and Pineapples. *Int. J. Food Sci. Technol.* **2021**, *56*, 1777–1787. [CrossRef]
66. Hill, K.; Ostermeier, R.; Töpfl, S.; Heinz, V. Pulsed Electric Fields in the Potato Industry. In *Pulsed Electric Fields Technology for the Food Industry*; Food Engineering Series; Springer: Cham, Switzerland, 2022; pp. 325–335. [CrossRef]
67. Shorstkii, I.; Sosnin, M.; Smetana, S.; Toepfl, S.; Parniakov, O.; Wiktor, A. Correlation of the Cell Disintegration Index with Luikov's Heat and Mass Transfer Parameters for Drying of Pulsed Electric Field (PEF) Pretreated Plant Materials. *J. Food Eng.* **2022**, *316*, 110822. [CrossRef]
68. Cemazar, M.; Sersa, G.; Frey, W.; Miklavcic, D.; Teissié, J. Recommendations and Requirements for Reporting on Applications of Electric Pulse Delivery for Electroporation of Biological Samples. *Bioelectrochemistry* **2018**, *122*, 69–76. [CrossRef]
69. Zhang, L.; Wang, L.J.; Jiang, W.; Qian, J.Y. Effect of Pulsed Electric Field on Functional and Structural Properties of Canola Protein by Pretreating Seeds to Elevate Oil Yield. *LWT—Food Sci. Technol.* **2017**, *84*, 73–81. [CrossRef]
70. Taha, A.; Ahmed, E.; Ismaiel, A.; Ashokkumar, M.; Xu, X.; Pan, S.; Hu, H. Ultrasonic Emulsification: An Overview on the Preparation of Different Emulsifiers-Stabilized Emulsions. *Trends Food Sci. Technol.* **2020**, *105*, 363–377. [CrossRef]
71. Hu, H.; Taha, A.; Khalifa, I. Effects of Novel Processing Methods on Structure, Functional Properties, and Health Benefits of Soy Protein. In *Phytochemicals in Soybeans*; CRC Press: Boca Raton, FL, USA, 2022; pp. 301–318. [CrossRef]
72. Rahman, M.M.; Lamsal, B.P. Ultrasound-Assisted Extraction and Modification of Plant-Based Proteins: Impact on Physicochemical, Functional, and Nutritional Properties. *Compr. Rev. Food Sci. Food Saf.* **2021**, *20*, 1457–1480. [CrossRef]
73. Liu, H.H.; Kuo, M.I. Ultra High Pressure Homogenization Effect on the Proteins in Soy Flour. *Food Hydrocoll.* **2016**, *52*, 741–748. [CrossRef]
74. Wang, C.; Wang, J.; Zhu, D.; Hu, S.; Kang, Z.; Ma, H. Effect of Dynamic Ultra-High Pressure Homogenization on the Structure and Functional Properties of Whey Protein. *J. Food Sci. Technol.* **2020**, *57*, 1301–1309. [CrossRef]
75. Cao, H.; Jiao, X.; Fan, D.; Huang, J.; Zhao, J.; Yan, B.; Zhou, W.; Zhang, H.; Wang, M. Microwave Irradiation Promotes Aggregation Behavior of Myosin through Conformation Changes. *Food Hydrocoll.* **2019**, *96*, 11–19. [CrossRef]
76. Solaesa, Á.G.; Villanueva, M.; Muñoz, J.M.; Ronda, F. Dry-Heat Treatment vs. Heat-Moisture Treatment Assisted by Microwave Radiation: Techno-Functional and Rheological Modifications of Rice Flour. *LWT* **2021**, *141*, 110851. [CrossRef]
77. Thirumdas, R.; Sarangapani, C.; Annapure, U.S. Cold Plasma: A Novel Non-Thermal Technology for Food Processing. *Food Biophys.* **2014**, *10*, 1–11. [CrossRef]
78. Ji, H.; Dong, S.; Han, F.; Li, Y.; Chen, G.; Li, L.; Chen, Y. Effects of Dielectric Barrier Discharge (DBD) Cold Plasma Treatment on Physicochemical and Functional Properties of Peanut Protein. *Food Bioprocess. Technol.* **2018**, *11*, 344–354. [CrossRef]
79. Akiyama, H.; Katsuki, S.; Redondo, L.; Akiyama, M.; Pemen, A.J.M.; Huiskamp, T.; Beckers, F.J.C.M.; van Heesch, E.J.M.; Winands, G.J.J.; Voeten, S.J.; et al. Pulsed Power Technology. In *Bioelectrics*; Springer: Tokyo, Japan, 2017; pp. 41–107. ISBN 978-4-43156-095-1.
80. Barbosa-Cánovas, G.V.; Altunakar, B. Pulsed Electric Fields Processing of Foods: An Overview. In *Pulsed Electric Fields Technology for the Food Industry*; Food Engineering Series; Springer: Boston, MA, USA, 2006; pp. 3–26.
81. Buchmann, L. Emerging Pulsed Electric Field Process Development for Bio-Based Applications. Ph.D. Thesis, ETH Zurich, Zurich, Switzerland, 2020.
82. Schoenbach, K.H.; Neumann, E.; Heller, R.; Vernier, P.T.; Teissie, J.; Beebe, S.J. Introduction. In *Bioelectrics*; Springer: Tokyo, Japan, 2017; pp. 1–40. ISBN 978-4-43156-095-1.
83. Platt, C. *Encyclopedia of Electronic Components*; O'Reilly Media Inc.: Newton, MA, USA, 2012; Volume 2012, ISBN 978-1-44933-389-8.

84. Sharma, P.; Bhatti, T.S. A Review on Electrochemical Double-Layer Capacitors. *Energy Convers. Manag.* **2010**, *51*, 2901–2912. [CrossRef]
85. Toepfl, S.; Heinz, V.; Knorr, D. Overview of Pulsed Electric Field Processing for Food. In *Emerging Technologies for Food Processing*; Elsevier Ltd.: Amstrerdam, The Netherlands, 2005; pp. 69–97. ISBN 978-0-12676-757-5.
86. Nishino, A. Capacitors: Operating Principles, Current Market and Technical Trends. *J. Power Sources* **1996**, *60*, 137–147. [CrossRef]
87. Maged, M.; Eiss, A.H.A. Pulsed Electric Fields for Food Processing Technology. In *Structure and Function of Food Engineering*; InTech: London, UK, 2012.
88. Alexander, C.K.; Sadiku, M. *Fundamentals of Electric Circuits*, 6th ed.; McGraw-Hill Education: New York, NY, USA, 2017; ISBN 978-0-07802-822-9.
89. Aoude, C.; Lammerskitten, A.; Parniakov, O.; Zhang, R.; Grimi, N.; El Zakhem, H.; Vorobiev, E. Equipment and Recent Advances in Pulsed Electric Fields. In *Innovative and Emerging Technologies in the Bio-Marine Food Sector*; Elsevier: Amsterdam, The Netherlands, 2021; pp. 149–172. [CrossRef]
90. Heinz, V.; Toepfl, S. Pulsed Electric Fields Industrial Equipment Design. In *Pulsed Electric Fields Technology for the Food Industry*; Food Engineering Series; Springer: Cham, Switzerland, 2022; pp. 489–504. [CrossRef]
91. Huppertz, T.; Vasiljevic, T.; Zisu, B.; Deeth, H. Novel Processing Technologies: Effects on Whey Protein Structure and Functionality. In *Whey Proteins: From Milk to Medicine*; Elsevier Inc.: Amsterdam, The Netherlands, 2018; ISBN 978-0-12812-124-5.
92. Abd El-Salam, M.H.; El-Shibiny, S.; Salem, A. Factors Affecting the Functional Properties of Whey Protein Products: A Review. *Food Rev. Int.* **2009**, *25*, 251–270. [CrossRef]
93. Onwulata, C.I.; Tunick, M.H.; Qi, P.X. *Extrusion Texturized Dairy Proteins. Processing and Application*, 1st ed.; Elsevier Inc.: Amsterdam, The Netherlands, 2011; Volume 62, ISBN 978-0-12385-989-1.
94. Mazurkiewicz, J.; Kołoczek, H.; Tomasik, P. Effect of the External Electric Field on Selected Tripeptides. *Amino Acids* **2015**, *47*, 1399–1408. [CrossRef]
95. Sui, Q.; Roginski, H.; Williams, R.P.W.; Versteeg, C.; Wan, J. Effect of Pulsed Electric Field and Thermal Treatment on the Physicochemical and Functional Properties of Whey Protein Isolate. *Int. Dairy J.* **2011**, *21*, 206–213. [CrossRef]
96. Bekard, I.; Dunstan, D.E. Electric Field Induced Changes in Protein Conformation. *Soft Matter* **2014**, *10*, 431–437. [CrossRef]
97. Sharma, P.; Oey, I.; Everett, D.W. Thermal Properties of Milk Fat, Xanthine Oxidase, Caseins and Whey Proteins in Pulsed Electric Field-Treated Bovine Whole Milk. *Food Chem.* **2016**, *207*, 34–42. [CrossRef]
98. Iametti, S.; Gregori, B.; Vecchio, G.; Bonomi, F. Modifications Occur at Different Structural Levels During the Heat Denaturation of Beta-Lactoglobulin. *Eur. J. Biochem.* **1996**, *237*, 106–112. [CrossRef] [PubMed]
99. Rodrigues, R.M.; Vicente, A.A.; Petersen, S.B.; Pereira, R.N. Electric Field Effects on β-Lactoglobulin Thermal Unfolding as a Function of PH—Impact on Protein Functionality. *Innov. Food Sci. Emerg. Technol.* **2019**, *52*, 1–7. [CrossRef]
100. Odriozola-Serrano, I.; Bendicho-Porta, S.; Martín-Belloso, O. Comparative Study on Shelf Life of Whole Milk Processed by High-Intensity Pulsed Electric Field or Heat Treatment. *J. Dairy Sci.* **2006**, *89*, 905–911. [CrossRef]
101. Liu, Y.Y.; Zeng, X.A.; Deng, Z.; Yu, S.J.; Yamasaki, S. Effect of Pulsed Electric Field on the Secondary Structure and Thermal Properties of Soy Protein Isolate. *Eur. Food Res. Technol.* **2011**, *233*, 841–850. [CrossRef]
102. Chen, Y.; Wang, T.; Zhang, Y.; Yang, X.; Du, J.; Yu, D.; Xie, F. Effect of Moderate Electric Fields on the Structural and Gelation Properties of Pea Protein Isolate. *Innov. Food Sci. Emerg. Technol.* **2022**, *77*, 102959. [CrossRef]
103. Kinsella, J.E.; Whitehead, D.M. Proteins in Whey: Chemical, Physical, and Functional Properties. *Adv. Food Nutr. Res.* **1989**, *33*, 343–438. [CrossRef]
104. Kostić, A.T.; Barać, M.B.; Stanojević, S.P.; Milojković-Opsenica, D.M.; Tešić, Ž.L.; Šikoparija, B.; Radišić, P.; Prentović, M.; Pešić, M.B. Physicochemical Composition and Techno-Functional Properties of Bee Pollen Collected in Serbia. *LWT—Food Sci. Technol.* **2015**, *62*, 301–309. [CrossRef]
105. Yu, L.J.; Ngadi, M.; Raghavan, G.S.V. Effect of Temperature and Pulsed Electric Field Treatment on Rennet Coagulation Properties of Milk. *J. Food Eng.* **2009**, *95*, 115–118. [CrossRef]
106. Rodrigues, R.M.; Martins, A.J.; Ramos, O.L.; Malcata, F.X.; Teixeira, J.A.; Vicente, A.A.; Pereira, R.N. Influence of Moderate Electric Fields on Gelation of Whey Protein Isolate. *Food Hydrocoll.* **2015**, *43*, 329–339. [CrossRef]
107. Jin, S.; Yin, Y.; Wang, Y. Effects of Combined Pulsed Electric Field and Heat Treatment on Texture Characteristics of Whey Protein Gels. *Nongye Jixie Xuebao/Trans. Chin. Soc. Agric. Mach.* **2013**, *44*, 142–146. [CrossRef]
108. Tian, Y.; Zhang, Z.; Zhang, P.; Taha, A.; Hu, H.; Pan, S. The Role of Conformational State of PH-Shifted β-Conglycinin on the Oil/Water Interfacial Properties and Emulsifying Capacities. *Food Hydrocoll.* **2020**, *108*, 105990. [CrossRef]
109. Golovanov, A.P.; Hautbergue, G.M.; Wilson, S.A.; Lian, L.Y. A Simple Method for Improving Protein Solubility and Long-Term Stability. *J. Am. Chem. Soc.* **2004**, *126*, 8933–8939. [CrossRef] [PubMed]
110. Kramer, R.M.; Shende, V.R.; Motl, N.; Pace, C.N.; Scholtz, J.M. Toward a Molecular Understanding of Protein Solubility: Increased Negative Surface Charge Correlates with Increased Solubility. *Biophys. J.* **2012**, *102*, 1907–1915. [CrossRef] [PubMed]
111. Zayas, J.F. Solubility of Proteins. In *Functionality of Proteins in Food*; Springer: Berlin/Heidelberg, Germany, 1997; pp. 6–75.
112. Melchior, S.; Calligaris, S.; Bisson, G.; Manzocco, L. Understanding the Impact of Moderate-Intensity Pulsed Electric Fields (MIPEF) on Structural and Functional Characteristics of Pea, Rice and Gluten Concentrates. *Food Bioprocess. Technol.* **2020**, *13*, 2145–2155. [CrossRef]

113. Wu, L.; Zhao, W.; Yang, R.; Chen, X. Effects of Pulsed Electric Fields Processing on Stability of Egg White Proteins. *J. Food Eng.* **2014**, *139*, 13–18. [CrossRef]
114. Lucey, J.A. Formation and Physical Properties of Milk Protein Gels. *J. Dairy Sci.* **2002**, *85*, 281–294. [CrossRef]
115. Schmitt, C.; Bovay, C.; Rouvet, M.; Shojaei-Rami, S.; Kolodziejczyk, E. Whey Protein Soluble Aggregates from Heating with NaCl: Physicochemical, Interfacial, and Foaming Properties. *Langmuir* **2007**, *23*, 4155–4166. [CrossRef]
116. McClements, D.J. Protein-Stabilized Emulsions. *Curr. Opin. Colloid Interface Sci.* **2004**, *9*, 305–313. [CrossRef]
117. Taha, A.; Ahmed, E.; Hu, T.; Xu, X.; Pan, S.; Hu, H. Effects of Different Ionic Strengths on the Physicochemical Properties of Plant and Animal Proteins-Stabilized Emulsions Fabricated Using Ultrasound Emulsification. *Ultrason. Sonochem.* **2019**, *58*, 104627. [CrossRef]
118. O'Sullivan, J.; Murray, B.; Flynn, C.; Norton, I. The Effect of Ultrasound Treatment on the Structural, Physical and Emulsifying Properties of Animal and Vegetable Proteins. *Food Hydrocoll.* **2016**, *53*, 141–154. [CrossRef]
119. Sun, W.W.; Yu, S.J.; Zeng, X.A.; Yang, X.Q.; Jia, X. Properties of Whey Protein Isolate-Dextran Conjugate Prepared Using Pulsed Electric Field. *Food Res. Int.* **2011**, *44*, 1052–1058. [CrossRef]

MDPI
St. Alban-Anlage 66
4052 Basel
Switzerland
Tel. +41 61 683 77 34
Fax +41 61 302 89 18
www.mdpi.com

*Foods* Editorial Office
E-mail: foods@mdpi.com
www.mdpi.com/journal/foods

www.ingramcontent.com/pod-product-compliance
Lightning Source LLC
Chambersburg PA
CBHW061800210926
51850CB00006B/601
*9783036568768*